Fundamental Concepts of Bioinformatics

Dan E. Krane
Wright State University, Department of Biological Sciences

Michael L. Raymer
Wright State University, Department of Computer Science

San Francisco ▪ Boston ▪ New York
Cape Town ▪ Hong Kong ▪ London ▪ Madrid ▪ Mexico City
Montreal ▪ Munich ▪ Paris ▪ Singapore ▪ Sydney ▪ Tokyo ▪ Toronto

Senior Project Manager: Peggy Williams
Editorial Assistant: Michael McArdle
Production Editors: Larry Lazopoulos, Jamie Sue Brooks
Production Service: Matrix Productions, Inc.
Composition: Omegatype Typography, Inc.
Text Design: Carolyn Deacy
Cover Design: Jennifer Dunn
Copyeditor: Loretta Palagi
Proofreader: Sally Scott
Manufacturing Supervisor: Vivian McDougal
Marketing Manager: Josh Frost
Cover and Text Printer: Phoenix Color

Library of Congress Cataloguing-in-Publication Data

Krane, Dan E.
Fundamentals of bioinformatics / Dan E. Krane, Michael L. Raymer.
p. cm.
ISBN 0-8053-4633-3 (pbk. : alk. paper)
1. Bioinformatics. I. Raymer, Michael L. II. Title.
QH324.2 .K72 2002
570'.285—dc21
2002012574

ISBN 0-8053-4633-3

www.aw.com/bc

4 5 6 7 8 9 10 – PBT – 08 07 06 05 04 03

Contents

1 Molecular Biology and Biological Chemistry 1

Preface

This book arose primarily from our own need for a text that we would be happy to recommend to our own undergraduate students interested in bioinformatics. We believe strongly that the best work in this new field arises from the interaction of individuals who are well versed in two disciplines, biology and computer science, that often have little in common in terms of language, approaches to problem solving, and even physical location within universities and colleges.

There is no particular shortage of books that catalog (and occasionally explain) websites with useful bioinformatics tools for biologists interested in analyzing their own data. There are also a few written for computer scientists that describe strategies for making algorithms more computationally efficient. Yet, a fundamental problem exists in the way that texts prepare students for work in bioinformatics. Most aim to train life sciences students to use existing web-based programs without fostering an understanding of the relative importance of different variables or how those programs might be customized for specific applications. The smaller number of books that are designed to teach computer scientists to write programs using established algorithms typically fail to convey an understanding of what constitutes biological significance or the limitations of molecular data gathering. While the very nature of bioinformatics requires an understanding of the objectives and limitations of both disciplines, students are rarely exposed to both.

The demand for bioinformaticians is very high, and the trend in bioinformatics education (as well as much of higher education in general) has been to move away from islands of information. Many graduate programs in bioinformatics are responding to the demands of the marketplace and their students by providing extensive remedial training in either biology or computer science for those who have graduated with degrees in the other area. These graduate programs in bioinformatics tend to be at least one, and usually two years longer than traditional graduate experiences. Our goal with this book is to reach students while it is still easy for them to become comfortable thinking about problems and arriving at solutions both as biologists and as computer scientists. By focusing on the actual algorithms at the heart of bioinformatics, our aim is to train computer scientists and biologists to have a common language and basis of understanding. After a general introduction to molecular biology and chemistry (primarily for the benefit of computer scientists and biology majors who have not yet encountered this material), the text revolves around algorithms that students could use to solve problems with pencil and paper. With that level of appreciation in place, it is not hard to illustrate how computers can provide invaluable assistance for larger and more complex data sets. Along the way, students are also introduced to existing databases and encouraged to use real data to test their skills. While mathematical and statistical calculations are included, all are fairly simple and understandable without a mastery of calculus. Sample programs, as well as examples of state-of-the-art web-based software based upon those same algorithms are also provided.

While new algorithms and methods are constantly being brought to bear on the problems of molecular biology, there remains a fundamental set of techniques and concepts that computer scientists and biologists use to address data-driven problems. A firm understanding of these fundamentals will allow students from both backgrounds to advance in their studies of computational molecular biology and bioinformatics. Several features of this text are specifically designed to provide students from varying backgrounds with their first taste of bioinformatics:

- A language-independent, pen-and-paper approach to problem solving allows students to understand the algorithms underlying various bioinformatics techniques, without becoming mired down in the syntactic details of implementing programs in a specific language or operating system.
- A hands-on, problem-solving approach is used to present the material, and the end-of-chapter questions are closely related to example problems presented in the text. Solutions for odd-numbered questions are provided in the text, while solutions for the remaining questions are available at the instructor's web site.
- Each chapter addresses the issues involved in biological data collection and bench experimentation, as well as the algorithmic issues involved in designing analysis techniques. Thus, students from any background will develop an appreciation for the parameters and complexities involved in both of these key aspects of bioinformatics.
- Appendix 1 provides non-programmers with their first programming experience using Perl. General concepts that are applicable to all structured programming languages are emphasized, so that students will be well equipped to learn new languages and programming methods after this brief introduction.
- Appendix 3 provides example algorithms in Perl that illustrate key concepts from each chapter of the text. These examples avoid Perl-specific constructs, instead focusing on general problem-solving approaches and algorithmic techniques that can be applied to a wide range of bioinformatics problems.

The work of assembling this book has been a greater challenge than either of us appreciated at its start. We never intended for it to be an encyclopedic review of the published literature in bioinformatics, but still found ourselves wanting to cite many excellent papers that we could not for reasons of time and space. We have also benefited from a great deal of support and help. The authors express their heartfelt thanks to David Paoletti for his excellent work in designing the algorithms for Appendix 3, as well as the various students who provided critical feedback on the initial versions of the text, figures, and example problems as we have used them for our own undergraduate bioinformatics courses here at Wright State University. We are likewise grateful to Michele Sordi, whose encouragement and guidance helped this text to germinate from an idea into reality, and to Peggy Williams for her creative suggestions, support and ubiquitous wit. Much of the credit for the finished product is due our production editor, Larry Lazopoulos, and to the first rate production staff at Matrix Productions, including Merrill Peterson and Michele Ostovar. Lastly, we are deeply appreciative to our wives, Carissa M. Krane and Delia F. N. Raymer, for their patience and understanding during those frequent times when the demands of writing took precedence over the matters of our families.

CHAPTER 1

Molecular Biology and Biological Chemistry

Biology has at least 50 more interesting years.

James D. Watson, December 31, 1984

The Genetic Material

Nucleotides

Orientation

Base pairing

The central dogma of molecular biology

Gene Structure and Information Content

Promoter sequences

The genetic code

Open reading frames

Introns and exons

Protein Structure and Function

Primary structure

Secondary, tertiary, and quaternary structure

The Nature of Chemical Bonds

Anatomy of an atom

Valence

Electronegativity

Hydrophilicity and hydrophobicity

Molecular Biology Tools

Restriction enzyme digests

Gel electrophoresis

Blotting and hybridization

Cloning

Polymerase chain reaction

DNA sequencing

Genomic Information Content

C-value paradox

Reassociation kinetics

The most distinguishing characteristic of living things is their ability to store, utilize, and pass on information. Bioinformatics strives to determine what information is biologically important and to decipher how it is used to precisely control the chemical environment within living organisms. Since that information is stored at a molecular level, the relatively small number of tools available to molecular biologists provides our most direct insights into that information content. This chapter provides a brief introduction or review of the format in which genetic information is maintained and used by living organisms as well as the experimental techniques that are routinely used to study it in molecular biology laboratories. Since that information is most relevant in terms of the effects that it has on the chemistry of life, that too is briefly reviewed.

The Genetic Material

DNA (deoxyribonucleic acid) is the genetic material. This is a profoundly powerful statement to molecular biologists. To a large extent, it represents the answer to questions that have been pondered by philosophers and scientists for thousands of years: "What is the basis of inheritance?" and "What allows living things to be different from nonliving things?" Quite simply, it is the information stored in DNA that allows the organization of inanimate molecules into functioning, living cells and organisms that are able to regulate their internal chemical composition, growth, and reproduction. As a direct result, it is also what allows us to inherit our mother's curly hair, our father's blue eyes, and even our uncle's too-large nose. The various units that govern those characteristics at the genetic level, be it chemical composition or nose size, are called **genes.** Prior to our understanding of the chemical structure of DNA in the 1950s, what and how information was passed on from one generation to the next was largely a matter of often wild conjecture.

Nucleotides

Genes themselves contain their information as a specific **sequence** of nucleotides that are found in DNA molecules. Only four different bases are used in DNA molecules: guanine, adenine, thymine, and cytosine (**G, A, T,** and **C**). Each base is attached to a phosphate group and a deoxyribose sugar to form a nucleotide. The only thing that makes one nucleotide different from another is which nitrogenous base it contains (Figure 1.1). Differences between each of the four nitrogenous bases is fairly obvious even in representations of their structures such as those in Figure 1.1, and the enzymatic machinery of living cells routinely and reliably distinguishes between them. And, very much like binary uses strings of zeros and ones and the English alphabet uses combinations of 26 different letters to convey information, all of the information within each gene comes simply from the order in which those four nucleotides are found along lengthy DNA molecules. Complicated genes can be many thousands of nucleotides long, and

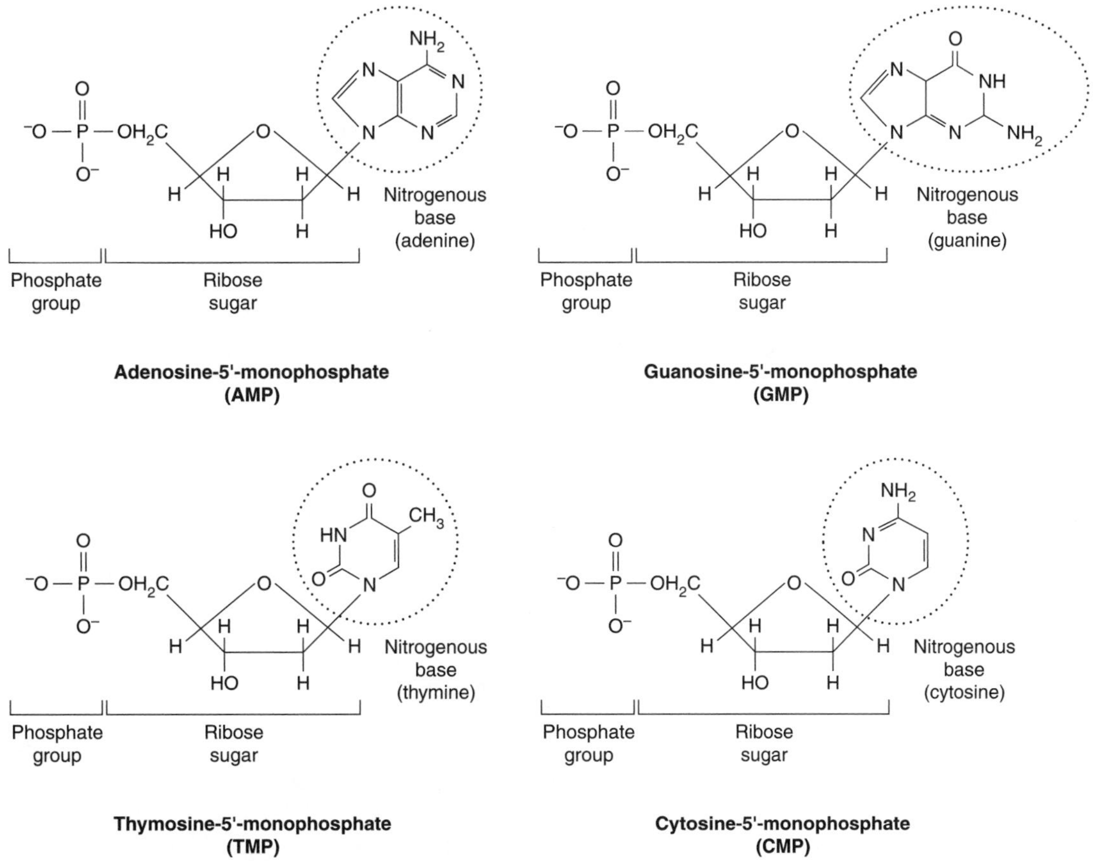

Adenosine-5'-monophosphate (AMP)

Guanosine-5'-monophosphate (GMP)

Thymosine-5'-monophosphate (TMP)

Cytosine-5'-monophosphate (CMP)

FIGURE 1.1 *Chemical structure of the four nucleotides used to make DNA. Each nucleotide can be considered to be made of three component parts: (1) a phosphate group, (2) a central deoxyribose sugar, and (3) one of four different nitrogenous bases.*

all of an organism's genetic instructions, its **genome,** can be maintained in millions or even billions of nucleotides.

Orientation

Strings of nucleotides can be attached to each other to make long **polynucleotide** chains or, when considered on a very large scale, **chromosomes.** The attachment between any two nucleotides is always made by way of a **phosphodiester bond** that connects the phosphate group of one nucleotide to the deoxyribose sugar of another (Figure 1.2). (Ester bonds are those that involve links made by oxygen atoms—phosphodiester bonds have a total of two ester bonds, one on each side of a phosphorous atom.)

All living things make these phosphodiester bonds in precisely the same way. Notice in Figure 1.2 that each of the five carbon atoms in a deoxyribose sugar has

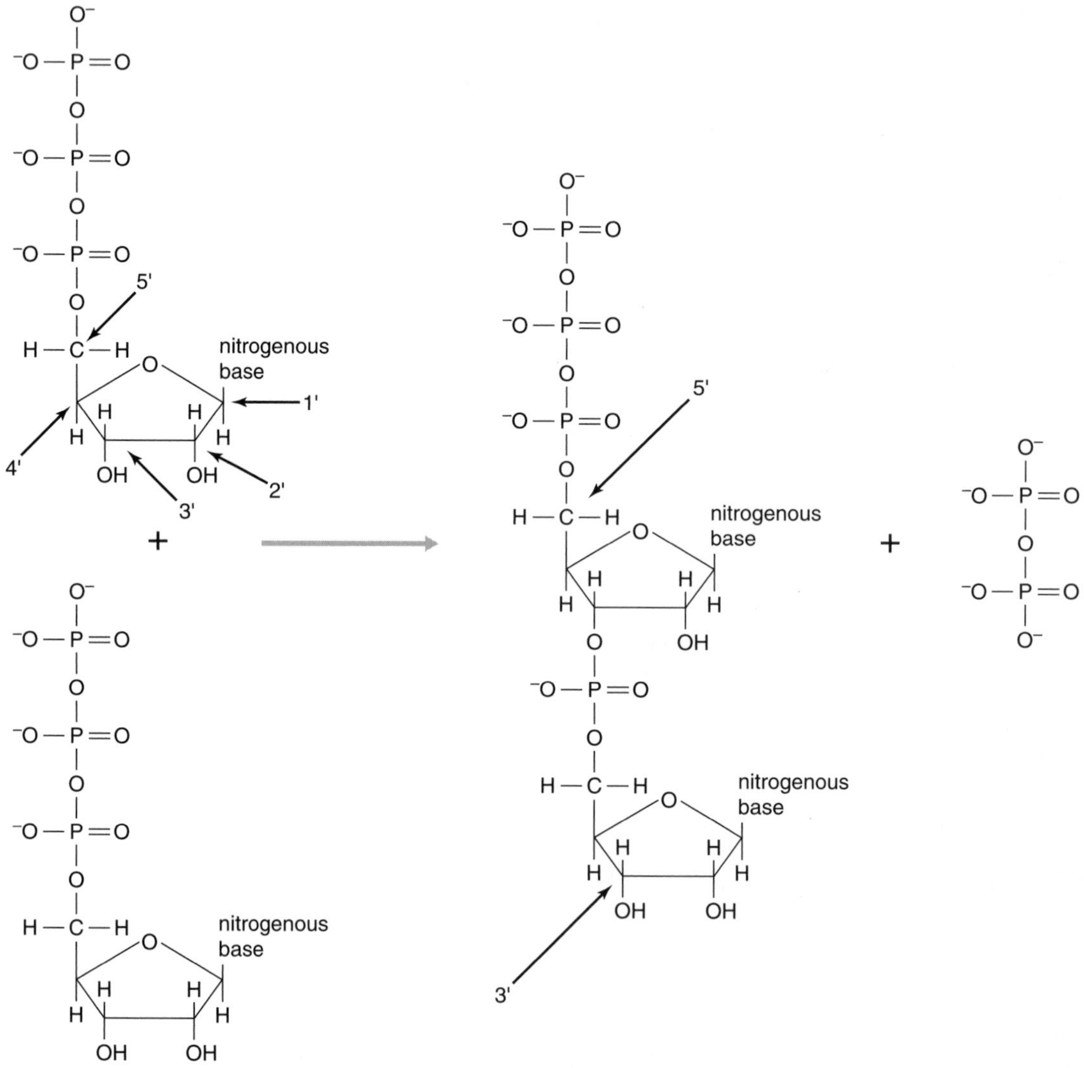

FIGURE 1.2 *The making of a phosphodiester bond. Nucleotides are added to growing DNA and RNA molecules only at their 3' ends.*

been assigned a specific numeric designation (1' through 5') by organic chemists. The phosphate group(s) of any single, unattached nucleotide are always found on its 5' carbon. Those phosphate groups are used to bridge the gap between the 5' carbon of an incoming deoxyribose sugar and the 3' carbon of a deoxyribose sugar at the end of a preexisting polynucleotide chain. As a result, one end of a string of nucleotides always has a 5' carbon that is not attached to another nucleotide, and the other end of the molecule always has an unattached 3' carbon. The difference between the 5' and 3' ends of a polynucleotide chain may seem subtle.

However, the orientation it confers to DNA molecules is every bit as important to cells as is our knowing that in written English we read from left to right and from top to bottom to understand the information content.

Base Pairing

A common theme throughout all biological systems and at all levels is the idea that structure and function are intimately related. Watson and Crick's appreciation that the DNA molecules within cells typically exist as double-stranded molecules was an invaluable clue as to how DNA might act as the genetic material. What they reported in their classic 1953 paper describing the structure of DNA is that the information content on one of those strands was essentially redundant with the information on the other. DNA could be replicated and faithfully passed on from one generation to another simply by separating the two strands and using each as a template for the synthesis of a new strand.

As we have already discussed, the information content in a DNA molecule comes from the specific sequence of its nucleotides. While the information content on each strand of a double-stranded DNA molecule is redundant it is not exactly the same—it is **complementary.** For every G on one strand, a C is found on its complementary strand and vice versa. For every A on one strand, a T is found on its complementary strand and vice versa. The interaction between G's and C's and between A's and T's is both specific and stable. The nitrogenous base guanine with its two-ringed structure is simply too large to pair with a two-ringed adenine or another guanine in the space that usually exists between two DNA strands. By the same token, the nitrogenous base thymine with its single-ringed structure is too small to interact with another single-ringed cytosine or thymine. Space is not a barrier to interaction between G's and T's or A's and C's but their chemical natures are incompatible, as will be described later in this chapter. Only the pairing between the nitrogenous bases G and C (Figure 1.3a) and the pairing between the nitrogenous bases A and T (Figure 1.3b) have both the right spacing and interaction between their chemical groups to form stable **base pairs.** In fact, the chemical interaction (specifically, three hydrogen bonds that form between G's and C's and two hydrogen bonds that form between A's and T's) between the two different kinds of base pairs is actually so stable and energetically favorable that it alone is responsible for holding the two complementary strands together.

Although the two strands of a DNA molecule are complementary they are not in the same 5'/3' orientation. Instead, the two strands are said to be **antiparallel** to each other, with the 5' end of one strand corresponding to the 3' end of its complementary strand and vice versa. Consequently, if one strand's nucleotide sequence is 5'-GTATCC-3', the other strand's sequence will be 3'-CATAGG-5'. By convention, and since most cellular processes involving DNA occur in the 5' to 3' direction, the other strand's sequence would typically be presented as: 5'-GGATAC-3'. Strictly speaking, the two strands of a double-stranded DNA molecule are *reverse* complements of each other. Sequence features that are 5' to a particular reference point are commonly described as being "upstream" while those that are 3' are described as being "downstream."

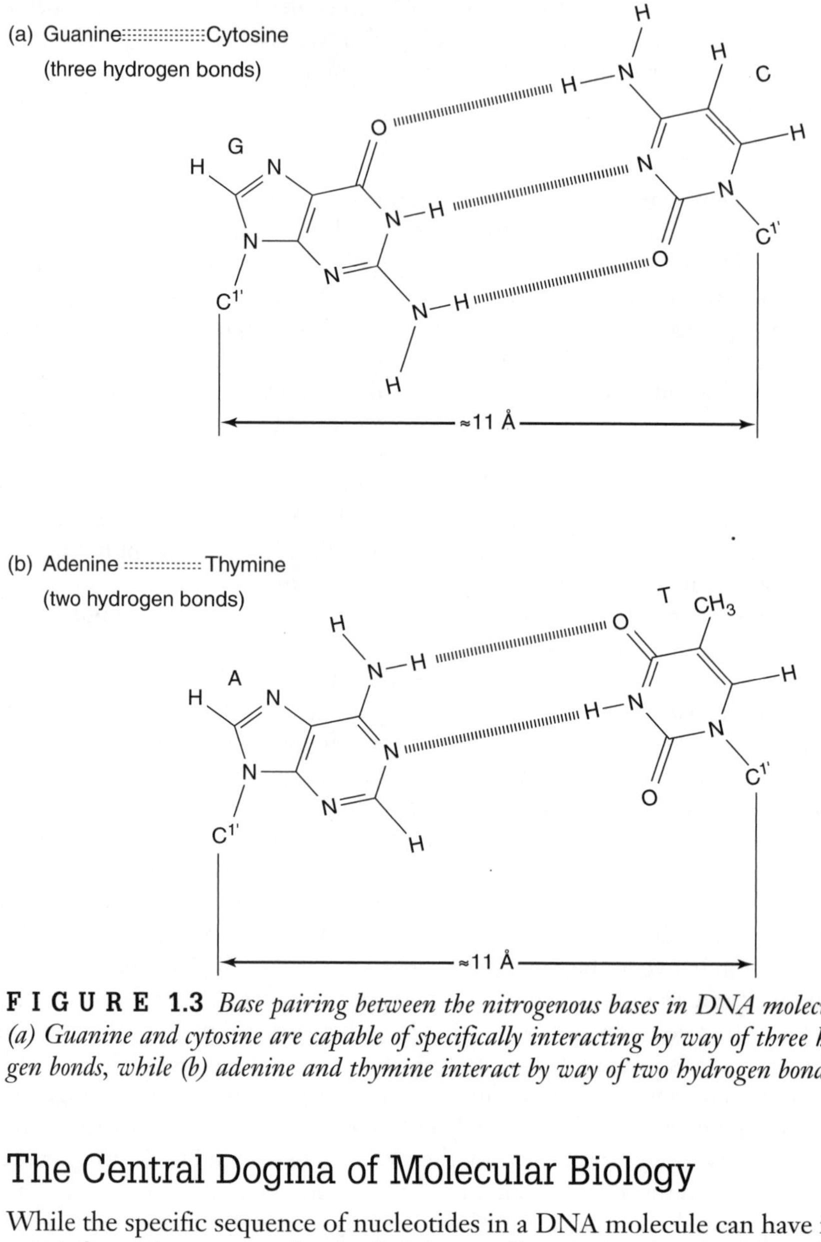

FIGURE 1.3 *Base pairing between the nitrogenous bases in DNA molecules. (a) Guanine and cytosine are capable of specifically interacting by way of three hydrogen bonds, while (b) adenine and thymine interact by way of two hydrogen bonds.*

The Central Dogma of Molecular Biology

While the specific sequence of nucleotides in a DNA molecule can have important information content for a cell, it is actually proteins that do the work of altering a cell's chemistry by acting as biological catalysts called **enzymes.** In chemistry catalysts are molecules that allow specific chemical reactions to proceed more quickly than they would have otherwise occurred. Catalysts are neither consumed nor altered in the course of such a chemical process and can be used to catalyze the same reaction many times. The term *gene* is used in many different ways, but one of its narrowest and simplest definitions is that genes spell out the instructions needed to make the enzyme catalysts produced by cells. The

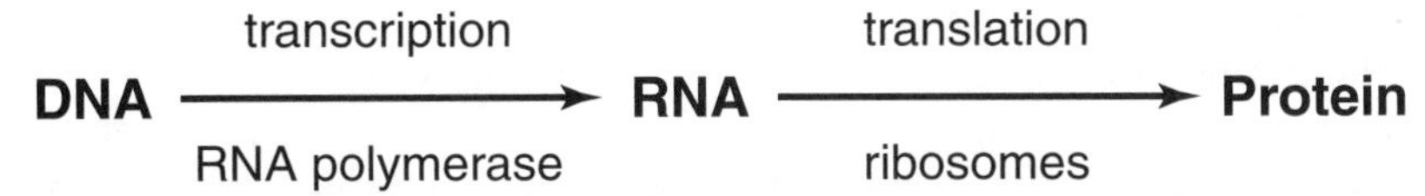

FIGURE 1.4 *The central dogma of molecular biology. Information in cells passes from DNA to RNA to proteins. RNA is made from DNA molecules during transcription by RNA polymerases. Proteins are made from the information content of RNA molecules as they are translated by ribosomes. DNA polymerases also make copies of DNA molecules during the replication process of cell division.*

process by which information is extracted from the nucleotide sequence of a gene and then used to make a protein is essentially the same for all living things on Earth and is described by the grandly named **central dogma** of molecular biology. Quite simply, information stored in DNA is used to make a more transient, single-stranded polynucleotide called RNA (ribonucleic acid) that is in turn used to make proteins (Figure 1.4). The process of making an RNA copy of a gene is called **transcription** and is accomplished through the enzymatic activity of an **RNA polymerase.** There is a one-to-one correspondence between the nucleotides used to make RNA (G, A, U, and C where "U" is an abbreviation for uracil) and the nucleotide sequences in DNA (G, A, T, and C, respectively). The process of converting that information from nucleotide sequences in RNA to the amino acid sequences that make a protein is called **translation** and is performed by a complex of proteins and RNA called **ribosomes.** Protein synthesis and structure are discussed at the end of this chapter.

Gene Structure and Information Content

Formatting and its interpretation are important considerations for any information storage system, be it a written text or a cell's DNA molecule. All cells go about interpreting their genetic instructions in the same way and rely on specific signals to "punctuate" their genes. Much of the "language" of DNA and the rules of its interpretation were worked out very early in the history of life on earth and, because of their central importance, have changed very little over the course of billions of years. As a result, both prokaryotic (bacteria) and eukaryotic (more complicated organisms like yeast, plants, pets, and people) organisms all use not only the same "alphabet" of nucleotides but also use essentially the same format and approach for storing and utilizing their genetic information.

Promoter Sequences

Gene expression, the process of using the information stored in DNA to make an RNA molecule and then a corresponding protein, can have significant energetic and opportunity costs for a cell. Organisms that express unneeded proteins are less likely to survive and reproduce relative to competitors that regulate their gene expression more appropriately. As a result, all cells place particular emphasis on controlling gene expression at its very start by making two crucial distinctions. First, they must reliably distinguish between those parts of an organism's genome that correspond to the beginnings of genes and those that do not. Second, they must be able to determine which genes code for proteins that are needed at any particular time.

Since RNA polymerases are responsible for the initiation of gene expression through their synthesis of RNA copies of genes, it is reasonable that the burden of making those two distinctions falls on them. Certainly not every nucleotide in a genome can correspond to the start of a gene any more than every letter on a printed page can correspond to the beginning of a sentence with useful information content. By the same token, RNA polymerases cannot simply look for any *one* particular nucleotide, like A, when looking for the start of a gene because each nucleotide occurs by chance so frequently throughout a cell's DNA. However, particular combinations of nucleotides are not as likely to occur by chance, and the greater the number of nucleotides involved, the smaller a chance occurrence becomes. The probability (P) that a string of nucleotides will occur by chance alone can be determined by the relatively simple formula $P = (1/4)^n$ if all nucleotides are present at the same frequency and where n is the string's length. Prokaryotic RNA polymerases actually scan along DNA looking for a specific set of approximately 13 nucleotides (1 nucleotide that serves as a transcriptional start site, 6 that are 10 nucleotides 5' to the start site, and 6 more that are 35 nucleotides 5' to it) that mark the beginning of genes. Those nucleotides, taken as a whole and in the proper positions relative to each other, are called **promoter sequences.** Given that most prokaryotic genomes are only a few million nucleotides long, these promoter sequences, which should occur only by chance about once in every 70 million nucleotides, allow RNA polymerases to uniquely identify the beginnings of genes with great statistical confidence. Eukaryotic genomes tend to be several orders of magnitude larger than those of prokaryotes and, as a result, eukaryotic RNA polymerases tend to recognize larger and more complex promoter sequences so that they too can reliably recognize the beginning of genes.

Two French biochemists, F. Jacob and J. Monod, were the first to obtain direct molecular insights into how cells distinguish between genes that should be transcribed and those that should not. Their work on prokaryotic gene regulation earned them a Nobel Prize in 1965 and revealed that the expression of structural genes (those that code for proteins involved in cell structure or metabolism) was controlled by specific regulatory genes. The proteins encoded by these regulatory genes are typically capable of binding to a cell's DNA near the promoter of the genes whose expression they control in some circumstances but not in others. It is the ability of these regulatory proteins to bind or not bind to specific nucleotide sequences in a fashion that is dependent on their ability to sense a cell's chemical environment that allows living things to respond appropriately to their environment. When the binding of these proteins makes it easier for an RNA polymerase to initiate transcription, **positive regulation** is said to have occurred. **Negative regulation** describes those situations where binding of the regulatory protein prevents transcription from occurring. Eventually, most prokaryotic structural genes were found to be turned on or off by just one or two regulatory proteins. Eukaryotes, with their substantially more complicated genomes and transcriptional needs, use larger numbers (usually seven or more) and combinations of regulatory proteins to control the expression of their structural genes.

The Genetic Code

While nucleotides are the building blocks that cells use to make their information storage and transfer molecules (DNA and RNA, respectively), amino acids are the units that are strung together to make the proteins that actually do most of the work of altering a cell's chemical environment. The function of a protein is intimately dependent on the order in which its amino acids are linked by ribosomes during translation and, as has already been discussed, that order is determined by the instructions transcribed into RNA molecules by RNA polymerases. However, although only four different nucleotides (nt) are used to make DNA and RNA molecules, 20 different amino acids (each with its own distinctive chemistry) are used in protein synthesis (Figure 1.5a) ($1 \text{ nt} \neq 1 \text{ aa}; 4^1 < 20$). There cannot be a simple one-to-one correspondence between the nucleotides of genes and the amino acids of the proteins they encode. The 16 different possible pairs of nucleotides also fall short of the task ($2 \text{ nt} \neq 1 \text{ aa}; 4^2 < 20$). However, the four nucleotides can be arranged in a total of 64 different combinations of three ($4^3 = 64$). As a result, it is necessary for ribosomes to use a **triplet code** to translate the information in DNA and RNA into the amino acid sequence of proteins. With only three exceptions, each group of three nucleotides (a **codon**) in an RNA copy of the coding portion of a gene corresponds to a specific amino acid (Table 1.1). The three codons that do not instruct ribosomes to insert a specific amino acid are called **stop codons** (functionally equivalent to a period at the end of a sentence) because they cause translation to be terminated. This same genetic code seems to have been in place since the earliest history of life on earth and, with only a few exceptions, is universally used by all living things today.

Notice in Table 1.1 that 18 of the 20 different amino acids are coded for by more than one codon. This feature of the genetic code is called **degeneracy.** It is therefore possible for mistakes to occur during DNA replication or transcription that have no effect on the amino acid sequence of a protein. This is especially true of mutations (heritable changes in the genetic material) that occur in the third (last) position of a codon. Each amino acid can be assigned to one of essentially four different categories: nonpolar, polar, positively charged, and negatively charged (Figure 1.5b). A single change within a triplet codon is usually not sufficient to cause a codon to code for an amino acid in a different group. In short, the genetic code is remarkably robust and minimizes the extent to which mistakes in the nucleotide sequences of genes can change the functions of the proteins they encode.

Open Reading Frames

Translation by ribosomes starts at translation-initiation sites on RNA copies of genes and proceeds until a stop codon is encountered. Just as three codons of the

(a) Side chain

R

H, H, N, C_α, C, O, OH, H

Amino group

Carboxyl group

FIGURE 1.5 *(a) Chemical structure of a generic amino acid. The amino group, alpha carbon, and carboxyl groups are identical for all 20 amino acids while each has its own distinctive R group. (b) Chemical structure of the 20 different amino acids complete with their distinctive R groups. Amino acids are grouped according to the properties of their side chains, shown in black. Standard three-letter and one-letter abbreviations for each of the amino acids are shown in parentheses.*

(b)

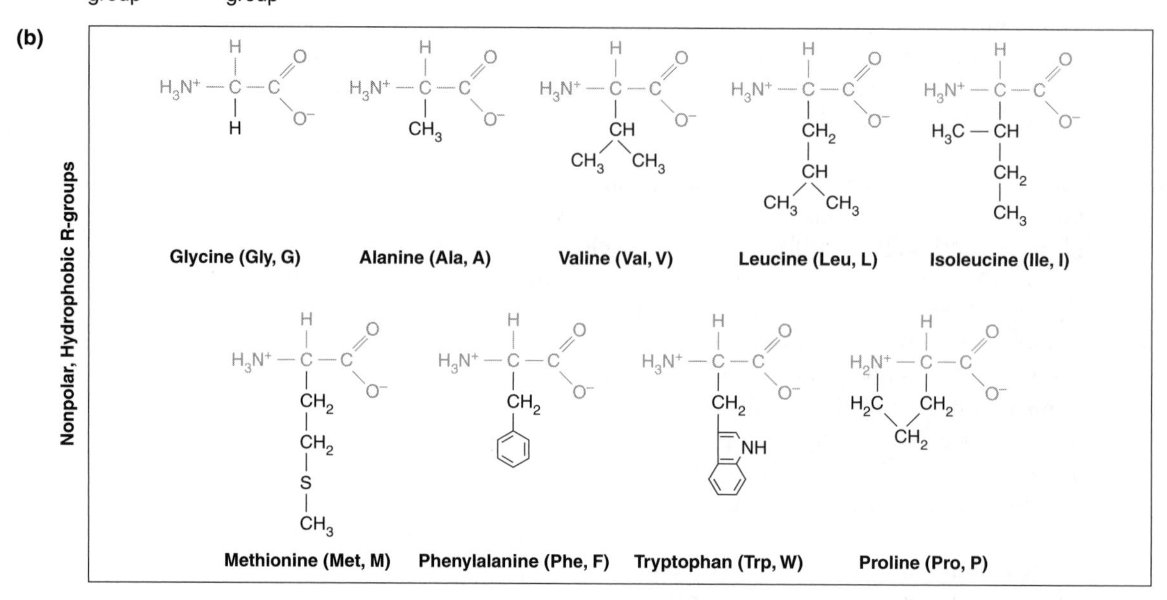

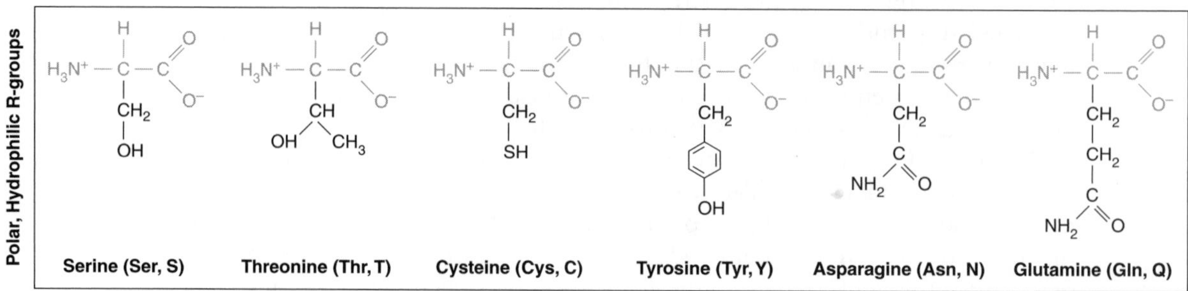

Electrically charged

Acidic

Aspartic acid (Asp, D)

Glutamic acid (Glu, E)

Basic

Lysine (Lys, K)

Arginine (Arg, R)

Histidine (His, H)

TABLE 1.1 A summary of the coding assignments of the 64 triplet codons. *Standard three-letter abbreviations for each of the most commonly used 20 amino acids are shown. The universality of the genetic code encompasses animals (including humans), plants, fungi, archaea, bacteria, and viruses. Small variations in the code exist in mitochondria and certain microbes. For instance, in a limited number of bacterial genes, a special UGA codon, normally a termination codon, is used as a codon for an unusual, 21st naturally occurring, amino acid selenocysteine. A 22nd naturally occurring amino acid, pyrrolysine, is coded for by UAG (a stop codon for most organisms) in some bacterial and eukaryotic species.*

(5') . . . pNpNpN . . . (3') in mRNA					
	Middle Base of Codon →				
Base at 5' End of Codon ↓	U	C	A	G	Base at 3' End of Codon ↓
U	phe (UUU)	ser	tyr	cys	U
	phe	ser	tyr	cys	C
	leu	ser	termination	termination	A
	leu	ser	termination	trp	G
C	leu	pro	his	arg	U
	leu	pro	his	arg	C
	leu	pro	gln	arg	A
	leu	pro	gln	arg	G
A	ile	thr	asn	ser	U
	ile	thr	asn	ser	C
	ile	thr	lys	arg	A
	met (and initiation)	thr	lys	arg	G
G	val	ala	asp	gly	U
	val	ala	asp	gly	C
	val	ala	glu	gly	A
	val	ala	glu	gly	G

genetic code are reserved as stop codons, one triplet codon is always used as a **start codon.** Specifically, the codon AUG is used both to code for the amino acid methionine as well as to mark the precise spot along an RNA molecule where translation begins in both prokaryotes and eukaryotes. Accurate translation can only occur when ribosomes examine codons in the phase or **reading frame** that is established by a gene's start codon. Unless a mistake involving some multiple of three nucleotides occurs, alterations of a gene's reading frame change every amino acid coded downstream of the alteration, and such alterations typically result in the production of a truncated version of the protein due to ribosomes encountering a premature stop codon.

Most genes code for proteins that are hundreds of amino acids long. Since stop codons occur in a randomly generated sequence at about every 20th triplet codon (3 codons out of 64), one of the reading frames of the RNA copies of most genes has unusually long runs of codons in which no stop codons occur. These strings of codons uninterrupted by stop codons are known as **open reading frames** (ORFs) and are a distinguishing feature of many prokaryotic and eukaryotic genes.

Introns and Exons

The messenger RNA (mRNA) copies of prokaryotic genes correspond perfectly to the DNA sequences present in the organism's genome with the exception that the nucleotide uracil (U) is used in place of thymine (T). In fact, translation by ribosomes almost always begins while RNA polymerases are still actively transcribing a prokaryotic gene.

Eukaryotic RNA polymerases also use uracil in place of thymine, but much more striking differences are commonly found between the mRNA molecules seen by ribosomes and the nucleotide sequences of the eukaryotic genes that code for them. In eukaryotes the two steps of gene expression are physically separated by the nuclear membrane, with transcription occurring exclusively within the nucleus and translation occurring only after mRNAs have been exported to the cytoplasm. As a result, the RNA molecules transcribed by eukaryotic RNA polymerases can be modified before ribosomes ever encounter them. The most dramatic modification that is made to the primary RNA transcripts of most eukaryotic genes is called **splicing** and involves the precise excision of internal sequences known as **introns** and the rejoining of the **exons** that flank them (Figure 1.6). Splicing is far from a trivial process and most eukaryotic genes have a large number of sometimes very large introns. An extreme example is the gene associated with the disease cystic fibrosis in humans, which has 24 introns and is over 1 million nucleotides (1 mega base pair or 1 Mb) long even though the mRNA seen by ribosomes is only about 1,000 nucleotides (1 kilo base pair or 1 kb) long. Failure to appropriately splice the introns out of a primary eukaryotic RNA transcript typically introduces frame shifts or premature stop codons that render useless any protein translated by a ribosome. Regardless of the tissue or even the organism being considered, the vast majority of eukaryotic introns conform to what is known as the "GT–AG rule," meaning that the first two nucleotides in the DNA sequence of all introns begin with the dinucleotide GT and end with the dinucleotide AG. Pairs of nucleotides occur too often just by chance to be a sufficient signal for the enzyme complexes responsible for splicing in eukaryotes, **spliceosomes,** and approximately six additional nucleotides at the 5' and 3' ends of introns are also scrutinized—sometimes differently in some cell types relative to others. This **alternative splicing** allows a huge increase in the diversity of proteins that eukaryotic organisms can use and is accomplished by often subtle modifications of spliceosomes and accessory proteins that are responsible for recognizing intron/exon boundaries.

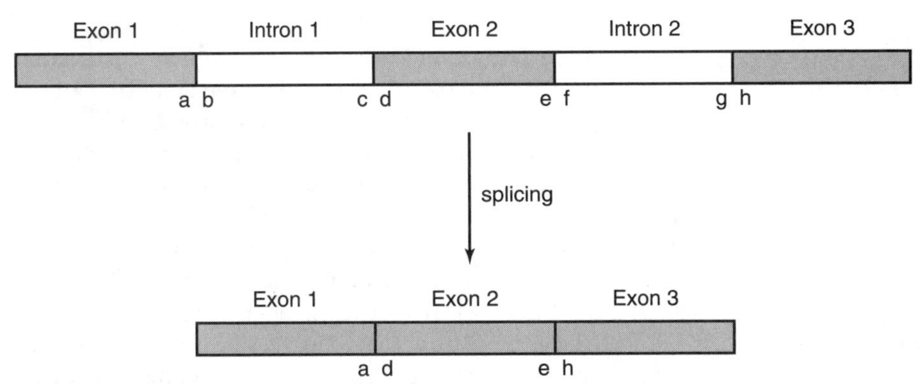

FIGURE 1.6 *Splicing of the primary transcripts of eukaryotic genes results in the removal of introns and the precise joining of exons. Intron sequences are quickly degraded once they are removed while the spliced mRNA is exported out of the nucleus for translation by ribosomes.*

Protein Structure and Function

Proteins are the molecular machinery responsible for performing most of the work of both prokaryotic and eukaryotic cells. The tasks undertaken by proteins are incredibly diverse. **Structural proteins,** such as collagen, provide rigidity and support in bones and connective tissues. Other proteins called **enzymes** act as biological catalysts, like the digestive enzyme pepsin that helps to break down and metabolize food. Proteins are also responsible for transportation of atoms and small molecules throughout an organism (e.g., hemoglobin), signaling and intercellular communication (e.g., insulin), absorbing photons to enable vision (e.g., rhodopsin), and myriad other functions.

Primary Structure

Following the genetic instructions contained in messenger RNA, proteins are translated by ribosomes as linear polymers (chains) of amino acids. The 20 amino acids have similar chemical structures (Figure 1.5a), varying only in the chemical group attached in the R position. The constant region of each amino acid is called the *backbone,* while the varying R group is called the *side chain.* The order in which the various amino acids are assembled into a protein is the sequence, or **primary structure,** of the protein. As with DNA, the protein chain has directionality. One end of the protein chain has a free amino (NH) group, while the other end of the chain terminates in a carboxylic acid (COOH) group. The individual amino acids in the protein are usually numbered starting at the **amino terminus** and proceeding toward the **carboxy terminus.**

After translation, a protein does not remain in the form of a simple linear chain. Rather, the protein collapses, folds, and is shaped into a complex globular structure. The order in which the various amino acids are assembled into a protein

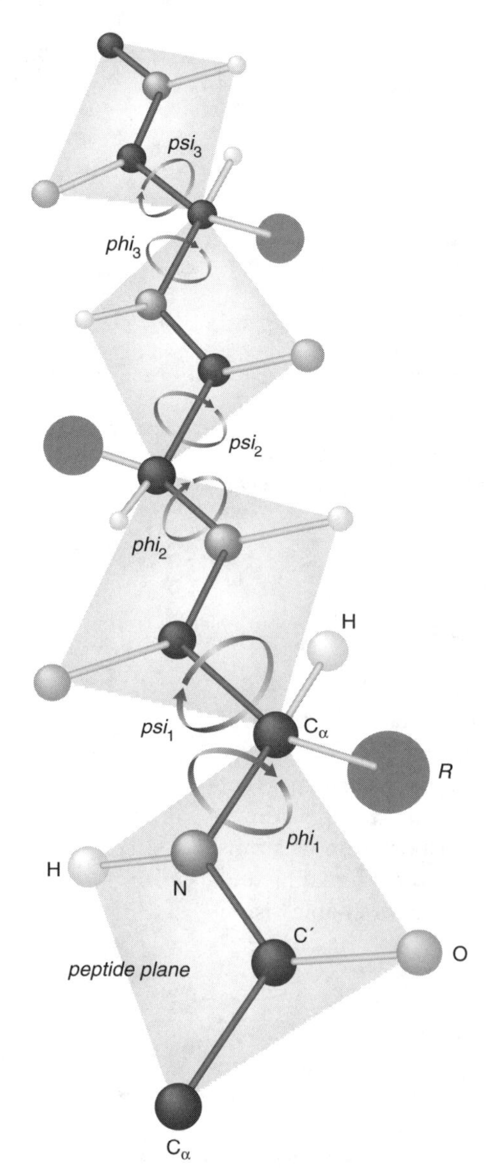

FIGURE 1.7 *Rigid and mobile regions of the protein backbone. Most of the backbone is rigid. The chemical bonds to the alpha carbons are rotatable. The angles of rotation for each alpha carbon's bonds are called phi (ϕ) and psi (ψ).*

largely determines the structure into which it will fold. The unique structure into which a particular protein will fold is called the **native structure** of the protein. The native structures of proteins give them unique properties that allow them to perform their particular roles in the context of a living organism.

The chemistry of a protein backbone forces most of the backbone to remain planar (Figure 1.7). The only "movable" segments of the protein backbone are the bonds from the nitrogen to the alpha carbon (the carbon atom to which the side chain is attached) and the bond between the alpha carbon and the carbonyl carbon (the carbon with a double bond to an oxygen atom). These two chemical bonds allow for circular (or "dihedral") rotation, and are often called phi (ϕ) and psi (ψ), respectively. Thus, a protein consisting of 300 amino acids will have 300 phi and psi angles, often numbered ϕ_1, ψ_1 through ϕ_{300}, ψ_{300}. All of the various conformations attainable by the protein come from rotations about these 300 pairs of bonds.

Secondary, Tertiary, and Quaternary Structure

Careful examination of proteins whose structures are known reveals that a very small number of patterns in local structures are quite common. These structures, formed by regular intramolecular hydrogen bonding (described below) patterns, are found in nearly every known protein. The location and direction of these regular structures make up the **secondary structure** of the protein. The two most common structures are the α-helix and the β-sheet (Figure 1.8). Often, the secondary structures are the first portions of the protein to fold after translation. Alpha (α) helices are characterized by phi and psi angles of roughly –60°, and exhibit a spring-like helical shape with 3.6 amino acids per complete 360° turn. Beta (β) strands are characterized by regions of extended (nearly linear) backbone conformation with $\phi \approx -135°$ and $\psi \approx 135°$. Beta strands assemble into one of two types of beta sheets, as illustrated in Figure 1.9 on page 16. In anti-parallel sheets, adjacent strands run in opposite directions as you move along the protein backbone from amino to carboxy terminus. In parallel sheets, the strands run in the same direc-

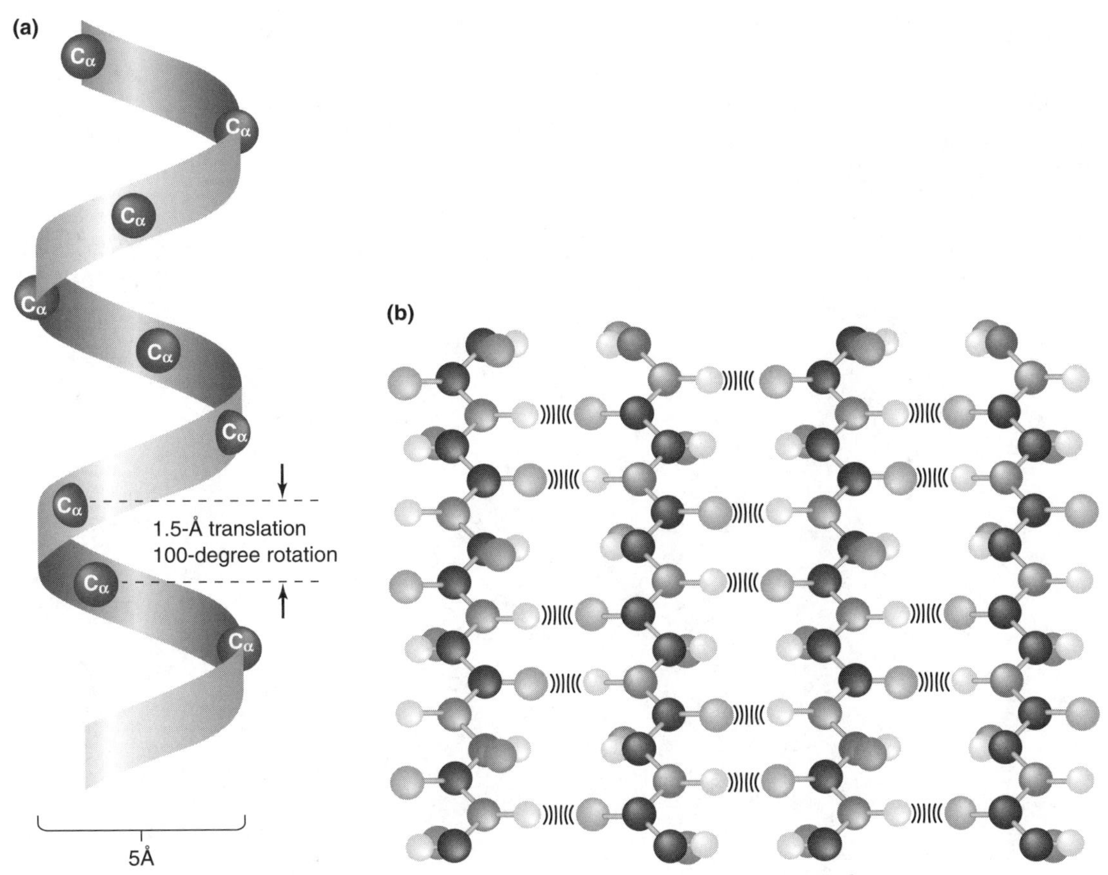

FIGURE 1.8 *Elements of secondary structure: (a) the alpha (α) helix and (b) the beta (β) sheet.*

tions. To make this possible, the strands of parallel beta sheets are often composed of amino acids that are nonlocal in the primary structure of the protein (Figure 1.9).

The regions of secondary structure in a protein pack together and combine with other less structured regions of the protein backbone to form an overall three-dimensional shape, which is called the **tertiary structure** of the protein. Often an active enzyme is composed of two or more protein chains that come together into a single large complex. When this occurs, the overall structure formed by the interacting proteins is commonly referred to as the **quaternary structure** of the enzyme.

The Nature of Chemical Bonds

As seen already, descriptions of nucleotides and proteins require at least a familiarity with the idea of chemical and hydrogen bonds. Much, if not all, of what we

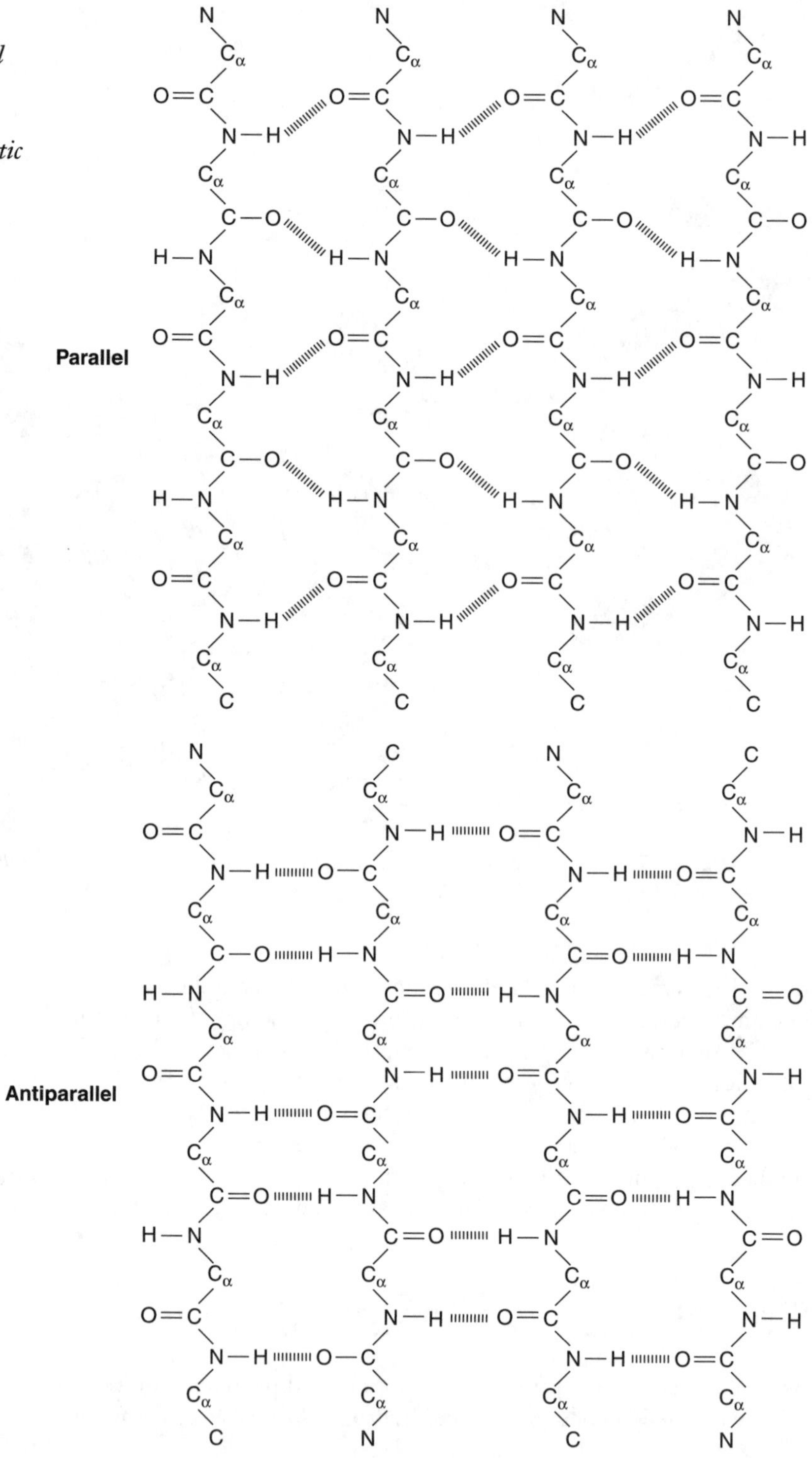

FIGURE 1.9
Parallel and anti-parallel beta sheets. Note that the hydrogen bonds (dotted lines) give greater energetic stability in anti-parallel sheets.

consider to be essential to life can be reduced to a set of chemical reactions and the characteristics of the enzymes that control the rate at which they occur. Even a passing understanding of basic chemistry gives deep insights into the way in which enzymes function and how molecules like proteins and DNA interact. Local differences in hydrophobicity and hydrophilicity, for instance, are fundamentally important to the functioning of most enzymes. L. Pauling won a Nobel Prize in 1962 for a book he wrote that made sense of such differences at a subatomic level. The following section describes the essence of his approach.

Anatomy of an Atom

By definition, **elements** are things that cannot be further reduced by chemical reactions. Elements themselves are made of individual atoms, which, in turn, are also made of smaller, subatomic particles. These smaller component parts of elements and atoms can only be separated by physical reactions, not chemical ones. Nuclear physicists have discovered hundreds of subatomic particles. Only three, however, are stable and particularly important to the discussion of the chemistry of living things. Those three subatomic particles are neutrons (weighing 1.7×10^{-24} gram and having no charge), protons (also weighing 1.7×10^{-24} gram and possessing one positive charge), and electrons (having only 1/2000th the mass of a proton or neutron and possessing a single negative charge). The number of protons in the nucleus of an atom determines what element it is. Generally, for every proton in an atomic nucleus there is an electron in orbit around it to balance the electrical charges. Electrons move in orbits at the speed of light and a relatively long way off from the nucleus. As a result, atoms are mostly empty space.

It also takes more and more energy for an electron to be moved away from the positive charges of atomic nuclei. Similarly, it takes more energy to carry a rock to the 20th floor of a building than it does to carry it to the 19th. The further an electron is from the nucleus of its atom, the more potential energy it must have. Packets of energy can be parceled out in a number of ways at the level of atoms, and one of the more common is through light (photons). Light plus an electron often results in an electron that is residing at an orbital with a higher energy level. Electrons often release light (a packet of energy) when they go from high to low orbitals. The amount of energy required for such a transition is narrowly defined and is known as a quantum (hence the term *quantum leap*). Electrons do not have predictable orbits in the same way that planets do. In fact, the best estimates that can be made about the position of an atom's electrons are based on confidence about where the electron is most likely to be at a given time. *Orbitals* are the three-dimensional space in which an electron spends 90% of its time.

Valence

Because the negative charges of electrons are repulsive to each other, only two can share an orbital at any given time. Electrons with the lowest amounts of energy are found in an orbital closest to the nuclei of atoms known as **1*s***. It is a spherical orbital and, again, it holds only two electrons. The second highest energy level for electrons has a total of four orbitals (**2*s*,** and **2*p*;** the 2*s* orbital is also spherical and the three 2*p* orbitals are dumbbell shaped).

The chemical properties of an atom depend on its outermost shell of electrons. Since atoms are mostly empty space, nuclei never meet in normal chemical reactions—only electrons way out at the edge of the atoms ever have an opportunity to interact. Although the number of protons in an atom never changes during a chemical reaction, the relative positions (and sometimes even the number) of electrons do.

Although maintaining a balance of charges (i.e., one electron for every proton in an atom) is Nature's highest priority, there is also a strong tendency to keep an atom's outermost shell of orbitals completely full or completely empty. These potentially conflicting tendencies can be resolved by allowing the electron orbitals of atoms to overlap. The sharing of electrons that results from the overlapping of those orbitals is typically part of a long-term association between the two atoms and is the basis of **covalent bonding.** Since the atoms of some elements such as helium, ${}_{2}He$ (the subscript number before an atomic symbol such as He states the number of protons in an atom's nucleus), have no unpaired electrons in their outermost orbital they are not chemically reactive and are never covalently bound to other atoms. In the same way, ${}_{10}Ne$ (in which both the 1*s* orbital and all four of the level-2 orbitals are filled with a total of 10 electrons) and ${}_{18}Ar$ are also unreactive. Atoms with similar valences have similar chemical properties: Carbon, ${}_{6}C$ (in which each of the four level-2 orbitals has a single electron), and silicon, ${}_{14}Si$ (in which each of the four level-3 orbitals has a single electron), react very similarly and are both capable of making four covalent bonds. As a result, the number of unpaired electrons in an atom's outermost orbital, its **valence,** takes on a special significance and represents its bonding capacity: ${}_{1}H = 1$, ${}_{8}O = 2$, ${}_{7}N = 3$, ${}_{6}C = 4$. The shape and size of compounds (a complex of two or more covalently bound atoms) are largely governed by the valences of the atoms that comprise them.

Electronegativity

The chemistry of living things is complicated by the fact that different nuclei have different affinities for electrons. The higher an atom's affinity for electrons, the higher its **electronegativity.** The relative electronegativity of an atom is a function of how many electrons it needs to acquire or to donate in order to completely fill or empty its outermost shell of orbitals. For instance, ${}_{1}H$ and ${}_{6}C$ both have outermost shells of electrons that are half full. Since their electronegativities are essentially the same, atoms are shared evenly in the covalent bonds between hydrogen and carbon atoms. This is substantially different from what occurs in the covalent bonds between hydrogen and carbon with oxygen. Since ${}_{8}O$ must either gain just two electrons or lose six, it is much more electronegative than hydrogen or carbon. Electrons involved in the covalent bonds of water (H_2O), for instance, tend to spend more time in the vicinity of the oxygen atom than the hydrogen atom. **Polar bonds** such as these result in a slight separation of charge that makes the oxygens of water molecules slightly negative and the hydrogens slightly positive. The slight separation of charges that result from polar covalent bonds allows for an important type of interaction between molecules called **hydrogen bonding.** Every water molecule is typically loosely associated with a network of other water molecules because the slight positive charges of their hydrogen atoms give them an affinity for the slight negative charges of the oxygens in their neighbors. Much less energy is required to break the association caused by hydrogen bonding than by covalent bonding because no electrons are shared between atoms in hydrogen bonds.

Hydrophilicity and Hydrophobicity

Chemists have found that most chemicals can be easily placed in one of just two categories: those that interact with water and those that do not. Molecules with polar bonds, like water itself, have some regions of positive and negative charge on their surfaces that are capable of forming hydrogen bonds with water. This makes them **hydrophilic** (literally, "water friendly") and allows them to be easily dissolved in watery solutions like the interior of a living cell. Other molecules that have atoms joined by only nonpolar covalent bonds are **hydrophobic** (literally, "afraid of water") and have much less basis of interaction with water molecules. In fact, their physical presence actually gets in the way of water molecules interacting with each other and prevents them from offsetting their partial charges. As a result, molecules such as fats that are composed primarily of carbon–carbon and carbon–hydrogen bonds are actually excluded from watery solutions and forced into associations with each other such as those observed in a cell's lipid bilayer membrane.

Molecular Biology Tools

Recognizing the information content in the DNA sequences of prokaryotic genomes is invariably easier than the equivalent task in more complicated eukaryotic genomes. For example, while the mRNA copies of both prokaryotic and eukaryotic genes have long ORFs, the DNA sequences of eukaryotic genes themselves often do not due to the presence of introns (see Chapter 6). The problem of identifying protein coding information within eukaryotic DNA sequences is further compounded by the fact that what may be an intron in one kind of eukaryotic cell may be an exon in another (described in greater detail in Chapter 6). These problems and others associated with deciphering the information content of genomes are far from insurmountable once the rules used by cells are known. In a quickly growing number of cases, it is bioinformaticians who recognize these rules from patterns they observe in large amounts of sequence data. It is the surprisingly small number of tools commonly used by molecular biologists, however, that both generates the raw data needed for such analyses and tests the biological significance of possible underlying rules. A set of roughly six different laboratory techniques, taken together, defines the entire discipline of molecular biology. These techniques are briefly described in this section.

Restriction Enzyme Digests

The Nobel Prize–winning work of Wilkins, Watson, and Crick in 1953 told the story of how DNA could act as the genetic material. Subsequent experiments confirmed this hypothesis, but it was not until nearly 20 years later that H. Smith and others made a serendipitous discovery that allowed researchers to manipulate DNA molecules in a specific fashion and to begin to decipher DNA's actual information content. In the course of studying what causes some bacterial cells to better defend themselves against viral infections, Smith and his colleagues found that bacteria produced enzymes that introduce breaks in double-stranded DNA molecules whenever they encounter a specific string of nucleotides. These proteins, **restriction enzymes,** can be isolated from bacterial cells and used in research laboratories as precise "scissors" that let biologists cut (and later "paste" together) DNA molecules. The very first of these proteins to be characterized was given the name *Eco*RI (*Eco* because it was isolated from *Escherichia coli;* R because it restricted DNA; I because it was the first such enzyme found in *E. coli*). *Eco*RI was found to cleave DNA molecules between G and A nucleotides whenever it encountered them in the sequence 5'-GAATTC-3' (Figure 1.10). Since then over 300 types of restriction enzymes have been found in other bacterial species that recognize and cut DNA molecules at a wide variety of specific sequences. Notice that *Eco*RI, like many restriction enzymes, cleaves (or digests) double-stranded DNA molecules in a way that leaves a bit of single-stranded DNA at the end of each fragment. The nucleotide sequences of those single-stranded regions (5'-AATT-3' in the case of *Eco*RI) are naturally complementary to each other. The resulting potential for base pairing makes these **sticky ends** capable of holding two DNA fragments together until another special enzyme called **ligase** can permanently link (or ligate) them together again by rebuilding the phosphodiester bonds that were broken by the restriction enzyme. Restriction enzymes that do not give rise to sticky ends create **blunt ends** that can be ligated to other blunt-ended DNA molecules.

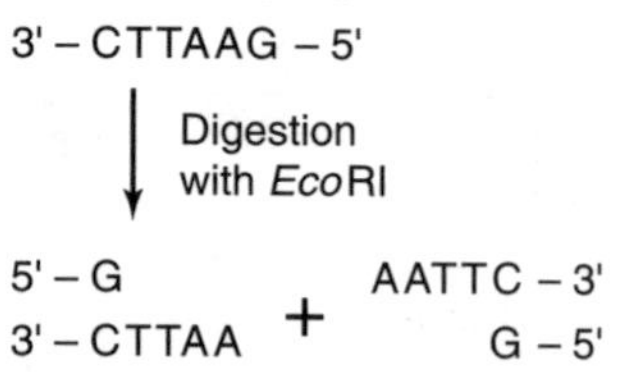

5' – GAATTC – 3'
3' – CTTAAG – 5'

↓ Digestion with *Eco*RI

5' – G + AATTC – 3'
3' – CTTAA + G – 5'

FIGURE 1.10
Digestion with the restriction enzyme Eco*RI.* Eco*RI introduces staggered breaks in DNA molecules whenever it encounters its recognition site (5'-GAATTC-3'). The single-stranded overhanging regions that result are capable of base pairing with each other and are referred to as* sticky ends.

The string of nucleotides recognized by *Eco*RI, its **restriction site,** should occur randomly in DNA sequences only once every $(1/4)^n$ base pairs where n equals 6 or, on average, once every 4,096 base pairs. Some restriction enzymes have smaller restriction sites (such as *Hinf*I, which finds and restricts at 5'-GATC-3' on average once every 256 base pairs) while others have larger sites (such as *Not*I, which finds and restricts at 5'-GCGGCCGC-3' on average once every 65,536 base pairs). Simply cutting a DNA molecule and determining how many fragments are made and the order in which the breaks occur when multiple restriction enzymes are used provide some limited insight into the specific organization and sequence of that DNA molecule. Such experiments are termed **restriction mapping.** Restriction enzymes also allowed the isolation and experimental manipulation of individual genes for the very first time.

Gel Electrophoresis

When dealing with a genome that is millions of base pairs long (such as *E. coli*'s) or even billions of base pairs long (such as the human genome), complete digestion with even a very specific restriction enzyme such as *Not*I can yield hundreds of thousands of DNA fragments. Separating all of those different fragments from each other is commonly accomplished by **gel electrophoresis,** another of the tools of molecular biology. In gel electrophoresis, DNA (or RNA or protein) fragments are loaded into indentations called wells at one end of a porous gel-like matrix typically made either from agarose or acrylamide. When an electric field is applied across these gels, the charged molecules naturally migrate toward one of the two electrodes generating the field. DNA (and RNA) with its negatively charged phosphate backbone is drawn toward the positively charged electrode. Very simply, small molecules have an easier time working their way through the gel's matrix than larger ones, and separation of the molecules on the basis of their size occurs (Figure 1.11). Larger molecules remain closer to the wells than smaller molecules, which migrate more quickly.

Blotting and Hybridization

Finding the single piece of DNA that contains a specific gene among hundreds or thousands is very much akin to the idea of finding a needle in a haystack even when the DNA fragments are size fractionated. Molecular biologists routinely

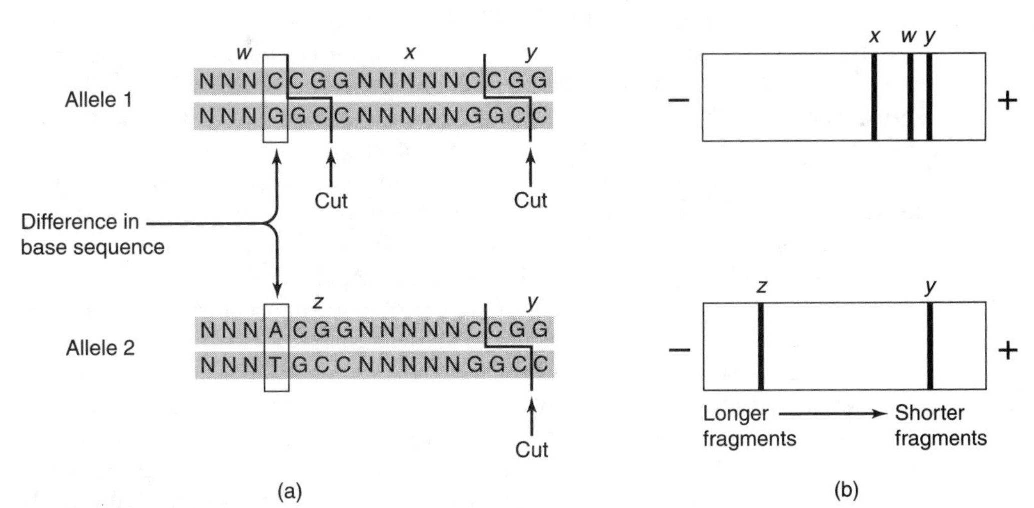

FIGURE 1.11 *Gel electrophoresis allows DNA fragments to be separated on the basis of their size. (a) Differences in DNA sequences can cause differences in the places where restriction enzymes break double-stranded DNA molecules. (b) The differences in the sizes of the restriction fragments that result can be easily detected by gel electrophoresis. Allele 1 has three bands (corresponding to regions* w, x, *and* y, *while allele 2 gives rise to two bands corresponding to regions* z *and* y.

employ another technique, **blotting and hybridization,** to draw attention to the one fragment they wish to study. In blotting, polynucleotides are transferred from the fragile gel that was used to separate them onto a more solid support such as a piece of nitrocellulose paper or a nylon membrane. This blotting process is mechanically simple and entails placing the membrane in contact with the gel and then using capillary action to pull the DNA up from the gel and onto the membrane. While water molecules can pass through the membrane (and be drawn into absorbent paper towels or a weak vacuum above it), DNA molecules are too large and remain associated with the membrane in the same relative positions that they had moved to during gel electrophoresis. Ultraviolet light or even simple baking can then be used to permanently attach the DNA fragments to the membrane.

A membrane prepared in this way is then ready for the second step in this detection process. Hybridization occurs when a labeled fragment of single-stranded DNA called a **probe** is allowed to base pair with the nucleic acids that have been transferred to a membrane. Typically 20 or more nucleotides in length, probes can be chosen on the basis of their being likely to find only one fragment of DNA on the membrane to which they can base pair. Probes can be chemically synthesized from scratch or can be fragments of DNA that have already been isolated in other experiments and even from related genes in different organisms. Any of a number of means can be used to label or tag a probe ranging from radioactivity to fluorescent dyes and even attaching enzymes that catalyze unusual reactions. These probes are allowed to wash over a membrane (often for several hours or even overnight) as part of a watery mix that also contains salt, pH buffers, and detergent. The stringency of the hybridization, particularly salt concentration and temperature, can be manipulated to allow probes to bind to sequences with less than perfect matches. At the end of the hybridization procedure, unbound probe is washed off and the membrane is examined to see where base pairing between the probe and its target sequence has occurred.

A variant of these membrane-based hybridization systems is the powerful **microarray** or DNA chip technology. Here, thousands and even tens of thousands of nucleotide sequences are each affixed to individual positions on the surface of a small silica (glass) chip. Fluorescently labeled copies of the RNA transcripts (cDNAs, described further in Chapter 6) from an organism being studied can then be washed over that surface and allowed to hybridize to complementary nucleotides. After washing, a laser is used to excite the fluorescent tags and then photodetectors quantify the amount of signal associated with each spot of known sequence. A popular application of this methodology results in the determination of relative RNA levels associated with huge numbers of known and predicted genes in a single experiment for a variety of organisms using commercially available microarrays. Quite literally, accurate measurements of *every* single gene (and even every processing variant of every gene) can be precisely assessed. The sensitivity of DNA chip technology is truly remarkable in that it can confirm the presence of as little as one transcript being present in every tenth cell of an experiment. Significant computational efforts are associated with the generation of such chips (ensuring the distinctiveness of each of the thousands of bound probes alone is challenging) as well as the interpretation of their results. (Variation among replicate experiments, evaluation of differences between test and control conditions, and determination of expression associations are all complicated by an abundance of data.)

Cloning

While cells manipulate and extract information from single DNA molecules on a routine basis, molecular biologists typically require quantities of material that are almost visible to the naked eye (many millions of molecules) for most of their analyses. DNA sequencing reactions (described below) in particular require higher purity and larger amounts of DNA than can be practically obtained through restriction enzyme digestion of genomic DNA and gel electrophoresis. A fairly simple solution to this problem has been to invoke the assistance of cells in the generation of sufficient quantities and qualities of specific DNA molecules for such purposes. In essence, **cloning** involves the insertion of specific DNA fragments into chromosome-like carriers called **vectors** that allow their replication in (and isolation from) living cells. Since all the copies of the fragment are identical, they are known as **molecular clones** and they can be purified for immediate study or stored in collections known as libraries for future analyses.

Once a restriction fragment that contains a sequence of particular interest has been generated as described above, its sticky ends can be used to help ligate it into a vector that has been cut with a restriction enzyme that has complementary sticky ends. The first vectors to be used were derived from bacterial viruses and from small extra-chromosomal pieces of DNA in prokaryotic cells called plasmids. These vectors are easy to manipulate in the laboratory and are especially useful for cloning relatively small pieces of DNA (ranging in size from dozens to 25,000 nucleotides in length). Newer alternatives derived from bacterial and yeast chromosomes are better suited for very large fragments of DNA (ranging from 100,000 to 1,000,000 base pairs long), but are not as amenable to handling and characterization. All vectors must have several features in common to be useful to molecular biologists. Those features include sequences that allow them to be replicated inside of living cells, sequences that confer a novel ability to their host cell so their presence can be detected, and distinguishing physical traits (such as size or shape) that allow them to be separated from the host cell's DNA.

A collection of genes, each of which is cloned into a vector, is known as a **genetic library.** An ideal genomic library would contain one copy of every segment of an organism's DNA. For example, if a 4,600,000-nucleotide-long genome (such as *E. coli*'s) were completely digested with a restriction enzyme such as *Eco*RI, then a total of more than 1,000 DNA fragments with an average length of 4,096 base pairs would each need to be cloned to make a complete genomic library. The number of clones (genome size divided by average fragment length) in such a perfect genomic library defines a **genomic equivalent.** Unfortunately, making a genomic library cannot be accomplished by simply digesting the genomic DNA of a single cell and making clones of each fragment. The cloning process is not efficient and it is usually necessary to harvest DNA from hundreds or thousands of cells to clone a single fragment. Further, the random nature of the cloning process ensures that some fragments will be cloned multiple times while others are not represented at all in one genomic equivalent. Increasing the number of clones in a genomic library increases the likelihood that it will contain at least one copy of any given segment of DNA. A genomic library with four to five genomic equivalents has, on average, four to five copies of every DNA segment and a 95% chance of containing at least one copy of any particular portion of the organism's genome. Details of this calculation are provided in Chapter 6. These realities have two practical implications: (1) Vectors that allow the cloning of larger fragments are better for making genomic libraries because fewer clones are needed to make a genomic equivalent, and (2) cloning the last 5% of a genome is often as difficult as cloning the first 95%.

In many cases a useful alternative to a genomic library is a **cDNA library.** The portions of a genome that are typically of greatest interest are those that correspond to the regions that code for proteins. One thing that all protein coding regions have in common is the fact that they are all converted into mRNAs before they are translated by ribosomes. Those mRNAs can be separated from all the other polynucleotides within a cell by means of a special enzyme called **reverse transcriptase,** which converts them back into complementary DNA (cDNA) sequences and then clones those cDNAs as part of a library. Simply showing up in a cDNA library is often enough to attach significance to a portion of a genome since cells usually only make mRNA copies of genes that are functionally important. Further, the relative abundance of cDNAs within a library from any given organism or cell type gives an indication as to how much a particular gene is expressed. A disadvantage to cDNA sequences, though, is that they typically contain only the information that is used by ribosomes in making proteins and not the important regulatory sequences and introns usually associated with genes. As a result, complete understanding of a gene's structure and function usually comes only after characterization of both its genomic and cDNA clone. The creation of screening libraries to determine which clones contain sequences of interest is accomplished by similar kinds of blotting and hybridization strategies used to distinguish between one DNA fragment and another.

Polymerase Chain Reaction

Molecular cloning provides a means of organizing and indefinitely maintaining specific portions of a genome in a way that also allows large quantities of that region to be isolated and used in more detailed analyses. When little information about the sequence of the region is known and large quantities of a region are needed, a powerful alternative to cloning is the use of the **polymerase chain reaction (PCR)** method. Developed by K. Mullis in 1985, PCR relies on an understanding of two idiosyncrasies associated with DNA polymerases (the enzymes responsible for replicating DNA during cell division). First, like RNA polymerases, all DNA polymerases add new nucleotides onto just the 3' (and never the 5') end of a DNA strand during synthesis; hence, there is a definite directionality to DNA synthesis. Second, while it is the job of a DNA polymerase to make double-stranded DNA molecules by using the information inherent to a single-stranded DNA molecule, DNA polymerases can only begin DNA synthesis by adding nucleotides onto the end of an existing DNA strand (Figure 1.12). PCR takes advantage of those two quirks of DNA polymerases to drive the replication of very specific regions of a genome that are of interest to a molecular biologist. One double-stranded copy of such a region can be replicated into two double-stranded copies after one round of amplification. Those two copies can each be duplicated to give rise to four copies during a second round of amplification. In just a couple of hours and after 20 to 30 such rounds of exponential amplification, a specific region of DNA is usually present in enormously higher quantities (theoretically, 2^{20} to 2^{30} or 1,048,576 to 1,073,741,824 copies, assuming that only one copy was present at the start of the process) than other DNA sequences present at the start of the process (Figure 1.12). These amplified DNA molecules are produced much more quickly and efficiently than those obtained from clones yet they can be used in many of the same ways. The amplifying nature of PCR gives it the additional advantage of being able to start with much smaller quantities of material (such as those typically associated with museum or even fossil and forensic specimens) than are usually amenable to cloning experiments.

DNA synthesis occurs only at specific segments of a genome during PCR amplification because of the specific primers that are added to the reaction mixture at the very start of the amplification process. Like the probes used in hybridization experiments, PCR primers are typically 20 or more nucleotides in length to ensure that each can bind specifically to only one target sequence within an organism's genome. The specific sequences used to make primers in the first place typically come from DNA sequence analyses of similar regions in closely related organisms and at some point usually require the more laborious process of cloning and screening described earlier.

DNA Sequencing

The ultimate molecular characterization of any piece of DNA comes from determining the order or sequence of its component nucleotides. All DNA sequencing strategies involve the same three steps: (1) the generation of a complete set of subfragments for the region being studied whose lengths differ from each other by a single nucleotide, (2) labeling of each fragment with one of four different tags that are dependent on the fragment's terminal nucleotide, and (3) separating those fragments by size in a way (usually some form of acrylamide gel electrophoresis) that allows the sequence to be read by detecting the order in which the different tags are seen.

A. M. Maxam and W. Gilbert developed the first successful DNA sequencing strategy in the late 1970s. However, the **Maxam-Gilbert method** relied on chemical degradation to generate the DNA subfragments needed for sequencing, so it quickly fell out of favor when a safer and more efficient DNA polymerase-based method was developed by F. Sanger a few years later. The Sanger approach is sometimes referred to as a **chain-termination method** because the subset of

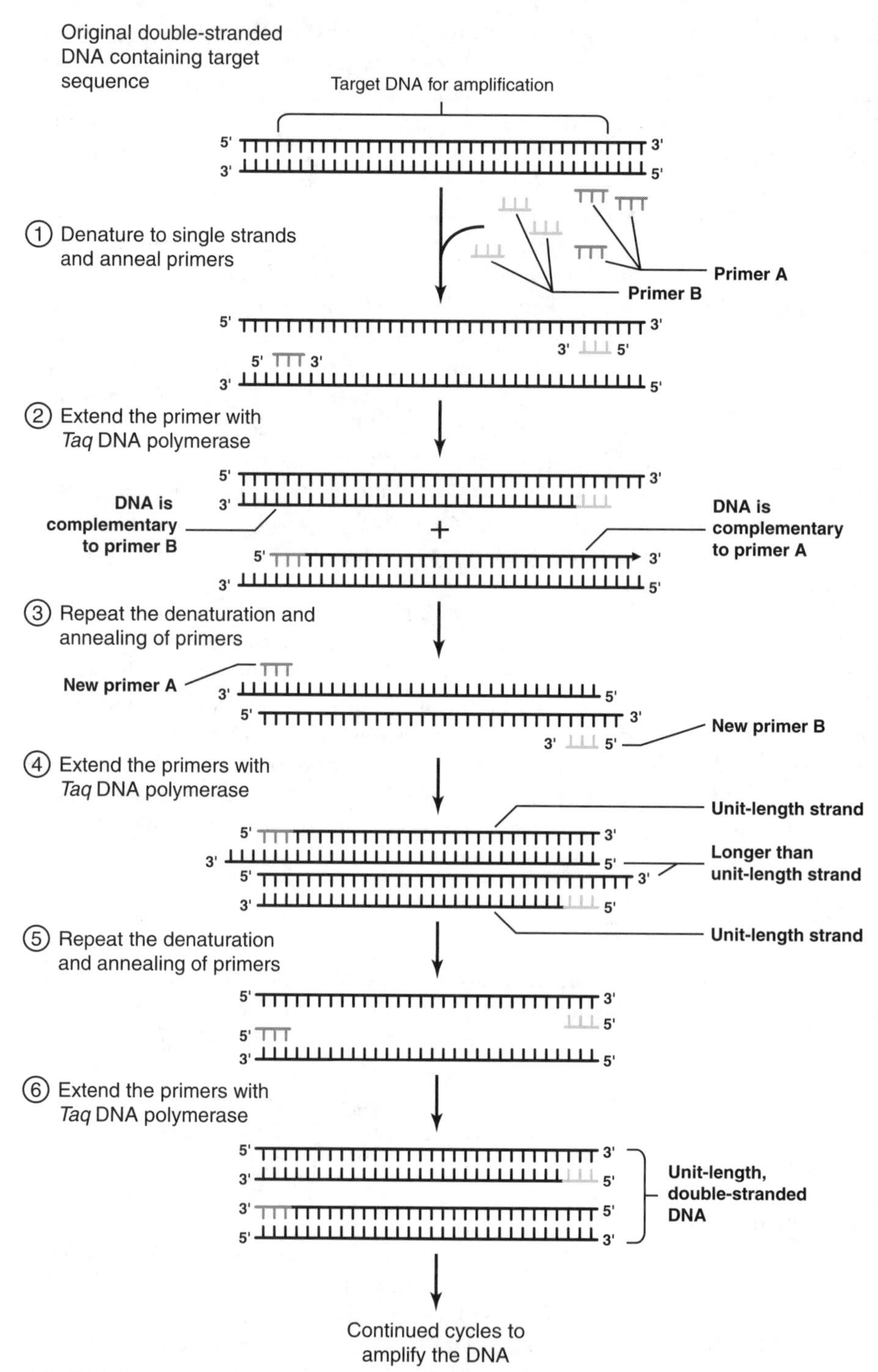

FIGURE 1.12 *The polymerase chain reaction.*

fragments needed for sequencing is generated by the incorporation of modified nucleotides that prevent a DNA polymerase from adding any additional bases to the chain (Figure 1.13). Those chain-terminating nucleotides differ from their normal counterparts in that they are missing their 3' hydroxyl group (see Figure 1.2) onto which the next nucleotide in a growing DNA strand is usually attached. They usually also have a tag such as a fluorescent dye that allows for their detection when they are size fractionated.

In an ideal sequencing reaction the modified and normal nucleotides are mixed in a ratio that allows DNA polymerases to randomly incorporate a chain terminator only once in every 500 or so nucleotides so that a complete set of subfragments from a region up to 1,000 base pairs long can be sequenced at one time. Improvements in the methodology and particularly in the automation of DNA sequencing now make it possible for a single analyst to generate tens and even hundreds of thousands of base pairs of DNA sequence data in a single day—quite a contrast to the 50 base pairs of sequence data for which Gilbert (along with Sanger) shared a Nobel Prize in 1980. Still, the short size (roughly 1,000 base pairs) of each piece of sequence information relative to the overall size of a genome (billions of base pairs in many eukaryotes) and even to genes (sometimes hundreds of thousands of nucleotides long) can make assembling sequences of useful size a computationally challenging task.

Genomic Information Content

As mentioned earlier, an organism's genome can be millions or billions of base pairs long. It was possible to obtain interesting insights into how complex a genome was and how much useful information it contained long before it was possible to determine the order in which its nucleotides were arranged. Even now that automated sequencing has made the sequencing of complete genomes feasible, those earliest approaches used to characterize genomes as a whole still provide useful insights into the quantity and complexity of their genetic information.

C-Value Paradox

In 1948 the discovery was made that the amount of DNA in every cell of a given organism is the same. These measures of a cell's total DNA content are referred to as **C values** (Figure 1.14). Interestingly, while genome size within a species is constant, large variations across species lines have been observed but not in a way that correlated well with organismal complexity. The absence of a perfect correlation between complexity and genome size is often called the **C-value paradox** (Figure 1.14). Total DNA amounts often differ by 100-fold or more even between very similar species. The clear (but difficult to prove) implication is that a large portion of the DNA in some organisms is expendable and does not contribute significantly to an organism's complexity.

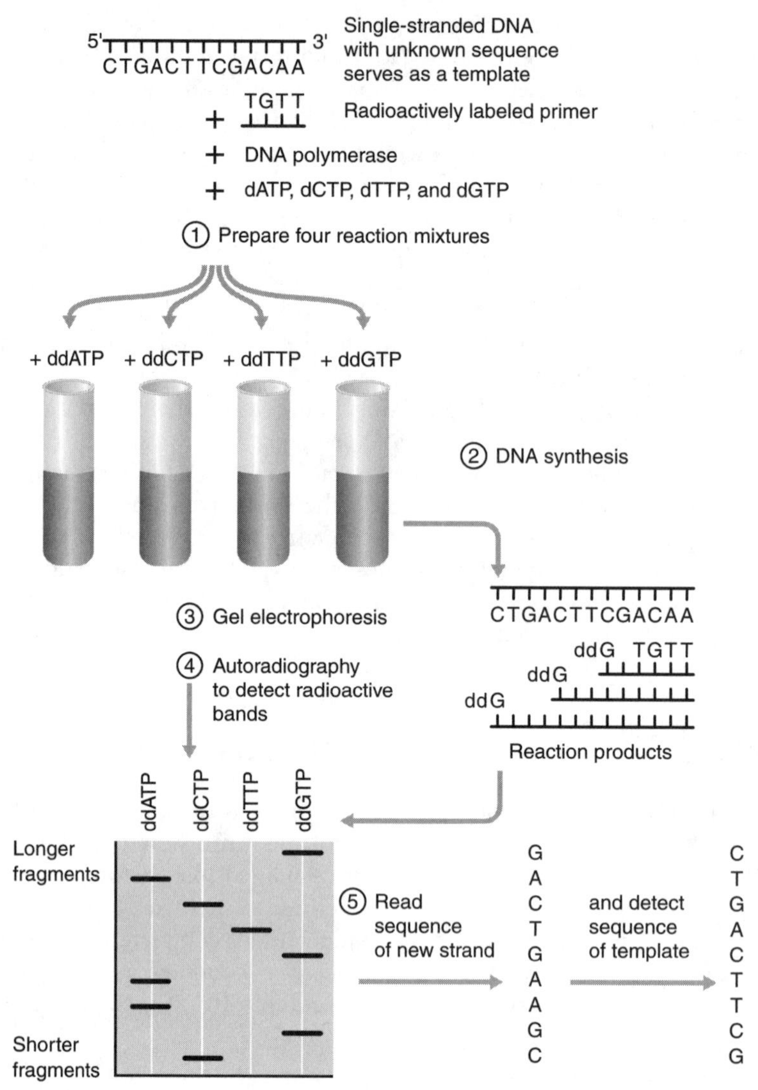

FIGURE 1.13 *The Sanger dideoxy method of DNA sequencing.*

Reassociation Kinetics

When the complementary strands of double-stranded DNA are separated (denatured) by heat or alkali treatment, they can readily reform (renature) a conventional double-stranded structure when conditions are returned to those typically encountered inside a cell. Quite a bit can be learned about the structure of genomes simply by examining the way in which their denatured DNA renatures. In the simplest terms, the more unique a sequence in a genome, the more time it will take for each strand to find and hybridize to its complement. Studies

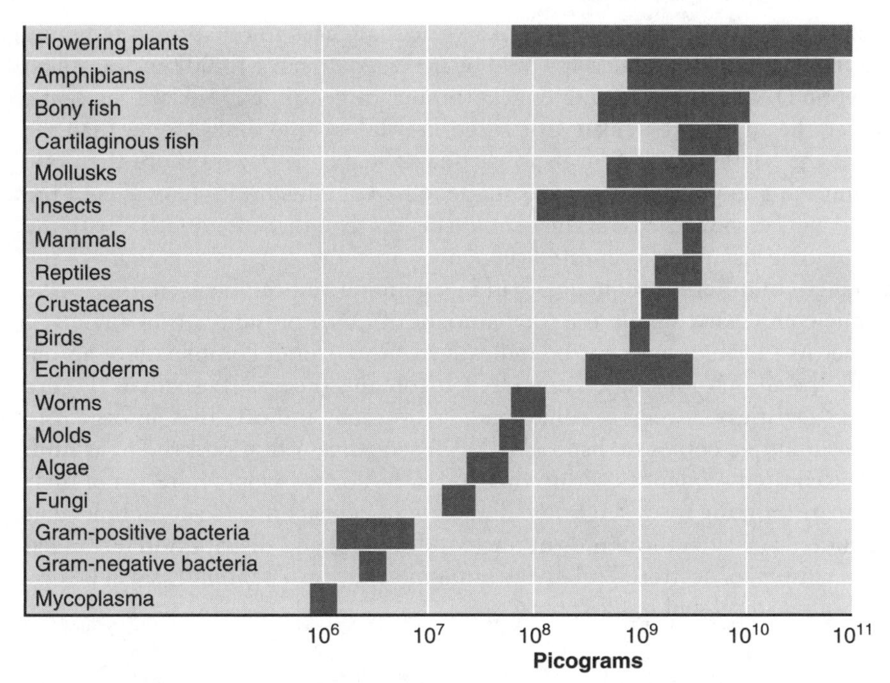

FIGURE 1.14 *The DNA contents of the haploid genomes of a variety of different organisms. C values are generally correlated to morphological complexity in simpler eukaryotes but vary significantly among more complex eukaryotes. The range of DNA content within a phylum is indicated by the shaded areas.*

by R. Britten and others in the 1960s revealed that the time course of DNA renaturation could be conveniently described by a **cot equation.** The cot equation relates the fraction of single-stranded DNA remaining (c/c_0) after t seconds of renaturing multiplied by the amount of denatured genomic DNA at the start of the experiment (c_0). A specific value, $c_0t_{1/2}$, derived from such experiments can be obtained for any organism and is directly proportional to the number of nucleotides of nonrepeated sequence within the organism's DNA. Measuring the time ($t_{1/2}$) required for one-half of the single-stranded DNA to renature ($c/c_0 = 0.5$) allows an experimental determination of the total amount of unique genetic information encoded in a genome. Since the $c_0t_{1/2}$ is the product of the concentration and time required to proceed halfway, a greater $c_0t_{1/2}$ implies a slower reaction and reflects a situation in which there are fewer copies of any particular sequence within a given mass of DNA.

For example, if the c_0 of DNA is 10 picograms (pg), it will contain 2,000 copies of each sequence in a bacterial genome whose size is 0.005 pg, but will contain only 2 copies of each sequence present in a eukaryotic genome of size 5 pg. So, the same absolute amount of DNA (c_0) will provide a concentration of each eukaryotic sequence that is 2,000/2 = 1,000 times lower than that of each bacterial sequence. Since the rate of renaturation depends on the concentration of complementary sequences, for the eukaryotic sequences to renature at the same rate as the bacterial sequences it would be necessary to have 1,000 times as much eukaryotic DNA. If the starting concentrations of genomic DNA are the same, however, the $c_0t_{1/2}$ of the eukaryotic reaction will be 1,000 times longer than the $c_0t_{1/2}$ of the bacterial reaction if no particular sequence occurs more than one time in each genome. If an organism's genome contains multiple copies of the same sequence, then its $c_0t_{1/2}$ values should be less than those for another with the same genome size but no repeated sequences. In short, $c_0t_{1/2}$ is a measure of the total length of different sequences within a genome and can be used to describe genomic complexity. While the total amount of DNA present within any given genome (its C value) may not be indicative of the overall complexity of an organism, the amount of single-copy DNA it contains (its $c_0t_{1/2}$) usually is (Figure 1.14). Disparities between C values and $c_0t_{1/2}$ usually indicate that an organism contains multiple copies of disposable DNA sequences often referred to as **junk DNA.** Repeated sequences within this junk DNA differ widely in terms of their complexity (ranging from single and dinucleotide repeat units to repeat units that are hundreds or even thousands of nucleotides long) and distribution (arranged in local clusters or scattered relatively randomly) within a genome, as will be discussed in greater detail in Chapter 6.

Chapter Summary

DNA is an information storage molecule for cells. The specific order of its four different nucleotides is transcribed by RNA polymerases into mRNAs that are then translated by ribosomes into proteins. Twenty different amino acids are used to make proteins, and the specific order and composition of those building blocks play an important role in establishing and maintaining the structure and function of enzymes. Molecular biologists have a fairly limited set of tools to study DNA and its information content. Restriction enzymes cut DNA molecules when they encounter specific strings of nucleotides. Electrophoresis allows such DNA fragments to be separated on the basis of their size and charge. Blotting and hybridization techniques allow specific DNA fragments to be found within a mixture of other DNA fragments, while cloning allows specific molecules to be propagated and used over and over again. PCR is a popular and versatile alternative to cloning that allows specific DNA fragments to be amplified and characterized. Ultimate characterization of a DNA molecule comes from determining the order of its nucleotides and can be accomplished by DNA sequencing techniques. Reassociation kinetics have revealed that a cell's DNA content (its C value) does not always correspond directly to an organism's information content due to the large amounts of "junk DNA" found in complex organisms.

Readings for Greater Depth

Numerous textbooks give excellent overviews of molecular biology. For a concise description of genes and our understanding of them, try P. Portin, 1993, The concept of the gene: Short history and present status, *Q. Rev. Biol.* **68:** 173–223.

Discovery of the structure of DNA won Watson and Crick a Nobel Prize. Their classic single-page paper describing their insight is J. D. Watson and F. H. C. Crick, 1953, Genetical implications of the structure of deoxyribonucleic acid, *Nature* **171:** 964–967.

Watson himself relates the dramatic and often unscientific race to discover the structure of DNA in a very informative and often humorous book that has also been made into a major motion picture: J. D. Watson, 1968, *The Double Helix: A Personal Account of the Discovery of the Structure of DNA.* Atheneum, New York.

Reassociation experiments and "junk DNA" are both described in R. J. Britten, D. E. Graham, and B. R. Neufeld, 1974, Analysis of repeating DNA sequences by reassociation. *Methods Enzymol.* **29:** 363–418.

A general text that describes the structure of genes and the experimental techniques used to study them is B. Lewin, 2000, *Genes VII,* Oxford University Press, New York.

Questions and Problems

* **1.1** Deoxyribonucleic acid (DNA) differs from ribonucleic acid (RNA) in two ways: (1) RNA uses the nitrogenous base uracil in place of DNA's thymine, and (2) the hydroxyl (OH) group attached to the 2' carbon of the deoxyribose sugar of RNA is replaced with just a hydrogen (H) in DNA. Sketch the chemical structures of the deoxyribose sugar used by DNA and the ribose sugar used by RNA.

1.2 What is the complementary sequence to the following string of nucleotides? Be sure to label the 5' and 3' ends of the sequence that you write. 5'-GGATCGTAGCCTA-3'.

* **1.3** Diagram the "central dogma" of molecular biology complete with labels that indicate the portions that correspond to transcription and translation and indicate what enzymes are responsible for those important steps.

1.4 Organic molecules that contain hydroxyl groups (—OH) are called alcohols. Would you expect such molecules to be hydrophobic or hydrophilic? Why?

* **1.5** Examine the chemical structures of the amino acid R groups shown in Figure 1.5b. What atom(s) is found in the R groups that are in the hydrophilic amino acids that generally is absent in the nonpolar group?

1.6 How frequently would you expect to find the sequence of nucleotides provided in Question 1.2 in a DNA molecule simply as a result of random chance? Assume that each of the four nucleotides occurs with the same frequency.

* **1.7** How many nucleotides long would a DNA sequence need to be in order for it to not be found by chance more than once in a genome whose size is 3 billion base pairs long?

1.8 Distinguish between positive and negative regulation of gene expression.

* **1.9** What sequence of amino acids would the following RNA sequence code for if it were to be translated by a ribosome?: 5'-AUG GGA UGU CGC CGA AAC-3'. What sequence of amino acids would it code for if the first nucleotide were deleted and another "A" were added to the 3' end of the RNA sequence?

1.10 A circular piece of DNA known to be 4,000 bp long is cut into two pieces when treated with the restriction enzyme *Eco*RI: One piece is 3,000 bp long and the other is 1,000 bp long. Another restriction enzyme, *BamHI*, cuts the same DNA molecule into three pieces of the following lengths: 2,500, 1,200, and 300 bp. When both *Eco*RI and *BamHI* are used to cut the DNA molecule together, fragments of the following sizes are generated: 1,600, 1,200, 900, 200, and 100 bp. Use this information to make a restriction enzyme map of this circular DNA molecule.

* **1.11** How does a cDNA library differ from a genomic library?

CHAPTER 2

Data Searches and Pairwise Alignments

It is a capital mistake to theorize before one has data.

Sir Arthur Conan Doyle (1859–1930)

In a very real sense, any alignment between two or more nucleotide or amino acid sequences represents an explicit hypothesis regarding the evolutionary history of those sequences. As a direct result, comparisons of related protein and nucleotide sequences have facilitated many recent advances in understanding the information content and function of genetic sequences. For this reason, techniques for aligning and comparing sequences, and for searching sequence databases for similar sequences, have become cornerstones of bioinformatics.

This chapter describes the methods by which alignments between two or more related nucleotide or polypeptide sequences can be found, evaluated, and used to search through databases of sequence information for genes or proteins relevant to a particular research problem. Closely related sequences are typically easy to align and, in fact, alignment can provide a strong indicator of how closely related the sequences are. Sequence alignments provide important information for solving many of the key problems in bioinformatics including determining the function of a newly discovered genetic sequence; determining the evolutionary relationships among genes, proteins, and entire species; and predicting the structure and function of proteins.

Dot Plots

One of the simplest methods for evaluating similarity between two sequences is to visualize regions of similarity using **dot plots.** To construct a simple dot plot, the first sequence to be compared is assigned to the horizontal axis of a plot space and the second is then assigned to the vertical axis. Dots are then placed in the plot space at each position where both of the sequence elements are identical. Adjacent regions of identity between the two sequences give rise to diagonal lines of dots in the plot (Figure 2.1a).

Such plots quickly become overly complex and crowded when large, similar sequences are compared, as seen in Figure 2.1b. Sliding windows that consider more than just one position at a time are an effective way to deal with this problem. Figure 2.1c illustrates this method for a window size of 10 and also invokes a similarity cutoff of 8. First, nucleotides 1–10 of the *X*-axis sequence are compared with nucleotides 1–10 of the sequence along the *Y* axis. If 8 or more of the 10 nucleotides (nt) in the first comparison are identical, a dot is placed in position (1,1) of the plot space. Next the window is advanced one nucleotide on the *X* axis, so that nucleotides 2–11 of the *X*-axis sequence are now compared with 1–10 of the sequence along the *Y* axis. This procedure is repeated until each 10 nt subsequence of the *X* axis has been compared to nts 1–10 of the *Y* axis. Then the *Y*-axis window is advanced by one nucleotide, and the process repeats until all 10 nt subsequences of both sequences have been compared.

As illustrated in Figure 2.1c, the sliding window can significantly reduce the noise in the dot plot, and make readily apparent the regions of significant similarity between the two sequences. Window sizes and cutoff scores can both be varied easily depending on the similarity of the two sequences being compared. The ultimate objective is typically to choose criteria that draw attention to regions of significant similarity without allowing noise levels to be distracting. A trial and error approach is often best when first analyzing new data sets.

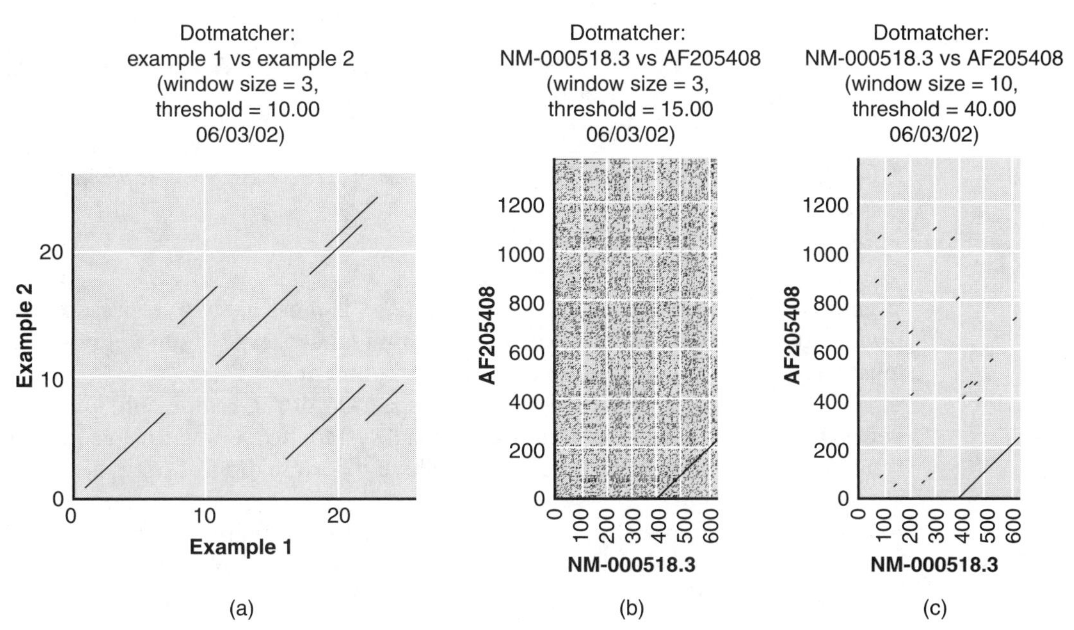

FIGURE 2.1 *(a) A simple dot plot for two DNA sequences that share extensive regions of similarity. (b) A complete dot plot comparing nucleotide sequences from the beta (β) globin genes of human* (Homo sapiens) *and orangutan* (Pongo pygmaeus). *(c) The same two beta (β) globin gene sequences using a window of 10 nucleotides and a similarity cutoff of 8.*

Simple Alignments

An **alignment** between two sequences is simply a pairwise match between the characters of each sequence. A true alignment of nucleotide or amino acid sequences is one that reflects the evolutionary relationship between two or more **homologs** (sequences that share a common ancestor). Homology is not a matter of degree—at any given position in an alignment, sequences (and individual positions) either share a common ancestor or they do not. In contrast, the overall similarity between two sequences can be described as a fractional value. As described in greater detail in Chapter 3, three kinds of changes can occur at any given position within a sequence: (1) a mutation that replaces one character with another, (2) an insertion that adds one or more positions, or (3) a deletion that deletes one or more positions. Insertions and deletions have been found to occur in nature at a significantly lower frequency than mutations. Since there are no homologs of inserted or deleted nucleotides in compared sequences, **gaps** in alignments are commonly added to reflect the occurrence of this type of change.

In the simplest case, where no internal gaps are allowed, aligning two sequences is simply a matter of choosing the starting point for the shorter sequence. Consider the following two short sequences of nucleotides: AATCTATA and AAGATA. These two sequences can be aligned in only three different ways when no gaps are allowed, as shown in Figure 2.2. To determine which of the three

```
AATCTATA        AATCTATA         AATCTATA
AAGATA            AAGATA             AAGATA
```

FIGURE 2.2 *Three possible simple alignments between two short sequences.*

alignments shown in Figure 2.2 is optimal (e.g., is most likely to represent the true relationship between the sequences assuming that they are homologous), we must decide how to evaluate, or score, each alignment.

While dot plots are useful for visual inspection of the regions of similarity between two sequences, a numeric scoring system for evaluating sequence similarity has obvious advantages for objective determination of optimal alignments. In the simple, gap-free alignments shown in Figure 2.2, the scoring function is determined by the amount of credit an alignment receives for each aligned pair of identical residues (the **match score**) and the penalty for aligned pairs of nonidentical residues (the **mismatch score**). The score for a given alignment is:

$$\sum_{i=1}^{n} \begin{cases} \text{match score; if } seq1_i = seq2_i \\ \text{mismatch score; if } seq1_i \neq seq2_i \end{cases}$$

where n is the length of the longer sequence. For example, assuming a match score of 1 and a mismatch score of 0, the scores for the three alignments shown in Figure 2.2 would be 4, 1, and 3, from left to right.

Gaps

Consideration of the possibility of insertion and deletion events significantly complicates sequence alignments by vastly increasing the number of possible alignments between two or more sequences. For example, the two sequences in Figure 2.2, which can be aligned in only three different ways without gaps, can be aligned in 28 different ways when two internal gaps are allowed in the shorter sequence. Figure 2.3 shows just three of those possible alignments.

```
AATCTATA        AATCTATA        AATCTATA
AAG-AT-A        AA-G-ATA        AA--GATA
```

FIGURE 2.3 *Three possible gapped alignments between two short sequences.*

Simple Gap Penalties

In scoring an alignment that includes gaps, an additional term, the **gap penalty,** must be included in the scoring function. A simple alignment score for a gapped alignment can be computed as follows:

$$\sum_{i=1}^{n} \begin{cases} \text{gap penalty; if } seq1_i = \text{'–' or } seq2_i = \text{'–'} \\ \text{match score; if no gaps and } seq1_i = seq2_i \\ \text{mismatch score; if no gaps and } seq1_i \neq seq2_i \end{cases}$$

For example, assuming a match score of 1, a mismatch score of 0, and a gap penalty of –1, the scores for the three gapped alignments shown in Figure 2.3 would be 1, 3, and 3, from left to right. The first alignment in Figure 2.3 with its score of 1 is still the one that is least likely to represent the true evolutionary relationship between these sequences if these criteria are the ones on which we rely.

Origination and Length Penalties

Using simple gap penalties, it is not uncommon to find a number of equally optimal alignments between two sequences. One method to further distinguish between alignments is to differentiate between those that contain many isolated gaps and those that contain fewer, but longer, sequences of gaps. As mentioned earlier, mutations are rare events but insertions and deletions appear to be even less common. In a very real sense, any given pairwise alignment represents a hypothesis about the evolutionary path two sequences have taken since they last shared a common ancestor. When competing hypotheses are being considered, the one that invokes the fewest number of unlikely events is, by definition, the one that is most likely to be correct. Consider two arbitrary sequences of lengths 12 and 9. Any alignment between these two sequences will necessarily have three gaps in the shorter sequence. Assuming the two sequences are truly homologous from beginning to end, the difference in length can be accounted for by nucleotide insertions in the longer sequence, nucleotide deletions in the shorter sequence, or a combination of the two. Since there is no way to determine, without knowing the original precursor sequence, whether a gap was caused by an insertion in one sequence or a deletion in the other, such events are commonly referred to as insertion/deletion, or **indel,** events.

Since multiple nucleotide insertions and deletions are not uncommon relative to single-nucleotide indels, it is statistically more likely that the difference in length between the two sequences is the result of a single 3-nt indel than by multiple insertions or deletions. Thus we can bias our alignment scoring function to reward alignments that are more likely from an evolutionary perspective by assigning a smaller gap penalty for alignments that extend an existing sequence of gaps than for originating a new gap sequence by inserting a gap between two nongap positions. The gap penalty is thus broken into two parts: an **origination penalty** for starting a new series of gaps in one of the sequences being aligned, and a **length penalty** that depends on the number of sequential missing characters. By assigning a length penalty that is smaller than the origination penalty, the scoring function rewards alignments that place gaps together in sequential positions. Consider the three alignments in Figure 2.3. Using an origination penalty of –2, a length penalty of –1, a match score of +1 and a mismatch score of 0, the scores for the three alignments shown are –3, –1, and +1, from left to right. Note that the last two alignments, both of which were scored +3 using a uniform gap penalty, now receive differing scores. The rightmost alignment, which unites the two gap positions into a single indel that is two characters long, is preferred over the middle alignment, which contains two indels that are each one character long.

Scoring Matrices

Just as the gap penalty can be broken down to reward the most evolutionarily likely alignments, the mismatch penalty can be used to provide further discrimination between similar alignments. In our previous examples, each nongap position in which the aligned nucleotides (or amino acids) did not match resulted in the same penalty to the alignment score. Again relying on the assumption that two sequences being aligned are truly homologous, we can immediately observe that some substitutions are more common than others. For example, consider two protein sequences, one of which has an alanine in a given position. A substitution to another small, hydrophobic amino acid, such as valine, would be less likely to have an impact on the function of the resulting protein than a substitution to a large, charged residue such as lysine. Intuitively, it seems that the more conservative substitution would be more likely to maintain a functional protein and less likely to be selected against than a more dramatic substitution. Thus, in scoring an alignment, we might want to score positions in which an alanine is aligned with a valine more favorably than positions in which an alanine is aligned with a bulky or charged amino acid like lysine.

Once the alignment score for each possible pair of nucleotides or residues has been determined, the resulting **scoring matrix** is used to score each nongap position in the alignment. For nucleotide sequence alignments, scoring matrices are generally quite simple. For example, BLAST—a commonly used tool for aligning and searching nucleotide sequences that is described in greater detail later in this chapter—defaults to a very simple matrix that assigns a score of +5 if the two aligned nucleotides are identical, and –4 otherwise. Figure 2.4 illustrates several alternative scoring matrices for nucleotide alignments. The rightmost matrix provides a mild reward for matching nucleotides, a mild penalty for **transitions**—substitutions in which a purine (A or G) is replaced with another purine or a pyrimadine (C or T) replaces another pyrimadine—and a more severe penalty for **transversions,** in which a purine is replaced with a pyrimidine (C or T) or vice versa.

Several criteria can be considered when devising a scoring matrix for amino acid sequence alignments. Two of the most common are based on observed chemical/physical similarity and observed substitution frequencies. For example, in similarity-based matrices, pairing two different amino acids that both have aromatic functional groups might receive a significant positive score, while pairing

Identity Matrix

	A	T	C	G
A	1	0	0	0
T	0	1	0	0
C	0	0	1	0
G	0	0	0	1

BLAST Matrix

	A	T	C	G
A	5	–4	–4	–4
T	–4	5	–4	–4
C	–4	–4	5	–4
G	–4	–4	–4	5

Transition Transversion Matrix

	A	T	C	G
A	1	–5	–5	–1
T	–5	1	–1	–5
C	–5	–1	1	–5
G	–1	–5	–5	1

FIGURE 2.4 *Scoring matrices for aligning DNA sequences.*

an amino acid that has a nonpolar functional group with one that has a charged functional group might result in a scoring penalty. Scoring matrices have been derived based on residue hydrophobicity, charge, electronegativity, and size. Another similarity-based matrix for amino acids is based on the genetic code: A pair of residues is scored according to the minimum number of nucleotide substitutions necessary to convert a codon from one residue to the other (see Table 1.1 in Chapter 1). One problem with similarity-based matrices results from the difficulty of combining these various physical, chemical, and genetic scores into a single meaningful matrix.

A more common method for deriving scoring matrices is to observe the actual substitution rates among the various amino acid residues in nature. If a substitution between two particular amino acids is observed frequently, then positions in which these two residues are aligned are scored favorably. Likewise alignments between residues that are not observed to interchange frequently in natural evolution are penalized. One commonly used scoring matrix based on observed substitution rates is the **point accepted mutation (PAM)** matrix. The scores in a PAM matrix are computed by observing the substitutions that occur in alignments between similar sequences (see Box 2.1). First, an alignment is constructed between sequences with very high (usually >85%) identity. Next, the **relative mutability,** m_j, for each amino acid, j, is computed. The relative mutability is simply the number of times the amino acid was substituted by any other amino acid. For example, the relative mutability of alanine, m_a, is computed by counting the number of times alanine is aligned with non-alanine residues. Next, A_{ij}, the number of times amino acid j was replaced by amino acid i, is tallied for each amino acid pair i and j. For example, A_{cm} is the number of times methionine residues were replaced with cysteine in any pair of aligned sequences. Finally, the substitution tallies (the A_{ij} values) are divided by the relative mutability values, normalized by the frequency of occurrence of each amino acid, and the log of each resulting value is used to compute the entries, R_{ij}, in the PAM-1 matrix. The resulting matrix is sometimes referred to as a **log odds matrix,** since the entries are based on the log of the substitution probability for each amino acid.

The normalization of each matrix entry is done such that the PAM matrix represents substitution probabilities over a fixed unit of evolutionary change. For

BOX 2.1 Construction of a PAM Matrix

1. Construct a multiple sequence alignment. Multiple alignments are discussed further later in this chapter. Below is an example of a simplified multiple alignment:

```
ACGCTAFKI    GCGCTLFKI
GCGCTAFKI    ASGCTAFKL
ACGCTAFKL    ACACTAFKL
GCGCTGFKI
```

2. From the alignment, a phylogenetic tree is created, indicating the order in which the various substitutions shown by the alignment might have taken place. Phylogenetic trees are discussed in detail in Chapters 4 and 5. For now, however, simply note that the tree suggests which substitutions occurred in the sequences involved in the multiple alignment. Consider the following phylogenetic tree, which shows various substitutions among the amino acids in the previous multiple alignment:

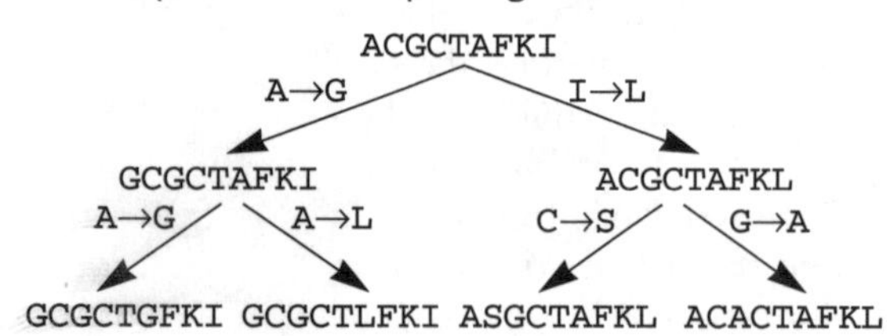

3. For each amino acid type, the frequency with which it is substituted by every other amino acid is calculated. It is assumed that substitutions are equally likely in each direction, so a substitution such as A→G would also count as a G→A substitution. For example, to determine the frequency, $F_{G,A}$, of A to G substitutions, we count all A→G and G→A branches in the tree. For the tree above, $F_{G,A} = 3$.

4. Compute the relative mutability, m_i, of each amino acid. The relative mutability is the number of times the amino acid is substituted by any other amino acid in the phylogenetic tree. This number is then divided by the total number of mutations that *could* have affected the residue. This denominator is calculated as the total number of substitutions across the entire tree times two, multiplied by the frequency of the amino acid, times a scaling factor of 100. The value 100 is used so that the PAM-1 matrix will represent 1 substitution per 100 residues. For example, consider the A residues in the phylogenetic tree above. There are a total of 4 mutations involving A. We divide this value by the number of mutations in the entire tree times two ($6 \times 2 = 12$), times the relative frequency of A residues (10 A's out of 63 total residues in the alignment = 0.159), times 100. Thus $m_A = 4/12 \times 0.159 \times 100 = 0.0209$.

5. Compute the mutation probability, M_{ij}, for each pair of amino acids.

   $M_{ij} = \dfrac{m_j F_{ij}}{\sum_i F_{ij}}$. For our example,

   $$M_{G,A} = \frac{0.0209 \times 3}{4} = 0.0156.$$

   The denominator, ΣF_{ij}, is simply the total number of substitutions involving A in the phylogenetic tree.

6. Finally, each M_{ij} is divided by the frequency of occurrence, f_i, of residue i, and the log of the resulting value becomes the entry R_{ij} in the PAM matrix. The frequency of occurrence is simply the number of occurrences of the residue in the multiple alignment, divided by the total number of residues. For G, the frequency of occurrence from our multiple alignment is $f_G = 0.1587$ (10 G residues divided by 63 total residues). So the entry in our PAM matrix for $R_{G,A}$ would be $\log(0.0156/0.1587) = \log(0.0982) \approx -1.01$.

7. By repeating for each pair of amino acids, we obtain the off-diagonal entries, R_{ij}, of the PAM matrix for our alignment data and phylogenetic tree. The diagonal entries are computed by taking $M_{jj} = 1 - m_j$, and then following step 6 to obtain R_{jj}.

PAM-1, this unit is 1 substitution (or accepted point mutation) per 100 residues, or one **PAM unit.** In other words, the probabilities in the PAM-1 matrix answer the following question: "Suppose I start with a given polypeptide sequence M at time t, and observe the evolutionary changes in the sequence until 1% of all amino acid residues have undergone substitutions at time $t + n$. Let the new sequence at time $t + n$ be called M'. What is the probability that a residue of type j in M will be replaced by i in M'?" The answer to this question can be obtained from entry R_{ij} of the PAM-1 matrix. By multiplying the PAM-1 matrix by itself, we can approximate the substitution rates over multiple PAM units. The particular PAM matrix that is most appropriate for a given sequence alignment depends on the length of the sequences and on how closely the sequences are believed to be related. It is more appropriate to use the PAM-1 matrix to compare sequences that are closely related, whereas the PAM-1000 matrix might be used to compare sequences with very distant relationships. In practice, the PAM-250 matrix is a commonly used compromise.

Another popular scoring matrix, the BLOSUM matrix, is also derived by observing substitution rates among similar protein sequences. For BLOSUM, ungapped alignments of related proteins are grouped using statistical clustering techniques (described in Chapter 4), and substitution rates between the clusters are calculated. This clustering approach helps to avoid some statistical problems that can occur when the observed substitution rate is very low for a particular pair of amino acids. Like the PAM matrices, various BLOSUM matrices can be constructed to compare sequences with different degrees of relatedness. The significance of the numbering for BLOSUM matrices, however, can be thought of as the inverse of the PAM numbers. In other words, lower numbered PAM matrices are appropriate for comparing more closely related sequences, while lower numbered BLOSUM matrices are used for more distantly related sequences. As a rule of thumb, a BLOSUM-62 matrix is appropriate for comparing sequences of approximately 62% sequence similarity, while a BLOSUM-80 matrix is more appropriate for sequences of about 80% similarity.

Dynamic Programming: The Needleman and Wunsch Algorithm

Once a method for scoring alignments is selected, an algorithm to find the best alignment or alignments between two sequences can be developed. The most obvious method, exhaustive search of all possible alignments, is generally not feasible. For example, consider two modest-sized sequences of 100 and 95 nucleotides. If we were to devise an algorithm that computed and scored all possible alignments, our program would have to test ~55 million possible alignments, just to consider the case where exactly five gaps are inserted into the shorter sequence. As the lengths of the sequences grow, the number of possible alignments to search quickly becomes **intractable,** or impossible to compute in a reasonable amount of time. We can overcome this problem by using **dynamic programming,** a method of breaking a problem apart into reasonably sized subproblems, and using these partial results to compute the final answer. S. Needleman and C. Wunsch were the first to apply a dynamic programming approach to the problem of sequence alignment. Their algorithm, which is similar to the one presented below, is one of the cornerstones of bioinformatics.

The key to understanding the dynamic programming approach to sequence alignment lies in observing how the alignment problem is broken down into subproblems. Suppose we have the following two sequences to align: CACGA and CGA. Assume for now that we are using uniform gap and mismatch penalties. There are three possibilities for the first position in our alignment: (1) We can place a gap in the first sequence (not likely in this case since the first sequence is longer), (2) place a gap in the second sequence, or (3) place a gap in neither sequence. For the first two cases the alignment score for the first position will equal the gap penalty, while the rest of the score will depend on how we align the remaining parts of each sequence. For the last case the alignment score for the first position will equal the match bonus, since we are aligning two C's. Again, the rest of the score will depend on how we align the remaining sequences. This breakdown of the problem is illustrated in Figure 2.5.

If we knew the score for the best alignment we could achieve between ACGA and GA, we could immediately compute the score for the first row in the table. Likewise, if we knew the score for the best alignment we could achieve between the remaining sequences for the second and third rows of the table, then we could compute the best score obtainable for each of the three choices for the initial position of our alignment. With all three scores computed, choosing one of the three possibilities would be a simple matter of selecting the one that leads to the best alignment score.

Suppose we begin our alignment by choosing the first row in Figure 2.5, and aligning the initial C in each sequence. We would go on to compute the score for aligning the sequence ACGA and GA. As we continue our search, progressing through all possible sequence alignments, we are likely to encounter the same

First Position	Score	Sequences Remaining to be Aligned
C C	+1	ACGA CGA
– C	–1	CACGA GA
C –	–1	ACGA CGA

FIGURE 2.5 *Three possibilities for aligning the first position in the sequences CACGA and CCGA. The match bonus is +1, the mismatch score is 0, and the gap penalty is –1.*

question many times: What is the best possible score for aligning ACGA and GA? The dynamic programming method uses a table to store partial alignment scores, so that we can avoid recomputing them on the fly.

The dynamic programming algorithm computes optimal sequence alignments by filling in a table of partial sequence alignment scores until the score for the entire sequence alignment has been calculated. The algorithm utilizes a table in which the horizontal and vertical axes are labeled with the two sequences to be aligned. Figure 2.6 illustrates the partial alignment score table for the two sequences ACAGTAG and ACTCG, where the gap penalty is –1, the match bonus is +1, and the mismatch score is 0. An alignment of the two sequences is equivalent to a path from the upper left corner of the table to the lower right. A horizontal move in the table represents a gap in the sequence along the left axis. A vertical move represents a gap in the sequence along the top axis, and a diagonal move represents an alignment of the nucleotides from each sequence.

		A	C	T	C	G
	0	–1	–2	–3	–4	–5
A	–1	1				
C	–2					
A	–3					
G	–4					
T	–5					
A	–6					
G	–7					

FIGURE 2.6 *A partial scores table for aligning sequences ACTCG and ACAGTAG. The gap penalty is –1, the match score is +1, and the mismatch score is 0. The first entry to be filled in is position (2,2) (score shown in gray).*

At the outset of the algorithm, the first row and column of the table are initialized with multiples of the gap penalty, as shown in Figure 2.6. We begin filling in the table with position (2,2), the second entry in the second row. This position represents the first column of our alignment. Recall that we have three possibilities for this first position: a gap in the first sequence, a gap in the second sequence, or an alignment of the nucleotides from each sequence (no gap). Likewise, we can fill the first position in the table with one of three possible values:

1. We can take the value from the left (2,1) and add the gap penalty, representing a gap in the sequence along the left axis;
2. We can take the value from above (1,2) and add the gap penalty, representing a gap in the sequence along the top axis; or
3. We can take the value from the diagonal element above and to the left (1,1) and add the match bonus or mismatch penalty for the two nucleotides along the axes, representing an alignment of the two nucleotides.

To fill in the table, we take the maximum value of these three choices. For the example in Figure 2.6, we obtain the values –2, –2, and 1, respectively, for the three options, and so we select the maximal value, 1. This is equivalent to an alignment between the initial A in each of the two sequences. Once we have position (2,2) filled, we can fill in the rest of row 2 in a similar manner, followed by row 3, and likewise for the rest of the table. Figure 2.7 illustrates the completed table. By way of example, consider position (2,3) in the table. The three choices for filling in this position are:

		A	C	T	C	G
	0	–1	–2	–3	–4	–5
A	–1	1	0	–1	–2	–3
C	–2	0	2	1	0	–1
A	–3	–1	1	2	1	0
G	–4	–2	0	1	2	2
T	–5	–3	–1	1	1	2
A	–6	–4	–2	0	1	1
G	–7	–5	–3	–1	0	2

FIGURE 2.7 *The completion of the partial scores table from Figure 2.6.*

1. We can take the value from the left (1) and add the gap penalty (–1), resulting in the value 0;

2. We can take the value from above (–2) and add the gap penalty, resulting in the value –3; or

3. We can take the value from the diagonal element above and to the left (–1) and add the mismatch score (0), since the two nucleotides for this position (A and C) do not match, resulting in the value –1.

Since we take the maximum of these three choices, we place the value 0 in position (2,3).

Once the table has been completed, the value in the lower right represents the score for the optimal gapped alignment between the two sequences. For the example shown in Figure 2.7, we can see that the score for the optimal alignment will be 1. Note that we have determined this score without having to exhaustively score all possible alignments between the two sequences. Furthermore, now that we have the partial scores table, we can reconstruct the optimal alignment(s) between the two sequences. Sometimes we will find more than one alignment that achieves the optimal score, as is the case in this example.

To reconstruct the alignment from the scores table, we need to find a path from the lower rightmost entry in the table to the upper leftmost position. To form this path, we can move from our current position in the table to any other position that could legally have produced the score in our current position. As an example, look at the lower rightmost position in Figure 2.7. The score in this position is 2. Of the three choices for forming this score, only one could have produced a score of 2: taking the diagonal element, which has score 0, and adding the match penalty. Since this is the only possible way to obtain the 2 in the current column, we draw an arrow to the diagonal element, as shown in Figure 2.8. From this new position (7,5) there is likewise only one position from which the score could have been obtained, and again it is from the diagonal element. As before, we draw an arrow to this position. We continue this process until all possible paths are completed back to position (1,1). These paths now represent all the optimal alignments between the two sequences.

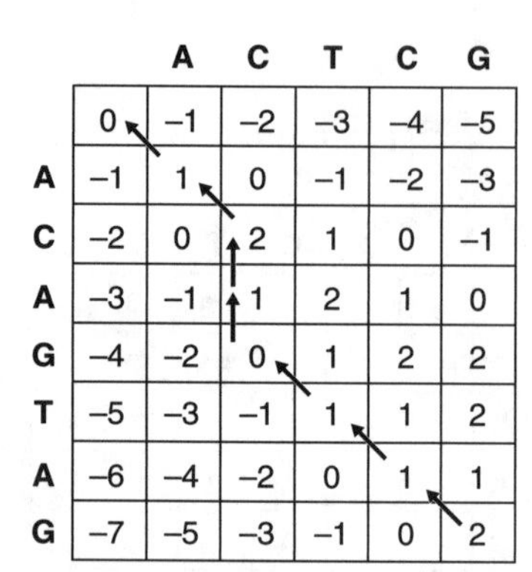

		A	C	T	C	G
	0	–1	–2	–3	–4	–5
A	–1	1	0	–1	–2	–3
C	–2	0	2	1	0	–1
A	–3	–1	1	2	1	0
G	–4	–2	0	1	2	2
T	–5	–3	–1	1	1	2
A	–6	–4	–2	0	1	1
G	–7	–5	–3	–1	0	2

FIGURE 2.8 *Arrows indicating all allowed paths from the lower right corner of the partial scores table to the upper left corner. Each path represents one of several equally optimal alignments.*

To convert a path to an alignment, we simply recall our original interpretation of the partial scores array. A vertical move represents a gap in the sequence along the top axis, a horizontal move represents a gap in the sequence along the left axis, and a diagonal move represents an alignment of the nucleotides from each sequence at the current position. For example, consider the path highlighted in Figure 2.8. The sequence of moves in this path, from lower right to upper left, is ↖↖↑↑↖↖↖. Using these moves we reconstruct the alignment "backwards," from right to left. The first arrow is diagonal, so we align the last two nucleotides:

```
G
G
```

The next two arrows are also diagonal so we align the next two positions as well:

```
TCG
TAG
```

The next arrow is vertical, so we place a gap in the top sequence aligned with the next nucleotide from the left sequence:

```
-TCG
GTAG
```

Continuing in this manner, from right to left, we obtain the following optimal alignment, with an alignment score of 2:

```
AC--TCG
ACAGTAG
```

If other paths were present (Figure 2.8), we would know there were other optimal alignments, which also would receive a score of 2. By following all the paths in the partial scores table, we can reconstruct all possible optimal alignments between the two sequences.

Global and Local Alignments

Semiglobal Alignments

The basic alignment algorithm discussed so far performs a **global alignment.** That is, it compares two sequences in their entirety; the gap penalty is assessed regardless of whether gaps are located internally within a sequence, or at the end of one or both sequences. This is not always the most desirable way to align two sequences. For example, suppose we wish to search for the short sequence ACGT within the longer sequence AAACACGTGTCT. Of the several possible alignments between these two sequences, the one we are most interested in is:

```
AACACGTGTCT
---ACGT----
```

This is the most interesting alignment because it demonstrates that the shorter sequence appears in its entirety within the longer sequence. When searching for the best alignment between a short sequence and an entire genome, for example, we might wish to avoid penalizing for gaps that appear at one or both ends of a sequence. **Terminal gaps** are usually the result of incomplete data acquisition and do not have biological significance—hence it is appropriate to treat them differently than internal gaps. This approach is sometimes referred to as a **semiglobal alignment.** Fortunately, we can conduct such an alignment by making a few small changes to our basic dynamic programming algorithm.

Recall that in our original algorithm, a vertical move in the table was equivalent to a gap in the sequence on the topmost axis. It should be clear that by moving vertically to the bottom of the table, and then horizontally to the rightmost edge, we would produce the following alignment for the sequences in Figure 2.6:

```
-------ACTCG
ACAGTAG-----
```

From the top left corner of the table, each downward move adds an additional gap to the beginning of the first sequence in the alignment. Since each gap adds a gap penalty to the overall alignment score (remember, we are using simple, uniform gap penalties), we initialize the first column of the table with multiples of the gap penalty. Now, suppose we wish to allow initial gaps in the first sequence with no penalty. We can make this change quite easily by simply initializing the first column of the table to all zeros. Now each vertical move in the first column does not add a gap penalty to the alignment score. Likewise, by initializing the first row of the table to all zeros, we can allow initial gaps in the second sequence without penalty.

To allow gaps at the end of a sequence without penalty, we must change our interpretation of the table slightly. Suppose we have the following alignment:

```
ACACTGATCG
ACACTG----
```

If we use this alignment to construct a path through a partial alignment scores table, after the first six nucleotides are aligned, we would find ourselves at the bottom row of the table, as illustrated in Figure 2.9. To reach the lower right corner of the table, we need to make four horizontal moves in this row. Under our previous scoring strategy, each of these moves would incur a gap penalty for the alignment. We can easily rectify this by allowing horizontal moves in the bottom row of the table with no gap penalty. Likewise, we can allow end gaps in the first sequence by allowing vertical moves in the last column of the table without assessing a gap penalty. By initializing the first row and column of our partial scores table with zero values, and by allowing free horizontal and vertical moves in the last row and column of the table, respectively, we now have a modification of our Needleman and Wunsch algorithm that searches for semiglobal alignments.

		A	C	A	C	T	G	A	T	C	G
	0	0	0	0	0	0	0	0	0	0	0
A	0	1	0	1	0	0	0	1	0	0	0
C	0	0	2	1	2	1	0	0	1	1	0
A	0	1	1	3	2	2	1	1	0	1	1
C	0	0	2	2	4	3	2	1	1	1	1
T	0	0	1	2	3	5	4	3	2	1	1
G	0	0	0	1	2	3	6	6	6	6	6

FIGURE 2.9 *The leftmost column and top row of this partial alignment scores table have been initialized to the value zero. Additionally, horizontal moves in the bottom row and vertical moves in the rightmost column are penalty free. This allow gaps at the start or end of either sequence to remain unpenalized.*

The Smith-Waterman Algorithm

Sometimes even semiglobal alignments do not afford the flexibility needed in a sequence search. For example, suppose you have a long sequence of DNA, and you would like to find any subsequences that are similar to any part of the yeast genome. For this sort of comparison, a semiglobal alignment will not suffice, since each alignment will be penalized for every nonmatching position. Even if there is an interesting subsequence that matches part of the yeast genome, all of the non-matching residues are likely to produce an abysmal alignment score. The appropriate tool for this sort of search is a **local alignment,** which will find the best matching subsequences within the two search sequences. Consider, by way of example, the following two sequences: AACCTATAGCT and GCGATATA. Using our semiglobal alignment algorithm, and using a gap penalty of –1, a match bonus of +1, and a mismatch score of –1, we will obtain the following alignment:

```
AAC-CTATAGCT
-GCGATATA---
```

In one view, this is a fairly poor alignment; four of the first five positions in the alignment are mismatches or gaps, as are the last three positions. However, this alignment does reveal that there is a matching region in the center of the two sequences: the subsequence TATA. With minimal modifications, our dynamic programming method can be used to identify subsequence matches while ignoring mismatches and gaps before and after the matching region. The resulting algorithm was first introduced by F. Smith and M. Waterman in 1981, and is a fundamental technique in bioinformatics.

To perform a local alignment, we modify our global alignment algorithm by allowing a fourth option when filling in the partial scores table. Specifically, we can place a zero in any position in the table if all of the other methods result in scores lower than zero. Once the table is completed in this manner, we simply find the maximum partial alignment score in the entire table, and work backwards, as before, constructing our alignment until we reach a zero. The resulting local alignment will represent the best matching subsequence between the two sequences being compared. Figure 2.10 illustrates the partial scores matrix for the two sequences in our previous example. Recall that the global alignment illustrated the matching subsequence TATA. The maximal value in the partial alignment scores table in Figure 2.10 is 4. Starting with this position, and working backward until we reach a value of 0, we obtain the following alignment:

```
TATA
TATA
```

		A	A	C	C	T	A	T	A	G	C	T
	0	0	0	0	0	0	0	0	0	0	0	0
G	0	0	0	0	0	0	0	0	0	1	0	0
C	0	0	0	1	1	0	0	0	0	0	2	1
G	0	0	0	0	0	0	0	0	0	1	0	1
A	0	1	1	0	0	0	1	0	1	0	0	0
T	0	0	0	0	0	1	0	2	1	0	0	1
A	0	1	1	0	0	0	2	0	3	2	1	0
T	0	0	0	0	0	1	1	3	2	2	1	2
A	0	1	1	0	0	0	2	2	**4**	3	2	1

FIGURE 2.10 *By allowing the option of placing a zero in the partial scores table at any position where a positive value cannot be obtained, the dynamic programming method can be modified to search for local alignments. Here, the best local alignment between the sequences AACCTATAGCT and GCGATATA is represented by the maximal value (4) in the table.*

The local alignment algorithm has identified exactly the subsequence match that we identified from our previous semiglobal alignment. When working with long sequences of many thousands, or even millions, of nucleotides, local alignment methods can identify subsequence matches that would be impossible to find using global or semiglobal alignments.

Database Searches

While sequence alignments can be an invaluable tool for comparing two known sequences, a far more common use of alignments is to search through a database of many sequences to retrieve those that are similar to a particular sequence. If, for example, we had identified a region of the human genome that we believe is a previously unidentified gene, we might compare our putative gene with the millions of other sequences in the GenBank database at the National Center for Biological Information (NCBI). The search results, consisting of other sequences that align well with (and thus are similar to) our sequence, might give us an indication of the functional role of our newfound gene along with valuable clues regarding its regulation and expression and its relationship to similar genes in humans and other species.

In performing database searches, the size and sheer number of sequences to be searched (at the time of the writing of this text, there were more than 13 million sequences in GenBank) often precludes the obvious and direct approach of aligning a query sequence with each sequence in the database and returning the sequences with the highest alignment scores. Instead, various indexing schemes and heuristics must be used to speed the search process. Many of the commonly used database search algorithms are not guaranteed to produce the best match from the database, but rather have a high probability of returning most of the sequences that align well with the query sequence. Nevertheless, the efficiency of these tools in finding sequences similar to a query sequence from the vast repositories of available sequence data has made them invaluable tools in the study of molecular biology.

BLAST and Its Relatives

One of the most well known and commonly used tools for searching sequence databases is the BLAST algorithm, introduced by S. Altschul *et al.* in the early 1990s. The original BLAST algorithm searches a sequence database for maximal ungapped local alignments. In other words, BLAST finds subsequences from the database that are similar to subsequences in the query sequence. Several variations of the BLAST algorithm are available for searching protein or nucleotide sequence databases using protein or nucleotide query sequences. To illustrate the basic concepts of BLAST searches, we will discuss the BLASTP algorithm, which searches for protein sequence matches using PAM or BLOSUM matrices to score the ungapped alignments.

To search a large database efficiently, BLASTP first breaks down the query sequence into **words,** or subsequences of a fixed length (4 is the default word length). All possible words in the query sequence are calculated by sliding a window equal in size to the word length over the query sequence. For example, a protein query sequence of AILVPTV would produce four different words: AILV (4 characters, starting with the first character), ILVP (starting with the second character), LVPT, and VPTV. Once all of the words in the query sequence have been determined, words composed mostly of common amino acids will be discarded. The sequences in the database are then searched for occurrences of the search words. Each time a word match is found in the database, the match is extended in both directions from the matching word until the alignment score falls below a given threshold. Since the alignment is ungapped, the extension only involves adding additional residues to the matching region and recalculating the score according to the scoring matrix. The choice of the threshold value for continuing the extension is an important search parameter, because it determines how likely the resulting sequences are to be biologically relevant homologs of the query sequence. Figure 2.11 shows a simplified overview of the BLASTP search process for a simple polypeptide sequence.

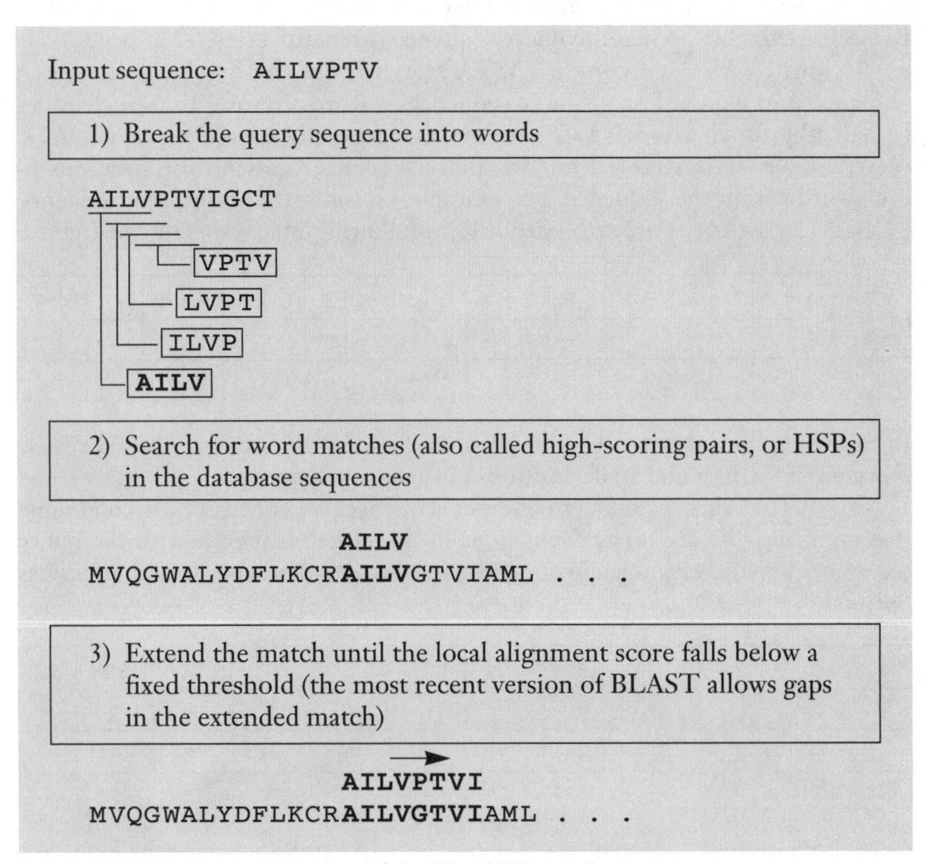

FIGURE 2.11 *Overview of the BLASTP search process.*

Numerous sequence alignment and database search algorithms have been developed for various specific types of sequence searches. As mentioned previously, BLASTP searches protein sequence databases for polypeptide sequences. Other variations of BLAST, including BLASTN and BLASTX, allow searching of nucleotide sequence databases and translating from nucleotide sequences to protein sequences prior to searching, respectively. BLAST 2.0, the most recent version of BLAST, inserts gaps to optimize the alignment. PSI-BLAST, another member of the BLAST family, summarizes results of sequence searches into **position-specific scoring matrices,** which are useful for protein modeling and structure prediction.

FASTA and Related Algorithms

The FASTX algorithms are another commonly used family of alignment and search tools. FASTA and its relatives perform gapped local alignments between sequences. Since FASTX searches perform several detailed comparisons between the query sequence and each sequence in the database, FASTX searches generally require significantly more execution time than the BLAST searches. However, the FASTX algorithms are considered by some to be more sensitive than BLAST, particularly when the query sequence is repetitive.

As with BLAST searches, a FASTA search begins by breaking the search sequence into words. For genomic sequences a word size of 4 to 6 nucleotides is generally used, while 1 to 2 residues are generally used for polypeptides. Next, a table is constructed for the query sequence showing the locations of each word within the sequence. For example, consider the amino acid sequence `FAMLGFIKYLPGCM`. For a word size of 1, the following table would be constructed:

Word	A	C	D	E	F	G	H	I	K	L	M	N	P	Q	R	S	T	V	W	Y
Pos.	2	13			1	5		7	8	4	3		11							9
					6	12				10	14									

In this table, the column for phenylalanine (F) contains entries 1 and 6 because F occurs in the first and sixth positions of the query sequence.

To compare this sequence to a target sequence, we construct a second table that compares the amino acid positions in the target sequence with the query sequence. For the target sequence `TGFIKYLPGACT` this table would appear as follows:

1	2	3	4	5	6	7	8	9	10	11	12
T	G	F	I	K	Y	L	P	G	A	C	T
	3	–2	3	3	3	–3	3	–4	–8	2	
	10	3				3		3			

Consider position 2, a glycine (G) residue. Looking at the table for the query sequence, we can quickly see that glycines are present in positions 5 and 12 of the query sequence. The distance between 5 and 12 and the position of the first glycine in the target sequence (position 2) produces the two entries 3 and 10. For the second glycine, in position 9, we likewise subtract 9 from 5 and 12, obtaining entries –4 and 3. Amino acids, such as threonine (T), that are not found in the query sequence are not included in this table.

Note the large number of instances of the distance 3 in the second table. This suggests that by offsetting the target sequence by 3, we might obtain a reasonable alignment between the two sequences. In fact, we would obtain the following:

```
FAMLGFIKYLPGCM
    ||||||||
   TGFIKYLPGACT
```

By comparing the offset tables for two sequences, areas of identity can be found quickly. Once these areas are found, they are joined to form larger sequences, which are then aligned using a full Smith-Waterman alignment. However, because the alignment is constrained to a known region of similar sequence, FASTA is much faster than performing a complete dynamic programming alignment between the query sequence and all possible targets.

Alignment Scores and Statistical Significance of Database Searches

While a database search will always produce a result, the sequences found cannot be assumed to be related to the search sequence without more information. The primary indicator of how similar the search results are to a query sequence is the alignment score. Alignment scores, however, vary among the different database search algorithms, and are not, of themselves, a sufficient indicator that two sequences are related. Given a database search result with an alignment score S, an appropriate question to ask is "Given a set of sequences *not related* to the query sequence (or even random sequences), what is the probability of finding a match with alignment score S simply by chance?" To answer this question, database search engines generally provide a P score or an E score along with each search result. While they answer slightly different questions, the two scores are closely related, and often have very similar values. Given a database result with an alignment score S, the E score is the expected number of sequences of score $> = S$ that would be found by random chance. The P score is the probability that *one or more* sequences of score $> = S$ would have been found randomly. Low values of E and P indicate that the search result was unlikely to have been obtained by random chance, and thus is likely to bear an evolutionary relationship to the query sequence. While E values of 10^{-3} and below are often considered indicative of statistically significant results, it is not uncommon for search algorithms to produce matches with E values on the order of 10^{-50}, indicating a very strong likelihood of evolutionary relationship between the query sequence and the search results.

Multiple Sequence Alignments

While all of the alignment algorithms discussed thus far are designed to perform pairwise alignments, it is often necessary to simultaneously align a number of sequences. For example, when observing a number of sequences in order to determine substitution frequencies, a **multiple sequence alignment,** which aligns the gaps among all the sequences as much is possible, is often preferable to a set of pairwise alignments. Multiple sequence alignments are also vital for the creation of scoring matrices, such as the PAM and BLOSUM matrices, discussed earlier in this chapter.

The most straightforward techniques for performing multiple alignments are logical extensions of the dynamic programming methods we have discussed so far. For aligning *n* sequences, an *n*-dimensional array is used instead of the two-dimensional array used in the Needleman-Wunsch algorithms, but the algorithm is otherwise the same. Unfortunately, the computational complexity of multiple alignment methods grows rapidly with the number of sequences being aligned. Even using supercomputers or networks of workstations, multiple sequence alignment is an intractable problem for more than 20 or so sequences of average length and complexity. As a result, alignment methods using heuristics have been developed. These methods, including the well-known CLUSTAL algorithm, cannot guarantee an optimal alignment, but can find near-optimal alignments for larger numbers of sequences than would be possible with full dynamic programming techniques.

The CLUSTAL algorithm, first described by D. G. Higgins and P. M. Sharp in 1988, begins by aligning closely related sequences and then adds increasingly divergent sequences to produce a complete multiple sequence alignment. First, the algorithm constructs a phylogenetic tree (see Chapters 4 and 5) to determine the degrees of similarity among the sequences being aligned. Using this tree as a guide, closely related sequences are aligned two at a time using dynamic programming for the pairwise alignments.

Selection of a scoring matrix for alignments can present a significant problem in multiple sequence alignments. Some matrices, such as PAM-1 and BLOSUM-90, are appropriate for closely related sequences, while others, such as PAM-1000 and BLOSUM-35, might be more appropriate for very divergent sequences. Use of an inappropriate scoring matrix will generally result in a poor alignment. In CLUSTALW, the most recent version of the CLUSTAL algorithm, sequences are weighted according to how divergent they are from the most closely related pair of sequences, and the gap opening and gap extension penalties, as well as the scoring matrix selection, are based on the weight of each sequence.

An additional consideration for scoring multiple sequence alignments is that it is now possible for two sequences in the alignment to have a gap in the same position. A common strategy is to assign a score of zero for aligned gap positions.

Note also that, like pairwise methods, multiple sequence alignments are based solely on nucleotide or amino acid similarity between sequences. The goal of a multiple alignment is generally to align regions of similar structural or functional importance among sequences. While sequence similarity is an important indicator of related function, it is often the case that a molecular biologist has additional knowledge about the structure or function of a particular protein or gene. Information such as the locations of secondary structure elements, surface loop regions, and active sites are often used to tune multiple alignments by hand in order to produce biologically meaningful results.

Chapter Summary

An alignment between two or more genetic or amino acid sequences represents a hypothesis about the evolutionary path by which the two sequences diverged from a common ancestor. While the true evolutionary path cannot be inferred with certainty, sequence alignment algorithms can be used to identify alignments with a low probability of occurrence by chance. The selection of a scoring function has a significant bearing on the results of a sequence alignment. Various techniques are available to bias scoring functions to discover evolutionarily likely alignments, including the use of scoring matrices such as the PAM and BLOSUM matrices. The algorithm first described by Needleman and Wunsch for global sequence alignment and also the local alignment method of Smith and Waterman have become the cornerstones on which numerous database search algorithms, including the BLASTX and FASTX tools, have been built. These algorithms use indexing, heuristics, and fast comparison techniques to allow an entire database of sequences to be rapidly compared with a query sequence.

Readings for Greater Depth

The PAM matrix was first described in M. Dayhoff, R. M. Schwartz, and B. C. Orcutt, 1978, A model of evolutionary change in proteins, in *Atlas of Protein Sequence and Structure*, Vol. 5, pp. 345–352, National Biomedical Research Foundation, Silver Spring, MD.

Another approach to scoring alignments, the BLOSUM matrix, was introduced in S. Henikoff and J. G. Henikoff, 1992, Amino acid substitution matrices from protein blocks, *Proc. Nat. Acad. Sci. U.S.A.* **89:** 10915–10919.

Needleman and Wunsch's plenary algorithm for global sequence alignment was detailed in S. B. Needleman and C. D. Wunsch, 1970, A general method applicable to the search for similarities in the amino acid sequences of two proteins, *J. Mol. Biol.* **48:** 443–453.

Smith and Waterman's ubiquitous algorithm for local alignments is succinctly described in F. F. Smith and M. S. Waterman, 1981, Identification of common molecular subsequences, *J. Mol. Biol.* **147:** 195–197.

The well-known BLAST engine for database searching is described in S. F. Altschul, W. Gish, W. Miller, E. W. Myers, and D. J. Lipman, 1990, A basic local alignment search tool, *J. Mol. Biol.* **215:** 403–410.

FAST, an alternative sequence search and alignment method, was first detailed in D. J. Lipman and W. R. Pearson, 1985, Rapid and sensitive protein similarity search, *Science* **227:** 1435–1441.

The CLUSTAL algorithm for multiple sequence alignment was first described in D. G. Higgins and P. M. Sharp, 1988, CLUSTAL: A package for performing multiple sequence alignment on a microcomputer, *Gene* **73:** 237–244. CLUSTALW, a commonly used variant of the original algorithm, is described in J. D. Thompson, D. G. Higgins, and T. J. Gibson, 1994, CLUSTALW: Improving the sensitivity of progressive multiple sequence alignment through sequence weighting, positions-specific gap penalties and weight matrix choice, *Nucl. Acids Res.* **22:** 4673–4680.

MUSCA, an alternative multiple sequence alignment method from IBM's bioinformatics group, is described in L. Parida, A. Floratos, and I. Rigoutsos, 1999, An approximation algorithm for alignment of multiple sequences using motif discovery. *J. Combinatorial Optimization* **3:** 247–275.

MultAlin, another multiple sequence alignment method, is described in F. Corpet, 1988, Multiple sequence alignment with hierarchical clustering, *Nucl. Acids Res.* **16:** 10881–10890.

Questions and Problems

* **2.1** What are some situations in which a molecular biologist might wish to perform a pairwise sequence alignment? A multiple sequence alignment? A sequence database search?

2.2 For the following two sequences, construct a simple dot plot using graph paper. Place each sequence along one axis, and place a dot in the plot for each identical pair of nucleotides:

```
GCTAGTCAGATCTGACGCTA
GATGGTCACATCTGCCGC
```

Does your dot plot reveal any regions of similarity?

* **2.3** For the two sequences in Question 2.2, construct a dot plot using a sliding window of size 4 and a similarity cutoff of three nucleotides. Does this plot reveal any regions of similarity between the two regions?

2.4 Determine the alignment score for each of the following sequence alignments:

a. Global alignment: match score = +1, mismatch score = 0, gap penalty = –1

```
TGTACGGCTATA
TC--CGCCT-TA
```

b. Global alignment: match score = +1, mismatch score = 0, gap penalty = –1

```
--TCTGTACGCGATCATGT
TAGC-GTCCGATAT-A---
```

c. Global alignment: match score = +1, mismatch score = –1, origination penalty = –2, length penalty = –1

```
AGATAGAAACTGATATATA
AGA-A-A-ACAGAG-T---
```

d. Global alignment: match score = +1, mismatch score = –1, origination penalty = –2, length penalty = –1

```
AGATAGAAACTGATATATA
AG---AAAACAGAGT----
```

e. Semiglobal alignment: match score = +1, mismatch score = –1, origination penalty = –2, length penalty = –1

```
AGATAGAAACTGATATATA
AG---AAAACAGAGT----
```

* **2.5** Using the Needleman and Wunsch dynamic programming method, construct the partial alignment score table for the following two sequences, using the following scoring parameters: match score = +1, mismatch score = 0, gap penalty = –1.

```
ACAGTCGAACG
ACCGTCCG
```

What is the optimal global alignment between these sequences?

2.6 Using the same scoring parameters as in Question 2.5, use the modified Needleman and Wunsch method to compute the optimal semiglobal alignment for the two sequences in Question 2.5.

* **2.7** Using the Smith-Waterman method, construct the partial alignment scoring table for a local alignment of the following two sequences:

```
ACGTATCGCGTATA
GATGCTCTCGGAAA
```

CHAPTER 3

Substitution Patterns

Nature, red in tooth and claw.

Alfred Tennyson (1809–1892)

Comparisons of the nucleotide sequences from two or more organisms such as those described in the previous chapter frequently reveal that changes have accumulated at the level of their DNA even when the sequences are from functionally equivalent regions. In fact, it is not uncommon to find sequences that have become so different over the course of evolution that reliable sequence alignments are difficult to obtain. Analyses of both the number and nature of substitutions that have occurred are of central importance to the study of molecular evolution. Because the process of natural selection is very effective at removing harmful changes, such analyses also provide powerful clues to bioinformaticians interested in recognizing and characterizing the portions of genes that are most functionally important. This chapter explains how and why the rate of nucleotide substitutions differs within and between genes as well as across species boundaries.

Patterns of Substitutions within Genes

Alterations in DNA sequences can have dire consequences for living cells. Most genes are very close to being in an optimal state for an organism in its typical environment, and the axiom "if it's not broken, don't fix it" definitely applies. It should not be surprising then for cells to have developed elaborate mechanisms that ensure the accuracy of DNA replication and repair. Still, **mutations** (both exchanges of one nucleotide for another and insertion/deletion events) do occur and cells do not always pass on a perfect copy of their genetic instructions. Those who study molecular evolution divide those mistakes into three categories: (1) those that are disadvantageous or **deleterious,** (2) those that are advantageous, and (3) those that are effectively **neutral** (have no effect on the fitness of an organism). The relative abundance of changes that fall into each category is still an open question, but two things are well accepted: (1) Advantageous changes are in a substantial minority, and (2) some changes in nucleotide sequences have greater consequences for an organism than do others. Not all portions of a gene are created equal—at least not from the perspective of how closely they are scrutinized during the process of natural selection.

Mutation Rates

The number of substitutions two sequences have undergone since they last shared a common ancestor, K, can often be determined simply by counting the differences between them. When K is expressed in terms of the number of substitutions per site and coupled with a divergence time (T) it is easily converted into a rate (r) of substitution. Because substitutions are assumed to accumulate simultaneously and independently in both sequences, the substitution rate is obtained by simply dividing the number of substitutions between two homologous sequences by $2T$ as shown in this equation:

$$r = K/(2T)$$

Note that in order to estimate substitution rates data must always be available from at least two species. This simple equation can be quite powerful: If evolutionary rates between several species are similar, substitution rates can give insights into the dates of evolutionary events for which no other physical evidence is available. Comparisons of substitution rates within and between genes are even more commonly used to determine the roles of different genomic regions as described below.

Functional Constraint

Changes to genes that diminish an organism's ability to survive and reproduce are typically removed from the gene pool by the process of **natural selection.** Since proteins are responsible for carrying out most of the important work of cells, it should not be surprising that changes to the nucleotide sequence of genes that also change the catalytic or structural properties of proteins are especially subject to natural selection. Portions of genes that are especially important are said to be under **functional constraint** and tend to accumulate changes very slowly over the course of evolution. Many changes to the nucleotide sequence of a gene have no effect on the amino acid sequence or expression levels of proteins and are much less subject to correction by natural selection—changes of this type accumulate relatively quickly.

Numerous analyses have confirmed that different portions of genes do accumulate changes at widely differing rates that reflect the extent to which they are functionally constrained. An example of those differences is shown in Table 3.1 for the changes that have accumulated within the beta-like globin gene of four mammals since they last shared a common ancestor roughly 100 million years ago. Recall from our discussion of gene structure in Chapter 1 that a typical eukaryotic gene is made up of some nucleotides that specify the amino acid sequence of a protein (coding sequences) and other nucleotides that do not code

TABLE 3.1 Average pairwise divergence among different regions of the human, mouse, rabbit, and cow beta-like globin genes.

Region	Length of Region (bp) in Human	Average Pairwise Number of Changes	Standard Deviation	Substitution Rate (substitutions/site/10^9 years)
Noncoding, overall	913	67.9	14.1	3.33
Coding, overall	441	69.2	16.7	1.58
5' Flanking sequence	300	96.0	19.6	3.39
5' Untranslated sequence	50	9.0	3.0	1.86
Intron 1	131	41.8	8.1	3.48
3' Untranslated sequence	132	33.0	11.5	3.00
3' Flanking sequence	300	76.3	14.3	3.60

Note: No adjustment is made for the possibility that multiple changes may have occurred at some sites.

for amino acids in a protein (noncoding sequences). The roughly two-times higher rate of change in the noncoding sequences of beta-like globin genes suggests that, taken as a whole, they are not as functionally constrained as the adjacent coding sequences (3.33×10^{-9} changes/site/year vs. 1.58×10^{-9} changes/site/year). Noncoding sequences can be subdivided into many different categories including introns, leader regions and trailer regions that are transcribed but not translated, and 5' and 3' flanking sequences that are not transcribed (Figure 3.1). Each of those regions also tends to accumulate changes at different rates that are generally correlated with the extent to which their nucleotides are functionally constrained. For instance, it should not be surprising that one of the lowest substitution rates seen in Table 3.1 (1.86×10^{-9} substitutions/site/year) is associated with the 5' sequences of the gene that are transcribed but not translated because so many of the nucleotides in that region are functionally important for the appropriate translation of beta-like globin proteins. In contrast, the nucleotides that are downstream of the beta-like globin gene's polyadenylation signal appear to play little if any functional role and are free to accumulate substitutions at a relatively fast rate (3.60×10^{-9} substitutions/site/year).

The results reported in Table 3.1 come from a fairly small data set—an analysis of roughly 1.3 kb containing the beta-like globin gene in each of four different mammalian genomes. It is entirely reasonable to expect that other genes accumulate substitutions at different rates and even for beta-like globin genes to be under different levels of functional constraint in different species. However, Table 3.1 does illustrate a general trend: Changes accumulate most rapidly in introns and flanking sequences, next most rapidly in other regions that are transcribed but not translated, and least rapidly within coding sequences. Data from

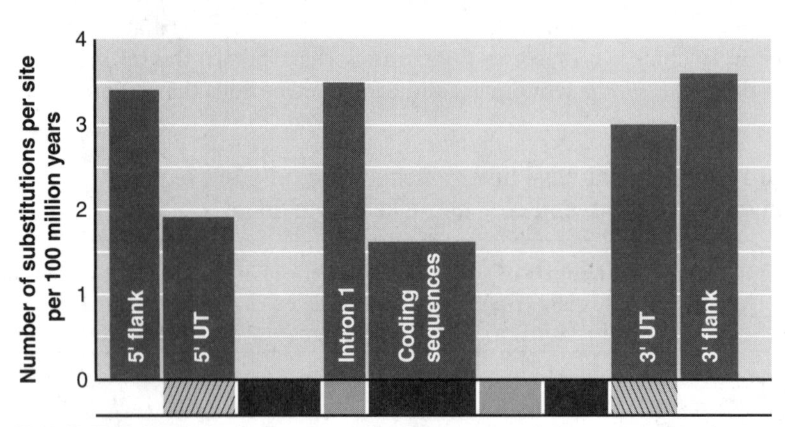

FIGURE 3.1 *Structure and relative rate of change within the beta-like globin gene in four mammals. Three hundred base pairs of the 5' and 3' flanking sequences are shown as open boxes; the 5' transcribed but untranslated (5' UT) sequence is represented as a forward slash-filled box; the 3' transcribed but untranslated (3' UT) sequence is shown as a backward slash-filled box; exons are shown as black boxes; and introns are shown as gray boxes. Relative rates of change are taken from Table 3.1.*

the beta-like globin genes also give a general feel for the time frames in which nucleotide changes accumulate. While a 0.35% change in a nucleotide sequence per million years (the approximate rate for introns and flanking sequences) may seem inordinately slow from a human perspective, it is relatively fast when considered from the perspective of molecular evolution.

Synonymous vs. Nonsynonymous Substitutions

As was pointed out in Chapter 1, 18 of the 20 different amino acids are encoded by more than one triplet codon (Table 1.1). For example, four different codons (GGG, GGA, GGU, and GGC) all code for the amino acid glycine. Any change at the third position of a codon for glycine (changing GGG to GGC, for instance) results in a codon that still causes ribosomes to insert a glycine at that point in the primary structure of a protein. Changes such as these at the nucleotide level of coding sequences that do not change the amino acid sequence of the protein are reasonably called **synonymous substitutions.** In contrast, changes to the second position of a glycine codon would result in changes to the amino acid sequence of a protein and are therefore known as **nonsynonymous substitutions** (for example, GCG codes for the amino acid alanine).

If natural selection distinguishes primarily between proteins that function well and those that do not, then synonymous substitutions should be observed more frequently than nonsynonymous ones. However, of the 47 substitutions that have accumulated within the human and rabbit beta-like globin genes in the past 100 million years, 27 are synonymous substitutions and 20 are nonsynonymous substitutions. This is especially remarkable in light of the fact that there were almost three times as many opportunities for *nonsynonymous* substitutions within the coding sequence of that gene.

From the example of mutations to the glycine codons described above, it should be clear that not all positions within triplet codons are as likely to result in nonsynonymous substitutions. In fact, the nucleotides in triplet codons can be placed in one of three different categories on that basis. **Nondegenerate sites** are codon positions where mutations always result in amino acid substitutions (i.e., UUU codes for phenylalanine, CUU codes for leucine, AUU codes for isoleucine, and GUU codes for valine). **Twofold degenerate sites** are those codon positions where two different nucleotides result in the translation of the same amino acid, but the two other nucleotides code for a different amino acid (e.g., GAU and GAC both code for aspartic acid, whereas GAA and GAG both code for glutamic acid). **Fourfold degenerate sites** are codon positions where changing a nucleotide to any of the three alternatives has no effect on the amino acid that ribosomes insert into proteins (such as the third codon position of glycine described above). Again, if natural selection works primarily at the level of mutations that alter protein function, it should not be surprising that nucleotide changes should accumulate most rapidly at fourfold degenerate sites and least rapidly at nondegenerate sites. That is exactly what is observed for the substitutions that have accumulated in the coding sequence of the human and rabbit beta-like globin genes (Table 3.2). The substitution rate at fourfold degenerate sites

TABLE 3.2 Divergence between different kinds of sites within the coding sequence of the human and rabbit beta-like globin genes.

Region	Number of Sites (bp)	Number of Changes	Substitution Rate (substitutions/ site/10^9 years)
Nondegenerate	302	17	0.56
Twofold degenerate	60	10	1.67
Fourfold degenerate	85	20	2.35

Note: Sequences used are available from GenBank (accession numbers V00497 and V00879, respectively). No adjustment is made for the possibility that multiple changes may have occurred at some sites. A divergence time of 100 million years is assumed.

often approaches that of 3' flanking sequences and other regions that are relatively free of selective constraint.

Indels and Pseudogenes

All of the changes discussed to this point in this chapter have focused on those occurring within transcriptionally active genes. In such a setting there is a very strong bias against insertion and deletion (indel) events because of their tendency to alter the reading frame used by ribosomes. That bias against frameshift mutations in coding regions is so strong that DNA replication and repair enzymes in general appear to have evolved in a way that makes indels roughly 10 times less likely to occur in any region of the genome than simple exchanges of one base for another.

It is also interesting to consider substitution patterns in genes that once were under selective constraint but have become transcriptionally inactive. As will be discussed in greater detail in Chapter 6, genes with new functions are commonly derived from genes with existing, useful functions. Duplication of an entire gene allows one copy to provide the necessary function of the original and the other copy to accumulate substitutions in a way that is free of selective constraint. On occasion, the evolving copy of such a gene undergoes some changes that give it an important new function and it again becomes important to the fitness of the organism. More often, however, one of the copies of a duplicated gene becomes a **pseudogene** when it acquires mutations that make it nonfunctional and transcriptionally inactive. Mammalian genomes are littered with such pseudogenes, and their sequences tend to accumulate substitutions at a very fast rate—at an average of almost four substitutions per site per 100 million years and just a bit faster than the 3' flanking sequences of genes within a species.

Substitutions vs. Mutations

Natural selection has an insidious effect on the data that are typically available for analysis by bioinformaticians. With only very rare exceptions, the only alleles available for characterization in naturally occurring populations of organisms are those that have not had a detrimental effect on fitness. The point here is that while all changes to the nucleotide sequence of a gene may be possible, not all are seen. An appreciation of this data-sampling problem gives rise to an interesting and subtle distinction between the use of the words *mutation* and *substitution* in molecular evolution studies. Mutations are changes in nucleotide sequences that occur due to mistakes in DNA replication or repair processes. **Substitutions** are mutations that have passed through the filter of selection on at least some level. Estimates of substitution rates are common in the field of molecular evolution. At the same time it is very difficult to estimate mutation rates reliably because natural selection can be so subtle and pervasive. This becomes a particularly important issue because comparisons between substitution and mutation rates give the best indication as to how functionally constrained a sequence actually is. Synonymous (and pseudogene) substitution rates (K_s), such as those shown in Table 3.3, are generally considered to be fairly reflective of the actual mutation rate operating within a genome because they are not subject to natural selection. Nonsynonymous substitutions rates (K_a) are not because they are subject to natural selection.

Fixation

Most naturally occurring populations of organisms harbor a substantial amount of genetic variation. Humans, for instance, differ from each other at an average of 1 base pair out of every 200. Different versions of any given gene within a species of organism are known as **alleles.** Differences among alleles cover a broad spectrum ranging from those that are relatively innocuous (i.e., a single difference in the nucleotide sequence of a 3' flanking sequence) to those that have very dramatic consequences (i.e., the presence of a premature stop codon that causes a truncated, nonfunctioning protein to be produced). Change in the relative frequencies of these different alleles is the essence of evolution.

With the exception of those that are introduced through migration or transfer across species boundaries, new alleles arise from mutations occurring to an existing allele within a single member of a population. As a result, new versions of genes typically begin at very low frequencies (q):

$$q = 1/2N$$

where N is the number of reproductively active diploid organisms within the population. As mentioned earlier, mutations that make organisms less likely to survive and reproduce tend to be removed from the gene pool through the process of natural selection and their frequencies eventually return to 0 (unless new mutations re-create the allele). The apparent rarity of mutations that dramatically increase the fitness of organisms suggests that most genes are at least fairly close to an optimal state for an organism's typical environment. However, when advantageous alleles do arise, their frequencies should move progressively toward 1.

Knowing that substitution rates are fairly low and that changes that alter fitness achieve a frequency of either 0 or 1 relatively quickly, what accounts for the

TABLE 3.3 Ratios of synonymous differences per synonymous site (K_s) and nonsynonymous differences per nonsynonymous site (K_a) for a variety of mammalian genes.

Gene	Codons (in human)	Human/mouse		Human/cow		Human/rabbit		Mouse/cow		Mouse/rabbit		Cow/rabbit		Averages	
		K_s	K_a	K_s	K_a	K_s	K_a	K_s	K_a	K_s	K_a	K_s	K_a	K_s	K_a
Erythropoietin	194	0.481	0.063	0.242	0.068	0.394	0.070	0.495	0.076	0.480	0.058	0.342	0.071	0.406	0.068
Growth hormone	217	0.321	0.100	0.236	0.106	0.220	0.113	0.380	0.046	0.396	0.027	0.244	0.048	0.299	0.073
Prolactin receptor	621	0.304	0.082	0.249	0.122	0.321	0.072	0.358	0.124	0.413	0.088	0.300	0.114	0.324	0.100
Prolactin	226	0.364	0.098	0.368	0.085	0.395	0.064	0.382	0.112	0.307	0.131	0.521	0.064	0.390	0.092
Serum albumin	610	0.528	0.062	0.329	0.067	0.324	0.075	0.477	0.065	0.500	0.065	0.327	0.067	0.414	0.067
Alpha globin	143	0.584	0.022	0.236	0.025	0.204	0.038	0.505	0.025	0.539	0.041	0.242	0.048	0.385	0.033
Beta globin	148	0.324	0.033	0.271	0.046	0.294	0.015	0.263	0.062	0.392	0.039	0.333	0.059	0.313	0.042
Prothrombin	608	0.033	0.687	0.033	1.040	0.075	1.602	0.196	0.887	0.037	1.442	0.078	0.318	0.075	0.996
Apolipo-protein E	317	0.199	0.148	0.132	0.117	0.108	0.114	0.187	0.160	0.165	0.144	0.125	0.126	0.153	0.135
Carbonic anhydrase I	336	0.255	0.159	0.203	0.149	0.207	0.138	0.338	0.113	0.284	0.115	0.187	0.117	0.246	0.132
P53	392	0.372	0.059	0.351	0.061	0.382	0.045	0.457	0.067	0.412	0.054	0.378	0.056	0.392	0.057
Histone 2A	115	0.967	0.057	1.110	0.057	0.174	0.034	0.298	0.006	1.176	0.025	1.192	0.025	0.820	0.033
Column averages		0.394	0.131	0.313	0.162	0.258	0.198	0.361	0.145	0.425	0.186	0.356	0.093	0.351	0.152

relatively high levels of variation seen within naturally occurring populations of organisms? Quite simply, much of the variation that is observed among individuals must have little beneficial or detrimental effect and be essentially **selectively neutral.** The probability (P) that any truly neutral variant of a gene will eventually be lost from a population is simply a matter of chance and is equal to $1 - q$, where q is the relative frequency of the allele in the population. By the same token, the probability that a particular neutral allele will be **fixed** (occur within a population at a frequency of 1) is equal to q, the current frequency of the gene in the population. Even though the chance of fixation for a new version of a gene might be small, neutral mutations can persist in populations for very long periods of time (even by molecular evolution standards), and the mean time for fixation for a new neutral mutation is effectively equal to the amount of time required for $4N$ generations to pass.

All of this has direct, practical implications for bioinformaticians interested in determining the functionality of genes. The variation between and especially within species seen in comparative sequence analyses such as those described above is almost always limited to those portions of a gene that are free of functional constraint. Analysis of such variation helps to distinguish between those regions of genomes that contain genes and those that do not (discussed in greater depth in Chapter 6). Comparative sequence analysis also goes a long way toward obviating the need for the time-consuming and experimentally grueling process of **saturation mutagenesis** in which molecular biologists make all possible changes to the nucleotide sequence of a gene to determine which alter its function. To a large extent, nature is performing a perpetual saturation mutagenesis experiment, and the majority of the variation available for us to observe corresponds to changes that do not significantly alter the function of genes.

Estimating Substitution Numbers

The number of substitutions (K) observed in an alignment between two sequences is typically the single most important variable in any molecular evolution analysis. If an optimal alignment suggests that relatively few substitutions have occurred between two sequences, then a simple count of the substitutions is usually sufficient to determine a value for K. However, even before the nucleotide sequences of any genes were available for analysis, T. Jukes and C. Cantor (1969) realized that alignments between sequences with many differences might cause a significant underestimation of the actual number of substitutions since the sequences last shared a common ancestor.

Jukes-Cantor Model

Where substitutions were common, there were no guarantees that a particular site had not undergone multiple changes such as those illustrated in Figure 3.2. To address that possibility, Jukes and Cantor assumed that each nucleotide was just as likely to change into any other nucleotide. Using that assumption, they

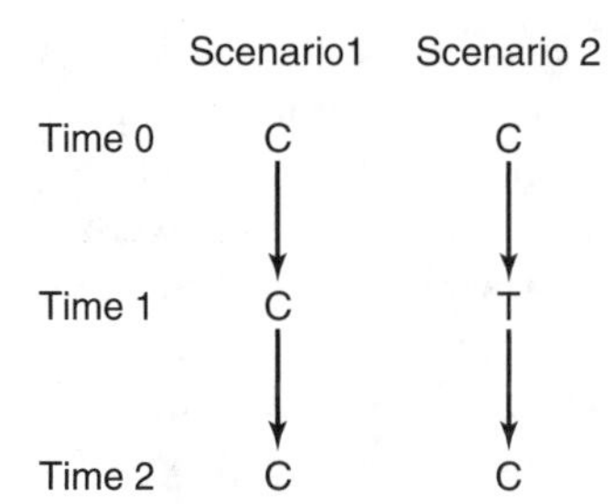

FIGURE 3.2 *Two possible scenarios where multiple substitutions at a single site would lead to underestimation of the number of substitutions that had occurred if a simple count were performed.*

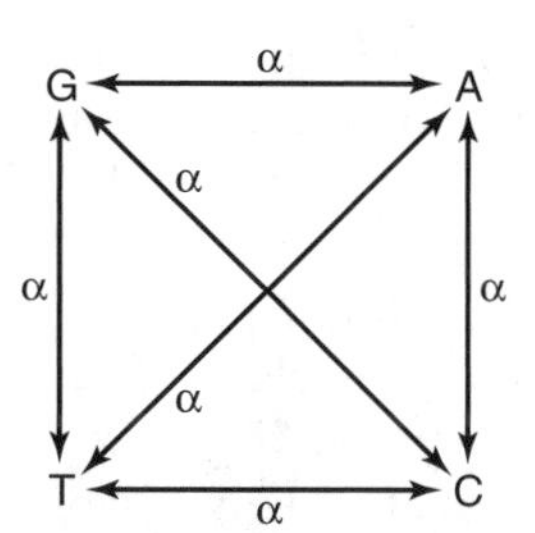

FIGURE 3.3 *Diagram of the Jukes-Cantor model of nucleotide substitution. For their model, Jukes and Cantor assumed that all nucleotides changed to each of the three alternative nucleotides at the same rate, α.*

created a mathematical model (diagrammed in Figure 3.3) in which the rate of change to any one of the three alternative nucleotides was assumed to be α and the overall rate of substitution for any given nucleotide was 3α. In that model, if a site within a gene was occupied by a C at time 0, then the probability (P) that that site would still be the same nucleotide at time 1 would be $P_{C(1)} = 1 - 3\alpha$. Because a reversion (back mutation) to C could occur if the original C changed to another nucleotide in that first time span, at time 2 the probability, $P_{C(2)}$, would be equal to $(1 - 3\alpha)P_{C(1)} + \alpha\,[1 - P_{A(1)}]$. Further expansion suggested that at any given time (t) in the future, the probability that that site would contain a C was defined by the following equation:

$$P_{C(t)} = 1/4 + (3/4)e^{-4\alpha t}$$

Ten years later nucleotide sequence data became available for the first time and made it clear that Jukes's and Cantor's assumption of a global uniformity in substitution patterns was an oversimplification. Even still, their model continues to provide a useful framework for taking into account the actual number of substitutions per site (K) when multiple substitutions are possible. Through manipulations of the Jukes-Cantor equation above, we can derive a fairly simple equation that yields an estimate of the true number of substitutions that have occurred between two sequences when only a pairwise counting of differences is available:

$$K = -3/4 \ln[1 - (4/3)(p)]$$

In this equation p is the fraction of nucleotides that a simple count reveals to be different between two sequences. This equation is totally consistent with the idea that when two sequences have few mismatches between them, p is small and the chance of multiple substitutions at any given site is also small. It also suggests that when the observed number of mismatches is large, the actual number of substitutions per site will be larger than what is actually counted directly.

Transitions and Transversions

Nucleotides can be divided into two separate categories on the basis of the structure of their nitrogenous bases (see Figure 1.1). Guanine and adenine are called **purines** because their nitrogenous bases have a two-ring structure. In contrast, **pyrimidines** like cytosine, thymine, and uracil all have nitrogenous bases with only a one-ring structure. The first actual nucleotide sequence data, which became available in the 1970s, made it clear that exchanging one nucleotide for another within or between these classes occurred at significantly different rates. Specifically, **transitions** (exchanging one purine for another or exchanging one pyrimidine for another) occurred at least three times as frequently as **transversions** (exchanging a purine for a pyrimidine or vice versa).

Kimura's Two-Parameter Model

In 1980, M. Kimura developed a two-parameter model that took into account the different rates of transitions and transversions. In his model transitions were assumed to occur at a uniform rate of α and transversions at a different, uniform rate of β (diagrammed in Figure 3.4). With these parameters, if a site within a gene was occupied by a C at time 0, then the probability (P) that that site would still be the same nucleotide at time 1 would be $P_{CC(1)} = 1 - \alpha - 2\beta$. Back mutations could still occur between time 1 (t_1) and time 2 (t_2) and the probability that the site would still contain a C, $P_{CC(2)}$, is the sum of the probabilities associated with four different scenarios: (1) C remained unchanged at t_1 and at t_2; (2) C changed at t_1 and reverted by a transition to C at t_2; (3) C changed to a G at t_1 and reverted by a transversion to C at t_2; and (4) C changed to an A at t_1 and reverted by a transversion back to C at t_2 (Figure 3.5). Thus, for this model, the probability that the site is still occupied by a C at t_2 is

$$P_{CC(2)} = (1 - \alpha - 2\beta)P_{CC(1)} + \beta P_{GA(1)} + \beta P_{AC(1)} + \alpha P_{TC(1)}$$

As with the Jukes-Cantor model, further expansion suggested that at any given time (t) in the future, the probability that that site would contain a C is defined by this equation:

$$P_{CC(t)} = 1/4 + (1/4)e^{-4\beta t} + (1/2)e^{-2(\alpha + \beta)t}$$

The symmetry of the substitution scheme in both the Jukes-Cantor model and Kimura's two-parameter model results in all four nucleotides having an equal probability of being the same between time 0 and any point in the future ($P_{GG(t)} = P_{AA(t)} = P_{TT(t)} = P_{CC(t)}$). And, just as with the Jukes-Cantor one-parameter model, further manipulation of this equation and those for the other three nucleotides results in a useful equation that yields an estimate of the true number of substitutions that have occurred between two sequences when only a pairwise counting of differences is available:

$$K = 1/2 \ln[1/(1 - 2P - Q)] + 1/4 \ln[1/(1 - 2Q)]$$

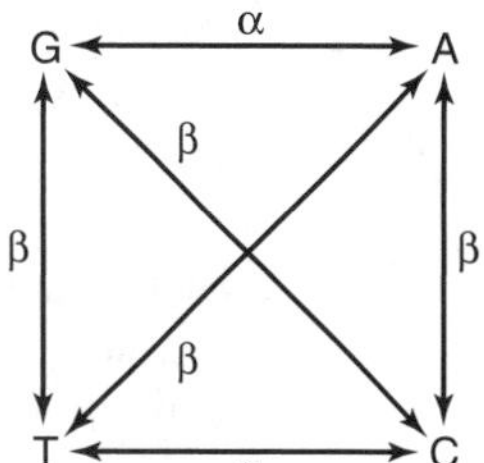

FIGURE 3.4 *Diagram of Kimura's two-parameter model of nucleotide substitution. Kimura assumed that nucleotide substitutions occurred at essentially two different rates: α for transitions (i.e., changes between G and A or between C and T), and β for transversions (changes between purines and pyrimidines).*

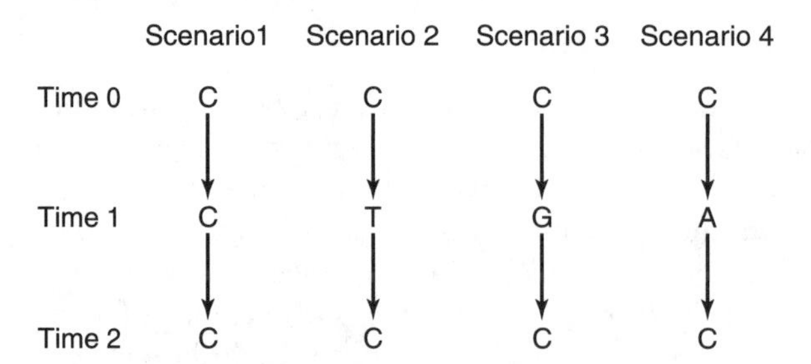

FIGURE 3.5 *Four possible routes by which a site appears to have been unchanged after two time intervals have passed.*

In this equation P is the fraction of nucleotides that a simple count reveals to be transitions, and Q is the fraction that are transversions. If no distinction is made between transitions and transversions (i.e., $p = P + Q$), this equation reduces precisely to the equivalent Jukes-Cantor equation for estimating the number of substitutions that have occurred between two sequences.

Models with Even More Parameters

The large amounts of sequence data that have been generated since the early 1980s have revealed that Kimua's assumption that nucleotides change at two different rates is almost as much of an oversimplification as Jukes's and Cantor's assumption that all nucleotides had an equal probability of changing into any of the other three. Since each of the four nucleotides can change into any of the other three, 12 different types of substitutions are possible: G → A; G → T; G → C; A → G; A → T; A → C; T → G; T → A; T → C; C → G; C → A; and C → T.

Matrices such as the one shown in Table 3.4 can be generated that attach probabilities to each of those different types of mutation, which can then be used in a mathematically cumbersome 12-parameter model. A 13th parameter can also be invoked that compensates for differences in matrices like the one in Table 3.4 that are caused by substitution biases associated with regional genomic GC context. Although intimidating if being analyzed by hand, such complex models can be fairly easily implemented computationally. However, such complex models invoke more assumptions than the one- and two-parameter models described earlier (i.e., that the nucleotide sequences being studied were at an equilibrium state at the time of their divergence). Sampling errors due to the comparatively small number of instances of each kind of change are also compounded during the analysis. Consequently, simulation studies suggest that one- and two-parameter models often give more reliable results than more complex alternatives—and are themselves virtually indistinguishable when closely related sequences are studied.

TABLE 3.4 Relative frequencies of nucleotide substitutions in Alu-Y (Sb) sequences throughout the human genome.

	To				
From	A	T	C	G	Row Totals
A	—	4.0	4.6	9.8	18.4
	—	(1.5)	(1.7)	(3.6)	(6.7)
T	3.3	—	10.4	2.7	16.4
	(1.2)	—	(3.8)	(1.0)	(6.0)
C	7.2	17.0	—	6.2	31.1
	(5.0)	(33.2)	—	(4.5)	(42.6)
G	23.6	4.6	6.0	—	34.2
	(37.7)	(3.2)	(3.7)	—	(44.7)
Column totals	34.1	26.3	21.0	9.0	
	(44.0)	(37.8)	(9.2)	(18.7)	

Note: Members of the *Alu* repeat family are approximately 260 base pairs in length. They are derived from one or a small number of ancestral sequences that have been duplicated almost 1 million times during primate evolution.

The relative frequencies of substitutions observed involving each of the four nucleotides within 403 Alu-Y (Sb) repeat sequences scattered throughout the human genome excluding those involving CpG dinucleotides. Values in parentheses were obtained when substitutions at CpG dinucleotides were not excluded. A total of 7,433 substitutions (2,713 of which were at sites other than CpG dinucleotides) were accumulated by the 403 Alu-Y (Sb) repeats included within this analysis since they were propagated roughly 19 million years ago.

Substitutions between Protein Sequences

The proportion (p) of different amino acids between two protein sequences can be calculated simply as

$$p = n/L$$

where n is the number of amino acids that differ between the two sequences, and L is the number of positions at which differences could be observed in the aligned sequences. However, refining estimates of the number of substitutions that have occurred between the amino acid sequences of two or more proteins is generally more difficult than the equivalent task for noncoding DNA sequences. Just as with DNA sequences, back mutations can result in significant undercounting of substitutions. And, in addition to the fact that some substitutions occur more frequently than others, the substitutional path from one amino acid to another is not always the same length. For instance, the CCC codon for proline can be converted to the CUC codon for leucine with just one mutation, but it is not possible to convert it to a codon for isoleucine (such as AUC) without at least two mutations. The problem is complicated even more by the fact that most amino acid substitutions do not have an equivalent effect on protein function, and effects can differ greatly from one context to another. One solution to these problems is to weight each amino acid substitution differently by using empirical data from a variety of different protein comparisons to generate a matrix such as the PAM matrix described in Chapter 2.

Variations in Evolutionary Rates between Genes

Just as variations in evolutionary rates are readily apparent in comparisons of different regions within genes, striking differences in the rates of evolution between genes have also been observed. If stochastic factors (like those arising from small population sizes due to sampling error) are ruled out, the difference in rates must be attributable to one or some combination of two factors: (1) differences in mutation frequency and/or (2) the extent to which natural selection affects the locus. While some regions of genomes do seem to be more prone to random changes than others, synonymous substitution rates rarely differ by more than a factor of 2. That difference is far from sufficient to account for the roughly 200-fold difference in nonsynonymous substitution rates observed between the different mammalian genes listed in Table 3.3. As was the case for variation in observed substitution rates within genes, variation of substitution rates between genes must be largely due to differences in the intensity of natural selection at each locus.

Specific examples of two classes of genes, histones and apolipoproteins, illustrate the effects of different levels of functional constraint (Table 3.3). Histones are positively charged, essential DNA binding proteins present in all eukaryotes. Almost every amino acid in a histone such as histone H2A interacts directly with specific chemical residues associated with negatively charged DNA. Virtually any change to the amino acid sequence of histone H2A affects its ability to interact with DNA. As a result, histones are one of the slowest evolving groups of proteins known, and it is actually possible to replace the yeast version of histone H2A with its human homolog with no effect despite hundreds of millions of years of independent evolution. Apolipoproteins, in contrast, accumulate nonsynonymous substitutions at a very high rate. They are responsible for nonspecifically interacting with and carrying a wide variety of lipids in the blood of vertebrates. Their lipid-binding domains are composed predominantly of hydrophobic amino acids. Any similar amino acid (i.e., leucine, isoleucine, and valine) appears to function in those positions just as well as another so long as it too is hydrophobic.

Also, while amino acid substitutions within many genes are generally deleterious, we should point out that natural selection actually favors variability within populations for some genes. The genes associated with the human leukocyte antigen (HLA), for instance, are actually under evolutionary pressure to diversify. As a result, the rate of nonsynonymous substitutions within the HLA locus is actually greater than that of synonymous substitutions (mean number of synonymous substitutions, K_s = 3.5%, and nonsynonymous substitutions, K_a = 13.3%, for the 57-amino-acid-long sequence of the antigen recognition site within five variants of the human HLA-A locus). The HLA locus contains a large multigene family whose protein products are involved with the immune system's ability to recognize foreign antigens. Within human populations, roughly 90% of individuals receive different sets of HLA genes from their parents, and a sample of 200 individuals can be expected to have 15 to 30 different alleles.

Such high levels of diversity in this region are favored by natural selection because the number of individuals vulnerable to infection by any single virus is likely to be substantially less than it would have been had they all had similar immune systems. At the same time that host populations are pressured to maintain diverse immune systems, viruses too are under pressure to evolve rapidly. Error-prone replication coupled with diversifying selection causes the rate of nucleotide substitutions within the influenza *NS* genes to be 1.9×10^{-3} nucleotide substitutions per site per year—roughly a million times greater than the synonymous substitution rate for representative mammalian genes such as those in Table 3.3.

Molecular Clocks

As described above, the differences in the nucleotide and amino acid replacement rates between nuclear genes can be striking but is likely to be due primarily to differences in the selective constraint on each individual protein. However, rates of molecular evolution for loci with similar functional constraints can be quite uniform over long periods of evolutionary time. In fact, the very first comparative studies of protein sequences performed by Emile Zuckerkandl and Linus Pauling in the 1960s suggested that substitution rates were so constant within homologous proteins over many tens of millions of years that they likened the accumulation of amino acid changes to the steady ticking of a **molecular clock.**

The molecular clock may run at different rates in different proteins, but the number of differences between two homologous proteins appeared to be very well correlated with the amount of time since speciation caused them to diverge independently, as shown in Figure 3.6 on page 72. This observation immediately stimulated intense interest in using biological molecules in evolutionary studies. A steady rate of change between two sequences should facilitate not only the determination of phylogenetic relationships between species but also the times of their divergence in much the same way that radioactive decay was used to date geological times.

Despite its great promise, however, Zuckerkandl and Pauling's molecular clock hypothesis has been controversial. Classical evolutionists argued that the erratic tempo of morphological evolution was inconsistent with a steady rate of molecular change. Disagreements regarding divergence times have also placed in question the very uniformity of evolutionary rates at the heart of the idea.

Relative Rate Test

Most divergence dates used in molecular evolution studies come from interpretations of the notoriously incomplete fossil record and are of questionable accuracy. To avoid any questions regarding speciation dates, Sarich and Wilson (1973)

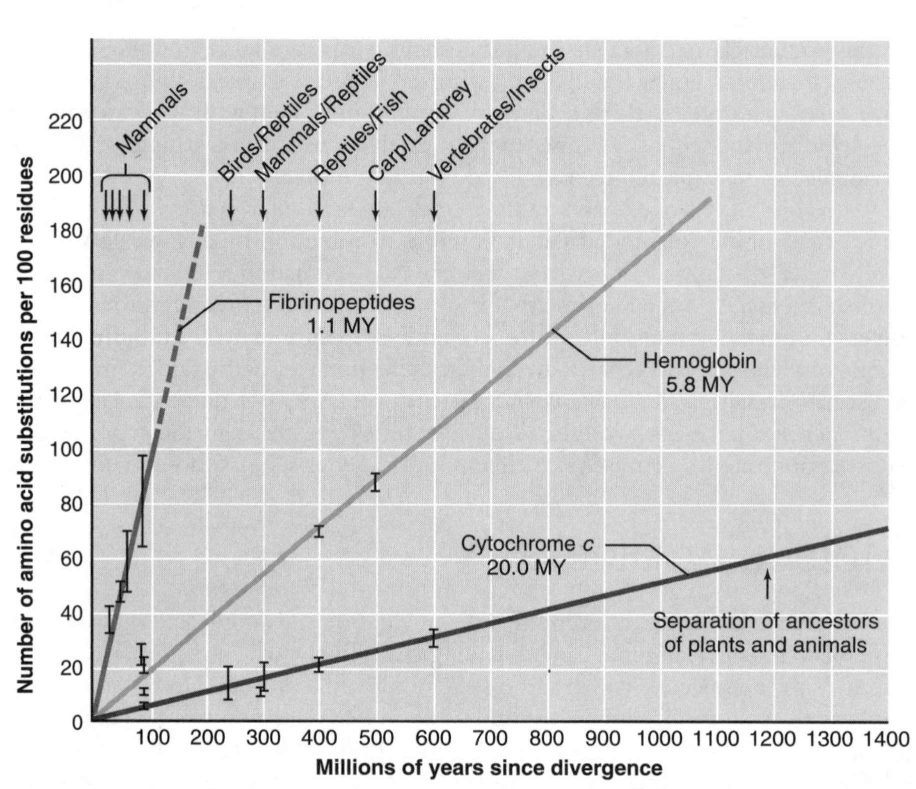

FIGURE 3.6 *Numbers of amino acids replaced and species divergence times are well correlated for a number of proteins.*

devised a simple way to estimate the overall rate of substitution in different lineages that does not depend on specific knowledge of divergence times. For example, to determine the **relative rate** of substitution in the lineages for species 1 and 2 in Figure 3.7, we need to designate a less related species 3 as an **outgroup.** Outgroups can usually be readily agreed on; for instance, in this example, if species 1 and 2 are humans and gorillas, respectively, then species 3 could be another primate such as a baboon. In the evolutionary relationship portrayed in Figure 3.7, the point in time when species 1 and 2 diverged is marked with the letter "A." The number of substitutions between any two species is assumed to be the sum of the number of substitutions along the branches of the tree connecting them such that

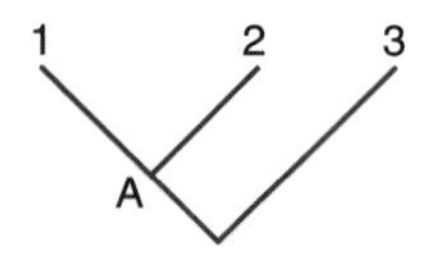

FIGURE 3.7 *Phylogenetic tree used in a relative rate test. Species 3 represents an outgroup known to have been evolving independently prior to the divergence of species 1 and species 2. "A" denotes the common ancestor of species 1 and species 2.*

$$d_{13} = d_{A1} + d_{A3}$$

$$d_{23} = d_{A2} + d_{A3}$$

$$d_{12} = d_{A1} + d_{A2}$$

where d_{13}, d_{23}, and d_{12} are easily obtained measures of the differences between species 1 and 3, species 2 and 3, and species 1 and 2, respectively. Simple algebraic manipulation of those statements allows the amount of divergence that has taken place in species 1 and in species 2 since they last shared a common ancestor to be calculated using these equations:

$$d_{A1} = (d_{12} + d_{13} - d_{23})/2$$

$$d_{A2} = (d_{12} + d_{23} - d_{13})/2$$

By definition, the time since species 1 and species 2 began diverging independently is the same, so the molecular clock hypothesis predicts that values for d_{A1} and d_{A2} should also be the same.

The data available for testing the molecular clock's premise that the rate of evolution for any given gene is constant over time in all evolutionary lineages are increasing exponentially. Substitution rates in rats and mice have been found to be largely the same. In contrast, molecular evolution in humans and apes appears to have been only half as rapid as that which has occurred in Old World monkeys since their divergence. Indeed, relative rate tests performed on homologous genes in mice and humans suggest that rodents have accumulated substitutions at twice the rate of primates since they last shared a common ancestor during the time of the mammalian radiation 80 to 100 million years ago. Even a casual review of Table 3.3 shows that the rate of the molecular clock varies among taxonomic groups. Such departures from constancy of the clock rate pose a problem in using molecular divergence to date the times of existence of recent common ancestors. Before such inferences can be made, it is necessary to demonstrate that the species being examined have a uniform clock such as the one observed within rodents.

Causes of Rate Variation in Lineages

Several possible explanations have been put forward to account for the differences in evolutionary rates revealed by the relative rate tests. For instance, generation times in monkeys are shorter than those in humans, and the generation time of rodents is much shorter still. The number of germ line DNA replications should be more closely correlated with substitution rates than simple divergence times. Differences may also be due in part to a variety of other differences between two lineages since the time of their divergence, such as average repair efficiency, metabolic rate (and the related rate of generation of oxygen free radicals), and the necessity to adapt to new ecological niches and environments. Such parameters tend to be very difficult to quantify in a useful way: We know that at the time of divergence both groups of organisms had similar attributes, and we know the extent of their differences at the present time but typically have very little information about their relative differences at all other times during the course of their evolution.

Evolution in Organelles

The average length of mammalian mitochondrial DNA (often abbreviated mtDNA) is approximately 16,000 base pairs, while the chloroplast DNA (often abbreviated cpDNA) of vascular plants ranges in size from 120,000 to 220,000 base pairs. The single, circular chromosomes of both organelles contain several protein and RNA encoding genes that are essential for their function. The relatively small size of their chromosomes and unusual pattern of inheritance (i.e., in mammals, mitochondria are contributed only by the mother and never the father) have fostered a considerable amount of interest in the way in which they accumulate substitutions.

The high concentration of mutagens (particularly oxygen free radicals) present within mitochondria as a result of their metabolic activity appears to subject mtDNA to an almost 10-fold higher rate of mutation than that found in nuclear DNA. That higher rate of change results in a correspondingly higher rate of both synonymous and nonsynonymous substitutions in mitochondrial genes. As a result, comparisons of mtDNA are often used to study relationships between closely related populations of organisms (but are often less useful for species that have diverged for more than 10 million years due to the expectation of multiple substitutions at each site). Chloroplast DNA seems to accumulate substitutions at a much slower pace than mtDNA, and values for K_s and K_a actually tend to be only one-fourth to one-fifth the rates observed for nuclear genes in the same species.

Chapter Summary

DNA, like any molecule, accumulates chemical damage with time. When that damage or a DNA replication error results in a change to the information content of a DNA molecule, a mutation is said to have occurred. Mutations do not all affect the fitness of an organism to the same degree. As a result, natural selection causes many to be lost from the gene pool, and the changes that remain are referred to as substitutions. Substitution rate can be used as a measure of the functional importance of a gene or other portion of a genome. Several models that take into account the possibility of multiple substitutions at any given site have been made to estimate the true number of substitutions that have occurred between two nucleotide or amino acid sequences and, generally speaking, those with the fewest parameters perform the best. Just as some genes accumulate substitutions more quickly than others, relative rate tests show that some organisms have a faster rate of substitution than others even when genes with similar functional constraints are considered.

Readings for Greater Depth

A description of the original Jukes and Cantor model of substitutions is found in T. H. Jukes and C. R. Cantor, 1969, Evolution of protein molecules, in H. N.

Munro (ed.), *Mammalian Protein Metabolism*, pp. 21–123, Academic Press, New York.

Kimura's two-parameter model is described in detail in M. Kimura, 1980, A simple method for estimating evolutionary rates of base substitutions through comparative studies of nucleotide sequences, *J. Mol. Evol.* **16**: 111–120.

The molecular clock hypothesis was first suggested in E. Zuckerkandl and L. Pauling, 1965, Evolutionary divergence and convergence in proteins, in V. Bryson and H. J. Vogel (eds.), *Horizons in Biochemistry*, pp. 97–166, Academic Press, New York.

A comprehensive set of algorithms associated with substitution modeling can be found in R. F. Doolittle, 1990, *Molecular Evolution: Computer Analysis of Protein and Nucleic Acid Sequences*, Academic Press, San Diego, CA.

The relative rate test of Sarich and Wilson is described in detail in V. M. Sarich and A. C. Wilson, 1973, Generation time and genomic evolution in primates, *Science* **179**: 1144–1147.

The neutral theory of evolution (genetic differences within populations are maintained by a balance between the effects of mutation and random genetic drift) was first espoused in M. Kimura, 1968, Evolutionary rate at the molecular level, *Nature* **217**: 731–736.

A fascinating and thought-provoking consideration of the societal (as well as biological) implications of evolutionary theory can be found in R. Dawkins, 1976, *The Selfish Gene*, Oxford University Press, New York.

Questions and Problems

* **3.1** Given that there are currently 6 billion reproductively active humans on earth and that the average human generation time is roughly 30 years, how long would it take for a single, neutral mutation that occurs within you to be fixed within the human population?

3.2 Using the same data presented in Question 3.1, what is the probability that a single, new, neutral mutation in you will be fixed? How much more likely is it that it will be lost?

* **3.3** The following sequence is that of the first 45 codons from the human gene for preproinsulin. Using the genetic code (refer back to Table 1.1), determine what fraction of mutations at the first, second, and third positions of these 45 codons will be synonymous.

```
ATG GCC CTG TGG ATG CGC CTC CTG CCC CTG CTG GCG CTG CTG GCC
CTC TGG GGA CCT GAC CCA GCC GCA GCC TTT GTG AAC CAA CAC CTG
TGC GGC TCA CAC CTG GTG GAA GCT CTC TAC CTA GTG TGC GGG GAA
```

At which position is natural selection likely to have the greatest effect and are nucleotides most likely to be conserved?

3.4 The sequences shown below represent an optimum alignment of the first 50 nucleotides from the human and sheep preproinsulin genes. Estimate the number of substitutions that have occurred in this region since humans and sheep last shared a common ancestor using the Jukes-Cantor model.

Human: ATGGCCCTGT GGATGCGCCT CCTGCCCCTG CTGGCGCTGC TGGCCCTCTG

Sheep: ATGGCCCTGT GGACACGCCT GGTGCCCCTG CTGGCCCTGC TGGCACTCTG

* **3.5** Using the number of substitutions estimated in Question 3.4 and assuming that humans and sheep last shared a common ancestor 100 million years ago, estimate the rate at which the sequence of the first 50 nucleotides in their preproinsulin genes has been accumulating substitutions.

3.6 Would the mutation rate be greater or less than the observed substitution rate for a sequence of a gene such as the one shown in Question 3.4? Why?

* **3.7** If the rate of nucleotide evolution along a lineage is 1.0% per million years, what is the rate of substitution per nucleotide per year? What would be the observed rate of divergence between two species evolving at that rate since they last shared a common ancestor?

3.8 Assume that the sequence of the first 50 nucleotides from the chicken preproinsulin gene can be optimally aligned with the homologous sequences in the human and sheep used in Question 3.4, as shown below.

Human: ATGGCCCTGT GGATGCGCCT CCTGCCCCTG CTGGCGCTGC TGGCCCTCTG

Sheep: ATGGCCCTGT GGACACGCCT GGTGCCCCTG CTGGCCCTGC TGGCACTCTG

Chicken: ATGGCTCTAT GGACACGCCT TCTGCCTCTA CTGGCCCTGC TAGCCCTCTG

What are the relative rates of evolution within the human and sheep lineages since the time of the mammalian radiation for this region?

CHAPTER 4

Distance-Based Methods of Phylogenetics

Nothing in biology makes sense except in the light of evolution.

Theodosius Dobzhansky (1900–1975)

The phylogenetic relationship between two or more sets of sequences is often extremely important information for bioinformatic analyses such as the construction of sequence alignments. It is not unreasonable to think of the kinds of molecular data discussed in the previous chapter as being something of a historical document that contains within it evidence of the important steps in the evolution of a gene. The very same evolutionary events (substitutions, insertions, deletions, and rearrangements) that are important to the history of a gene can also be used to resolve questions about the evolutionary history and relationships between entire species. In fact, the phylogenetic relationships among many kinds of organisms are difficult to determine in any other way. As a result, a variety of different approaches have been devised to reconstruct genealogies from molecular data not just for genes but for species of organisms as well. This chapter introduces the basic vocabulary of phylogenetics and focuses on cluster analysis, the oldest and most statistically based of those approaches commonly used to infer evolutionary relationships.

History of Molecular Phylogenetics

Taxonomists were naming and grouping organisms long before it was even suspected that evolutionary records might be retained within the sequence information of their genomes. Drawing heavily on studies in anatomy and physiology, the field of taxonomy has produced countless valuable insights, especially once Darwin's ideas caused Linnaeus's system of grouping and naming organisms to reflect evolutionary relationships. Those insights are largely responsible for such varied and dramatic accomplishments as the development of new crops for agriculture, the discovery of treatments for infectious diseases, and even the idea that all living things on this planet share a single common ancestor.

Consideration of similarities and differences at a molecular level seemed a natural addition to the tools commonly used by taxonomists when G. H. F. Nuttall demonstrated in 1902 and 1904 that the extent to which the blood of an organism generated an immune response when it was injected into a test organism was directly related to how evolutionarily related the two organisms were. Through such experiments he examined the relationship of hundreds of organisms and was among the first to correctly conclude that humans and apes shared a common ancestor with each other more recently than they did with other primates. Antibodies and their varying abilities to interact with other molecules are still used by some scientists today as a fast phylogenetic screening tool when working with organisms for which little or no DNA or protein sequence data are available.

It was not until the 1950s, though, that molecular data began to be used extensively in phylogenetic research. **Protein electrophoresis** allowed the separation and comparison of related proteins on the basis of somewhat superficial features such as their size and charge. Rates at which denatured genomes could cross-hybridize with each other gave indications of relatedness but worked best only when fairly closely related organisms were used. **Protein sequencing** also became possible for the first time though it was not until the 1960s that the complete amino acid sequences of proteins of even modest size could be easily generated. Taken together, these molecular approaches stimulated a significant change in terms of the types of organisms whose phylogenies could be studied. They also provided an abundance of parameters that could be measured and began to highlight the inadequacy of the largely intuitive approaches to data analysis that had sufficed for previous generations of taxonomists.

By the time that actual genomic information became available in the 1970s, first in the form of restriction enzyme maps and then as actual DNA sequence data, a flurry of interest in phylogenetic reconstruction had resulted in the generation of a variety of mathematically rigorous approaches for the molecular biologists who were generating exponentially increasing amounts of molecular data. For the first time it was possible to assign statistical confidences to phylogenetic groupings and it also became comparatively easy to formulate testable hypotheses about evolutionary processes.

Today, DNA sequence data are substantially more abundant than any other form of molecular information. Traditional taxonomic approaches based on anatomic differences continue to provide complementary data to evolutionary studies, and paleontological information provides irreplaceable clues about the actual time frames in which organisms accumulate differences and evolve. However, it is the ease with which molecular approaches such as PCR (described in Chapter 1) provide homologous sequence data that promises to deliver the raw materials needed to answer the most important outstanding questions regarding the history and relationships of life on this planet.

Advantages to Molecular Phylogenies

Because evolution is defined as genetic change, genetic relationships are of primary importance in the deciphering of evolutionary relationships. The greatest promise of the molecular clock hypothesis (see Chapter 3) is the implication that molecular data can be used to decipher the phylogenetic relationships between all living things. Quite simply, organisms with high degrees of molecular similarity are expected to be more closely related than those that are dissimilar. Before the tools of molecular biology were available to provide molecular data for such analyses, taxonomists were forced to rely on comparisons of **phenotypes** (how organisms looked) to infer their **genotypes** (the genes that gave rise to their physical appearance). If phenotypes were similar, it was assumed the genes that coded for the phenotypes were also similar; if the phenotypes were different, the genes were different. Originally the phenotypes examined consisted largely of gross anatomic features. Later, behavioral, ultrastructural, and biochemical characteristics were also studied. Comparisons of such traits were successfully used to construct evolutionary trees for many groups of plants and animals and, indeed, are still the basis of many evolutionary studies today.

However, relying on the study of such traits has limitations. Sometimes similar phenotypes can evolve in organisms that are distantly related in a process called **convergent evolution.** For example, if a naïve biologist tried to construct an evolutionary tree on the basis of whether eyes were present or absent in an organism, he might place humans, flies, and mollusks in the same evolutionary group, since all have light-detecting organs. In this particular case, it is fairly obvious that these three organisms are not closely related—they differ in many features other than the possession of eyes, and the eyes themselves are very different in their design. The point, though, is that phenotypes can sometimes be misleading about evolutionary relationships, and phenotypic similarities do not always reflect genetic similarities.

Another problem with relying on phenotypes to determine evolutionary relationships is that many organisms do not have easily studied phenotypic features suitable for comparison. For example, the study of relationships among bacteria has always been problematic, because bacteria have few obvious traits even when examined with a microscope, let alone ones that correlate with the degree of their genetic relatedness. A third problem arises when we try to compare distantly related organisms. What phenotypic features should be compared, for example, in an analysis of bacteria, worms, and mammals, which have so few characteristics in common?

Analyses that rely on DNA and protein sequences are often free of such problems because many homologous molecules are essential to all living things. Even though Chapter 3 warns that the relative rate of molecular evolution may vary from one lineage to another and molecularly inferred divergence times must be treated with caution, molecular approaches to generating phylogenies can usually be relied on to group organisms correctly. Many have argued that molecular phylogenies are more reliable even when alternative data (such as morphologic data) are available because the effects of natural selection are generally less pronounced at the sequence level. On those occasions when differences between molecular and morphological phylogenies are found, they usually create valuable opportunities to examine the effect of natural selection acting at the level of phenotypic differences.

Phylogenetic Trees

Central to most studies of phylogeny is the concept of a **phylogenetic tree**—typically a graphical representation of the evolutionary relationship among three or more genes or organisms. These trees truly can be pictures that are worth a thousand words, and it is possible for them to convey not just the relatedness of data sets but also their divergence times and the nature of their common ancestors.

Terminology of Tree Reconstruction

Sometimes also referred to as a dendrogram, phylogenetic trees are made by arranging **nodes** and **branches** (Figure 4.1). Every node represents a distinct taxonomical unit. Nodes at the tips of branches (**terminal nodes**) correspond to a gene or organism for which data have actually been collected for analysis, while **internal nodes** usually represent an inferred common ancestor that gave rise to two independent lineages at some point in the past. For example, in Figure 4.1, nodes I, II, III, IV, and V are terminal nodes that represent organisms for which sequence data are available. In contrast, the internal nodes A, B, C, and D represent **inferred ancestors** (of I and II, III and IV, A and B, and C and V, respectively) for which empirical data are no longer available. Computer programs often convey this basic information about the structure of a phylogenetic tree in a series of nested parentheses called the **Newick format.** For instance, the Newick format would describe the tree in Figure 4.1 as (((I, II), (III, IV)), V).

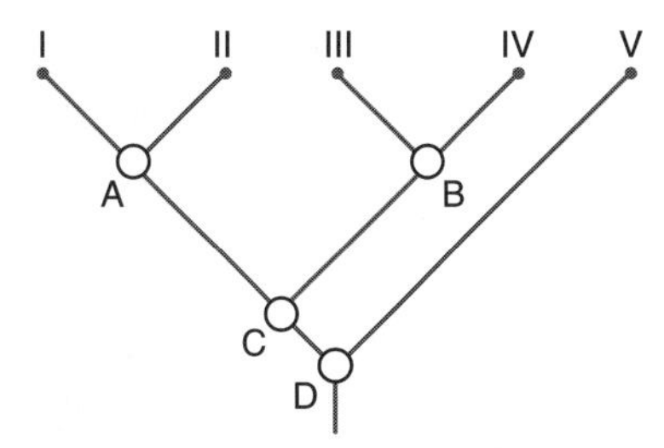

FIGURE 4.1 *A phylogenetic tree illustrating the evolutionary relationships among five species (I, II, III, IV, and V). Filled circles represent terminal nodes while open circles correspond to internal nodes. Inferred common ancestors to the species for which sequence data are available are labeled with letters (A, B, C, and D). The root of the tree corresponds to D.*

The internal nodes of most trees have only two lineages that descend from them and are said to be **bifurcating** as a result, although it is also possible for internal nodes to be **multifurcating** and have three or more descendant lineages. Multifurcating nodes can be interpreted in one of two ways: (1) An ancestral population simultaneously gave rise to three or more independent lineages, or (2) two or more bifurcations occurred at some point in the past but limitations in the data available make it impossible to distinguish the order in which they happened. Just as the branching patterns of a phylogenetic tree can be used to convey information about the sequence in which evolutionary events occurred, the length of branches is sometimes used to indicate the extent to which different data sets have diverged. **Scaled trees** are ones in which branch lengths are proportional to the differences between pairs of neighboring nodes. In the best of cases, scaled trees are also **additive,** meaning that the physical length of the branches connecting any two nodes is an accurate representation of their accumulated differences. In contrast, **unscaled trees** line up all terminal nodes and convey only their relative kinship without making any representation regarding the number of changes that separate them.

Rooted and Unrooted Trees

Another important distinction in phylogenetics is that between trees that make an inference about a common ancestor and the direction of evolution and those that do not (Figure 4.2). In **rooted trees** a single node is designated as a common ancestor, and a unique path leads from it through evolutionary time to any other node. **Unrooted trees** only specify the relationship between nodes and say nothing about the direction in which evolution occurred. Roots can usually be assigned to unrooted trees through the use of an **outgroup**—species that have unambiguously separated the earliest from the other species being studied. In the case of humans and gorillas when baboons are used as an outgroup, the root of the tree can be placed somewhere along the branch connecting baboons to the

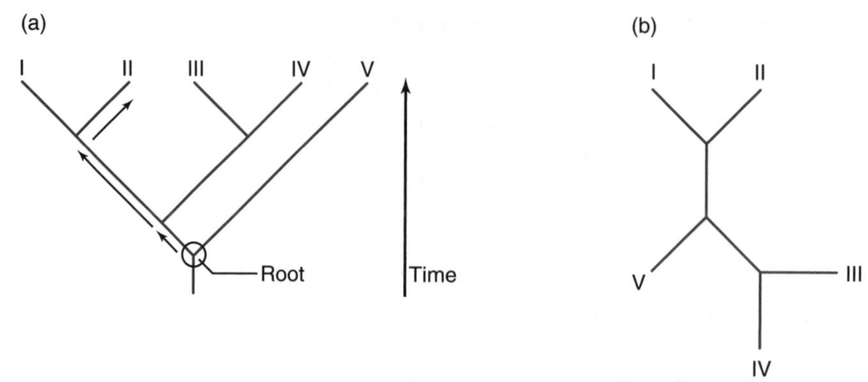

FIGURE 4.2 *(a) Rooted and (b) unrooted trees. Arrows indicate a unique path leading from the root to species D in the rooted tree. No inferences regarding the direction in which evolution occurred can be made from the unrooted tree.*

common ancestor of humans and gorillas. In a situation where only three species are being considered, three rooted trees are possible, but only one unrooted tree, as shown in Figure 4.3.

Contemplating all possible rooted and unrooted trees that might describe the relationship among three or four different species is not very difficult. However,

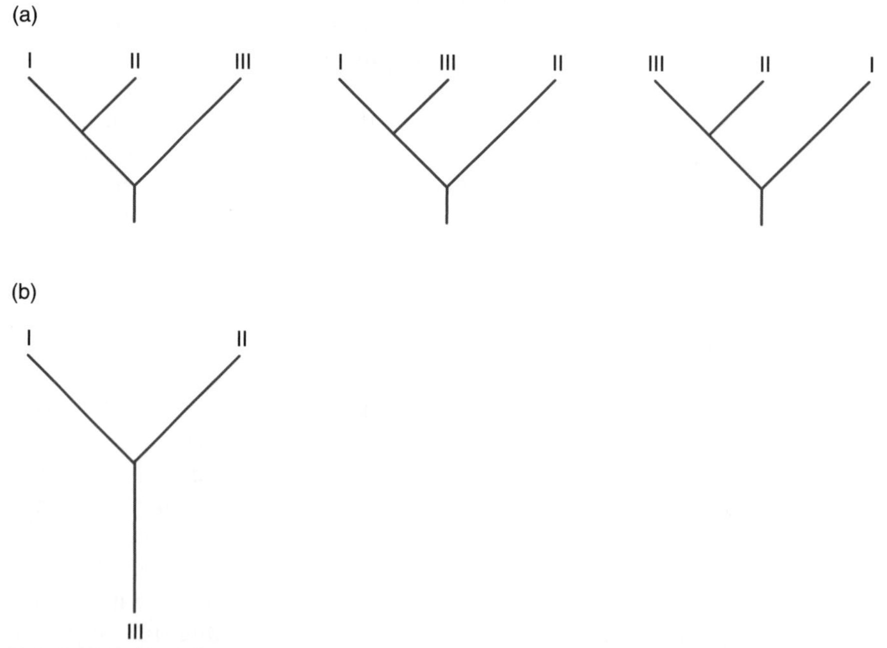

FIGURE 4.3 *All possible (a) rooted and (b) unrooted trees when only three species are considered.*

TABLE 4.1 Number of possible rooted and unrooted trees that can describe the possible relationships among fairly small numbers of data sets.

Number of Data Sets	Number of Rooted Trees	Number of Unrooted Trees
2	1	1
3	3	1
4	15	3
5	105	15
10	34,459,425	2,027,025
15	213,458,046,767,875	7,905,853,580,625
20	8,200,794,532,637,891,559,375	221,643,095,476,699,771,875

the number of possible trees quickly becomes staggering as more species are considered, as shown in Table 4.1. The actual number of possible rooted (N_R) and unrooted (N_U) trees for any number of species (n) can be determined with the following equations:

$$N_R = (2n - 3)!/2^{n-2}(n - 2)!$$
$$N_U = (2n - 5)!/2^{n-3}(n - 3)!$$

The reader should appreciate that the value for n can be profoundly large (conceivably every species or even every individual organism or copy of a gene that has ever been). Evaluating the relative merits of all possible trees is something that not even the fastest of computers can be expected to accomplish once more than a few dozen sequences or species are considered. Numerous shortcuts (some of which are described at the end of this chapter and also in Chapter 5) have been devised that focus only on those trees most likely to be reflective of the true relationship between the data sets.

Despite the staggering number of rooted and unrooted trees that can describe the relationships among even a small number of data sets (Table 4.1), only one of all possible trees can represent the true phylogenetic relationship among the genes or species being considered. Because the true tree is usually only known when artificial data are used in computer simulations, most phylogenetic trees generated with molecular data are referred to as **inferred trees.**

Gene vs. Species Trees

We would also like to point out that a phylogenetic tree based on the divergence observed within a single homologous gene is more appropriately referred to as a **gene tree** than a **species tree.** Such trees may represent the evolutionary history of a gene but not necessarily that of the species in which it is found. Species trees are usually best obtained from analyses that use data from multiple genes. For example, a recent study on the evolution of plant species used more than 100 different genes to generate a species tree for plants. While this may sound like an

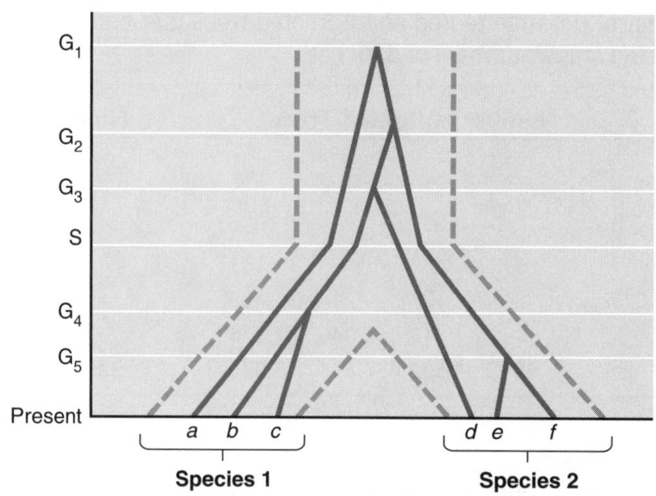

FIGURE 4.4 *Individuals may actually appear to be more closely related to members of a species other than their own when only one gene is considered. Gene divergence events (G_1 through G_5) often occur before as well as after speciation events (S). The evolutionary history of gene divergence resulting in the six alleles denoted* a *through* f *is shown in solid lines; speciation (i.e., population splitting) is shown by broken lines. Individual* d *would actually appear to be more closely related to individuals in species 1 if only this locus were considered even though it is a member of species 2.*

unnecessary demand for more data, it has proven to be important due to the fact that evolution occurs at the level of *populations* of organisms, not at the level of individuals. Divergence within genes typically occurs prior to the splitting of populations that occurs when new species are created. For the locus being considered in Figure 4.4, some individuals in species 1 may actually be more similar to individuals in species 2 than they are to other members of their own population. The differences between gene and species trees tend to be particularly important when considering loci where diversity within populations is advantageous such as the human leukocyte antigen (HLA) locus described in Chapter 3. If HLA alleles alone were used to determine species trees, many humans would be grouped with gorillas rather than other humans because the origin of the polymorphism they carry predates the split that gave rise to the two lineages.

Character and Distance Data

The molecular data used to generate phylogenetic trees fall into one of two categories: (1) **characters** (a well-defined feature that can exist in a limited number of different states) or (2) **distances** (a measure of the overall, pairwise difference between two data sets). Both DNA and protein sequences are examples of data that describe a set of discrete character states. Other examples of character data sets are those more commonly encountered in anatomically or behaviorally based taxonomy such as an organism's color or how long it takes to respond to a particular stimulus. As mentioned earlier, DNA sequence data are now so abundant that it is uncommon to find data sets that begin as distance measures such as those that can be generated by DNA–DNA hybridization studies among the genomes of organisms.

Character data can be fairly easily converted to distance data once criteria for determining the similarities among all possible character states have been established. For instance, a single value for the overall distance (D) between two genes from two species can be determined simply by generating an optimal pairwise alignment for the sequences, tallying the number of matching nucleotides (m), and dividing by the total number of sites at which matches could have been detected (t): $D = m/t$. Adjustments can be made for the possibility of multiple substitutions at any given site (see Chapter 3), and many biologists normalize distance values by expressing them in terms of "number of changes per 100 nucleotides." Protein distances can be calculated in the same way when amino acid sequences are aligned. However, a great deal of potentially important biological information can be lost in such conversions. For instance, it has long been appreciated that some kinds of substitutions are more likely to occur within nucleotide and protein sequences than others. Given that a C is replaced with a T almost three times as often as it is replaced with a G or an A within mammalian genomes, a reasonable argument could be made that C/T mismatches in pairwise alignments not be as heavily weighted as C/G or C/A mismatches when calculating the pairwise distance of DNA sequences. The situation can be much more complicated in protein comparisons where not only are some amino acids more likely to be replaced by others due to the chemical activity of their functional groups, but the number of substitutions at the DNA level to exchange one amino acid for another can also differ.

Mathematically based approaches to phylogenetic reconstruction (such as UPGMA, described later) generally discount the importance of such biological subtleties in data sets. **Pheneticists** tend to prefer these methods because they place a greater emphasis on the relationships among data sets than the paths they have taken to arrive at their current states. **Cladists,** however, are generally more interested in evolutionary pathways and patterns than relationships and tend to prefer the more biologically based approaches to tree generation (like maximum parsimony) discussed in Chapter 5. While disputes between the champions of the two approaches have often been surprisingly intense, it is fair to say that both approaches are widely used and work well with most data sets.

Distance Matrix Methods

Distinguishing which of all possible trees is most likely to be the true tree can be a daunting task and is typically left to high-speed computers. Pairwise distance matrices, tabular summaries of the differences between all data sets to be analyzed, are the raw material used by many popular phylogenetic tree reconstruction algorithms. A basic understanding of the logic behind these approaches should give the reader an understanding of exactly what information phylogenetic trees convey and what sort of molecular data are most useful for their generation. It should also make it easier to appreciate the circumstances that make using a variant approach advisable.

UPGMA

The oldest distance matrix method is also the simplest of all methods for tree reconstruction. Originally proposed in the early 1960s to help with the evolutionary analysis of morphological characters, the **unweighted-pair-group method with arithmetic mean (UPGMA)** is largely statistically based and, like all distance-based methods, requires data that can be condensed to a measure of genetic distance between all pairs of taxa being considered. In general terms, the UPGMA method requires a distance matrix such as one that might be created for a group of four taxa called A, B, C, and D. Assume that the pairwise distances between each of the taxa are given in the following matrix:

Species	A	B	C
B	d_{AB}	—	—
C	d_{AC}	d_{BC}	—
D	d_{AD}	d_{BD}	d_{CD}

In this matrix, d_{AB} represents the distance (perhaps simply the number of nonmatching nucleotides divided by the total number of sites where matches could have been found) between species A and B, while d_{AC} is the distance between taxa A and C, and so on.

UPGMA begins by clustering the two species with the smallest distance separating them into a single, composite group. In this case, assume that the smallest value in the distance matrix corresponds to d_{AB} in which case species A and B are the first to be grouped (AB). After the first clustering, a new distance matrix is computed with the distance between the new group (AB) and species C and D being calculated as $d_{(AB)C} = 1/2(d_{AC} + d_{BC})$ and $d_{(AB)D} = 1/2(d_{AD} + d_{BD})$. The species separated by the smallest distance in the new matrix are then clustered to make another new composite species. The process is repeated until all species have been grouped. If scaled branch lengths are to be used on the tree to represent the evolutionary distance between species, branch points are positioned at a distance halfway between each of the species being grouped (i.e., at $d_{AB}/2$ for the first clustering).

A practical example using UPGMA with actual sequence data will help illustrate the general approach just described. Consider the alignment between five different DNA sequences presented in Figure 4.5. Pairwise comparisons of the five different sequences yield a distance matrix such as the one shown in Table 4.2. Notice that because all five sequences are the same length and there are no gaps, the distance matrix in this case can simply be the number of nonmatching nu-

```
        10           20          30          40          50
A:  GTGCTGCACGG  CTCAGTATA  GCATTTACCC  TTCCATCTTC  AGATCCTGAA
B:  ACGCTGCACGG  CTCAGTGCG  GTGCTTACCC  TCCCATCTTC  AGATCCTGAA
C:  GTGCTGCACGG  CTCGGCGCA  GCATTTACCC  TCCCATCTTC  AGATCCTATC
D:  GTATCACACGA  CTCAGCGCA  GCATTTGCCC  TCCCGTCTTC  AGATCCTAAA
E:  GTATCACATAG  CTCAGCGCA  GCATTTGCCC  TCCCGTCTTC  AGATCTAAAA
```

FIGURE 4.5 *A five-way alignment of homologous DNA sequences.*

TABLE 4.2 A pairwise distance matrix that summarizes the number of nonmatching nucleotides between all possible pairs of sequences shown in Figure 4.5.

Species	A	B	C	D
B	9	—	—	—
C	8	11	—	—
D	12	15	10	—
E	15	18	13	5

cleotides observed in each pairwise comparison (i.e., the number of nonmatching nucleotides between A and B, d_{AB}, is 9).

The smallest distance separating any of the two sequences in the multiple alignment corresponds to d_{DE}, so species D and species E are grouped (Figure 4.6a). A new distance matrix is then made in which the composite group (DE) takes the place of D and E (Table 4.3). Distances between the remaining species and the new group are determined by taking the average distance between its two members (D and E) and all other remaining species [i.e., $d_{(DE)A} = 1/2(d_{AD} + d_{AE})$ so $d_{(DE)A} = 1/2(12 + 15) = 13.5$], as shown in Table 4.3.

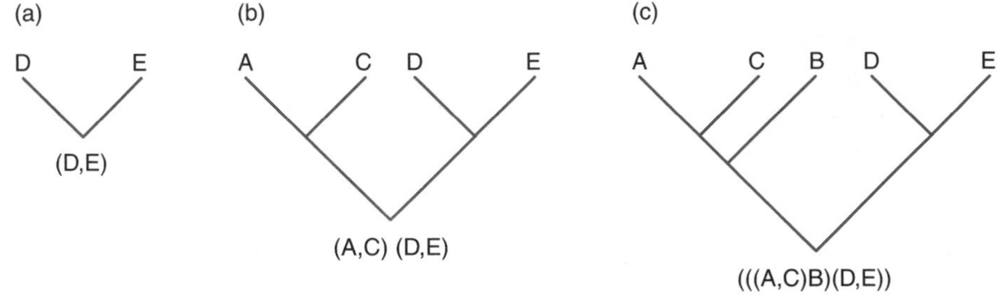

FIGURE 4.6 *A phylogenetic tree as it is constructed using the UPGMA method. (a) The first grouping (D,E) is between species D and E, which are connected by a single bifurcating branch. (b) The second grouping (A,C) is between species A and C, which are also connected to each other by a single bifurcating branch. (c) The last grouping ((A,C)B) unambiguously places the branching point for species B between that of the common ancestors of (A,C) and (D,E).*

TABLE 4.3 The distance matrix that results when species D and E of Table 4.2 are combined and considered as a single group.

Species	A	B	C
B	9	—	—
C	8	11	—
DE	13.5	16.5	11.5

TABLE 4.4 The distance matrix that results when species A and C of Table 4.3 are combined and considered as a single group.

Species	B	AC
AC	10	—
DE	16.5	12.5

The smallest distance separating any two species in this new matrix is the distance between A and C, so a new combined species, (AC), is created (Figure 4.6b). Another distance matrix using this new grouping then looks like the one shown in Table 4.4. In this last matrix the smallest distance is between species (AC) and B (d_{DE} = 10) so they are grouped as ((AC)B). One way to symbolically represent the final clustering of species is shown in Figure 4.6c. Alternatively, in the standard Newick format it could be described as (((A,C)B)(D,E)). In many cases, generating the distance matrix needed for the UPGMA method is the most tedious and time-consuming step in the entire process of inferring a phylogenetic tree. While small data sets can be analyzed by hand, computer programs can easily analyze data sets that are large in terms of both the number and the lengths of the sequences involved.

Estimation of Branch Lengths

Remember that in addition to describing the relatedness of sequences it is possible for the **topology** of phylogenetic trees to convey information about the relative degree to which sequences have diverged. Scaled trees that convey that information, often referred to as **cladograms,** do so by having the length of branches correspond to the inferred amount of time that the sequences have been accumulating substitutions independently.

Determining the relative length of each branch in a cladogram can also be easily calculated using the information in a distance matrix. If rates of evolution are assumed to be constant in all lineages, then internal nodes should simply be placed at equal distances from each of the species they give rise to on a bifurcating tree. For instance, using the sequences and distance matrix from Figure 4.5 and Table 4.2, respectively, the distance between species D and E (d_{DE}) is 5, and the pair of branches connecting each of those species to their common ancestor should each be $d_{DE}/2$ or 2.5 units long on a tree with scaled branch lengths. Similarly, A and C should be connected to their common ancestor by branches that are $d_{AC}/2$ or 4 units long, and the branch point between (AC) and (DE) should be connected to (AC) and (DE) by branches that are both $d_{(AC)(DE)}/2$ or 6.25 units long, as shown in Figure 4.7. This very simple approach to estimating branch lengths actually allows UPGMA to be one of only a very small number of approaches that intrinsically generate rooted phylogenetic trees.

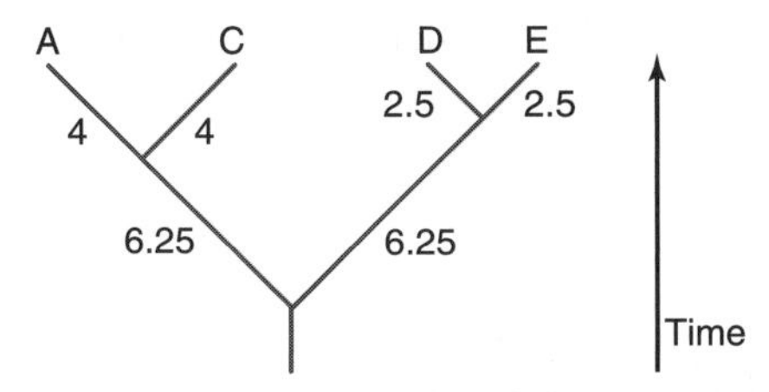

FIGURE 4.7 *A scaled tree showing the branch lengths separating four of the species depicted in Figure 4.6. Branch lengths are shown next to each branch. Branches are also drawn to scale to reflect the amount of differences between all species. If evolution has occurred at a constant rate for all of these lineages, then branch length also corresponds to divergence time.*

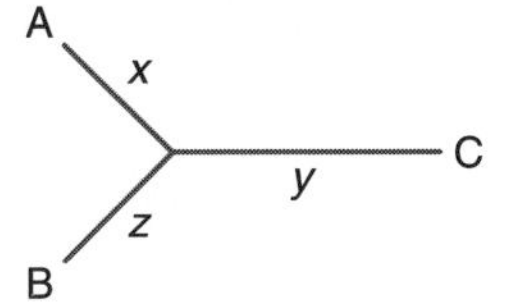

FIGURE 4.8 *The simplest tree whose branch lengths might have some meaningful information. Each of the three branches on this tree can be represented as a single variable* (x, y, *and* z)*.*

Determining branch lengths for a scaled tree is only slightly more complicated when it cannot be assumed that evolutionary rates are the same for all lineages. The simplest tree whose branch lengths might have some meaningful information is one with just three species and one branch point, such as the one shown in Figure 4.8. On such a tree, the length of each of the three branches can be represented by a single letter (x, y, and z) for which we know the following must be true:

$$d_{AC} = x + y$$
$$d_{AB} = x + z$$
$$d_{BC} = y + z$$

Simple algebraic manipulation of those equations can then be used to give equations for each branch length simply in terms of the values in a pairwise distance matrix:

$$x = (d_{AB} + d_{AC} - d_{BC})/2$$
$$y = (d_{AC} + d_{BC} - d_{AB})/2$$
$$z = (d_{AB} + d_{BC} - d_{AC})/2$$

Branch lengths for more complicated trees (ones with more than one branch point) can be estimated by continuing to consider just three branches at a time. Two of those branches, x and z in Figure 4.8, are the ones that connect the two most closely related species in the distance matrix, and one, y in Figure 4.8, connects their common ancestor to the common ancestor of all others in the distance matrix. For instance, for a distance matrix for five species (1, 2, 3, 4, and 5) in which UPGMA groups species 1 and 2 first, the values for d_{AC} and d_{BC} in the earlier equations are just the average values for the distance between those species and all other species taken together:

$$d_{AC} = (d_{13} + d_{14} + d_{15})/3$$
$$d_{BC} = (d_{23} + d_{24} + d_{25})/3$$

Transformed Distance Method

One strength of distance matrix approaches in general is that they work equally well with morphological and molecular data and even combinations of the two. They also take into consideration all of the data available for a particular analysis, whereas the alternative parsimony approaches described in Chapter 5 discard so-called uninformative sites.

One particular weakness of the UPGMA approach is that it assumes a constant rate of evolution across all lineages—something that the relative rate tests (Chapter 3) tell us is not always the case. Variations in rates of substitutions can be a serious problem for the UPGMA method and can easily cause it to produce trees with incorrect topologies. In fact, one indication that that is not the case for the set of five DNA sequences used to illustrate the UPGMA method (Figure 4.5) is that the branch lengths of the resulting tree are not additive (for example, the scaled tree shown in Figure 4.7 suggests that $d_{AE} = 4 + 6.25 + 6.25 + 2.5 = 19$ while the distance matrix in Table 4.2 tells us that the actual value for d_{AE} is 15).

Several distance matrix–based alternatives to UPGMA take the possibility of different rates of evolution within different lineages into account. The oldest and simplest of those is the **transformed distance method.** This approach, first described by J. Farris in 1977, takes full advantage of the power of an **outgroup**—a species that is known to have diverged from the common ancestor of all other species represented in a distance matrix prior to all other species being considered (the **ingroups**).

Using the same data set shown in Figure 4.5, assume that species D is known to be an outgroup to species A, B, and C. Also assume that the true relationship between these species is the one depicted in Figure 4.9 [(((A,B)C)D) in the Newick format]. If the numbers beside each of the branches in Figure 4.9 correspond to the number of mutations within the 50 base pairs of the sequence shown in Figure 4.5 that have accumulated along each lineage during each stage of its evolution, then the pairwise evolutionary distances are shown in Table 4.5 (an abbreviated version of Table 4.2). In this situation, D can be used as an external reference to transform the distances that separate the other species using this equation:

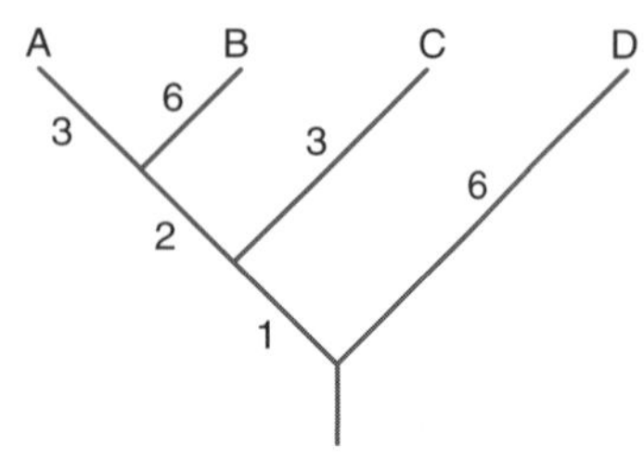

FIGURE 4.9 *The true phylogenetic relationship and branch lengths of four species: A, B, C, and D.*

$$d'_{ij} = (d_{ij} - d_{iD} - d_{jD})/2 + \bar{d}_D$$

where d'_{ij} is the transformed distance between species i and j, and $\bar{d}_D$ is the average distance between the outgroup and all ingroups (in this case: $d_D = (d_{AD} + d_{BD} + d_{CD})/3 = 37/3$). The term $\bar{d}_D$ is introduced simply to ensure that all transformed distance

TABLE 4.5 Pairwise evolutionary distances for the four species depicted in Figure 4.7, assuming that the tree is additive as it should be.

Species	A	B	C
B	9	—	—
C	8	11	—
D	12	15	10

TABLE 4.6 Transformed distance matrix for the three ingroups of Table 4.5 when D is used as an outgroup.

Species	A	B
B	10/3	—
C	16/3	16/3

values are positive values since negative distances are not possible from an evolutionary perspective. Using this value for $\bar{d}_D$ and the distances in Table 4.5, a transformed distance matrix for species A, B, and C can be generated (Table 4.6). The basic UPGMA approach can then be used with this new distance matrix; since d'_{AB} is the smallest value, A and B are the first to be clustered (A,B). Species C is added to the tree next and D, the outgroup, is added last to give a tree with the true topology: (((A,B)C)D).

The power of the transformed distance matrix approach comes from an insight that is so simple it is easy to miss: Ingroups only evolve separately from each other *after* they diverge, and any differences in the number of substitutions they have accumulated must have occurred since that time. In this situation, outgroups simply provide an objective frame of reference for comparing those rates of substitution.

As was the case in the example above, the transformed distance matrix approach is typically better at determining the true topology of a tree than the UPGMA method alone. It can also be applied in those situations where it is not possible to independently determine which species is an outgroup. Any ingroup can also provide a frame of reference suitable for use in transforming a distance matrix. The principal advantage of outgroups over ingroups for this purpose is that outgroups alone allow a root to be placed on the phylogenetic tree.

Neighbor's Relation Method

Another popular variant of the UPGMA method is one that emphasizes pairing species in such a way that a tree is created with the smallest possible branch lengths overall. On any unrooted tree, pairs of species that are separated from each other by just one internal node are said to be **neighbors.** The topology of a phylogenetic tree such as the one in Figure 4.10 gives rise to some useful algebraic relationships between neighbors. If the tree in Figure 4.9 is a true tree for which additivity holds, then the following should be true:

$$d_{AC} + d_{BD} = d_{AD} + d_{BC} = a + b + c + d + 2e = d_{AB} + d_{CD} + 2e$$

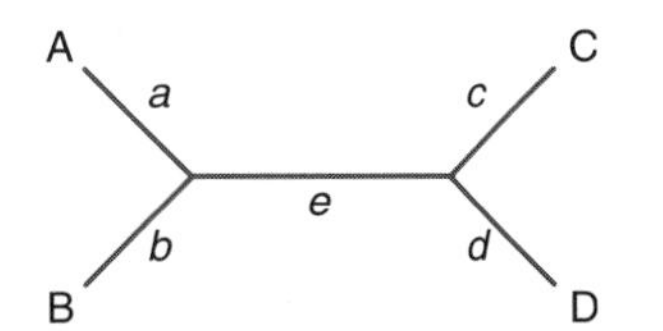

FIGURE 4.10 *A generic phylogenetic tree with four species (A, B, C, and D) and each branch uniquely labeled* (a, b, c, d, *and* e)*.*

where *a*, *b*, *c*, and *d* are the lengths of the terminal branches and *e* is the length of the single central branch. The following conditions, known together as the **four-point condition,** will also be true:

$$d_{AB} + d_{CD} < d_{AC} + d_{BD}$$

and

$$d_{AB} + d_{CD} < d_{AD} + d_{BC}$$

It is in this way that a **neighborliness approach** considers all possible pairwise arrangements of four species and determines which arrangement satisfies the four-point condition. An important assumption of the four-point condition is that branch lengths on a phylogenetic tree should be additive and, while it is not especially sensitive to departures from that assumption, data sets that are not additive can cause this method to generate a tree with an incorrect topology.

In 1977, S. Sattath and A. Tversky suggested a way to use the neighborliness approach for trees that described the relationship between more than four species. They begin by generating a distance matrix and using the values in that matrix to generate three values for the first four species: (1) $d_{AB} + d_{CD}$; (2) $d_{AC} + d_{BD}$; and (3) $d_{AD} + d_{BC}$. Whichever sum is the smallest causes both pairings to be assigned a score of 1, while the four other pairings are given a score of 0. This process is repeated for all possible sets of four species in the data set, and a running tally of the scores is maintained. The single pair of species with the highest score at the end of the analysis is grouped—these species are as neighbors—and a new distance matrix is generated as was done for UPGMA. The new distance matrix is then used in another scoring process until only three species remain (at which time the topology of the tree is unambiguously determined). While manageable when considering the relationship between only five or six species, this approach can become very computationally intensive for more complex trees.

Neighbor-Joining Methods

Other neighborliness approaches are also available including several variants called **neighbor-joining methods.** These methods start with a star-like tree with all species coming off a single central node regardless of their number. Neighbors are then sequentially found that minimize the total length of the branches on the tree. The principal difference between the neighbor-joining methods is the way in which they determine the sum of all branch lengths with each reiteration of the process. N. Saitou and M. Nei originally proposed in 1987 that the tree's total branch length be determined with this formula:

$$S_{12} = (1/(2(N-2))(\Sigma(d_{1k} + d_{2k}) + (1/2)d_{12} + (1/N-2)(\Sigma d_{ij})$$

where any pair of species can take positions 1 and 2 in the tree, N is the number of species represented in the distance matrix, k, is an accepted outgroup, and d_{ij} is the distance between species i and j. J. Studier and K. Keppler advocated a computationally faster algorithm in 1988:

$$Q_{12} = (N-2)d_{12} - \Sigma d_{1i} - \Sigma d_{2i}$$

All possible pairs of species are considered in each round of the process, and the pairing that makes a tree with the overall smallest branch lengths (the smallest values for S or Q) is grouped so that a new distance matrix can be generated.

Both the S and Q criteria have since been shown to be theoretically related. Both the neighbor-joining and neighbor's relation approach described above have also been shown to be theoretically equivalent in that they are fundamentally dependant on the four-point condition and the assumption of additivity. As a result, both neighbor-joining and neighbor's relation approaches tend to generate trees with very similar, if not identical, topologies.

Maximum Likelihood Approaches

Maximum likelihood approaches represent an alternative and purely statistically based method of phylogenetic reconstruction. With this approach, probabilities are considered for every individual nucleotide substitution in a set of sequence alignments. For instance, we know from Chapter 3 that transitions (exchanging one purine for another or one pyrimidine for another) are observed roughly three times as often as transversions (exchanging a purine for a pyrimidine or vice versa). In a three-way alignment where a single column is found to have a C, a T, and a G, it can be reasonably argued that a greater likelihood exists that the sequences with the C and T are more closely related to each other than they are to the sequence with a G. Calculation of probabilities is complicated by the fact that the sequence of the common ancestor to the sequences being considered is generally not known. They are further complicated by the fact that multiple substitutions may have occurred at one or more of the sites being considered and that all sites are not necessarily independent or equivalent. Still, objective criteria can be applied to calculating the probability for every site *and* for every possible tree that describes the relationship of the sequences in a multiple alignment. The number of possible trees for even a modest number of sequences (Table 4.1) makes this a very computationally intensive proposition, yet the one tree with the single highest aggregate probability is, by definition, the most likely to reflect the true phylogenetic tree.

The dramatic increase in the raw power of computers has begun to make maximum likelihood approaches feasible, and trees inferred in this way are becoming increasingly common in the literature. Note, however, that no one substitution model is as yet close to general acceptance and, because different models can very easily lead to different conclusions, the model used must be carefully considered and described when using this approach.

Multiple Sequence Alignments

As described in Chapter 2, sequence alignments are easiest when the sequences being aligned are very similar and have not experienced many insertion and deletion events. Simultaneous alignment of more than two sequences is a natural extension of the two-sequence case. As in pairwise alignments, sequences in a multiple alignment are customarily placed in a vertical list so that characters (or gaps) in corresponding positions occupy the same column. The major difficulty in aligning multiple sequences is computational. For instance, the insertion of a single nucleotide in one sequence requires that a gap be added to every other sequence in a multiple alignment and can easily wreak havoc with static scoring of gap insertion and length penalties. Several studies have also shown that the order in which sequences are added to a multiple alignment can significantly affect the end result.

Given that similar sequences can be aligned both more easily and with greater confidence, it is not surprising that the alignment of multiple sequences should take into consideration the branching order of the sequences being studied. If the phylogeny of the sequences being aligned is known before the alignment is made, sequences are generally added one at a time to the growing multiple alignment with the most related sequences being added first and the least related being added last. It is increasingly common, however, for analyses of the sequences themselves to be the way in which phylogenetic relationships are determined. In those cases, an integrated or unified approach is generally adopted that simultaneously generates an alignment and a phylogeny. This approach typically requires many rounds of phylogenetic analysis and sequence alignment and can be very computationally intensive. For instance, a common strategy of several popular multiple sequence aligning algorithms is to (1) generate a pairwise distance matrix based on all possible pairwise alignments between the sequences being considered, (2) use a statistically based approach such as UPGMA to construct an initial tree, (3) realign the sequences progressively in order of their relatedness according to the inferred tree, (4) construct a new tree from the pairwise distances obtained in the new multiple alignment, and (5) repeat the process if the new tree is not the same as the previous one.

Chapter Summary

The true relationship between homologous sequences is hardly ever known aside from computer simulation experiments. The numbers of alternative phylogenetic trees to choose among quickly become daunting even for relatively small numbers of sequences. A variety of approaches are available for inferring the most likely phylogenetic relationship between genes and species using nucleotide and amino acid sequence information. Distance-based approaches to phylogenetic reconstruction draw attention to just a few (and, often, just one) of those many possible trees by considering the overall similarities between available sequences and progressively grouping those that are most alike. Maximum likelihood approaches are computationally intensive but strive to draw attention to the phylogenetic relationship that is statistically most likely to represent the true relationship.

Readings for Greater Depth

The general utility of molecular data in phylogenetic studies is succinctly reviewed in T. A. Brown and K. A. Brown, 1994, Using molecular biology to explore the past, *BioEssays* **16**:719–726.

The simplest method for tree reconstruction, UPGMA, was first applied in R. R. Sokal and C. D. Michener, 1958, A statistical method for evaluating systematic relationships, *Univ. Kansas Sci. Bull.* **28**: 1409–1438.

The algorithm underlying the neighbor-joining approach as well as a comparison to the neighbor's relation method are described in detail in N. Saitou and M. Nei, 1987, The neighbor-joining method: A new method for reconstructing phylogenetic trees, *Mol. Biol. Evol.* **4**: 406–425.

A detailed description and review of the maximum likelihood approach (as well as references to papers describing different substitution models and algorithms) can be found in J. P. Huelsenbeck and K. A. Crandall, 1997, Phylogeny estimation and hypothesis testing using maximum likelihood, *Annu. Rev. Ecol. Syst.* **28**: 437–466.

Questions and Problems

* **4.1** What are some of the advantages of using molecular data to infer evolutionary relationships?

4.2 As suggested by the popular movie *Jurassic Park*, organisms trapped in amber have proven to be a good source of DNA from tens and even hundreds of millions of years ago. However, when using such sequences in phylogenetic analyses it is usually not possible to distinguish between samples that come from evolutionary "dead ends" and those that are the ancestors of organisms still alive today. Why would it not be possible to use information from ancient DNA in a standard UPGMA analysis?

* **4.3** Draw the phylogenetic tree that corresponds to the one described in the standard Newick format as (((A,B)C)(D,E))F.

4.4 Draw all possible rooted and unrooted trees for four species: A, B, C, and D.

* **4.5** What is the chance of randomly picking the one rooted phylogenetic tree that describes the true relationship between a group of eight organisms? Are the odds better or worse for randomly picking among all possible unrooted trees for those organisms?

4.6 Why is it easier to convert character data into distance data than vice versa? Which type of data would be preferred by a pheneticist or a cladist?

* **4.7** Using the five-way sequence alignment shown in Figure 4.5, construct a distance matrix by weighting transversions (A's or G's changing to C's or T's) twice as heavily as transitions (C's changing to T's, T's changing to C's, A's changing to G's, or G's changing to A's).

4.8 Use UPGMA to reconstruct a phylogenetic tree using the following distance matrix:

Species	A	B	C	D
B	3	—	—	—
C	6	5	—	—
D	9	9	10	—
E	12	11	13	9

* **4.9** Using the same distance matrix provided for Question 4.8 and assuming that species D is an outgroup, reconstruct a phylogenetic tree with the transformed distance method. Does the topology of this tree differ from the one you generated with UPGMA alone?

4.10 Using the same distance matrix provided for Question 4.8 and assuming that species F is an outgroup, reconstruct a phylogenetic tree with the transformed distance method. Does the topology of this tree differ from the one you generated using D as an outgroup?

* **4.11** Assume that the rate of evolution is not the same for all lineages in the tree you generated for Question 4.10. What are the relative lengths of each of the branches in the tree?

CHAPTER 5

Character-Based Methods of Phylogenetics

There is grandeur in this view of life, . . . from so simple a beginning endless forms most beautiful and most wonderful have been, and are being, evolved.

Charles Darwin (1809–1882)

While an understanding of the relationships between sequences and species has been the basis of profound advances in disciplines as varied as agriculture and medicine, bioinformaticians are generally more interested in using the relationships just as clues for other analyses such as those involved in multiple sequence alignments. Bioinformaticians tend to be more like the pheneticists described in the previous chapter than the pathway and evolutionary process–oriented cladists. At the same time, though, there is no denying that distance-based methods "look at the big picture" and pointedly ignore much potentially valuable information. All of that additional information is at the heart of what excites cladists and poses demanding (but solvable) computational problems that have intrigued many programmers. Of course, much can be learned about relationships from a consideration of the path taken in the process of divergence. Because the analyses employed by the distance- and character-based methods are so fundamentally different, agreement of their conclusions regarding relationships is considered to be particularly strong support for a phylogenetic tree—a very valuable commodity for bioinformaticians.

Parsimony

The concept of **parsimony** is at the very heart of all character-based methods of phylogenetic reconstruction. The term itself is borrowed from the English language and was popularly used in the United States during the 1930s and 1940s to somewhat derogatorily describe someone who was especially careful with the spending of their money. In a biological sense, it is used to describe the process of attaching preference to one evolutionary pathway over another on the basis of which pathway requires the invocation of the smallest number of mutational events. As described in the two previous chapters, phylogenetic trees represent theoretical models that depict the evolutionary history of three or more sequences. The two premises that underlie biological parsimony are quite simple: (1) Mutations are exceedingly rare events and (2) the more unlikely events a model invokes, the less likely the model is to be correct. As a result, the relationship that requires the fewest number of mutations to explain the current state of the sequences being considered is the relationship that is most likely to be correct.

Informative and Uninformative Sites

Before considering how it is that parsimony is used to infer phylogenetic relationships, it is necessary to consider what sites within a multiple sequence alignment might have useful information content for a parsimony approach. The short four-way multiple sequence alignment shown in Figure 5.1 contains positions that fall into two categories in terms of their information content for a parsimony analysis: those that have information (are **informative**) and those that do not (are **uninformative**). The relationship between four sequences can be described by

	Position					
Sequence	**1**	**2**	**3**	**4**	**5***	**6***
1	G	G	G	G	G	G
2	G	G	G	A	G	T
3	G	G	A	T	A	G
4	G	A	T	C	A	T

FIGURE 5.1 *An alignment of four homologous sequences, each of which is six nucleotides long. Asterisks (*) indicate the two columns within the alignment that correspond to informative sites. The four remaining sites are uninformative.*

only three different unrooted trees (as described in Chapter 4 and shown in Figure 5.2), and informative sites are those that allow one of those trees to be distinguished from the other two on the basis of how many mutations they must invoke. For the first position in the alignment in Figure 5.1, all four sequences have the same character (a "G") and the position is therefore said to be **invariant.** Invariant sites are clearly uninformative because each of the three possible trees that describe the relationship of the four sequences (Figure 5.2.1) invokes exactly the same number of mutations (0). Position 2 is similarly uninformative from a parsimony perspective because one mutation occurs in all three of the possible trees (Figure 5.2.2). Likewise, position 3 is uninformative because all three trees require two mutations (Figure 5.2.3), and position 4 is uninformative because all three trees require three mutations (Figure 5.2.4). In contrast, positions 5 and 6 are both informative because for both of them, one of the three trees invokes only one mutation and the other two alternative trees both require two (Figures 5.2.5 and 5.2.6).

In general, for a position to be informative regardless of how many sequences are aligned, it has to have at least two different nucleotides, and each of these nucleotides has to be present at least twice. All parsimony programs begin by applying this fairly simple rule to the data set being analyzed. Notice that four of the six positions being considered in the alignment shown in Figure 5.1 are simply discarded and not considered any further in a parsimony analysis. All of those sites would have contributed to the pairwise similarity scores used by a distance-based approach, and this difference alone can generate substantial differences in the conclusions reached by both types of approaches.

Unweighted Parsimony

Once uninformative sites have been identified and discarded, implementation of the parsimony approach in its simplest form can be straightforward. Every possible tree is considered individually for each informative site. A running tally is maintained for each tree that keeps track of the minimum number of substitutions required for each position. After all informative sites have been considered, the tree (or trees) that needs to invoke the smallest total number of substitutions

FIGURE 5.2 *Three possible unrooted trees that describe the possible relationships between the four sequences at all 6 positions shown in Figure 5.1. Each position is numbered as in Figure 5.1. The Newick format description of each of the trees is also shown. The two internal nodes of each tree contain a possible common ancestor's sequence, and lineages where mutations must have occurred are highlighted with arrows.*

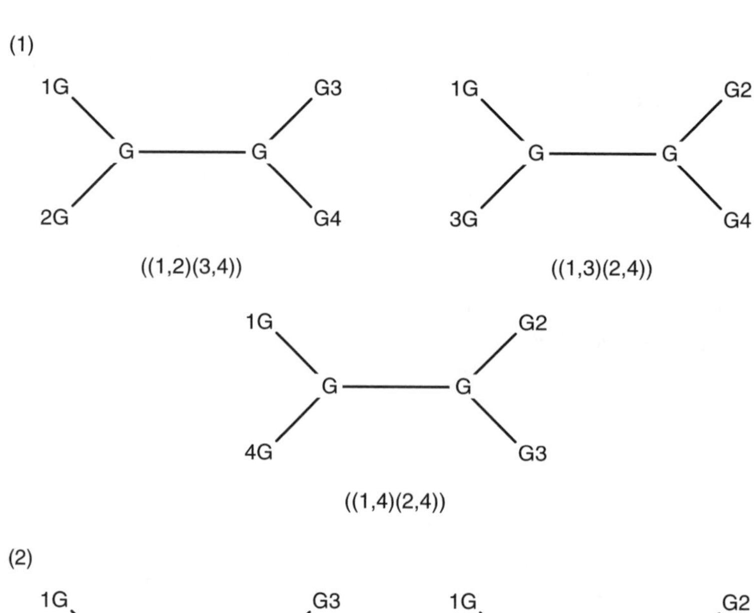

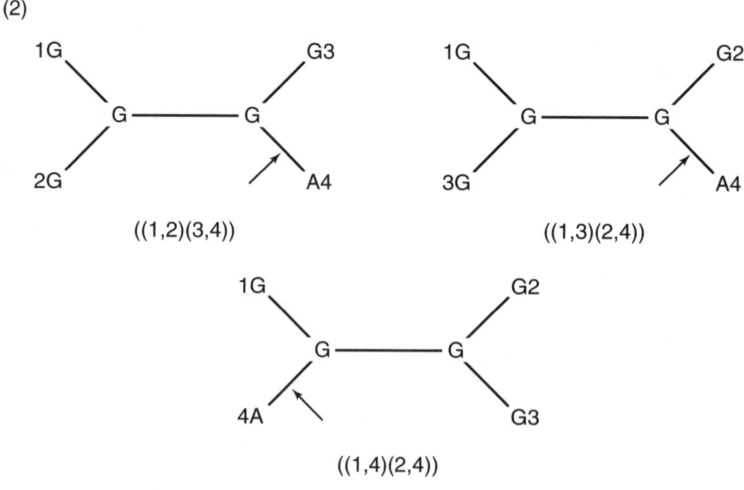

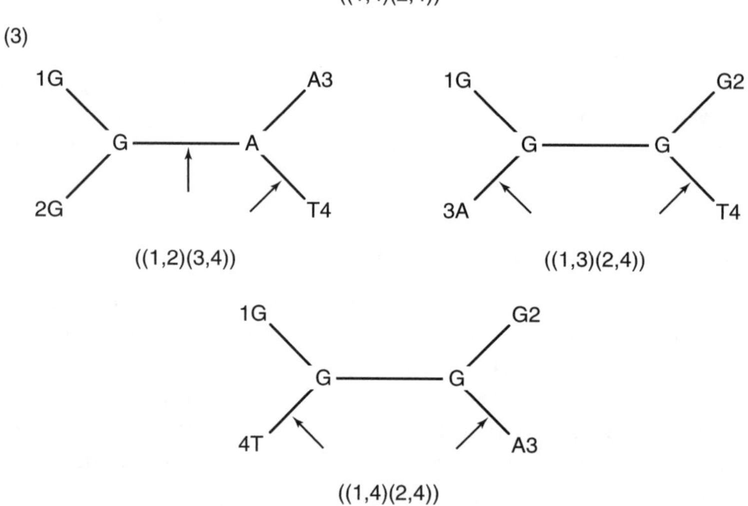

(4)

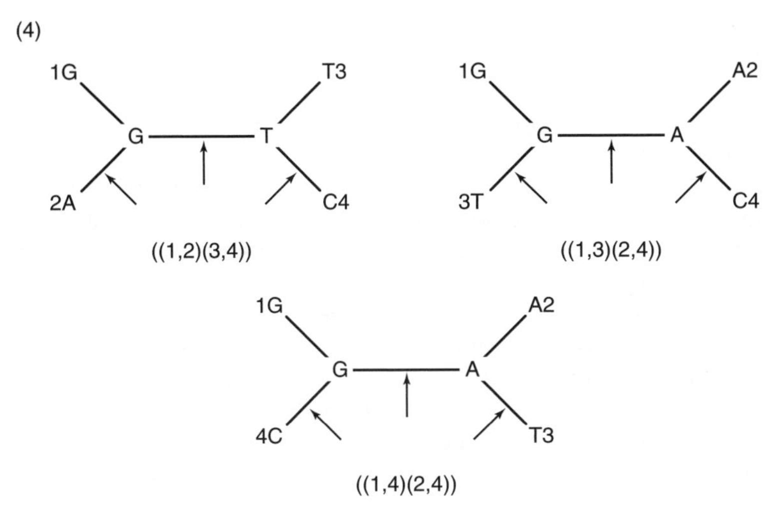
1G
T3
G
T
2A
C4
((1,2)(3,4))
1G
A2
G
A
3T
C4
((1,3)(2,4))
1G
A2
G
A
4C
T3
((1,4)(2,4))

(5)

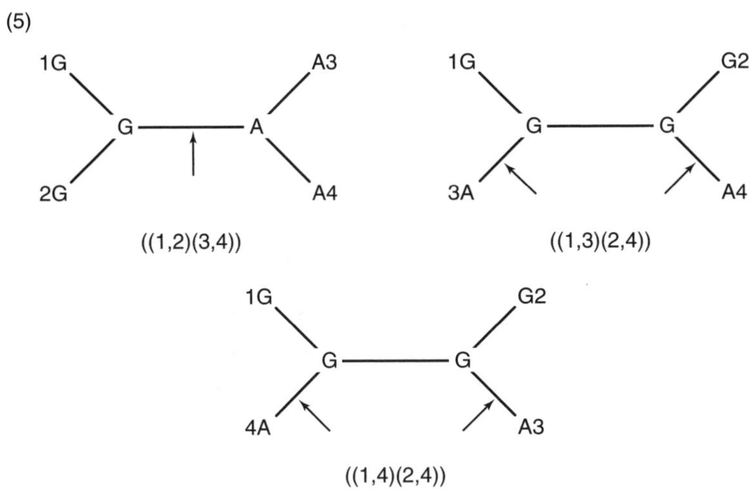
1G
A3
G
A
2G
A4
((1,2)(3,4))
1G
G2
G
G
3A
A4
((1,3)(2,4))
1G
G2
G
G
4A
A3
((1,4)(2,4))

(6)

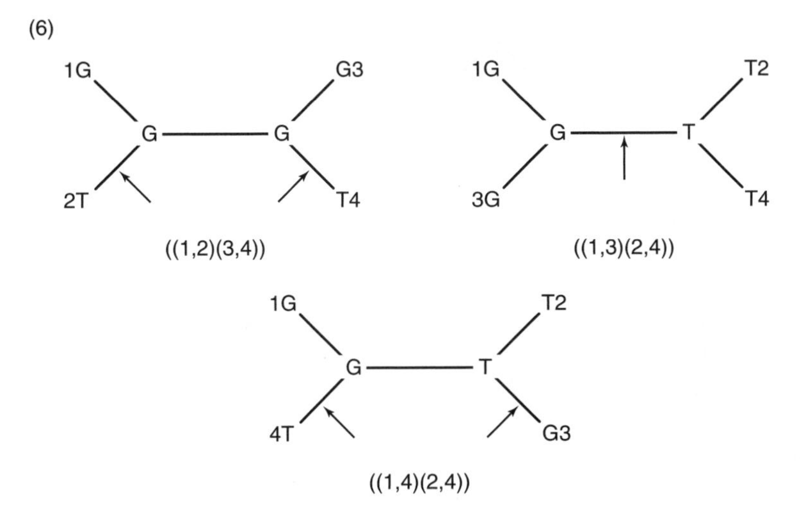
1G
G3
G
G
2T
T4
((1,2)(3,4))
1G
T2
G
T
3G
T4
((1,3)(2,4))
1G
T2
G
T
4T
G3
((1,4)(2,4))

is labeled the most parsimonious. The example presented in Figures 5.1 and 5.2 for a four-way sequence alignment is the simplest case possible. In an analysis involving only four sequences, each informative site can favor only one of the three alternative trees, and the tree supported by the largest numbers of informative sites must also be the most parsimonious.

Evaluation of data sets with five or more sequences is substantially more complicated due to three important differences. First, the number of alternative unrooted trees that need to be considered increases dramatically as the number of sequences in an alignment becomes greater than just four (Table 4.1). Even when only a small number of informative sites are being evaluated, applying the unmodified parsimony approach by hand is virtually unimaginable when as few as just seven or eight sequences are being considered. Second, the situation is further complicated by the fact that in analyses involving more than four sequences, individual informative sites can support more than one alternative tree (see Figures 5.3a and b), and the tree of maximum parsimony for the entire data

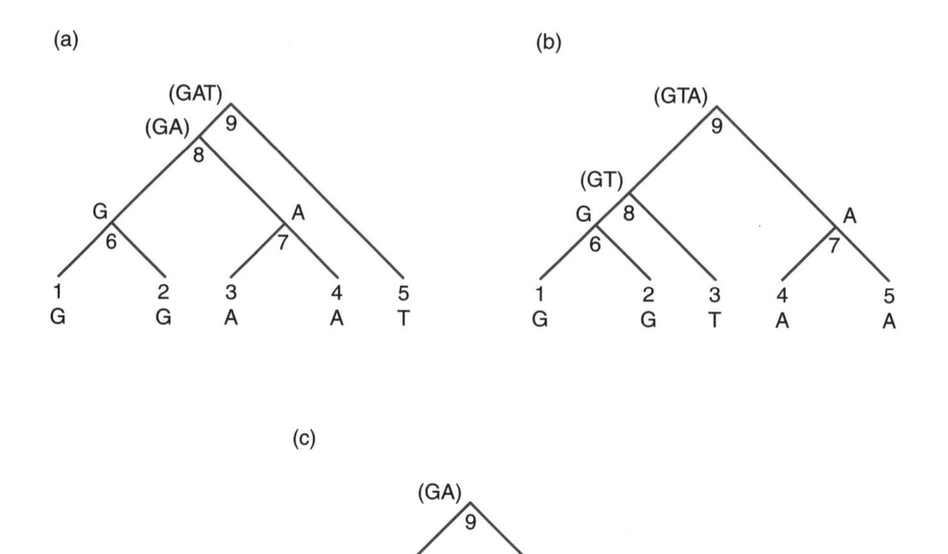

FIGURE 5.3 *Three of 15 unrooted trees that illustrate the relationship between five different taxonomic units. All three trees are equally parsimonious in that they each require a minimum of two substitutions. Terminal nodes are labeled 1 through 5 while the four internal nodes are labeled 6 through 9. The most likely candidates for inferred ancestral nucleotides according to the parsimony rule are shown at each internal node.*

set is not necessarily the one supported by the largest number of informative sites. Third, calculating the number of substitutions invoked by each alternative tree becomes considerably more difficult when five or more sequences are considered.

Figure 5.3 shows three of the possible 15 unrooted trees that describe the possible relationship between five different sequences. Determining the number of substitutions invoked by each tree requires inferring the most likely nucleotide at each of the four internal nodes from the nucleotides known to exist at each of the five terminal nodes. The parsimony rule makes inference of the ancestral nucleotide at node 6 in Figure 5.3a quite easy—the ancestral nucleotide must be a "G" or a substitution would have had to have occurred in the lineages leading to *both* terminal node 1 and terminal node 2. Identical reasoning can be used to justify assigning an "A" to node 7 in Figure 5.3a. The ancestral nucleotide at node 8 cannot be determined unambiguously for this tree but must be either a "G" or an "A" according to the parsimony rule. In the same way, the most parsimonious candidate nucleotides for node 9 are "G," "A," and "T." Consider also the alternative scenario involving nodes 6 and 8 in the tree shown in Figure 5.3c. In this case, nodes 1 and 2 suggest that the ancestral nucleotide at node 6 is "G" *or* "T." However, node 3 also casts a vote for "G" being the candidate nucleotide at node 8. By assigning a "G" as the ancestral nucleotide for nodes 6 and 8, only one substitution (along the lineage leading from node 6 to node 2) must be invoked for this portion of the tree. The three alternatives (assigning a "T" at node 6, a "T" at node 8, or a "T" at nodes 6 and 8) all require at least two substitutions for this portion of the tree.

From a mathematical (and programming) perspective the algorithm for assigning ancestral positions is as follows: The set of most likely candidate nucleotides at an internal node is the intersection of the two sets at its immediate descendant nodes if the intersection is not empty; otherwise it is the union of the sets at the descendant nodes. When a union is required to form a nodal set, a nucleotide substitution must have occurred at some point within the lineage leading to that position. The number of unions is therefore also the minimum number of substitutions required to account for the nucleotides at the terminal nodes since they all last shared a single common ancestor.

Of course, the method described above applies only to the informative sites and not to the uninformative sites discarded in the first step of a typical parsimony analysis. Calculating the minimum number of substitutions that any given tree invokes at an uninformative site is actually very easy. The minimum number is simply the number of different nucleotides present at the terminal nodes minus one. For example, if the nucleotides at a particular position in a five-way alignment are G, G, A, G, and T, then the minimum number of substitutions is $3 - 1 = 2$ regardless of the tree topology. Uninformative sites contribute an equal number of steps to all alternative trees and are often excluded entirely from parsimony analyses. However, when the total number of substitutions required at both informative and uninformative sites is reported for any particular tree, that value is referred to as the tree's **length.**

Weighted Parsimony

While the general principle that "mutations are rare events" is definitely sound, its suggestion that all mutations are equivalent should be an obvious oversimplification. We have already seen in Chapters 2, 3, and 4 that insertions and deletions are less likely than exchanging one nucleotide for another; long insertions and deletions are less common than short ones; some point substitutions are more likely than others (i.e., transitions vs. transversions); and mutations with functional consequences are less likely to occur than those that are inconsequential. If values can be attached to the relative likelihood of each of these kinds of mutations, those values can be translated into weights used by parsimony algorithms.

Unfortunately, it is not possible to arrive at a single set of likelihood-based weights that apply to all (or even a large fraction) of data sets. A very incomplete list of the problems that underlie determining a universal set of relative likelihoods includes (1) some sequences (such as noncoding sequences and especially those containing short, tandem repeats) are more prone to indels than others; (2) functional importance definitely differs from gene to gene and often from species to species even for homologous genes; and (3) subtle substitution biases (such as those that act to change the relative abundance of GC and AT base pairs or the relative abundance of one triplet codon over another that codes for the same amino acid) usually vary between genes and between species.

As a result, the best weighting schemes usually come from analyses of empirical data sets. And the best empirical data set available in terms of general equivalency is typically the one that is actually being analyzed. Assume, for instance, that for a particular multiple sequence alignment comparisons between each individual sequence and a consensus sequence suggest that transitions are three times more common than transversions. A parsimony analysis of the same set of sequences could then attach a value of 1 to every substitution that results in a transversion and a value of 0.33 to every one that results in a transition. The tree with the lowest score after all sites have been considered is then presented as the most likely at the end of the analysis.

Inferred Ancestral Sequences

A remarkable by-product of the parsimony approach is the inferred ancestral sequences that are generated during the course of analysis. Even though the common ancestor of sequences 1 and 2 in Figure 5.1 may have been extinct for tens of millions of years, the parsimony approach makes a fairly strong prediction that the nucleotide found at the fifth position of its sequence was a "G." This may seem trivial in the context of single nucleotides, but when considered over the course of entire genes or genomes, it can provide irreplaceable insights into evolutionary processes and pressures from long ago. When the structure and function of a protein are particularly well understood, amino acid replacements can even provide stunning insights into the day-to-day physiology and environment of truly ancient organisms. (The prototypical review of this kind of analysis remains the one by M. Perutz cited at the end of this chapter.)

Creationists often deride the theory of evolution because of the numerous gaps in the fossil record, and anthropologists frequently fall into the trap of responding by finding "missing links" (which in turn replaces the one original gap in the fossil record with two new ones—one on both sides of the missing link). Thanks to the inferred ancestors generated by parsimony analyses, the study of molecular evolution has no missing links, and intermediary states can be objectively inferred from the sequences of living descendants.

In technical terms, informative sites that support the internal branches in the inferred tree are deemed to be **synapomorphies** (a derived state that is shared by several taxa). All other informative sites are considered to be **homoplasies** (a character that has arisen in several taxa independently through convergence, parallelism, and reversals rather than being inherited from a common ancestor).

Strategies for Faster Searches

The underlying principle and rules of parsimony remain the same for both the simplest case of just a four-way alignment and for the more complex cases of deeper alignments. Still, using an unmodified parsimony approach quickly becomes impossible to do by hand for deep alignments even when the number of informative sites is relatively small. By the time 10 sequences are to be analyzed, more than 2 million trees must be considered (Table 4.1), and even the fastest computers cannot be expected to conduct an **exhaustive search** (an evaluation of every alternative unrooted tree) when as few as 12 sequences are aligned. Data sets are usually tens or even hundreds of times larger than these practical limitations allow, and a variety of algorithms have been developed that facilitate reliable determination of the most parsimonious trees without the need to consider all possible alternatives.

Branch and Bound

A clever and efficient means of streamlining the search for trees of maximum parsimony is the **branch and bound method** first proposed by Hardy and Penny in 1982. As the name suggests, the branch and bound method consists of two essentially different steps. The first step in applying the method is the determination of an upper bound, L, to the length of the most parsimonious tree for the data set. The value for L could be obtained from any randomly chosen tree that describes the relationship of all sequences being considered, but the method is most efficient if a reasonable approximation of what will be the most parsimonious tree is used to establish the upper bound (i.e., one obtained from a computationally fast method such as UPGMA described in Chapter 4).

The second step in the branch and bound method is the process of incrementally "growing" a tree by adding branches one at a time to a tree that describes the relationship of just a few of the sequences being considered (Figure 5.4). What

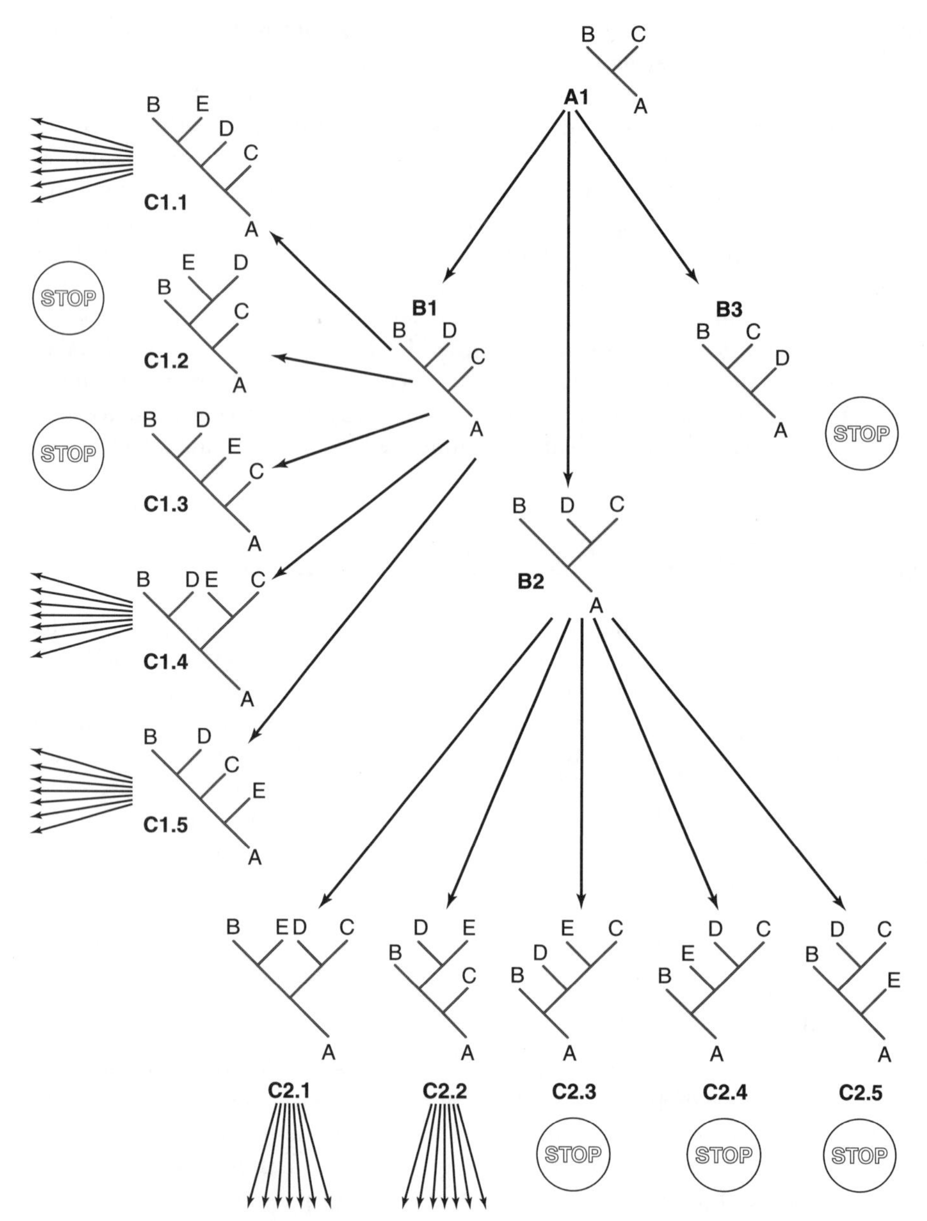

F I G U R E 5.4 *A diagrammatic representation of the branch and bound approach for maximum parsimony analysis. Only one tree can describe the relationship between three sequences (A, B, and C) and it is used as a starting point (tree A1). A fourth sequence (D) from the data set can be added to tree 1 in only three different ways to give rise to trees B1, B2, and B3. If the length of one of those trees is greater than that of the upper bound, then that tree is discarded and no additional branches are added to it in subsequent steps. An additional (fifth) branch is then added to all remaining trees and the process is repeated until a branch has been added for all sequences being analyzed.*

makes the process so effective is the fact that any tree made from a subset of the data that requires more substitutions than L must become longer still as branches corresponding to the remaining sequences are added to the tree. In other words, no tree that contains that particular branching pattern can possibly be the most parsimonious tree since a more parsimonious tree for the complete data set has already been found. If, during the course of the analysis, trees are found that require fewer substitutions than the tree that was used to establish the initial upper bound, the value of L can then be changed accordingly and the remainder of the analysis of the data set will be even more efficient.

Like exhaustive searches, the branch and bound method guarantees that no more parsimonious trees have been missed when the analysis is complete. It also has the substantial advantage of typically being several orders of magnitude faster than an exhaustive search. As a result, it can reasonably be used when as many as 20 sequences are being analyzed, but is typically not sufficient for analyses where more than 1×10^{21} unrooted trees are possible.

Heuristic Searches

Sequence information is increasingly abundant, however, and it is quite common for multiple alignments to be greater than 20 sequences deep. In such situations, algorithms that might not always find the most parsimonious tree must be employed. Several trial-and-error, self-educating techniques have been proposed but most of these **heuristic methods** are based on the same general assumption: Alternative trees are not all independent of each other. Because the most parsimonious trees should have very similar topologies to trees that are only slightly less parsimonious, heuristic searches all begin by constructing an initial tree and using it as a starting point for finding shorter trees.

As with the branch and bound approach, heuristic searches work best if the starting tree is a good approximation of the most parsimonious tree (i.e., one generated by UPGMA). Rather than building up all of the alternative trees branch by branch, though, most heuristic searches generate complete trees with topologies similar to the starting tree by **branch swapping** subtrees and grafting them onto other portions of the best tree found at that point of the analysis (Figure 5.5). Hundreds of new trees are typically generated from the starting tree in the first round of these analyses. All of the new trees that are shorter than the original are then pruned and grafted in the second round of the analysis. This process is repeated until a round of branch swapping fails to produce a tree that is equal to or shorter than the one generated in the previous round of pruning and grafting.

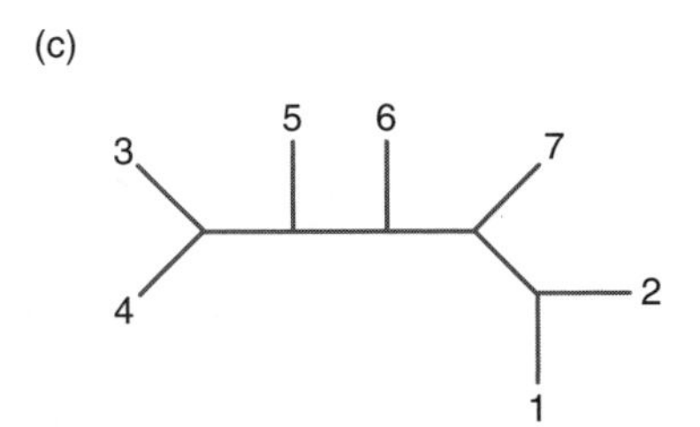

FIGURE 5.5 *An example of branch swapping for an unrooted tree that describes the relationship between seven sequences labeled 1 through 7. The branch containing sequences 1 and 2 is randomly chosen for pruning and then randomly placed between terminal nodes 6 and 7 to generate a new tree with a topology that is similar to the original.*

Heuristic algorithms deal with the impossibility of examining even a small fraction of the astronomical number of alternative unrooted trees for deep alignments by placing an emphasis on branch swapping of increasingly more parsimonious trees. This can result in the algorithm stalling in tree topologies that do not necessarily invoke the smallest number of substitutions. In other words, if the most parsimonious tree is actually not similar to the tree used at the very start of the branch swapping process, it might not be possible to arrive at it without making some rearrangements that first *increase* the number of substitutions. Of course, algorithms can be allowed to occasionally explore paths that increase the length of trees in the hope of getting past such local minima but such provisions can come at substantial cost in terms of required computations.

Since it is the depths of alignments and not their lengths that pose the greatest computational problems, an often better alternative is to divide deep alignments into several shallower batches of alignments. For instance, the relationship between a large number of homologous mammalian sequences can be fairly easily broken into parts: one part that contains several different primate sequences to ascertain the relationships at the tip of the primate trunk of the tree, another part that contains several different rodent sequences to determine the relationships at the tip of the rodent trunk of the tree, and another that contains representatives from a variety of other mammalian orders like artiodactyls (cows) and lagomorphs (rabbits) as well as a few primates and rodents to examine the older divergences and the final placement of the more detailed primate and rodent trunks. Prior knowledge of the general relationships of the sequences (i.e., that all primates are more closely related to each other than they are to any other mammal) is certainly helpful when such a strategy is adopted. It is not essential, however, since a heuristic algorithm could also be asked to consider separately any cluster of sequences that exceeds a particular threshold of pairwise similarity.

Consensus Trees

It is quite common for parsimony approaches to yield more than one equally parsimonious tree. Deep data sets of closely related sequences often produce hundreds or even thousands of trees—far too many to be useful as a summary of the underlying phylogenetic information. In such cases, a simple alternative is the presentation of a single **consensus tree** that summarizes a set of trees (i.e., all those that are most parsimonious). Branching points where all the trees being summarized are in agreement are presented in consensus trees as bifurcations. Points of disagreement between trees are collapsed into internal nodes that connect three or more descending branches. In a **strict consensus tree,** all disagreements are treated equally even if only one alternative tree is not consistent with hundreds of others that are in agreement regarding a particular branching point (Figure 5.6a). A commonly used alternative to so stringent a definition of "agreement between trees" is a **50% majority-rule consensus** where any internal node that is supported by at least half of the trees being summarized is portrayed as a simple bifurcation and those nodes where less than half of the trees agree are shown as multifurcations (Figure 5.6b). Of course, the threshold for what constitutes significant disagreement is a parameter that can be changed to any value from > 0% to 100% (a strict consensus tree).

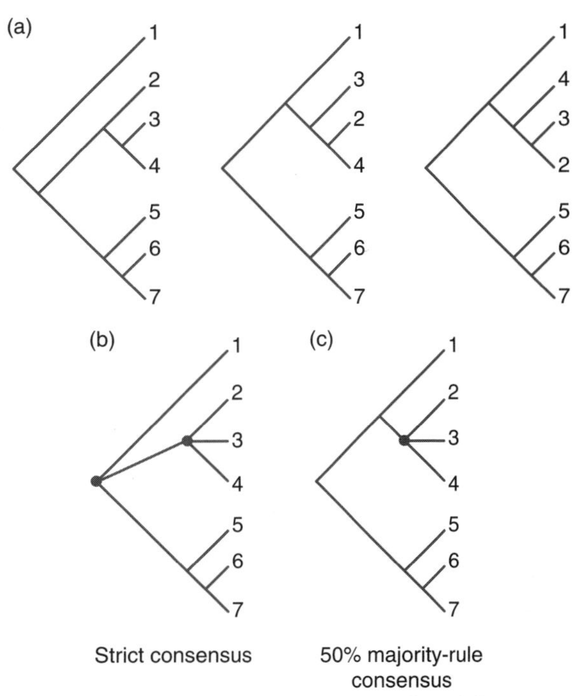

FIGURE 5.6 *Three trees (a) inferred from a single data set can be summarized in a single consensus tree using either (b) strict consensus criteria or (c) 50% majority-rule consensus criteria.*

Tree Confidence

All phylogenetic trees represent hypotheses regarding the evolutionary history of the sequences that make up a data set. Like any good hypothesis, it is reasonable to ask two questions about how well it describes the underlying data: (1) How much confidence can be attached to the overall tree and its component parts (branches)? and (2) How much more likely is one tree to be correct than a particular or randomly chosen alternative tree? While a variety of approaches have been proposed to address these two questions, a powerful resampling technique called bootstrapping has become the predominant favorite for addressing the first question, and a simple parametric comparison of two trees is typical of those used to address the second. Both are described below.

Bootstrapping

It is possible for portions of inferred trees to be determined with varying degrees of confidence. **Bootstrap tests** allow for a rough quantification of those confidence levels. The basic approach of a bootstrap test is straightforward: A subset of the original data is drawn (with replacement) from the original data set and a tree is inferred from this new data set, as illustrated in Figure 5.7. In a physical sense, the process is equivalent to taking the print out of a multiple alignment; cutting it up into pieces, each of which contains a different column from the alignment; placing all those pieces into a bag; randomly reaching into the bag and drawing out a piece; copying down the information from that piece before returning it to the bag; then repeating the drawing step until an artificial data set has been created that is as long as the original alignment. This whole process is repeated to create hundreds or thousands of resampled data sets, and portions of

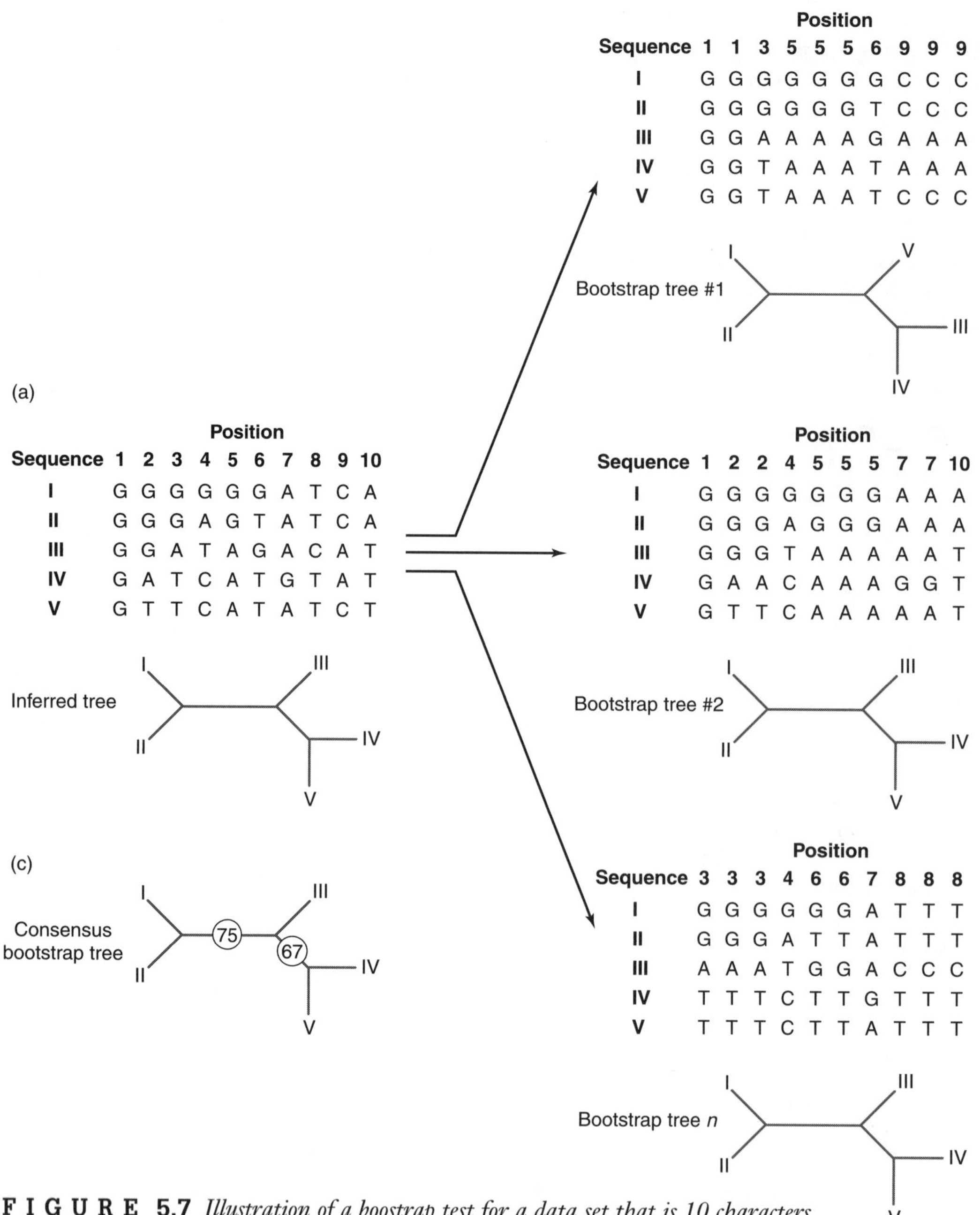

FIGURE 5.7 *Illustration of a boostrap test for a data set that is 10 characters long (positions are labeled 1 through 10) and five sequences (labeled I through V) deep. (a) The original data set and its most parsimonious tree is also shown. A random number generator chooses 10 times from the original 10 positions and creates the three resampled data sets with their corresponding trees (b). A consensus tree (c) for the three resampled data sets is shown with circled values indicating the fraction of bootstrapped trees that support the clustering of sequences I and II and sequences IV and V.*

the inferred tree that have the same groupings in many of the repetitions are those that are especially well supported by the entire original data set. Numbers that correspond to the fraction of bootstrapped trees yielding the same grouping are often placed next to the corresponding nodes in phylogenetic trees to convey the relative confidence in each part of the tree. Bootstrapping has become very popular in phylogenetic analyses even though some methods of tree inference can make it very time consuming to perform.

Despite their often casual use in the scientific literature, bootstrap results need to be treated with some caution. First, bootstrap results based on fewer than several hundred iterations (rounds of resampling and tree generation) are not likely to be reliable—especially when large numbers of sequences are involved. Simulation studies have also shown that bootstrap tests tend to underestimate the confidence level at high values and overestimate it at low values. And, since many trees have very large numbers of branches, there is often a significant risk of succumbing to "the fallacy of multiple tests," in which some results may appear to be statistically significant by chance simply because so many groupings are being considered. Still, some studies have suggested that commonly used solutions to these potential problems (i.e., doing thousands of iterations; using a correction method to adjust for estimation biases; collapsing branches to multifurcations wherever bootstrap values do not exceed a very stringent threshold) yield trees that are closer representations of the true tree than the single most parsimonious tree.

Parametric Tests

Since parsimony approaches often yield large numbers of trees that have the same minimum number of steps, it should not be surprising that alternative trees that invoke just a few additional substitutions can also be very common. Again, the underlying principle of parsimony suggests that the tree that invokes the smallest number of substitutions is the one that is most likely to depict the true relationship between the sequences. There is no limit to how many steps a most parsimonious tree might invoke, though, and data sets that are deep or involve dissimilar sequences can easily invoke many thousands of substitutions. In such cases it is reasonable to ask if a tree that is already so unlikely that it needs to invoke 10,000 substitutions is significantly more likely than an alternative that invokes 10,001. A common, related question is "How much more likely is the most parsimonious tree than a particular alternative that has previously been put forward to describe the relationship between these taxa?" One of the first parametric tests devised to answer such questions for parsimony analyses is that of H. Kishino and M. Hasegawa (1989). Their test assumes that informative sites within an alignment are both independent and equivalent and uses the difference in the minimum number of substitutions invoked by two trees, D, as a test statistic. A value for the variance, V, of D is determined by considering each of the informative sites separately as follows:

$$V = n/(n-1)\Sigma[D_i - (1/n)(\Sigma D_k)]^2$$

where n is the number of informative sites. A paired t-test with n–1 degrees of freedom can then be used to test the null hypothesis that the two trees are not different from each other:

$$t = (D/n)/[(V)^{1/2}n^{1/2}]$$

Alternative parametric tests are available not just for parsimony analyses but for distance matrix and maximum likelihood trees as well.

Comparison of Phylogenetic Methods

Neither the distance- nor the character-based methods of phylogenetic reconstruction make any guarantee that they yield the one true tree that describes the evolutionary history of a set of aligned sequences. This chapter and the one that precedes it have introduced just a few of the numerous variations on each approach that have been suggested, and massive simulation studies have been performed to compare the statistical reliability of virtually all tree-constructing methods. The results of these simulations are easy to summarize: Data sets that allow one method to infer the correct phylogenetic relationship generally work well with all of the currently popular methods. However, if many changes have occurred in the simulated data sets or rates of change vary among branches, then none of the methods works very reliably. As a general rule, if a data set yields similar trees when analyzed by the fundamentally different distance matrix and parsimony methods, that tree can be considered to be fairly reliable.

Molecular Phylogenies

Countless interesting and important examples of evolutionary relationships being deciphered from sequence analyses have accumulated during the past 30 years. Such studies often have important implications for medicine (i.e., a drug that effectively treats one kind of infection is likely to also prove effective in treating infections by related organisms), agriculture (i.e., it is often easiest to transfer disease resistance factors between closely related species), and conservation (i.e., is a population of organisms sufficiently distinct to qualify as a separate species and therefore deserves special protection?). Examples of just two such areas of analysis are described below.

The Tree of Life

One of the most striking cases where sequence data have provided new information about evolutionary relationships is in our understanding of the primary divisions of life. Many years ago, biologists divided all of life into two major groups, the plants and the animals. As more organisms were discovered and their features examined in more detail, this simple dichotomy became unworkable. It was later recognized that organisms could be divided into prokaryotes and eukaryotes on the basis of cell structure. More recently, several primary divisions of life have been recognized, such as the five kingdoms (prokaryotes, protista, plants, fungi, and animals) proposed by Whittaker. All along, though, negative evidence such as the *absence* of internal membranes (the primary distinction of prokaryotes) has been recognized as a notoriously bad way to group organisms taxonomically.

Beginning in the late 1970s, RNA and DNA sequences were used to uncover for the first time the primary lines of evolutionary history among all organisms. In one study, Carl Woese and his colleagues constructed an evolutionary tree of life based on the nucleotide sequences of the 16S rRNA, which all organisms (as well as mitochondria and chloroplasts) possess. As illustrated in Figure 5.8, their evolutionary tree revealed three major evolutionary groups: the Bacteria (the traditional prokaryotes as well as mitochondria and chloroplasts), the Eucarya

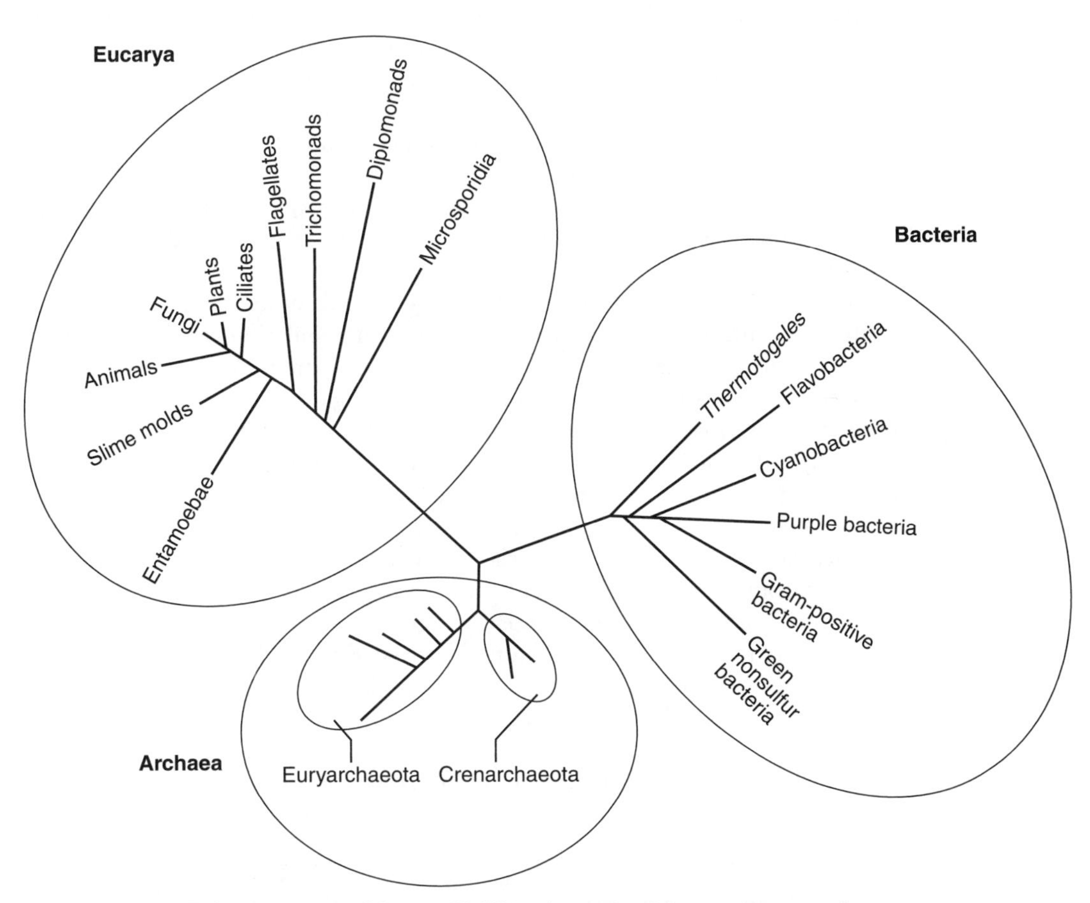

FIGURE 5.8 *An unrooted "tree of life" as determined by parsimony and distance matrix approaches using sequence data from 16S ribosomal RNA genes. The three main lines of descent are Eucarya, Bacteria, and Archaea. (Adapted from C. R. Woese, 1996. Phylogenetic trees: Whither microbiology?* Curr. Biol. ***6:*** *1060–1063.*

(eukaryotic organisms like plants, animals, and fungi), and the Archaea (including thermophilic bacteria and many other relatively little known organisms, some of which are only known through their rRNA sequences). Archaea and Bacteria, although both prokaryotic in that they had no internal membranes, were found to be as different genetically as Bacteria and Eucarya. The deep evolutionary differences that separate the Archaea and Bacteria were not obvious on the basis of phenotype, and the fossil record was absolutely silent on the issue. The differences became clear only after their nucleotide sequences were compared. Sequences of other genes, including 5S rRNAs, large rRNAs, and the genes coding for some fundamentally important proteins, have since been shown to also lend strong support to the idea that living organisms are best placed in three major evolutionary groups in this way.

Human Origins

Another field in which DNA sequences are being used to study evolutionary relationships is human evolution. In contrast to the extensive variation that is observed in size, body shape, facial features, skin color, etc., genetic differences among human populations are relatively small. Analysis of mtDNA sequences shows that the mean difference in sequence between two human populations is about 0.33%. Other primates exhibit much larger differences. For example, the two subspecies of orangutan differ by as much as 5%. This indicates that all human groups are closely related. Nevertheless, some genetic differences do occur among different human groups.

Surprisingly, the greatest differences are not found among populations located on different continents, but are seen among human populations residing in Africa. All other human populations show fewer differences than we find among the African populations. Many experts interpret these findings to mean that humans experienced their origin and early evolutionary divergence in Africa. After a number of genetically differentiated populations had evolved in Africa, it is hypothesized that a small group of humans may have migrated out of Africa and given rise to all other human populations. This hypothesis has been termed the "out-of-Africa theory." Sequence data from both mitochondrial DNA and the nuclear Y chromosome (the male sex chromosome) are consistent with this hypothesis.

Further interpretation of these data suggests that all people alive today have mitochondria that came from a "mitochondrial Eve" and that all men have Y chromosomes derived from a "Y chromosome Adam" roughly 200,000 years ago. Although the out-of-Africa theory is not universally accepted, DNA sequence data are playing an increasingly important role in the study of human evolution and indeed in the study of the evolution of many lineages.

Chapter Summary

Character-based methods of phylogenetic reconstruction are based primarily on the principle of parsimony—substitutions are rare events and the phylogeny that invokes the fewest number of substitutions is the one that is most likely to reflect the true relationship of the sequences being considered. In addition to providing insights into relationships between sequences, parsimony approaches also make potentially useful inferences regarding the sequences of long extinct ancestors to living organisms. Parsimony analyses can be very computationally intensive, however, especially when multiple alignments involving 20 or more sequences are considered. Data sets often yield several trees that are equally parsimonious, and consensus trees can be used to summarize them. Several methods are available for determining the robustness of parsimony trees including bootstrap and parametric tests, though no guarantees can be made that a tree created with either a character- or distance-based approach represents the true relationship between the sequences being considered.

Readings for Greater Depth

Every scientist should be able to point to a single paper that is their favorite. This one is ours. A thoroughly intimate understanding of the function of hemoglobin and an abundance of data that can be used to infer ancestral states are combined to produce a masterwork in the field of molecular evolution in M. Perutz, 1983, Species adaptation in a protein molecule, *Mol. Biol. Evol.* **1:** 1–28.

The most commonly used program for parsimony analysis has been and continues to be D. L. Swofford, 1993, *PAUP: Phylogenetic Analysis Using Parsimony*, Sinauer Associates, Sunderland, MA.

An efficient and popular method for calculating the substitution number associated with a specific tree is described in mathematical terms in W. Fitch, 1971, Toward defining the course of evolution: Minimum change for a specific tree topology, *Syst. Zool.* **20:** 406–416.

One of the first and largest simulation studies designed to put character- and distance-based methods of phylogenetic reconstruction to the test is described in J. Sourdis and M. Nei, 1988, Relative efficiencies of the maximum parsimony and distance-matrix methods in obtaining the correct phylogenetic tree. *Mol. Biol. Evol.* **5:** 298–311.

The possible existence of a mitochondrial Eve remains controversial among anthropologists while an African origin for humans is generally better accepted. The two hypotheses are related and both are discussed in E. Watson, P. Forster, M. Richards, and H. J. Bandelt, 1997, Mitochondrial footprints of human expansions in Africa. *Am. J. Hum. Genet.* **61:** 691–704.

Kishino's and Hasegawa's parametric test for parsimony analyses is described in H. Kishino and M. Hasegawa, 1989, Evolution of the maximum likelihood estimate of the evolutionary tree topologies from DNA sequence data, and the branching order in Hominoidae, *J. Mol. Evol.* **29:** 170–179.

Questions and Problems

* **5.1** What sites in the following alignment would be informative for a parsimony analysis? How many sites are invariant?

```
1  GAATGCTGAT  ATTCCATAAG  TCACGAGTCA  AAAGTACTCG
2  GGATGGTGAT  ACTTCGTAAG  TCCCGAGTCG  AAAGTACTCG
3  GGATGATGAT  ACTTCATAAG  TCTCAAATCA  AAGGTACTTG
4  GGATGCTGAC  ACTTCATAAG  TCGCGAGTCA  AAAGTACTTG
5  GGATGCTGAC  ACTCCGTAAG  TCCCGAGTCA  AATGTACTCG
```

5.2 Draw the 3 alternative unrooted trees for four taxa whose nucleotides at a position under consideration are T, T, C, and C. Label each of the internal nodes with the most likely candidates for inferred ancestral sequences.

* **5.3** How many of the 3 possible unrooted trees for the taxa used in the previous question are equally parsimonious in that they invoke a minimum of one substitution? How many invoke a minimum of two substitutions? Do any invoke more than a minimum of two substitutions?

5.4 Draw any 3 of the 105 alternative tree topologies that describe the relationship between six sequences. Using those 3 trees that you have drawn, infer the most likely ancestral sequences at each internal node if the terminal nodes are G, G, A, T, G, C. What is the minimum number of substitutions required by each of your 3 tree topologies?

* **5.5** Draw a strict consensus tree for the same 3 trees that you drew for Question 5.4. What would a strict consensus of all possible 105 trees look like?

5.6 What factor would have the greatest impact on the number of computations needed to complete a bootstrap analysis: doubling the depth or doubling the length of an alignment? Why?

CHAPTER 6

Genomics and Gene Recognition

Now we see through a glass, darkly.

New Testament,
1 Corinthians xiii, 12

In previous chapters, an analogy has been drawn between written texts and genomes. Letters correspond roughly to nucleotides, sentences are akin to genes, and the individual volumes in a set of encyclopedias are something like an organism's individual chromosomes. The analogy is a good one in many respects, but deciphering the information content within a genome is much more difficult than determining the meaning of a paragraph—even when the written language is unfamiliar. Unannotated genomic sequences have no obvious equivalent to the indentation that marks the beginning of a new paragraph or to the period we expect to find at the end of a sentence. The problem is further compounded in eukaryotes by the fact that our genomes are cluttered with a surprisingly large amount of "junk DNA" that appears to have little or no important information content at all. Still, like any useful information storage system, genomes do contain signals that allow cells to determine the beginnings and ends of genes and when it is appropriate to express them. One of the greatest challenges of bioinformatics to date has been the task of discovering those signals. This chapter describes the underlying biology that bioinformaticians use to begin making sense of the bewildering array of G's, A's, T's, and C's that constitute typical raw genomic sequence data. The challenges of incorporating these signals into useful gene-detecting algorithms are also addressed. Interestingly, the development of gene-finding tools has drawn attention to previously unsuspected biological mechanisms responsible for the regulation of gene expression.

Prokaryotic Genomes

Central to the very concept of being alive is the ability to respond to stimuli. As the simplest free-living organisms, prokaryotes represent an excellent opportunity to determine the molecular basis for those responses. From a prokaryotic perspective, appropriate responses to stimuli invariably involve at least some alteration of gene expression levels.

The newly acquired ability to analyze complete bacterial genomes provided particularly useful insights into the minimum requirements for life. It has become increasingly clear that much of the information content of prokaryotic genomes is dedicated simply to the maintenance of a cell's basic infrastructure such as its ability to make and replicate its DNA (requiring not more than 32 genes), its ability to make new proteins (requiring between at least 100 to 150 genes), and its ability to obtain and store energy (requiring a minimum of approximately 30 genes). Some very simple prokaryotes such as *Haemophilus influenzae* (the first to have its genome sequence completely determined) have been found to have relatively little beyond that minimal set of between 256 and 300 genes. Other prokaryotes with more genes use their additional information content to more efficiently utilize a wider array of resources that might be found in their environments.

As described in Chapter 1, the basic methodology of DNA sequencing has been essentially unchanged since the mid-1980s and rarely yields contiguous sets of data that are more than 1,000 nucleotides long. It is useful to consider just how daunting a task it is to determine the complete sequence of a typical prokaryote's genome. With a single, circular chromosome 4.60 million nucleotides in length, *Escherichia coli*'s genome requires a minimum of 4,600 sequencing reactions to be performed in order to obtain complete coverage. Substantially more than that minimum number of reactions is required in order to assemble **contigs** (continuous runs of nucleotides longer than what can be obtained from any single sequencing reaction) by detecting overlap between the data from multiple sequencing reactions (Figure 6.1). Also, what has become a standard approach to genome sequencing efforts usually begins with a random assortment of subclones (corresponding to subsets of the genome sequence data) from the genome of interest. There are no guarantees that each portion of the genome is represented at least one time unless it is also accepted that some regions are represented multiple times. From a statistical perspective, the chance of "covering" any given nucleotide in a 4.60-Mb genome with a single 1,000-bp-long clone is only 1,000/4,600,000. Similarly, the chance for a specific region to *not* be covered is 4,599,000/4,600,000. Assuming a large enough sample of subclones in a library, 95% coverage is likely to be achieved when N clones are sequenced where:

$$(4{,}599{,}000/4{,}600{,}000)^N = 0.05$$

As a result, if a genome is 4.60 million nucleotides long (one **genome equivalent**), it is necessary to begin with a set of subclones that actually contains more than 20 million nucleotides (a little more than four genome equivalents) in order to have at least a 95% chance of having every individual sequence represented at least once.

Despite the logistical difficulties associated with the experimental and computational aspects of genome sequencing, over 60 prokaryotic genomes have been completely sequenced since the mid-1990s. Some laboratories such as The Institute of Genetic Research (TIGR) have made bacterial genome sequencing

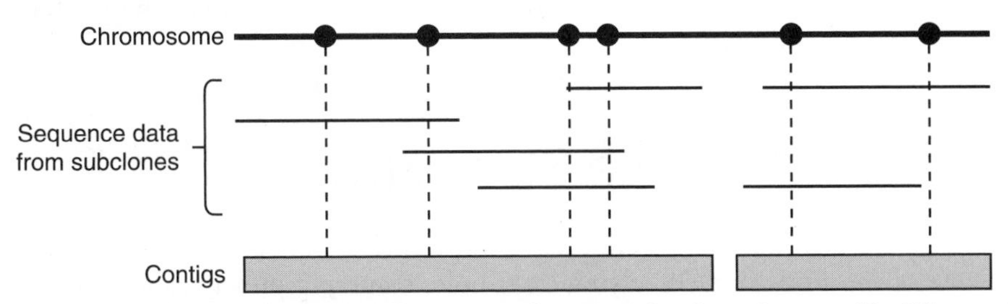

FIGURE 6.1 *Diagram illustrating the principle of contig assembly. The results of numerous sequencing reactions are examined for regions of significant overlap. Various unique points within the genome that may have been previously sequenced (often STSs, sequence tagged sites, or even ESTs, expressed sequence tags) are shown as dark circles on the chromosome and can be used to help order the sequence information. Contigs of the sequencing information are assembled and, in this case, the STSs and ESTs can be grouped into two contigs that are separated by a comparatively short gap.*

TABLE 6.1 Partial list of the many prokaryotic genomes that have been sequenced to date.

Organism	Genome size (Mb)	Gene number	Web site
Mycoplasma genitalium	0.58	470	http://www.tigr.org/tigr-scripts/CMR2/GenomePage3.spl?database=gmg
Helicobacter pylori	1.66	1,590	http://www.tigr.org/tigr-scripts/CMR2/GenomePage3.spl?database=ghp
Haemophilus influenzae	1.83	1,727	http://www.tigr.org/tigr-scripts/CMR2/GenomePage3.spl?database=ghi
Bacillus subtilis	4.21	4,100	http://www.pasteur.fr/Bio/SubtiList.html
Escherichia coli	4.60	4,288	http://www.genome.wisc.edu/k12.htm
Mesorhizobium loti	7.04	6,752	http://www.kazusa.or.jp/rhizobase/

Note: Genome sizes (in megabases) and predicted numbers of genes are based on complete genome sequence data maintained at the listed web sites.

into something of a cottage industry. TIGR's infrastructure of hundreds of high-throughput automated DNA sequencing machines that feed data directly to contig-assembling computers is now capable of completely characterizing several complete bacterial genome sequences every year. Examples of the types of prokaryotic organisms that have been studied by TIGR and other laboratories including their genome sizes and predicted numbers of genes are listed in Table 6.1. A more complete (and continuously updated) list as well as links to the actual sequence data themselves can be found at the TIGR web site (http://www.tigr.org). Bacterial genome sequencing has become so efficient that the U.S. Centers for Disease Control and Prevention (CDC) was able to perform complete genome comparisons of the anthrax strains used in the bioterrorism mailings in the United States in late 2001 within just weeks of their delivery to intended victims. (They consistently differed from the strain used by U.S. military labs at only four positions out of millions.)

Prokaryotic Gene Structure

The structure of prokaryotic genes is generally fairly simple, as shown in Figure 6.2. Just as you find yourself relying on simple, recurring punctuation marks as an aid to deciphering the information content of a written textbook, the proteins that are responsible for gene expression look for a recurring set of signals associated with every gene. Those genomic punctuation marks and the sometimes subtle variations between them allow distinctions to be made between genes that should be expressed and those that should not; the identification of the beginnings and ends of regions to be copied into RNA; and the demarcation of the beginnings and ends of regions of RNA that ribosomes should translate into proteins.

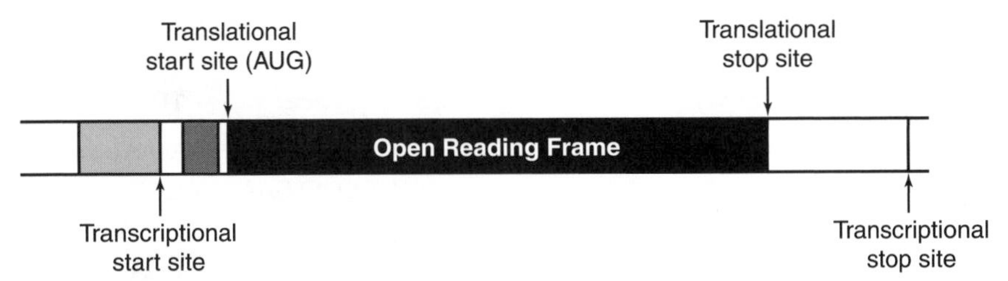

FIGURE 6.2 *The structure of a typical prokaryotic gene. A promoter region where RNA polymerase initiates transcription is shown as a gray box while an operator sequence (sometimes found to overlap or precede the promoter for some genes) to which a regulatory protein might bind is shown a darker gray box. Protein coding regions are shown as a black box, while transcription and translation start and stop sites are labeled with arrows.*

Like all other information content within genomes, these signals are conveyed in relatively short strings of nucleotides. They typically make up only a small fraction of the hundreds or thousands of nucleotides needed to code for the amino acid sequence of a protein.

Promoter Elements

As described in Chapter 1, the process of gene expression begins with transcription—the making of an RNA copy of a gene by an RNA polymerase. Prokaryotic RNA polymerases are actually assemblies of several different proteins that each play a distinct and important role in the functioning of the enzyme. All prokaryotic RNA polymerases rely on a protein called β' (beta-prime) for their ability to bind to DNA templates, β (beta) to link one nucleotide to another, α (alpha) to hold all subunits together, and σ (sigma) to be able to recognize the specific nucleotide sequences of promoters. The proteins β', β, and α are evolutionarily well conserved and are often very similar from one bacterial species to another. The subunit responsible for promoter recognition, σ, tends to be less well conserved, and several significantly different variants of it are often found in any given cell.

The ability to make RNA polymerases from significantly different σ factors is directly responsible for a cell's ability to turn on or off the expression of whole sets of genes. *E. coli* has 7 different σ factors though other closely related bacteria like *Bacillus subtilis* have as many as 10. A list of *E. coli*'s σ factors and the –35 and –10 sequences (see Chapter 1) that they help RNA polymerases bind to are shown in Table 6.2. When it is appropriate for *E. coli* to express those genes that are involved with responding to heat shock, RNA polymerases containing σ^{32} seek out and find those genes with "σ^{32} promoters." By the same token, the roughly 70% of *E. coli* genes that need to be expressed at all times during normal growth and development are transcribed by RNA polymerases containing σ^{70}.

TABLE 6.2 The seven σ factors known to be used by *E. coli.*

σ factor	Gene family	–35 sequence	–10 sequence
σ^{70}	General	TTGACA	TATAAT
σ^{32} (σ^{H})	Heat shock	TCTCNCCCTTGAA	CCCCATNTA
σ^{54} (σ^{N})	Nitrogen stress	CTGGCAC	TTGCA
σ^{28} (σ^{F})	Flagella synthesis	CTAAA	CCGATAT
σ^{38} (σ^{S})	Stationary phase genes	CGTCAA	CTNNTATAAT
σ^{20} (σ^{FecI})	Iron-dicitrate transport	TGGAAA	TGTAAT
σ^{24} (σ^{E})	Extracytoplasmic proteins	GAACTTC	TCTGA

Note: Consensus nucleotides associated with the –35 and –10 sequences of *E. coli's* promoters are shown (*N* = any nucleotide). The names of the σ factors are derived from their molecular weights measured in terms of kilodaltons (i.e., the molecular weight of σ^{70} is 70 kDa). Alternative, commonly used names for *E. coli's* σ factors are also shown. Some –35 and –10 sequences have not yet been well characterized (n.a.).

How well an RNA polymerase recognizes a gene's promoter is directly related to how readily it initiates the process of transcription. The –35 and –10 sequences recognized by any particular σ factor are usually described as a **consensus sequence**—essentially the set of most commonly found nucleotides at the equivalent positions of other genes that are transcribed by RNA polymerases containing the same σ factor. The better any given gene's –35 and –10 sequences match those consensus sequences, the more likely it is for an RNA polymerase to initiate transcription at that promoter.

The protein products of many genes are only useful when they are used in conjunction with the protein products of other genes. It is very common for the expression of genes with related functions to actually share a single promoter within prokaryotic genomes and to be arranged in an **operon.** This provides an elegant and simple means of ensuring that when one gene is transcribed, all others with similar roles are also transcribed. A classic example is that of the lactose operon—a set of three genes (beta-galactosidase, lactose permease, and lactose transacetylase) involved with the metabolism of the sugar lactose in bacterial cells. Transcription of the operon results in the synthesis of just one long **polycistronic** RNA molecule that contains the coding information needed by ribosomes to make all three proteins.

As mentioned in Chapter 1, individual regulatory proteins also facilitate bacterial gene expression responses to specific environmental circumstances at a much finer scale than different σ factors with affinities for a handful of different promoters could provide. The lactose operon, for instance, has a promoter that is recognized by RNA polymerases containing σ^{70} but is most efficiently expressed only when a cell's environment is rich in lactose and also poor in glucose (a preferred sugar that is more efficiently utilized than lactose). Responsiveness to lactose levels is mediated through a **negative regulator** called the lactose repressor protein (pLacI). When lactose levels are low, pLacI binds to a specific nucleotide sequence found only once in any given prokaryotic genome—the lactose operon's **operator sequence** located immediately downstream of the –10 sequence of the operon's promoter (Figure 6.3). When pLacI is bound to the operon's operator, it effectively acts as a roadblock that prevents RNA polymerases from transcribing any of the downstream coding sequences.

The protein encoded by the *LacI* gene (pLacI) is also capable of specifically binding to lactose. When lactose is bound to pLacI, the negative regulator's affinity for the operator is dramatically decreased (thereby making it is possible for the genes of the operon to be expressed). The lactose operon's sensitivity to glucose levels is accomplished through the action of a **positive regulator** called cyclic-AMP receptor protein (CRP). The –35 and –10 sequences of the lactose operon's promoter (5'-TTTACA-3' and 5'-TATGTT-3', respectively) are actually poor matches to those of the consensus sequences best recognized by RNA polymerases with the σ factor (see Table 6.2). As a result, the lactose operon is not expressed at high levels even when pLacI is not bound to the operator. The presence of CRP bound to a promoter is sufficient to make up for those deficiencies to an RNA polymerase, however. CRP does bind to the specific nucleotide sequences associated with the lactose operon's promoter but only when glucose levels within a cell are low.

There is much about prokaryotic promoter sequences and structures that is of interest to bioinformaticians. Computer programs can easily search a nucleotide sequence for a string of characters (though perfect matches are not required) known to be associated with the –35 and –10 sequences of different RNA polymerases. Attaching penalties to each nonmatching nucleotide within a putative promoter sequence allows different operons to be ranked in terms of which are most likely to be expressed at high levels in the absence of any positive regulators. Promoter recognition algorithms also allow the operons of an organism's newly sequenced genome to be organized in terms of their general expression

```
-81        -71        -61        -51        -41
   AACGCAATTAATGTGAGTTAGCTCACTCATTAGGCACCCC
-40        -30        -20        -10         -1
   AGGCTTTACACTTTATGCTTCCGGCTCGTATGTTGTGTGG
1           10         20         30         40
   AATTGTGAGCGGATAACAATTTCACACAGGAAACAGCTat
41          51         61         71         81
   gaccatgattacggattcactggccgtcgttttacaacgt
```

FIGURE 6.3 *Actual nucleotide sequence of the promoter region of* E. coli's *lac operon (GenBank accession #AE000141). Nucleotides associated with the –35 and –10 sequences are italicized; those associated with the binding site for pLacI are underlined; those for the binding site of CRP are double underlined. The transcriptional start site (+1) is shown in boldface italics as are the nucleotides associated with the operon's Shine-Delgarno sequence. The start codon for the beta-galactosidase gene and its downstream ORF are shown in lowercase letters.*

patterns. At the same time, many regulatory proteins (like CRP) have been discovered by first observing that a particular string of nucleotides other than –35 and –10 sequences is found to be associated with more than one operon's promoter.

Open Reading Frames

As described in Chapter 1, ribosomes translate a triplet genetic code in the RNA copy of a gene into the specific amino acid sequence of a protein. Of the 64 possible different arrangements of the four nucleotides in sets of three, three (UAA, UAG, and UGA) functionally act as periods to translating ribosomes in that they cause translation to stop. Most prokaryotic proteins are longer than 60 amino acids in length. (Within *E. coli*, the average length of a protein-coding region is 316.8 codons long, and less than 1.8% of the genes are shorter than 60 codons long.) Since stop codons are found in uninformative nucleotide sequences, approximately once every 21 codons (3 out of 64), a run of 30 or more triplet codons that does not include a stop codon (an **open reading frame** or an ORF) is in itself good evidence that the region corresponds to the coding sequence of a prokaryotic gene. In a statistical sense, if all codons can be expected to occur at the same frequency within a random DNA sequence, the chance of a sequence that is N codons long not containing a stop codon is $(61/64)^N$. A 95% confidence regarding the significance of the length of an ORF is equivalent to only a 5% likelihood of a "random" hit [$(61/64)^N = 0.05$] where N (the number of codons in a significantly long ORF) equals 60. Many gene mapping algorithms for prokaryotic organisms rely heavily on this single criterion.

Just as three codons of the genetic code are reserved as **stop codons,** one triplet codon is usually used as a **start codon.** Specifically, the codon AUG is used both to code for the amino acid methionine as well as to mark the precise spot along an RNA molecule where translation begins. (AUG is the first codon of 83% of *E. coli*'s genes; UUG and GUG make up the entirety of the remaining 17%.) If no likely promoter sequences are found upstream of a start codon at the beginning of the ORF before the end of the preceding ORF, then it is commonly assumed that the two genes are part of an operon whose expression is controlled at a further upstream promoter.

One other hallmark of prokaryotic genes that is related to their translation is the presence of the set of sequences around which ribosomes assemble at the 5' end of each open reading frame. Often found immediately downstream of transcriptional start sites and just upstream of the first start codon, ribosome loading sites (sometimes called **Shine-Delgarno sequences**) almost invariably include the nucleotide sequence 5'-AGGAGGU-3'. Point mutations in a gene's Shine-Delgarno sequence can prevent an mRNA from being translated. In some bacterial mRNAs where there are very few nucleotides between ORFs, translation between adjacent coding regions in a polycistronic mRNA is directly linked because ribosomes gain access to the start codon of the downstream ORF as they complete translation of the first ORF. The norm though is that each legitimate start codon has its own Shine-Delgarno sequence.

Conceptual Translation

In the 1960s and 1970s it was much easier to determine the amino acid sequence of a protein than the nucleotide sequence of the gene that coded for it. The advent of improved DNA sequencing methodologies in the 1980s and the success of numerous genome sequencing projects have changed that balance to the point where now the vast majority of protein sequences are inferred only from predicted gene sequences. The process of **conceptual translation** of gene sequences into the corresponding amino acid sequences of proteins using the genetic code (Table 1.1) is easily done by computers (see Appendix 3). Amino acid sequences in turn can be evaluated for structural tendencies such as the propensity to form alpha helices or beta sheets as described in the next chapter. Protein structure prediction from amino acid sequences rarely allows more than a guess at a particular gene's function, however. Comparisons with the amino acid sequences of proteins from other, better characterized organisms as well as the promoter sequences and genomic context of a gene often provide much more reliable clues about a protein's role. In the end, there is still no good substitute for the often laborious task of physically purifying and characterizing an enzyme using some of the biochemical approaches described in Appendix 2.

Termination Sequences

Just as RNA polymerases begin transcription at recognizable transcriptional start sites immediately downstream from promoters, the vast majority (greater than 90%) of prokaryotic operons also contain specific signals for the termination of transcription called **intrinsic terminators.** Intrinsic terminators have two prominent structural features: (1) a sequence of nucleotides that includes an inverted repeat (i.e., the sequence 5'-CGGATG|CATCCG-3' contains an inverted repeat centered at the "|" because "5'-CGGATG-3'" reads "5'-CATCCG-3'" on its complementary strand), and (2) a run of roughly six uracils immediately following the inverted repeat.

While RNA molecules are typically thought of as being single stranded, it is possible for them to adopt stable secondary structures due to intramolecular base pairing within their inverted repeats. The stability of an RNA secondary structure is directly related to the length of the (often imperfect) inverted repeats that base pair with each other and to the number of G's and C's relative to A's and T's within those repeats (as is described in greater detail in Chapter 8). For intrinsic terminators, each inverted repeat is typically 7 to 20 nucleotides long and rich in G's and C's.

The formation of secondary structures like the one shown in Figure 6.4 in an RNA molecule during its transcription has been experimentally proven to cause RNA polymerases to pause for an average of 1 minute (a very long time considering that prokaryotic RNA polymerases typically incorporate roughly 100 nucleotides a second otherwise). If the RNA polymerase pauses just as it is synthesizing a run of uracils in the new RNA, the unusually weak base pairing that occurs between uracils in the RNA and adenines in the DNA template allows the two polynucleotides to dissociate and effectively terminate transcription. The normal processivity (essentially, their molecular momentum) of RNA polymerases usually allows them to transcribe such runs of adenines in DNA templates. But, when coupled with the pause in synthesis caused by secondary structure in the RNA, the instability of uracil/adenine base pairing effectively and fairly precisely results in transcriptional termination.

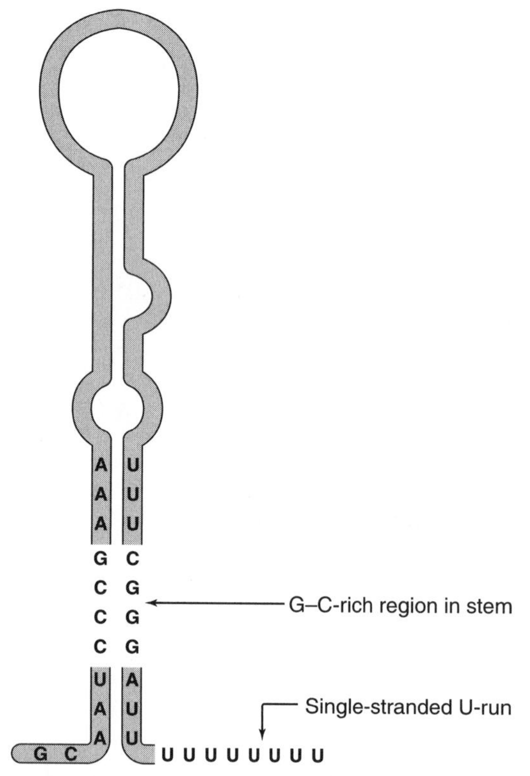

FIGURE 6.4 *Molecular structure of a prokaryotic intrinsic terminator of transcription. Intrinsic terminators include palindromic sequences that vary in length from 7 to 20 bp and tend to be rich in G's and C's. The stem-loop structure that results from intramolecular base pairing is immediately followed by a run of uracils in the RNA*

GC Content in Prokaryotic Genomes

Base pairing rules require that for every G within a double-stranded DNA genome there be a complementary C, but the only physical constraint on the fraction of nucleotides that are G/C as opposed to A/T is that the two values add to 100%. The abundance of G and C nucleotides relative to A and T nucleotides has long been recognized as a distinguishing attribute of bacterial genomes. Measuring genomic **GC content** has proven to be a particularly useful means of identifying bacterial species due simply to the fact that it can vary so dramatically across prokaryotic species with values ranging from 25% to 75% GC (with corresponding values for AT spanning 75% to 25%, respectively).

The GC content of each bacterial species seems to be independently shaped by mutational biases of its DNA polymerases and DNA repair mechanisms working over long periods of time. As a result, the relative ratios of G/C to A/T base pairs are generally uniform throughout any bacteria's genome. As more and more prokaryotic genomes are being completely sequenced, analyses of their GC content has revealed that much of bacterial evolution occurs through large-scale acquisitions of genes (corresponding to tens and even hundreds of thousands of nucleotides in length) from other organisms through a process called **horizontal gene transfer.** Since bacterial species often have widely differing genomic GC contents, genes that have been recently acquired in this way often have GC contents that differ significantly from those that have resided within the genome for longer time periods. Those differences in GC content in turn give rise to significantly different codon usage biases (not all triplet codons are used with the same frequency within a genome's coding sequences as described below) and even amino acid utilizations between genes that have been recently acquired and those that have been long time residents within its genome. In short, many bacterial genomes appear to be patchworks of regions with distinctive GC contents that reflect the evolutionary history of the bacteria as the essences of their ecological and pathogenic characters have changed.

Prokaryotic Gene Density

The density of genes within prokaryotic genomes is remarkably high. Completely sequenced bacterial and archaea chromosomes suggest that between 85% and 88% of the nucleotides are typically associated with the coding regions of genes. In the specific case of *E. coli*, a total of 4,288 genes with average coding sequence lengths of 950 bp are separated by an average of just 118 bp. The ability to reproduce quickly appears to be important to the evolutionary success of bacteria. It has been speculated that maximizing the coding efficiency of their chromosomes may be the result of a need to minimize the potentially rate-limiting step of DNA replication during cell division.

More recent studies that have used GC-content analyses as a means of deciphering the evolutionary history of bacterial genomes have suggested an alternative explanation. These studies indicate that deletions of large regions of chromosomes may be just as common as their acquisition. The only sequences likely to be left behind in the genomes of successful bacterial species (such as those whose genomes get sequenced) are the regions that are absolutely essential such as the coding regions of important genes. Regardless of the underlying cause, the fact remains that finding a gene in prokaryotic genomes is comparatively easy by considering a fairly small number of characteristics:

- Long open reading frames (60 or more codons in length),
- Matches to simple promoter sequences (small numbers of sigma factors help RNA polymerases recognize the –35 and –10 sequences of promoters),
- Recognizable transcriptional termination signal (inverted repeats followed by a run of uracils), and
- Comparisons with the nucleotide (or amino acid) sequences of known protein coding regions from other organisms.

Chances are extremely good that any randomly chosen nucleotide will be associated with the coding sequences or promoter of an important gene. Prokaryotic genomes have very little in the way of wasted space.

Eukaryotic Genomes

Eukaryotic organisms are exceptionally more complex than prokaryotes in virtually every respect. First, their internal membrane-bound compartments allow them to maintain a wide variety of chemical environments within each cell. Second, unlike prokaryotes, almost all eukaryotes live as multicellular organisms and each cell type usually has a distinctive pattern of gene expression even though every cell within the organism has the very same set of genetic instructions. Third, there seems to be relatively little constraint regarding the size of eukaryotic genomes and, as a result, they exhibit an enormously greater tolerance for what is arguably dispensable, "junk" DNA. The demands associated with these three factors alone have forced eukaryotic genomes themselves and the gene expression apparatus that interprets them to be much more complicated and flexible than what is found in even the most sophisticated of prokaryotes. Those two features, complexity and flexible rules of interpretation, have caused the analysis and annotation of eukaryotic genomes to be one of the most challenging ongoing problems for bioinformaticians.

Obtaining the complete nucleotide sequence of any eukaryote's genome is definitely a major undertaking. Unlike prokaryotes with their single copies of typically circular chromosomes, eukaryotic nuclei usually contain two copies of each of at least several different linear chromosomes. Most human cells, for instance, have two copies of a total of 22 different chromosomes (these autosomes are named 1 through 22) and two sex chromosomes (two X chromosomes in females or an X and a Y in males), with the shortest being 55,000,000 bp (55 mega base pairs or 55 Mb) long and the longest being 250,000,000 bp (250 Mb) long and an overall genome size of approximately 3,200,000,000 bp (3,200 Mb). The overall size of a eukaryotic genome is actually difficult for most people to contemplate let alone analyze and annotate. If the nucleotide sequence of the human genome were spelled out using the same font size seen in most encyclopedias, it would require 10 times the number of pages in the complete 32-volume set of the 2002 edition of the *Encyclopedia Britannica!*

The same problems encountered in the sequencing of prokaryotic genomes described earlier in this chapter definitely apply and are compounded significantly by the much larger size of even the simplest and most streamlined eukaryotic genome. The computational problems of finding overlaps between the very large number of contigs that are created in the course of standard genome sequencing approaches necessitate the use of more than just sequence similarities between clones. An extremely useful additional strategy has included establishing correspondence between physical maps (like the STSs and ESTs shown in Figure 6.1 and described in greater detail below) and genetic maps (inferred by examining the combination of traits seen in the progeny of mating experiments). Among the first eukaryotic genomes where these strategies proved to be helpful are those of yeast, *Drosophila melanogaster* (fruit flies), *Arabidopsis thaliana* (a mustard plant), and even humans. As presented in Table 6.3, the genomes of these eukaryotes are all orders of magnitude larger than those of the prokaryotes listed in Table 6.1. Comparisons of gene number estimates between prokaryotic and eukaryotic organisms are very difficult due to lack of confidence in the ability to accurately predict eukaryotic genes simply from an examination of DNA sequence information, as will be described below. Still, the best estimates of gene numbers listed in Table 6.3 are consistent with the idea that in most cases, eukaryotes are more complex than any prokaryotic counterpart.

TABLE 6.3 Partial list of eukaryotic genome sequencing projects.

Organism	Genome size (Mb)	Gene number	Web site
Saccharomyces cerevisiae (yeast)	13.5	6,241	http://genome-www.stanford.edu/Saccharomyces
Caenorhabditis elegans	100	18,424	http://www.sanger.ac.uk/Projects/C_elegans/
Arabidopsis thaliana (thale cress)	130	25,000	http://www.tair.org
Drosophila melanogaster (fruit flies)	180	13,601	http://flybase.bio.indiana.edu
Danio rerio (zebrafish)	1,700	na	http://zfish.uoregon.edu/
Homo sapiens (humans)	3,000	45,000	http://www.ncbi.nlm.nih.gov/genome/guide/

Note: Genome sizes (in megabases) and predicted numbers of genes are based on data maintained at the listed web sites.

Eukaryotic Gene Structure

The size of a eukaryotic genome alone has caused many people to liken the process of finding genes within their reams of sequence data to looking for a needle in a haystack. That old analogy actually falls very far short of conveying the enormity of the problem. A typical 2-gram needle in a 6,000-kilogram haystack would actually be 1,000 times easier to find, and that is only if genes were as different from the remainder of genomic DNA as a needle is from a piece of straw. Unfortunately, recognizing eukaryotic genes is substantially more difficult than the equivalent task in prokaryotic nucleotide sequences. The striking landmarks of prokaryotic open reading frames with statistically significant lengths used earlier are not found in eukaryotic genes due to the abundant presence of introns as described below. Eukaryotic promoters, like their prokaryotic counterparts, do have some conserved sequence features that gene-finding algorithms can search for but they tend to be much more diffuse and located far away from a gene's start codon.

In short, the problem of recognizing eukaryotic genes in genomic sequence data is a major challenge and promises to be one of the greatest open questions for bioinformaticians in the next several decades. The best attempts at solutions to date (programs such as Grail EXP and GenScan available on the web) have relied on **neural network** and **dynamic programming** techniques. Neural networks look for statistical similarities in characterized data sets (i.e., the sequences of known human genes) to predict the properties of related but uncharacterized data sets (i.e., raw sequence data from the human genome project). As described in Chapter 2, dynamic programming allows computers to efficiently explore all possible solutions to some kinds of complex problems. With correct predictions being made less than 50% of the time, current algorithms are a good start but not particularly reliable. Algorithms using these approaches scan sequences looking for a variety of features in appropriate orientations and relative positions.

Any one feature by itself might occur as the result of random chance, but the co-occurrence of several such as a likely promoter, a series of intron/exon boundaries, and a putative ORF with a favorable codon usage bias all lend weight to the prediction that a region may correspond to a gene. The biological significance of these features and what makes them distinctive are described below.

Promoter Elements

All of the information that a liver cell needs to be a liver cell is also present in a muscle or a brain cell. Regulation of gene expression is the only practical way to account for their differences and, as was the case for prokaryotes, the initiation of transcription represents one of the most efficient places to regulate expression. It should not be surprising then that eukaryotes have very elaborate mechanisms for regulating the initiation of transcription. At one level the added challenges of gene expression for eukaryotes are reflected simply in the number of different RNA polymerases they use. Unlike prokaryotes with a single RNA polymerase made from a handful of different proteins, all eukaryotic organisms employ three different kinds of RNA polymerase made of at least 8 to 12 proteins.

Each of the three eukaryotic RNA polymerases recognizes a different set of promoters, and each is used to transcribe different kinds of genes (Table 6.4). RNA polymerases I and III make RNA molecules that are themselves functionally important and needed at fairly constant levels in all eukaryotic cells at all times. RNA polymerase II is exclusively responsible for the transcription of the eukaryotic genes that code for proteins. The variety of promoter sequences RNA polymerase II recognizes definitely reflects the complexity of distinguishing between genes that should and should not be expressed at any given time in any given type of cell.

As with prokaryotes, eukaryotic promoters are the nucleotide sequences that are important for the initiation of transcription of a gene. Unlike prokaryotic operons where multiple genes share a single promoter, every eukaryotic gene has its own promoter. Most RNA polymerase II promoters contain a set of sequences known as a **basal promoter** where an RNA polymerase II **initiation complex** is assembled and transcription begins. The promoters of most RNA polymerase II transcribed genes also include several additional **upstream promoter elements** to which proteins other than RNA polymerase II specifically bind. Given a typical eukaryote's number of genes and cell types, it has been estimated that a minimum

TABLE 6.4 The three eukaryotic RNA polymerases.

RNA polymerase	Promoter location	Promoter complexity	Transcribed genes
RNA polymerase I	-45 to +20	Simple	Ribosomal RNAs
RNA polymerase II	Far upstream to -25	Very complex	Protein-coding genes
RNA polymerase III	+50 to +100	Simple	tRNAs, other small RNAs

Note: Positions of promoter sequences relative to the transcriptional start site, the complexity of sequences, and the types of genes they transcribe are listed.

of five upstream promoter elements are required to uniquely identify any particular eukaryotic protein coding gene and ensure that it is expressed in an appropriate fashion. Assembly of initiation complexes on a core promoter can occur in the absence of the proteins associated with any or all upstream elements but only in an inefficient way.

A fundamental difference between the initiation of transcription in prokaryotes and eukaryotes is that RNA polymerase II does not directly recognize the basal sequences of promoters. Instead, **basal transcription factors** including a TATA-binding protein (TBP) and at least 12 TBP-associated factors (TAFs) seek out and bind to the nucleotide sequences of the promoter in a specific order and then facilitate the binding of the catalytic unit of RNA polymerase II, as shown in Figure 6.5. As with prokaryotes, eukaryotic promoters include a TATA box (the consensus sequence in eukaryotes is actually 5'-TATAWAW-3' where "W" means either "A" or "T" is present at equal frequency) at –25 instead of –10 relative to the transcriptional start site (+1). In eukaryotes the +1 position is associated with an **initiator (Inr) sequence** (with a consensus of 5'-YYCARR-3' where "Y" means "C" or "T" and "R" means "G" or "A"). The +1 nucleotide is almost always the highly conserved A within the Inr sequence but other nucleotides can be consistently used for some genes.

Subtle differences in basal transcription factors are known to exist among different cell types in eukaryotic organisms and are likely to recognize slightly different sequences as promoters. These transcription factor differences ultimately play an important role in the tissue-specific expression of some genes. Still, the presence of a strong TATA box and Inr sequence in the appropriate relative orientation and position is a useful indicator to bioinformaticians that a eukaryotic genomic sequence is associated with a downstream protein-coding region.

Regulatory Protein Binding Sites

The initiation of transcription in eukaryotes is fundamentally different from what occurs in prokaryotes. In bacteria, RNA polymerases have a high affinity for promoters, and an emphasis is placed on negative regulation (such as that achieved by pLacI described earlier) by proteins that prevent gene expression from occurring at inappropriate times. In eukaryotes, RNA polymerases II and III do not assemble around promoters very efficiently, and the basal rate of transcription is very low regardless of how well a promoter matches consensus sequences. As a result, eukaryotes place a much greater emphasis on additional proteins acting as positive regulators. Some of these positive regulators are essentially **constitutive** in that they work on many different genes and do not seem to respond to external signals. Other proteins are more **regulatory** in that they play a role for a limited number of genes and do respond to external signals. The distinction between constitutive and regulatory activating proteins (often referred to as "transcription factors") in eukaryotes is far from precise but the concept still provides a useful framework for discussion.

The majority of transcription factors are sequence-specific DNA-binding proteins (Table 6.5). Some, like CAAT transcription factor and the CP family of

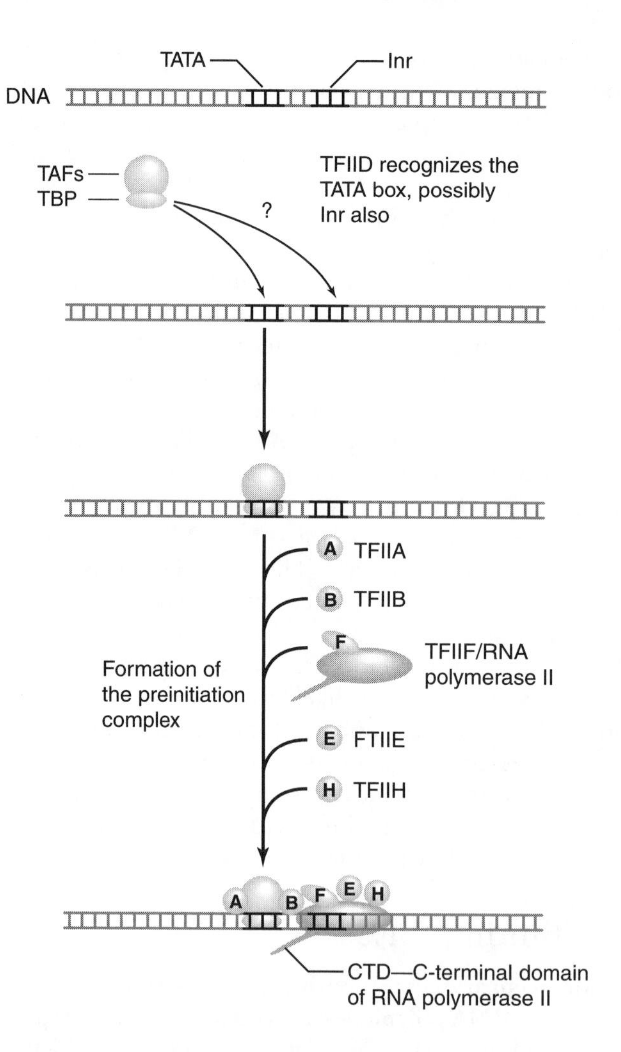

FIGURE 6.5 *Assembly of the RNA polymerase II preinitiation complex. Assembly of the preinitiation complex begins with recognition of the TATA box and possibly the Inr sequence by TBP (probably in conjunction with TAFs). Other components of the preinitiation complex assemble in the order shown and with known protein contacts in place. (Based on Roeder, 1996.)*

proteins, recognize consensus sequences relatively close to transcriptional start sites like the **CAAT box** found in the same orientation and in the vicinity of –80 in most eukaryotic genes. Others like Sp1 are called **enhancers** because they work equally well in either orientation and over a wider range relative to the start site (usually from –500 to +500). Enhancers also tend to work cumulatively when multiple binding sites are present. Some eukaryotic enhancers are known to work at distances as great as many tens of thousands of nucleotides upstream of tran-

TABLE 6.5 Examples of several eukaryotic transcription factors.

Protein factor	Consensus sequence	Role
Constitutive factors		
CAAT transcription factor	5'-GCCAATCT-3'	Ubiquitous
CP family	5'-GCCAATCT-3'	Ubiquitous
Sp1	5'-GGGCGG-3'	Ubiquitous
Oct-1	5'-ATGCAAAT-3'	Ubiquitous
Response factors		
Heat shock factor	5'-CNNGAANNTCCNNG-3'	Response to heat shock
Serum response factor in serum	5'-CCATATTAGG-3'	Response to growth factors
Cell-specific factors		
GATA-I	5'-GATA-3'	Only in erythroid cells
Pit-I	5'-ATATTCAT-3'	Only in pituitary cells
MyoDI	5'-CANNG-3'	Only in myoblast cells
NF-kB	5'-GGGACTTTCC-3'	Only in lymphoid cells
Developmental regulators		
Bicoid	5'-TCCTAATCCC-3'	Early embryo organization
Antennapedia	5'-TAATAATAATAATAA-3'	Embryonic head development
Fushi tarazu	5'-TCAATTAAATGA-3'	Embryonic segment pairing

Note: Examples include the sequences that the transcription factors specifically interact with and their roles. "N" means that all four nucleotides occur with roughly the same frequency.

scriptional start sites and have their effect by bending the DNA into a specific shape that brings other transcription factors into contact with each other to form structures called **enhanceosomes.** Still other transcription factors are available only under special circumstances and help mediate the response of eukaryotic cells to stimuli such as exposure to heat or allow genes to be expressed only in specific tissues or times during development.

Open Reading Frames

The nuclear membrane of eukaryotic cells provides a physical barrier that separates the process of transcription and translation in a way that never occurs in prokaryotes where translation by ribosomes typically begins as soon as an RNA polymerase has begun to make an RNA copy of a coding region.

Eukaryotes take advantage of the resulting delay in the initiation of translation that transport out of the nucleus necessitates to extensively modify their RNA polymerase II transcripts. Known as **hnRNAs** (heterogeneous RNAs) prior to processing, RNA polymerase II transcripts are capped, spliced, and polyadenylated as they are converted into mRNAs suitable for translation by ribosomes.

Capping refers to a set of chemical alterations (including methylation) at the 5' end of all hnRNAs. **Splicing** involves the wholesale and precise removal of often very large segments from the interior of hnRNAs. **Polyadenylation** describes the process of replacing the 3' end of an hnRNA with a stretch of approximately 250 A's that are not spelled out in the nucleotide sequence of a gene.

Each of the three kinds of modification just described can occur differently in different cell types. Variation in splicing in particular is a great boon to eukaryotic organisms in terms of their ability to meet the demands of tissue-specific gene expression without paying an inordinate cost in genome complexity. At the same time, though, splicing represents a serious problem for gene recognition algorithms. In short, the DNA sequences of eukaryotic genes do not have to possess the statistically significant long ORFs associated with DNA sequences of prokaryotic genes because the process of splicing typically removes countless stop codons that interrupt the ORFs in the DNA sequences of eukaryotic genes. If gene recognition algorithms can accurately model the splicing process, then the ORFs of eukaryotic mRNAs could be every bit as useful in gene recognition as they are in prokaryotes. The tissue-specific variability in splicing rules described below, however, makes this a very difficult problem to solve.

Introns and Exons

The genetic code (Table 1.1) was experimentally deciphered long before it was possible to determine the nucleotide sequence of genes. It came as quite a surprise in 1977, when the first eukaryotic genomic sequences were obtained, that many genes contained "intervening sequences" (**introns**) that interrupt the coding regions that were ultimately linked together in processed mRNAs (Figure 6.6).

At least eight distinctly different kinds of introns have since been found in eukaryotic cells though only one of those types, the one that conforms to a **GU-AG rule,** is predominantly associated with eukaryotic protein-coding genes. The GU-AG rule gets its name from the fact that the first two nucleotides at the 5' end of the DNA sequence of all of these introns are invariably 5'-GU-3' and the last two at the 3' end of the intron are always 5'-AG-3'. Additional nucleotides associated with these 5' and 3' splice junctions as well as an internal "branch site" located 18 to 40 bp upstream of the 3' splice junction are also scrutinized by the splicing machinery as indicated in the consensus sequences shown in Figure 6.6.

From an information storage perspective, it is worth noting that most of the sequences being scrutinized by the splicing apparatus actually lie within the intron itself and do not constrain the information content of the typically protein-coding **exon** sequences on either side that end up being linked as hnRNAs are processed into mRNAs. Introns must have a minimal length of about 60 bp simply to accommodate these splicing signals, although there appears to be no practical constraint on the upper bound of their length (a large number of human introns are many tens of thousands of base pairs long). Similarly, exon lengths also span a wide range of sizes in vertebrates with most being about 450 bp in length, although some are less than 100 bp long and others are known to be greater than 2,000 bp long.

(a)

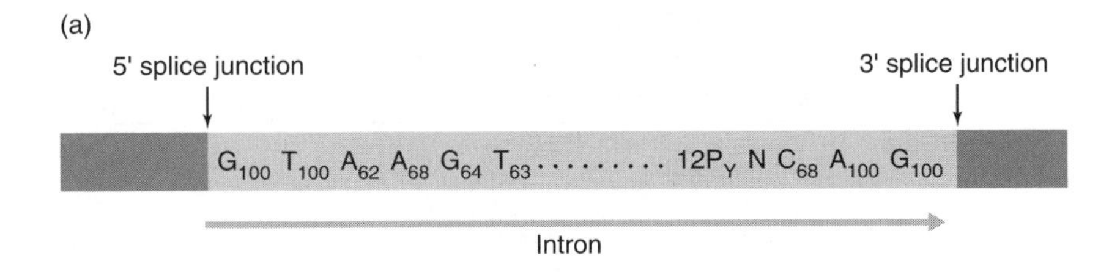

(b)

Correct splicing removes 3 introns by pairwise recognition of the junctions

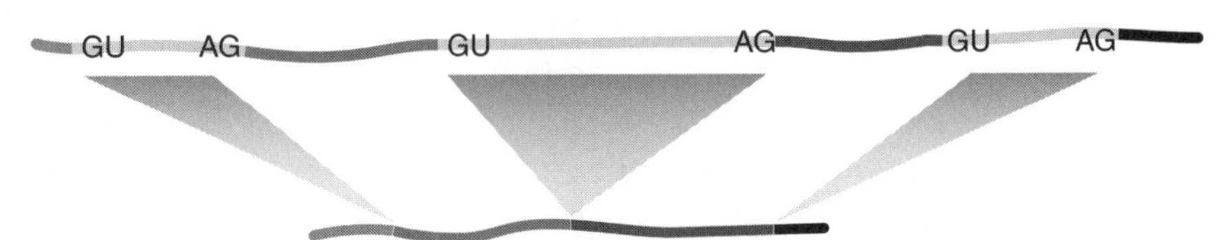

FIGURE 6.6 *A diagram representing the intron/exon structure of a eukaryotic gene and the mRNA it gives rise to after processing of an hnRNA transcript. (a) Conserved sequences associated with the 5' and 3' splice junctions of yeast introns are shown. Subscript numbers beside each consensus nucleotide indicate the frequency with which the nucleotide occurs in the introns of all known yeast genes. "Y" means that the consensus nucleotide is either a C or a T while "N" means that no one nucleotide is found at higher than expected frequency. (b) Splice junctions are recognized in pairwise recognition of the splice junctions by spliceosomes to give rise to mRNAs ready for translation by ribosomes.*

No strictly followed rules appear to govern the distribution of these introns though they are generally less common in simpler eukaryotes (i.e., the 6,000 genes of the yeast genome have a total of only 239 introns while some individual genes in humans are known to have 100 or more). Introns are a very common feature in the genes of most vertebrates, and almost 95% of all human genes have at least one (Table 6.6). Aside from the sequences required for splicing, the length and nucleotide sequences of introns appear to be under very little selective constraint. In contrast, the position of introns within any given gene does seem to be evolutionarily conserved in that they are often in identical positions in alignments of the sequences of homologous genes.

Alternative Splicing

All 5' splice junctions appear to be functionally equivalent to the splicing apparatus, as do all 3' splice junctions. Still, in usual circumstances splicing occurs only between the 5' and 3' sites of the *same* intron. The molecular basis for differentiating between splice junctions of the same and different introns appears to be more complex than simple scanning of the hnRNA sequence by the splicing machinery for adjacent sites and remains an important unanswered question for

TABLE 6.6 Intron/exon structure of some human genes.

Gene (GenBank Acc. #)	Intron #	Coding sequence (bp)	Intron length (bp)
Histone H3 (X83550)	0	410	0
α-Globin (J00153)	2	426	261
β-Globin (AF083883)	2	441	982
Insulin (J00265)	2	332	963
Keratin, type I (Y16787)	6	1,248	2,267
Alpha albumin (U51243)	13	1,797	20,273
Phosphofructokinase (AJ005577)	14	1,512	17,421
RECQL4 helicase (AB026546)	20	3,624	2,592
Factor VIII (M14113)	25	7,053	≈179,000
Cystic fibrosis TR (AH006034)	26	4,440	≈226,000
Hyperion protein (AJ010770)	49	11,733	≈160,000
Dystrophin (M18533)	78	11,057	≈2,400,000
Type VII collagen (NM_000094)	117	8,833	21,932

Note: The number of introns in each gene are shown as well as the total length of each gene's coding region and the total length of its intron sequences.

molecular biologists. The majority of eukaryotic genes appear to be processed into a single type of spliced mRNA (i.e., introns and exons are recognized in the same way in all cell types). However, it has been conservatively estimated that 20% of human genes give rise to more than one type of mRNA sequence due to **alternative splicing.** In one extreme example of alternative splicing, a single human gene has been shown to generate up to 64 different mRNAs from the same primary transcript.

As mentioned earlier, the processing of hnRNAs into mRNAs does not happen in exactly the same way in different eukaryotic cell types and under all circumstances (Figure 6.7). For instance, exons 2 and 3 of the mouse troponin T gene are mutually exclusive; exon 2 is used in smooth muscle while exon 3 is used in all other tissues as illustrated in Figure 6.7b. Smooth muscle cells possess a protein that binds to repeated sequences present on either side of the exon in the gene's hnRNA and apparently mask the splice junctions that must be recognized to include it in the mRNA.

It is also worth considering that the splicing apparatus itself is made from a variety of small nuclear RNAs (snRNAs) as well as several proteins—each of which may differ from one cell type to another. Much of the variability observed in the consensus sequences of the splice junctions and branch sites may actually reflect the specific recognition of different splicing signals in different eukaryotic tissues. Single genes do not necessarily give rise to single proteins in eukaryotic systems, and the development of useful computational splicing models remains a major challenge at the heart of gene recognition algorithms. A catalog of known splicing variants is maintained on the web at the intron database (http://www.introns.com/front.html).

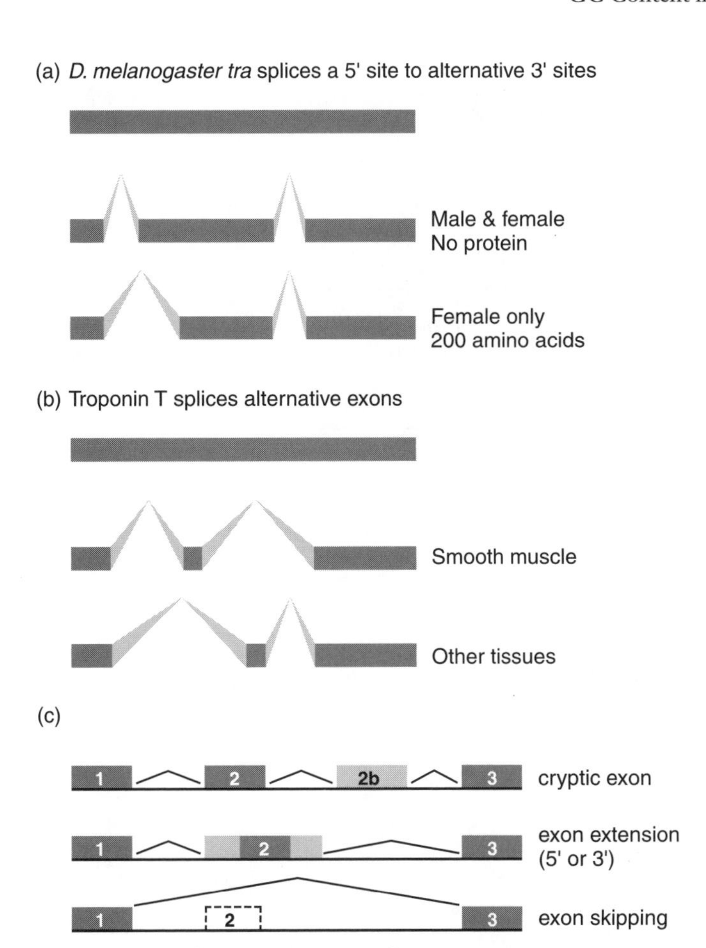

FIGURE 6.7 *Possible products of alternative splicing of a eukaryotic gene. (a) Alternative splicing of the* tra *gene in* D. melanogaster *is responsible for sex determination in fruit flies. (b) Alternative splicing of troponin T leads to distinctly different protein products in human smooth muscle cells relative to other tissues. (c) Many variants on the theme of alternative splicing are known.*

GC Content in Eukaryotic Genomes

Overall genomic GC content does not show the same variability between eukaryotic species as is observed in prokaryotes. It does seem to play a much more important role in gene recognition algorithms, though, for at least two reasons: (1) Eukaryotic ORFs are much harder to recognize, and (2) large-scale variation of GC content *within* eukaryotic genomes underlies useful correlations between genes and upstream promoter sequences, codon choices, gene length, and gene density.

CpG Islands

One of the first bioinformatics analyses ever performed on DNA sequence data was a statistical evaluation of the frequency that all possible pairs of nucleotides (GG, GA, GT, GC, AG, etc.) were observed in generic sequences from the human

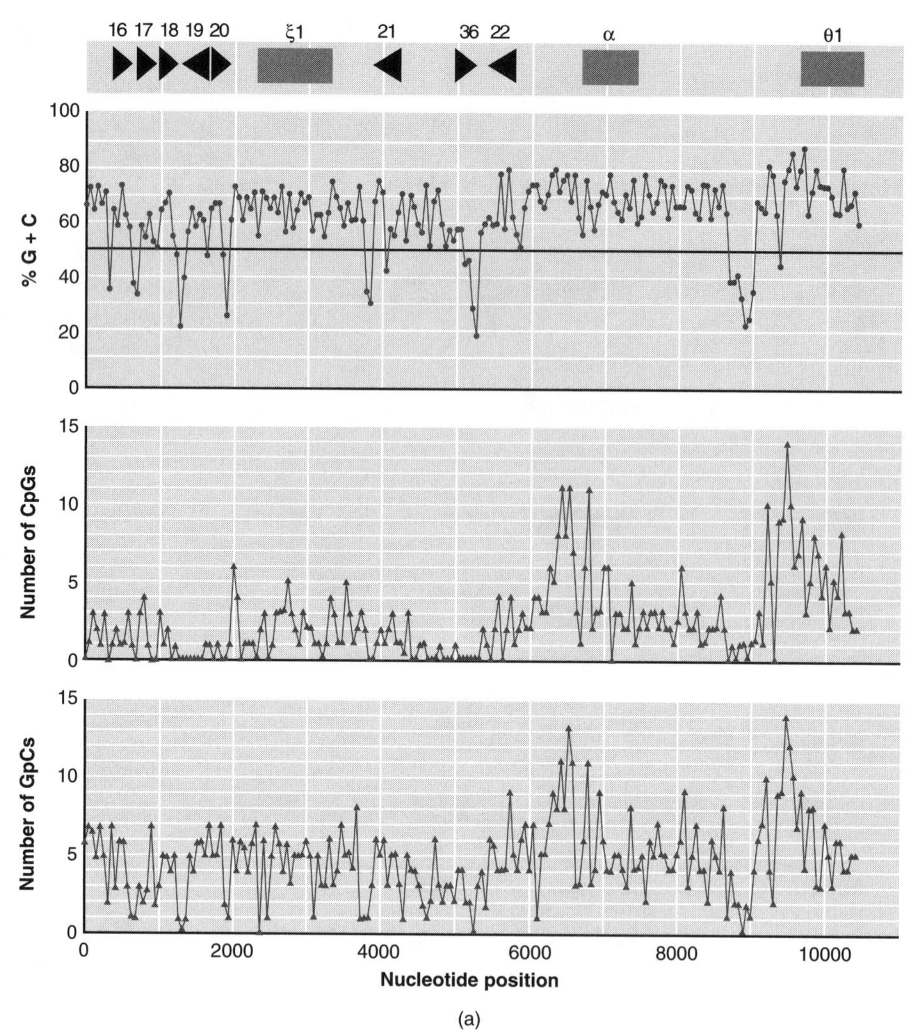

FIGURE 6.8 *Gene map, dinucleotide frequency, and GC content of the rabbit alpha- and beta-like globin gene clusters. (a) The rabbit alpha-like globin gene cluster: a set of tissue-specific genes in a GC-rich portion of the human genome. Regions corresponding to genes are shown as filled boxes while copies of the predominant SINE in rabbits (C repeats) are numbered and shown as filled arrows. The number of occurrences of the dinucleotide 5'-CG-3' in a sliding window of 200 bp is shown in the panel immediately below the gene map. A CpG island can be seen to be associated with the 5' ends of both the alpha- and theta$_1$-globin genes. The number of occurrences of the dinucleotide 5'-GC-3' in a 200-bp window such as those shown is generally higher than those for CpGs because only CpGs are hypermutable due to methylation. The bottom panel simply shows what fraction (%) of nucleotides within the same sliding window of 200 bp is either G or C. (b, opposite) The rabbit beta-like globin gene cluster: tissue-specific genes in a GC-poor portion of the human genome. Genes and repeated sequences are shaded as for the alpha-globin gene cluster of part (a). Occurrences of CpGs are placed above those of GpCs and overall GC content using the same 200-bp window as part (a).*

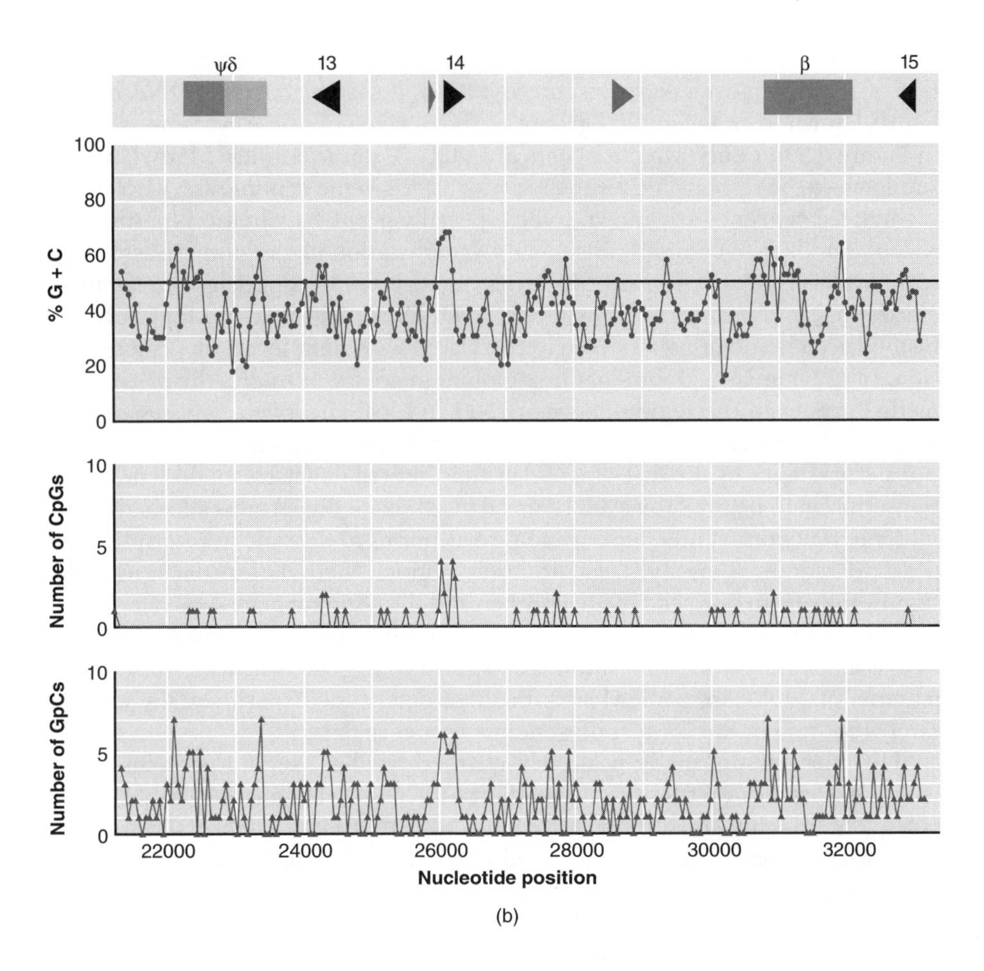

genome. A striking observation was made: CG dinucleotides (often called CpGs to reflect the phosphodiester bond that connects the two nucleotides) were found with only 20% of the frequency that should have occurred by chance (Figure 6.8). No other pair of nucleotides was found to be unusually over- or underrepresented.

An interesting exception to the general paucity of CpGs was found to be associated with stretches of 1–2 kb at the 5' ends of many human genes. These **CpG islands** typically span from about –1,500 to +500 and have densities of CpGs at the levels predicted by chance. Many of the individual CpGs appear to be involved with binding sites for known transcriptional enhancers such as Sp1 (see Table 6.5). Analysis of the complete human genome sequence indicates that there are approximately 45,000 such islands and that about half of them are associated with every known **housekeeping gene** (genes that are expressed at high levels in all tissues and at all times in development). Many of the remaining CpG islands appear to be associated with promoters of tissue-specific genes (like human α-globin shown in Figure 6.8a), although less than 40% of known tissue-specific genes have them (like human β-globin shown in Figure 6.8b). CpG islands are only rarely found in gene-free regions or with genes that have accumulated inactivating mutations.

CpG islands are intimately associated with an important chemical modification of DNA in most eukaryotes, **methylation.** A specific enzyme, DNA methylase, is known to attach methyl groups to the nitrogenous base cytosine as shown in Figure 6.9 but only when it occurs in 5'-CG-3' dinucleotides. Methylation itself seems to be responsible for the rarity of CpGs in the genome as a whole since methylated cytosines are known to be especially prone to mutation (particularly to TpGs and CpAs). High levels of DNA methylation in a region are associated with low levels of acetylation of histones (important DNA packaging proteins in eukaryotes) and vice versa. Low levels of DNA methylation and high levels of histone acetylation are also strongly correlated with high levels of gene expression. In the human γ-globin (gamma-globin) gene, for example, the presence of methyl groups in the region between –200 and +90 effectively suppresses transcription. Removal of the three methyl groups found upstream of the transcriptional start site or of the three methyl groups located downstream does not allow transcription to begin. However, removal of all six methyl groups allows the promoter to function. While there are some exceptions to this rule, transcription appears to require a methyl-free promoter region. Methylation patterns differ significantly from one cell type to another, and the γ-globin gene sequence is generally only free of methyl groups in erythroid cells. While the presence of CpG islands alone is a strong indication that a eukaryotic gene is nearby, methylation patterns of DNA are somewhat difficult to determine experimentally and are rarely reported in the context of genomic sequence data.

Histones are unusually positively charged, well-conserved proteins found within eukaryotes that have a high affinity for negatively charged DNA molecules. The roughly equal mixture (by mass) of DNA and closely associated histones within eukaryotic nuclei is called **chromatin.** Wrapping of DNA around histones and further organization of the histones themselves result in a final packaging ratio of approximately 1:10,000 for eukaryotic genomic DNA. As mentioned above, transcriptionally active regions are generally areas where the positive charge of histones is reduced by the addition of acetyl and methyl groups.

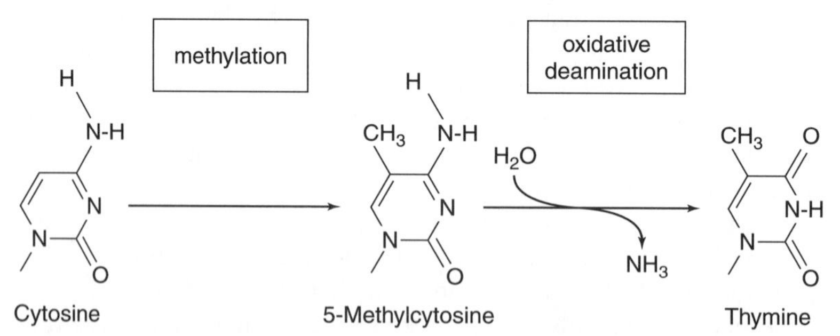

FIGURE 6.9 *Methylation of cytosine by DNA methylase in eukaryotes. Methylases add methyl groups (—CH_3) to the nitrogenous base cytosine only when they occur as part of a CpG dinucleotide. A common type of chemical damage to DNA, oxidative deamination, converts methylcytosine to thymine and, as a result, makes CpGs hypermutable relative to all other possible dinucleotides.*

The resulting low affinity of those histones for negatively charged DNA causes the chromatin to be less tightly packed and more accessible to RNA polymerases. Such areas of open chromatin are known as **euchromatin** in contrast to transcriptionally inactive, densely packed **heterochromatin.** Information stored in heterochromatin is not lost but is much less likely to be used in gene expression, much like the way information in a textbook is much less likely to be accessed if the book is packed away in a box in an attic. The packaging of DNA differs significantly from one cell type to another, and genomic sequence information aside from CpG islands gives only hints as to what a region's chromatin structure will be. Among those hints is what can be gleaned from isochore compartmentalization, which is described next.

Isochores

The genomes of vertebrates and plants (and perhaps of all eukaryotes) display evidence of a level of organization called **isochores** that is intermediate between that of genes and chromosomes. The working definition of isochores, "long regions of homogeneous base composition," has two operative parts. First, the genomic sequences of isochores are in excess of 1 million base pairs in length. Second, the GC content of an isochore is relatively uniform throughout (i.e., a sliding window of 1,000 bp moving across an isochore's entire length would have a GC content that rarely differed from the isochore's overall GC content by more than 1%), although it differs significantly as the transition is made from one isochore to the next.

Experiments performed on human chromosomes suggest that our genome is a mosaic (Figure 6.10) of five different classes of isochores: two that are poor in G's and C's (called L1 and L2 with 39% and 42% GC content on average, respectively) and three that are comparatively rich in G's and C's (called H1, H2, and H3 with 46%, 49%, and 54% GC content on average, respectively). The H (high density on buoyant density gradients) isochores of humans and other eukaryotes are particularly rich in genes and an excellent place for genome sequencing efforts to start. The GC-richest isochore, H3, for instance has at least 20 times the density of genes found in the AT-richest isochore, L1 (the lowest density on buoyant density gradients).

Perhaps even more interesting is the fact that the types of genes found in GC-rich and GC-poor isochores is also very different. Even though the human H3 isochore represents a relatively small fraction of our total genome (between 3% and 5%), it contains within it approximately 80% of all of our housekeeping genes. In contrast, the L1 and L2 isochores (together comprising about 66% of the human genome) contain about 85% of our tissue-specific genes. Isochore compartmentalization is also known to be associated with several other important features of eukaryotic genomes including methylation patterns and chromatin structure (GC-rich isochores tend to have low levels of methylation of their CpGs and to be stored as transcriptionally active euchromatin); means of regulating gene expression (GC-rich regions tend to have promoter sequence elements closer to the transcriptional start site); intron and gene length (GC-rich

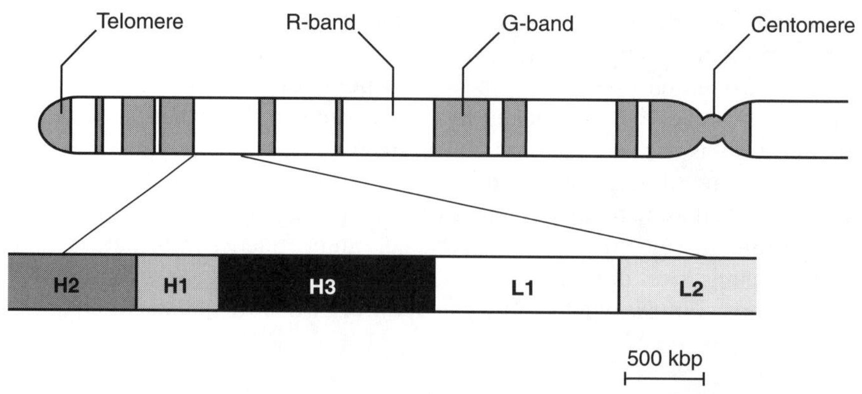

FIGURE 6.10 *The isochore compartmentalization of the human genome. This schematic shows a portion of a metaphase human chromosome with a blowup showing the mosaic nature of its underlying isochores. Metaphase chromosomes are often treated with dyes like Giemsa to generate characteristic banding patterns such as the G-band (Giemsa staining) and R-band (reverse Giemsa staining) bands. The five putative isochores of the human genome (L1, L2, H1, H2, and H3) have each been assigned a different shade of gray according to their relative GC contents with the GC-poorest isochore (L1) being white and the GC-richest being black. Human chromosomes are essentially random assortments of long fragments from different isochores as shown. L1 regions from all chromosomes together comprise approximately half of all of the human genome.*

regions tend to have shorter introns and genes); the relative abundance of short and long repeated sequences described at the end of this chapter (SINEs tend to be found in GC-rich isochores while LINEs predominate in GC-poor isochores); and the relative frequency of amino acids used to make the proteins of different genes (genes in GC-rich isochores tend to use amino acids that correspond to triplet codons that are rich in G's and C's). All of these features and the correlations between them can provide important clues to gene recognition algorithms.

Codon Usage Bias

Every organism seems to prefer to use some equivalent triplet codons (i.e., they code for the same amino acid) over others. For example, across the entire yeast genome, the amino acid arginine is specified by the codon AGA 48% of the time even though the other five functionally equivalent codons for arginine (CGT, CGC, CGA, CGG, and AGG) are each used with lower, relatively equal frequencies (approximately 10% each). Fruit flies show a similar but distinctly different codon usage bias for arginine with the codon CGC (33%) being preferred to the five alternatives (occurring with a frequency of about 13% each). A complete set of codon usage biases determined from completed genome projects can be found on the web at http://www.kazusa.or.jp/codon/CUTG.html. Some of the biological basis for these preferences may be related to avoidance of codons that are similar to stop codons as well as a preference for codons that allow efficient translation because they correspond to particularly abundant tRNAs within the organism. Regardless of the reason for the preferences, codon usage bias can differ significantly between eukaryotic species, and real exons generally reflect those preferences whereas randomly chosen strings of triplets do not.

Gene Expression

Given all the uncertainties of gene recognition in eukaryotes, the true test of any gene prediction effort is still the laboratory-based demonstration that a living cell actually transcribes the region into an RNA molecule. In addition to providing final verification, **transcriptomes** (the complete set of an organism's RNA sequences) are also an extremely valuable tool for bioinformaticians interested in finding genes in the first place. Some useful DNA sequence features for eukaryotic gene recognition algorithms are as follows:

- Known promoter elements (i.e., TATA and CAAT boxes),
- CpG islands,
- Splicing signals associated with introns,
- Open reading frames with characteristic codon utilization, and
- Similarity to the sequences of ESTs or genes from other organisms.

Even when the nucleotide sequence of only some of an organism's RNA transcripts is known, that information can be used to greatly facilitate gene recognition efforts by simply employing a series of data searches and pairwise sequence alignments like those described in Chapter 2. However, it is important to not lose sight of the fact that an organism's ability to alter its pattern of gene expression in response to changes in its environment is central to the very idea of being alive. That alone justifies the substantial amount of effort that has been put into developing the methodologies described below that allow a determination of what portions of a genome are actually transcribed.

cDNAs and ESTs

cDNAs (short for complementary DNAs) represent convenient ways of isolating and manipulating those portions of a eukaryotic genome that are transcribed by RNA polymerase II. cDNAs are made from the RNA isolated from eukaryotic cells in the fairly simple process illustrated in Figure 6.11. The resulting double-stranded DNA copies of processed mRNAs can be cloned into vectors and maintained as a **cDNA library.** The mere fact that a region is transcribed makes it of special interest to molecular biologists, and even small fragments of sequence information from individual cDNAs are enough to provide useful **ESTs** (expressed sequence tags) that can be used in contig assembly (as in Figure 6.1) or for gene mapping and recognition. When cDNA sequences are complete they can also be

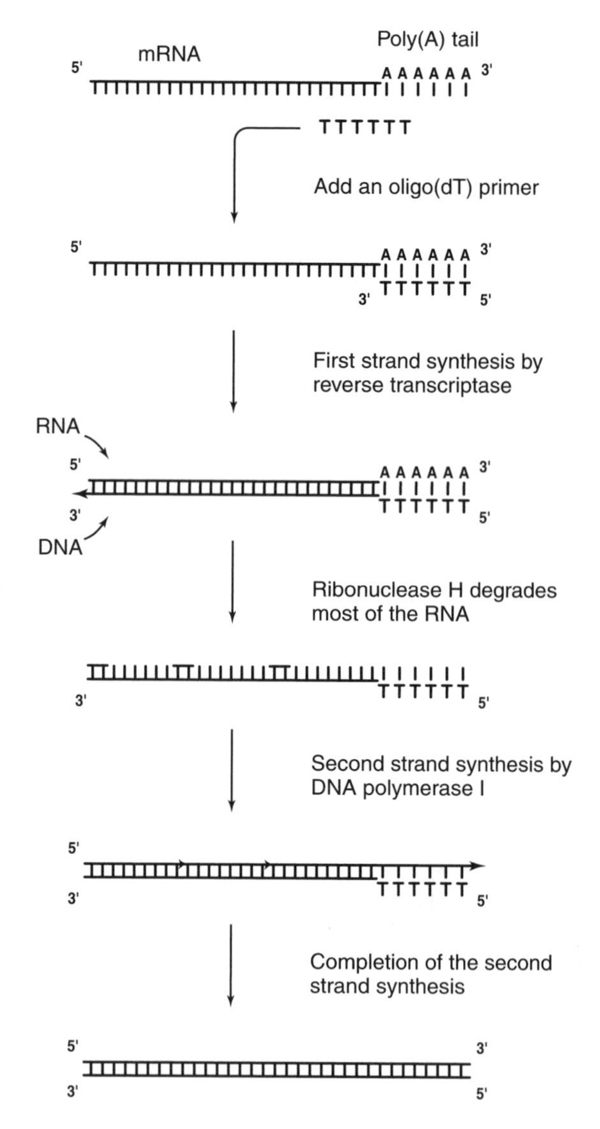

FIGURE 6.11 *A commonly used method for preparing cDNAs from a diverse population of mRNAs. Twenty-nucleotide-long oligo(dT) primers specifically hybridize to the poly(A) tail of processed mRNAs and are used by an enzyme (reverse transcriptase) to synthesize a DNA copy. Ribonuclease H is then used to specifically degrade most of the RNA component of the resulting RNA–DNA hybrids. Some remaining portions of the original RNA are then used as primers for a DNA polymerase. The resulting double-stranded DNA molecules can then be cloned into a vector and maintained as a cDNA library.*

compared to genomic sequences to provide unambiguous evidence of exon–intron boundaries.

Because a cell's mRNAs are derived from protein-coding genes, cDNAs give invaluable insights into both the population of genes being expressed at any given time as well as the mRNA's relative abundance. One means of assaying the complexity of the mRNA pool within a cell involves reassociation kinetics much like the $C_0t_{1/2}$ analyses described in Chapter 1. In essence, an excess amount of RNA from a cell is allowed to hybridize with cDNA copies prepared from the same organism. The resulting $\boldsymbol{R_0t_{1/2}}$ **values** (where R_0 is the starting concentration of RNA) typically indicate that, just like genomic DNA, RNA populations appear to have three components that differ in terms of their relative abundances. For instance, $R_0t_{1/2}$ analyses of chicken oviducts suggest that each cell contains roughly 100,000 copies of the ovalbumin mRNA (50% of all mRNA; $R_0t_{1/2}$ = 0.0015), 4,000 copies of 8 other mRNAs (15% of all mRNA; $R_0t_{1/2}$ = 0.04), and between 1 and 5 copies of 13,000 additional scarce mRNAs (35% of all mRNA; $R_0t_{1/2}$ = 30). In general, about 50% of the mRNA mass within a cell is found exclusively in that specific tissue (i.e., ovalbumin is expressed only in oviducts). Other, more sensitive methods such as those described below are needed to determine the extent of overlap of scarce mRNAs between different tissues.

Serial Analysis of Gene Expression

Determining the nucleotide sequence of every cDNA obtained from a single cell would certainly provide more detailed information about the expression patterns of eukaryotic genes. Sequencing several hundred thousand clones per cell (each with average lengths of over 1,000 bp) represents an impractical objective despite advances in high-throughput automated DNA sequencing. A sensitive alternative is **serial analysis of gene expression (SAGE).** In SAGE, cDNAs are made from a cell; the cDNAs are broken into small fragments (typically with restriction enzymes as described in Chapter 1) between 10 and 14 nucleotides long; and hundreds of those fragments are then randomly ligated (linked) into longer DNA molecules that are cloned and sequenced. Computers are then used to recognize the original small fragments in hundreds of clones by considering the recognition sequences of the restriction enzymes that were used as well as through comparisons to the sequences of known transcripts from the organism. The number of times that a tag from any particular transcript is observed should reflect the relative abundance of the corresponding transcript in the original cDNA pool.

Although computationally intensive, this approach has confirmed much of what was learned from the original $R_0t_{1/2}$ experiments. It has also provided important new details such as the fact that roughly 75% of the genes expressed in chicken liver and oviduct are the same. Specifically, approximately 12,000 genes are expressed in both liver and oviduct, about 5,000 additional genes are expressed only in liver, and a different set of 3,000 transcripts is found only in oviduct. Experiments using mammalian cells and tissues yield similar results and indicate that a set of roughly 10,000 housekeeping genes is constitutively expressed and codes for proteins whose functions are needed by every cell in an organism.

Microarrays

As mentioned in Chapter 1, a variant on the membrane-based Southern hybridization approach known as **microarrays** (Figure 6.12) is now routinely providing even more sensitive and detailed information about gene expression patterns than even the SAGE methodology allows. The small silica (glass) chips used in this approach are literally covered with thousands or tens of thousands of

(a)

(b)

FIGURE 6.12 *"Gene chip" microarray analyses allow changes in gene expression to be sensitively measured for very large numbers of genes. (a) An Affymetrix U95Aver2 GeneChip containing 12,625 rat genes represented as sets of 16 probe pairs. Each "cell" or spot contains a 25-mer oligonucleotide. The intensity of signal in each probe cell is proportional to the amount of hybridized, biotin-labeled cDNA bound to it. (b) Expression levels of a single gene from three different U95Aver2 GeneChips. The relative expression of this gene is represented by the signal value that is calculated using the intensities of the 16 probe pairs. PM, probe cells containing oligonucleotides that are "perfect matches" to the gene; MM, probe cells containing oligonucleotides with mismatches in the central nucleotide of the 25-mer. Conditions 2 and 3 result in an elevation in this gene's expression relative to condition 1. (Images kindly provided by Dr. Steven Berberich, Wright State University.)*

relatively short nucleotides of known sequence. In the case of one set of commercially available high-density oligonucleotide arrays (HDAs), expression patterns associated with 6,181 ORFs in the yeast genome can be analyzed simultaneously.

Each microarray chip is fabricated by high-speed robotics and has 20 spots containing 25-nucleotide-long probes (at a density of as many as 500,000 spots per 1.25 cm^2 with some chips containing more than 1 million) that can perfectly base pair with complementary nucleotides in each of the 6,181 expected transcripts. Each HDA also includes 20 spots that contain 25-mers that differ from the perfectly matching oligonucleotide at just one position. Target sequences for each gene are duplicated many times at many different places in each gene as a method of detecting errors.

An advantage to using small targets of 25 nucleotides is the ability to be very sensitive to hybridization conditions such that a single nucleotide change results in failure of the probe to bind to the target. Alternative chips with much longer targets (i.e., 250 nucleotides) can hybridize with smaller amounts of probe but are less sensitive to mismatches (MMs). At the most superficial level, the expression level of any gene is determined by subtracting the average amount of fluorescently labeled cDNA bound to the mismatch oligos from the average amount bound to its perfect match (PM) partner. The results are typically displayed as a grid in which every square represents a particular gene and relative levels (or changes) in expression are indicated by a color or gray scales such as those shown in Figure 6.12.

Gene expression profiling (sometimes also referred to as transcriptional profiling) has been applied to a wide variety of important biological problems including mapping of metabolic pathways, tissue typing, environmental monitoring, and answering a wide range of questions pertaining to medical diagnosis of disease states. Extremely large amounts of gene expression data that allow comparisons of gene expression patterns in diseased and normal states as well as between tissue types are maintained at web sites such as http://www.ncbi.nlm.nih.gov/UniGene/ddd.cgi?ORG=Hs.

In one recent application, gene expression patterns were used to distinguish between two frequently misdiagnosed lymphomas (diffuse large B-cell versus follicular). Microarrays with probes for 6,817 different human genes indicated that significant differences in expression at 30 genes existed between the two types of tumors. Considered together, the expression patterns of these 30 genes allowed correct classification of 71 of 77 tumors (91%)—a substantial improvement over alternative cytological indicators. Such improvements in diagnosis can be vitally important especially when treatment regimens are significantly different (as they are for these lymphomas).

The medical applications of gene expression profiling do not necessarily end with diagnosis either. In the case of the 58 patients with large B-cell lymphomas in the previous study, changes in gene expression patterns in response to treatment were also evaluated. A supervised machine learning prediction method was applied to the resulting data and was able to categorize treated patients into two categories with very different 5-year overall survival rates (70% versus 12%) with a high degree of confidence ($p = 0.00004$). The implications should be clear: The sooner it can be determined that a patient has not responded to a treatment, the sooner a new treatment can be attempted and the greater the likelihood for a positive outcome. Analyses such as these suggest that there is great promise for the development of much more individualized treatments than those that are currently available. The relatively new field of **pharmacogenomics** hopes to maximize the efficacy (and minimize the unwanted side effects) of treatments in just such a way using information about individuals' genetic makeup as well as how their gene expression patterns change in response to different therapies.

Transposition

As mentioned earlier in this chapter, prokaryotic genomes are exceptionally streamlined in terms of their information content. Even still, DNA transposons that are often present in multiple copies and often quite dispensable to their host represent an important component of bacterial genome anatomy. A single *E. coli* genome, for example, may contain as many as 20 different **insertion sequences** (ISs). Most of the sequence of an IS is dedicated to one or two genes that code for a transposase enzyme that catalyzes its transposition from one part of the genome to another in a conservative (the number of copies of the repeat does not change) or replicative (the copy number increases) fashion. Other kinds of bacterial transposons include composite transposons (pairs of IS elements that facilitate the transposition and sometimes the horizontal transfer of genes), Tn3-type transposons (which always transpose replicatively), and transposable phages (viruses that transpose replicatively as part of their normal infection cycle). The DNA transposons of prokaryotes are often distributed randomly throughout the genome, and their presence (let alone their position) is usually sufficiently variable to allow reliable distinctions to be made between strains of the same species.

It should not be surprising that eukaryotic genomes, with their abundance of noncoding and apparently dispensable DNA, also have DNA transposons, although current estimates suggest that there are fewer than 1,000 DNA transposon sequences in the human genome. Well-studied examples of eukaryotic transposons include the *Ac/Ds* elements of maize made famous by B. McClintock in the 1950s. Another important eukaryotic transposon is the 1,250-bp-long *mariner*, which was originally found in fruit flies but has since been discovered in a variety of eukaryotes including humans (suggesting that it might be a promising vehicle for both natural and engineered horizontal gene transfers between eukaryotic organisms). Much more common in eukaryotes though are transposons that are propagated by RNA intermediates known as **retrotransposons,** which are described below.

Repetitive Elements

DNA transposons present in multiple copies within a prokaryotic or eukaryotic genome qualify as "repeated DNA" or repetitive sequence elements. While uncommon in prokaryotes, repetitive elements that do not propagate through the action of a transposase make up a very large fraction of the genomes of the most complex eukaryotes. Those repeated sequences are typically divided into two categories that differ in their mode of propagation: tandemly repeated DNA and repeats that are interspersed throughout a genome.

Tandemly repeated (head to tail repeats such as 5'-CACACACA-3' where the repeat unit 5'-CA-3' is repeated four times) DNA can itself be subdivided into two categories: (1) satellite DNA and (2) mini/microsatellites. **Satellite DNA** gets its name from the fact that its very simple sequences with skewed nucleotide compositions give rise to DNA fragments with unusual densities relative to other genomic DNA. Buoyant density gradients can be used to separate human genomic DNA into four bands with different densities: one main band with an overall GC content of about 40.3% G+C and three lower density "satellite" bands above it. The repeating DNA sequences that give rise to these bands range in length from 5 to 200 bp in different eukaryotes and are typically present in millions of copies. Although some satellite DNA is scattered throughout eukaryotic genomes, most is located in the centromeres. Their very simple sequences are no more capable of containing useful information than a string of 997 repetitions of the word "really" in a student's 1,000-word-long report that says a book was *really* good.

Minisatellites and **microsatellites** are typically not abundant or long enough to give rise to their own bands on buoyant density gradients. Minisatellites form clusters up to 20,000 bp in length and have many tandem copies of sequences not more than 25 bp in length. The term *microsatellite* is typically used to describe clusters of shorter repeated sequences (usually four or fewer nucleotides in a row) that usually span less than 150 bp overall. Although microsatellites are relatively short, there are usually many of them fairly evenly distributed across a complex eukaryote's genome. In humans, for example, microsatellites with a CA repeat such as 5'-CACACACACACA-3' occur approximately once every 10,000 bp and make up 0.5% of the whole genome (15 Mb in all). Single base pair repeats (i.e., 5'-AAAAAAAA-3') make up another 0.3% of the human genome. DNA polymerases apparently lose their place during replication of these simple sequences and give rise to longer and shorter versions quite frequently. The resulting high level of variability in the lengths of microsatellites from one individual to another has made them useful genetic markers to geneticists (as well as to forensic scientists and for paternity/maternity testing).

Many repeated sequences within eukaryotic genomes appear to be scattered fairly randomly across a genome rather than being tandemly clustered together as is the case for satellites and mini/microsatellites. This other kind of repeated sequence appears to be propagated by the synthesis of an RNA intermediate in a process called **retrotransposition.** The basic mechanism involves three steps: (1) An RNA copy of the transposon is transcribed by an RNA polymerase just like a normal gene, (2) the RNA copy is converted into a DNA molecule by a special enzyme called reverse transcriptase, and (3) the reverse transcriptase inserts the DNA copy of the transposon into another site elsewhere in the genome.

The reverse transcriptase required for this process does not seem to be part of the normal complement of genes in a eukaryotic genome but rather to be acquired from infecting (or stranded) retroviruses (viruses like the one responsible for AIDs in humans where the genome is stored as RNA rather than DNA). Very common kinds of **retroposons** within mammalian genomes are **LINEs** (long interspersed nuclear elements) and **SINEs** (short interspersed nuclear elements). LINEs are likely to be stranded retroviruses and contain a reverse transcriptase-like gene that is essential for the propagation of both themselves and SINEs. Human L1 repeats are a classic example of LINEs in that full-length repeats are 6,100 bp long and present in approximately 3,500 copies scattered throughout the human genome. Full-length L1 repeats as well as hundreds of thousands of truncated fragments comprise a total of approximately 5% of every mammalian genome.

Each mammalian order (i.e., primates, rodents, artiodactyls) has its own, independently derived SINE. The human genome's *Alu* repeat is a good example in that it is has an average length of 258 bp and is present in more than 1,000,000 copies scattered across the human genome. Even though SINEs comprise up to 10% of some mammalian genomes, they are often considered the epitome of "junk DNA" since they are generally not associated with functionally constrained sequences. The term "junk DNA" is one that must be used with caution, however, in that many sequences that have been relegated to that category have subsequently been found to simply have been playing previously unappreciated roles. Regardless, from a bioinformatics perspective, many genome analyzing algorithms (including database search programs) begin by "masking" known repeated sequences such as SINEs and LINEs because they are known to have little useful information content for gene recognition or typical comparisons between sequences.

Eukaryotic Gene Density

The C-value paradox described in Chapter 1 made clear that much of a eukaryote's genome is dispensable decades before molecular biologists began to seriously contemplate determining the nucleotide sequence of complete genomes. The human genome project alone has amply confirmed that observation. Of the 3,000 Mb in the human genome, not more than 90 Mb (3%) corresponds to coding sequences, and approximately 810 Mb (27%) are associated sequences such as introns, promoters, and pseudogenes. The remaining 2,100 Mb (70%) of the human genome can be divided into two different kinds of "junk" that is under little or no selective constraint: unique sequences (1,680 Mb or 56%) and repetitive DNA (420 Mb or 14%). In short, genes are definitely "far between" even in gene-rich regions such as the H3 isochores in the most complex eukaryotes where there seems to be little evolutionary pressure to rein in genome size. The average distance *between* human genes is in the range of 65,000 bp (given a genome size of 3 billion bp and a total of about 45,000 genes)—in the ballpark of 10% of the *total* genome size of a simple prokaryote like *M. genitalium.* To someone who is illiterate, only the pictures within an encyclopedia have useful information content. It is possible that all the sequences between the genes whose functions we recognize may one day take on a greater significance to us in much the same way that letters and punctuation between pictures becomes more important as one learns to read.

Mutational analyses (such as genetically engineered "knockout" organisms) are making it increasingly clear that many genes code for proteins that are responsible for multiple (often unsuspected) functions. Genomic sequence analyses are also suggesting that many genes are present in multiple, redundant copies of subtle variants—in humans, perhaps as many as an average of three to four genes for every function. Less complex eukaryotes tend to have higher densities of genes on their chromosomes than more complex organisms like vertebrates. The yeast genome, for instance, has an average ORF length of approximately 1,400 bp with an average separation of roughly 600 bp.

Chapter Summary

Gene recognition in prokaryotes is comparatively simple and can rely heavily on searches for statistically significant, long open reading frames. Prokaryotic genomes are characterized by a notably high density of information content and, as a rule, are fairly easy to analyze. Eukaryotic genomes with their very low density of information and prodigious sizes represent a striking contrast. Eukaryotic gene recognition software must consider a wide variety of different features when looking for genes. These features include adherence to codon usage biases within an ORF; the presence of an upstream CpG island complete with other promoter sequences; and splice junctions and internal branch sites that are good matches to the consensus sequences for introns. Unfortunately, the rules associated with all of these features are cluttered with common exceptions and often vary from one organism to another and even from one genomic or cell-type context to another. The best gene recognition algorithms to date take advantage of all of these features as well as others but are still plagued with high rates of both false positives and negatives. Recent increases in both the number and types of sequences used for training and evaluation coupled with additional data (i.e., knowing in advance the sequences of most if not all of an organism's mRNAs) should provide the basis for significant improvements in the years to come.

Readings for Greater Depth

Deciphering the evolutionary history of prokaryotic genes and genomes often requires an appreciation of their GC content such as that described in J. G. Lawrence and H. Ochman, 1998, Molecular archaeology of the *Escherichia coli* genome, *Proc. Nat. Acad. Sci. U.S.A.* **95:** 9413.

There has been much speculation regarding the minimum number of genes required for a free-living organism such as that presented in A. R. Mushegian and E. V. Koonin, 1996, A minimal gene set for cellular life derived by comparison of complete bacterial genomes, *Proc. Nat. Acad. Sci. U.S.A.* **93:** 10,268–10,273.

The process of initiation of transcription by RNA polymerase II in eukaryotes is described in careful detail in R. G. Roeder, 1996, The role of initiation factors in transcription by RNA polymerase II. *Trends Biochem. Sci.* **21:** 327–335. It is also reviewed in A. Ishihama, 2000, Functional modulation of *Escherichia coli* RNA polymerase, *Annu. Rev. Microbiol.* **54:** 499–518.

A comprehensive review of isochores and the many important features associated with them is provided by the founder of this field of research in G. Bernardi, 1995, The human genome: Organization and evolutionary history. *Ann. Rev. Genet.* **29:** 445–476.

A general review of the applications of computer-based studies in genome sequence analyses can be found in T. F. Smith, 1998, Functional genomics—Bioinformatics is ready for the challenge, *Trends Genet.* **14:** 291–293.

Determining the sequence of a genome is just the start of the process of understanding how an organism uses its information content. The first steps in the functional analysis of the yeast genome are described in S. G. Oliver, M. K. Winson, D. B. Kell, and F. Baganz, 1998, Systematic functional analysis of the yeast genome, *Trends Biotechnol.* **16:** 373–378.

Questions and Problems

* **6.1** Use Figure 6.3b to determine at how many sites within the –35 and –10 promoter regions of the Lac operon's promoter there are differences relative to the consensus sequence for σ^{70} promoters. How should those nucleotides be changed if you were interested in increasing the levels of expression of the operon within *E. coli*?

6.2 The promoter of a predicted *E. coli* gene is found to contain the following nucleotide sequence:

5'-ACTGGACCCTTGAAGGCGACGTCGGCCTACCCGATCTCCACTGTATGGATCCGGA-3'

What can you infer about the circumstances under which the gene is most likely to be expressed?

* **6.3** Suggest two different experimental strategies that might be used to generate the sequence data that would connect the two contigs illustrated in Figure 6.1 into a single, longer contig.

6.4 Find the longest stretch of nucleotides in the following single-stranded RNA sequence that would be able to form a hairpin loop such as the one shown in Figure 6.4.

5'-GGGCGCGAAUAUCCCGGAGUCCGUAUGACCCCAUGCGGACUACGGGAUAUUCA-3'

* **6.5** What three nucleotide substitutions to the nucleotide sequence in Question 6.4 would allow it to form a more stable secondary structure? Assume that G/C base pairs are more energetically stable than A/T base pairs and that the phosphodiester backbone of DNA requires at least four nucleotides to be in the "looped" portion of a hairpin loop. Explain your choices.

6.6 Assuming that all codons and nucleotides occur with the same frequency, what is the probability that a stretch of 316 codons does not contain a single stop codon simply by chance? How large does an ORF need to be to achieve a statistical significance of 1%?

* **6.7** In which prokaryotic RNA polymerase subunit would you expect to find mutations that confer resistance to antibiotic drugs that are structurally similar to nucleotides?

6.8 How do the RNA polymerases of prokaryotes and eukaryotes differ?

* **6.9** How many of the nucleotides in the following sequence of a human gene's promoter are likely to be methylated in a tissue where the gene is transcriptionally inactive? Active?

```
5'-GGGCGCGAATATCCCGGAGTCCGTATGACCTACATATTCATGATCGCTAGCC-3'
3'-CCCGCGCTTATAGGGCCTCAGGCATACTGGATGTATAAGTACTACGGATCGG-5'
```

6.10 Using the information in Table 6.5, what transcription factors are likely to bind to the promoter sequence given in Question 6.9? What can you conclude about the expression pattern of the associated gene?

* **6.11** The length of the primary transcript of the human dystrophin gene is roughly four times larger than the entire genome of some prokaryotes (i.e., *M. genitalium*). Compare the fraction of those two sequences that correspond to coding information that is actually used by ribosomes.

6.12 Into how many different mRNAs can an hnRNA be processed if it has three exons and if all its 5' splice junctions can be used with any downstream 3' splice junction? Diagram the information content of each mRNA.

* **6.13** Using the genetic code shown in Table 1.1, determine how many of the 20 amino acids could not be present at a position in a protein that corresponds to where an intron had been spliced out of an hnRNA.

6.14 Using Entrez on the NCBI home page (http://www.ncbi.nlm.nih.gov/Entrez/), find an entry that includes the sequence of one of the genes mentioned in this chapter. (*Note:* You should obtain the genomic sequence, not one that corresponds to an mRNA.) Print the corresponding GenBank file. Use the file's annotations to identify as many gene features as possible (i.e., transcriptional start site, polyadenylation signal, exons and introns) and mark those features on the sequence portion of the file. Make a scale diagram of the gene using the style shown in Figure 6.2 with open boxes being used to represent intron sequences and filled boxes to represent exons.

* **6.15** Given that *Arabidopsis thaliana* genes use the codon GUU three times more frequently than any of the three alternative codons for the amino acid valine, find the reading frame in the *A. thaliana* sequence below that is most likely to correspond to the one used by ribosomes during translation of the corresponding mRNA. Use ORF lengths and codon usage biases to explain your answer.

```
5'-GAGCGGAAGUGUUCGAUGUACUGUUCCAGUCAUGUGUUCACC-3'
```

CHAPTER 7

Protein and RNA Structure Prediction

What a piece of work is man! How noble in reason! how infinite in faculties! in form and moving, how express and admirable!

Hamlet, *Act II, scene ii*

Proteins are the molecular machinery that regulates and executes nearly every biological function. Structural proteins, such as collagen, support and strengthen our connective tissues; mechanoenzymes, such as myosin in skeletal muscle, provide movement on both microscopic and macroscopic scales; other enzymes catalyze chemical reactions of all kinds, enabling and controlling digestion and metabolism, the immune system, reproduction, and a staggering array of other functions. Protein interactions with DNA and RNA enable the production of new proteins, and regulate their levels, responding as appropriate to internal and external environmental changes. One of the foremost tasks of molecular biology is to further the understanding of the relationships between the various genes in an organism's genome and the proteins they encode.

Proteins are synthesized as linear chains of amino acids, but they quickly fold into a compact, globular form *in vivo.* In his seminal work in the late 1960s, C. Anfinsen first demonstrated that unfolded, or **denatured,** proteins repeatedly assume the same conformation when allowed to refold. This **native structure** is essential for biological function. Only when folded into their native globular structure are most proteins fully biologically active. Understanding the forces that drive protein folding is perhaps the most significant unanswered question in biochemistry. No algorithm currently exists that will consistently predict the three-dimensional shape of a protein. The ramifications of such an algorithm would be so significant to molecular biology that the problem is considered a "grand challenge" problem—one of the most important problems to be addressed by contemporary computer science.

Amino Acids

Amino acids are the building blocks of proteins. Like DNA and RNA, proteins are synthesized as linear polymers (chains) composed of smaller molecules. Unlike DNA and RNA, in which there are four nucleotides from which to choose, proteins are constructed from 20 amino acids with a variety of sizes, shapes, and chemical properties.

Each amino acid has a **backbone** consisting of an amide (—NH_2) group, an **alpha carbon,** and a carboxylic acid, or carboxylate (—COOH) group. To the alpha carbon, a **side chain** (often denoted —R) is attached, as shown in Figure 7.1. The side chains vary with each amino acid, and these various side chains confer unique stereochemical properties on each amino acid.

The amino acids are often grouped into three categories. The **hydrophobic amino acids,** which have side chains composed mostly or entirely of carbon and hydrogen, are unlikely to form hydrogen bonds with water molecules. The **polar amino acids,** which often contain oxygen and/or nitrogen in their side chains, form hydrogen bonds with water much more readily. Finally, the **charged amino acids** carry a positive or negative charge at biological pH. Figure 7.2 shows the primary amino acid sequence of the prokaryotic protein superoxide dismutase.

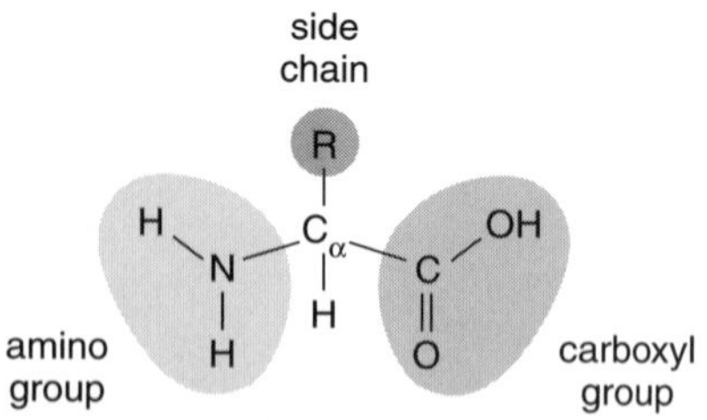

FIGURE 7.1 *The structure of an amino acid. The side chain, represented as R, distinguishes the different amino acids. The backbone atoms are constant for all 20 amino acids.*

(a)

```
DEFINITION     P.leiognathi bacteriocuprein superoxide dismutase gene, complete cds.
ACCESSION      J02658
SOURCE         P.leiognathi (ATCC 25521) DNA, clone pPhB-2.
BASE COUNT          277 a       140 c       179 g       245 t
ORIGIN         256 bp upstream of BglII site.
         1  agtaaaaatt tagcaattaa gtagtgttga tgaaatggta agagtaaaaa gtacacacgc
        61  tatgggatta atcttcttag cgaatgtttg agatattatc gataactata atcgtaaata
       121  tcagctatac ctttttgtta aaagcatgtt taatgcctgt ggaaaataaa aacaataagg
       181  ataaaatatg aacaaggcaa aaacgttact cttcaccgct ctagcttttg gtttatctca
       241  ccaagcgtta gcacaagatc tcacggttaa aatgaccgat ctacaaacag gtaagcctgt
       301  tggtacgatt gaactaagcc aaaataaata cggagtagta tttacacctg aactggcaga
       361  tttaacaccg gggatgcatg gcttccatat tcatcaaaat ggtagctgtg cttcatcaga
       421  aaaagacggc aaagttgttt taggtggcgc tgctggtgga cattacgatc ctgagcacac
       481  aaataaacac ggtttcccat ggactgatga taatcataaa ggtgatctgc cagcactgtt
       541  tgtgagtgca aatggtttag caactaaccc tgttttagcg ccacgtttaa cgttgaaaga
       601  actaaaaggt cacgcaatta tgatccatgc tggtggtgat aatcactctg atatgccaaa
       661  agcattaggt ggcggcggcg cacgtgtggc gtgtggtgtg atccaataat ttagtgagaa
       721  ccagcagcga atttgtcgct gttggtttta ttttaatcag attaagtttt ttagaaacag
       781  ccagttaatt gtaaaatatg taaaaatgtg aaattcaggt gaatttgaaa tcttctctta
       841  a
```

(b)

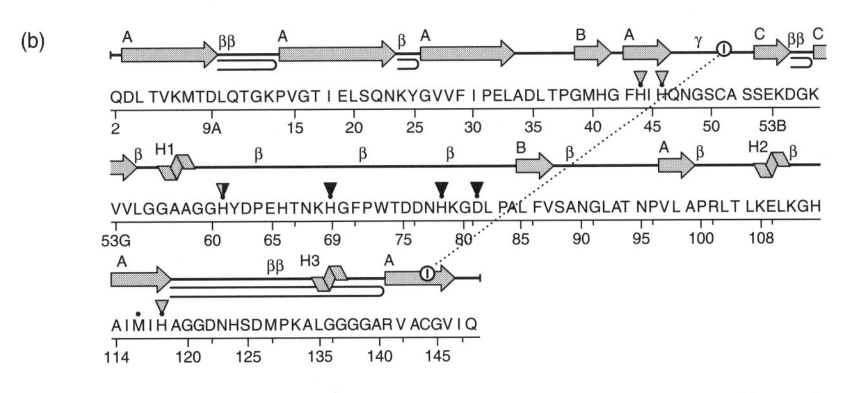

(c)

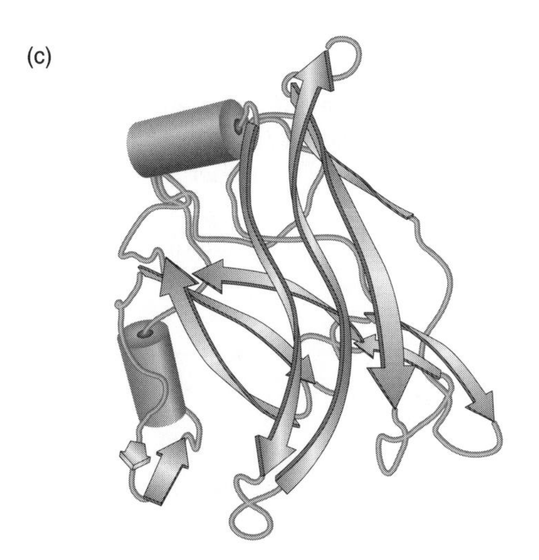

FIGURE 7.2 *(a) Excerpts from the genbank entry for the Photobacterium leiognathi bacteriocuprein superoxide dismutase gene (accession #J02658). (b) The primary and secondary structure for the corresponding protein (arrows and coils represent the positions of beta strands and helical regions, respectively. (c) The tertiary structure of the protein (PDB entry 1BZO, rendered using VMD molecular graphics software [Humphrey, W. Dalke, A. and Schulten, K., "VMD—Visual Molecular Dynamics", J. Molec. Graphics 1996, 14(1), 33–38, www.ks.uiuc.edu/Research/vmd]).*

BOX 7.1 pH, pKa, and pI

A very important concept in chemistry and biology both is that of concentration: a measure of how much of a solute is dissolved in a solvent. At the heart of the idea of concentration is the ratio of molecules of solute to solvent. The molecular weight (essentially the sum total of all the protons and neutrons of a componound) of NaCl is 58. By definition, in 58 grams of NaCl there is 1 *mole* of NaCl. The actual number of molecules in a mole is 6.02×10^{23}. As a result, the number of molecules in 58 g of NaCl is the same as the number of molecules of NaOH in 40 g (because the molecular weight of NaOH is 40). A solution that has 1 mole suspended in 1 liter is said to be 1 ***molar*** (or *M*).

It is important to remember that water is a molecule whose atoms have notably different electronegativities. Occasionally, the oxygen atoms in water actually strip away one of the hydrogen's electrons and an ionic situation results. In other words, H_2O becomes H+ and OH–. About 1 molecule in 554 million water molecules has disassociated in this way in a bottle of pure water. What concentration does that correspond to? 1×10^{-7} molar (or moles/liter of water). Although this dissociation is rare and is reversible, it is of great biological importance because the H^+ and OH^- are very reactive. As a result, the concentration of hydrogen ions has a special name and unit of measure: **pH,** where p stands for the negative of the log of the molar concentration. A solution of pure water where the concentration is 1×10^{-7} moles/liter is said to be pH 7 (or neutral).

Because living organisms are water-based systems, pH plays a role in nearly every biological reaction. Simply put, pH is a measure of the concentration of unassociated protons in a solution. When H+ ion concentrations are high, OH^- concentrations are low because free OH^- ions have more opportunities to reassociate with the abundant H^+ ions. By the same token, when H^+ ion concentrations are low, OH^- concentrations are high because free OH^- ions have fewer opportunities to reassociate with H^+ ions. Acidic solutions, which contain relatively large concentrations of free protons, have lower pH values, while basic solutions, which contain few free protons, have higher pH. The value of pH typically ranges from 0 (very acidic) to 14 (very basic). The environ-

The order of the amino acids in a protein's primary sequence plays an important role in determining its secondary structure and, ultimately, its tertiary structure. It is worthwhile for the aspiring bioinformatician to study the structure and properties of each amino acid carefully, since the properties of these small molecules provide proteins with both their structure and their biological function.

Polypeptide Composition

A chain of several amino acids is referred to as a **peptide.** Longer chains are often called **polypeptides** or proteins. When two amino acids are covalently joined,

ment inside of a cell is generally close to neutral (pH = 7).

Just like water, many of the amino acid side chains have dissociable protons (essentially hydrogen atoms). For example, the carboxylic acid group of aspartic acid can exist in a protonated (OH) or deprotonated (O^-) state, depending on the pH (i.e., the availability of free H^+ ions) of its environment. The **pKa** of an amino acid is a measure of the relative ease with which it releases its dissociable protons. When the pH of the solution is equal to the pKa of an amino acid, approximately half of the residues will be protonated and half will dissociate (release their protons). At one pH unit lower than the pKa of an amino acid, that amino acid will be approximately 90% protonated. At two pH units lower than the pKa of an amino acid, the amino acid will be approximately 99% protonated. For example, almost all aspartic acid (pKa = 4.3) amino acids within a polypeptide are deprotonated (negatively charged) at neutral pH (7.0) but only about half have a charge associated with them at the acidic pH of 4.3.

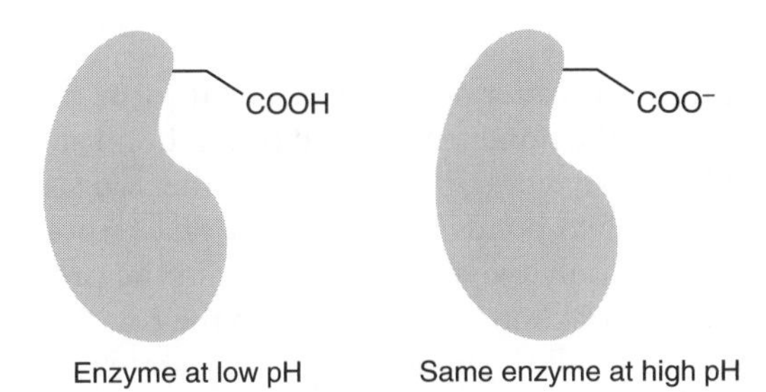

Protonation of a carboxylate group.

When the protons of an amino acid's side chain become dissociated, the amino acid becomes charged, having lost a proton, but no electrons. The more deprotonated a protein becomes, the more negative charge it accumulates. Likewise, as more residues become protonated, more positive charge is accumulated. The **isoelectric point (pI)** of a protein is the pH at which the net charge of the protein is neutral. Isoelectric points can reveal some valuable general information about the amino acid composition of a protein. For example, a protein with a greater amount of acidic amino acids than basic amino acids will have a pI of much less than 7. If the opposite is true, the pI will be much greater than 7. Human pepsin, a digestive enzyme, is very rich in glutamic and aspartic acids and has a pI of 1.2.

one of the amino acids loses a hydrogen (H^+) from its amine group, while the other loses an oxygen and a hydrogen (OH^-) from its carboxylate group, forming a carbonyl (C=O) group (and water, H_2O). The result is a **dipeptide**—two amino acids joined by a **peptide bond**—and a single water molecule, as shown in Figure 7.3. In a polypeptide, the amino acids are sometimes referred to as amino acid **residues,** because some atoms of the original amino acid are lost as water in the formation of the peptide bonds.

Like DNA and RNA molecules, polypeptides have a specific directionality. The **amino terminus** (or N terminus) of the polypeptide has an unbonded amide group, while the **carboxy terminus** (or C terminus) ends in a carboxylic acid group instead of a carbonyl. Protein sequences are usually considered to start at the N terminus and progress toward the C terminus.

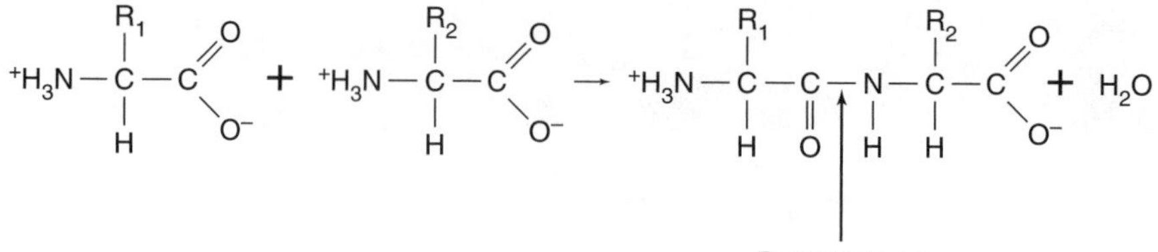

FIGURE 7.3 *Two amino acids are joined by dehydration synthesis (meaning a water molecule is released) to form a dipeptide.*

The sequence of amino acids that comprise a protein completely determines its three-dimensional shape, its physical and chemical properties, and ultimately its biological function. This sequence is called the **primary structure** of the protein. It is worthwhile to learn the one-letter codes for each amino acid (see Figure 7.2), because these codes allow protein sequences to be expressed as a string of characters, much like nucleotide sequences.

Secondary Structure

Backbone Flexibility, ϕ and ψ

The non-side-chain atoms of the amino acids in a polypeptide chain form the **protein backbone.** As noted in Chapter 1, the bond lengths and planar bond angles of the covalent bonds in the backbone are more or less fixed. In other words, the peptide group is rigid and planar—the two bonds between the alpha carbon (C_α) and the other backbone atoms are the only two rotatable bonds in the protein backbone. All flexibility in the protein backbone is derived from the rotation of these two bonds. The angle of rotation about the bond between the amide nitrogen and the alpha carbon is referred to as ϕ (phi), while the angle of rotation about the bond between the alpha carbon and the carbonyl carbon is called ψ (psi).

The backbone conformation of an entire protein can be specified in terms of the phi and psi angles of each amino acid, as shown in Figure 7.4. Not all values of phi and psi are physically realizable. Some phi/psi combinations result in a **steric collision** (physical overlap of the space occupied by atoms) between the side-chain atoms of one residue and the backbone of the next. The Ramachandran plot (Figure 7.5 on page 162) illustrates the allowable values for phi and psi for nonglycine residues. Because of their lack of a side chain (other than hydrogen), glycine residues have a much greater range of allowable phi and psi angles than most other residues.

As described in Chapter 1, most protein backbones contain elements of secondary structure, including alpha helices ($\phi \approx \psi \approx -60°$), and beta strands ($\phi \approx -135°$ and $\psi \approx 135°$), which associate with other beta strands to form parallel or anti-parallel beta sheets. Given a protein sequence (or primary structure), the first step in predicting the three-dimensional shape of the protein is determining what regions of the backbone are likely to form helices, strands, and **beta turns**—the

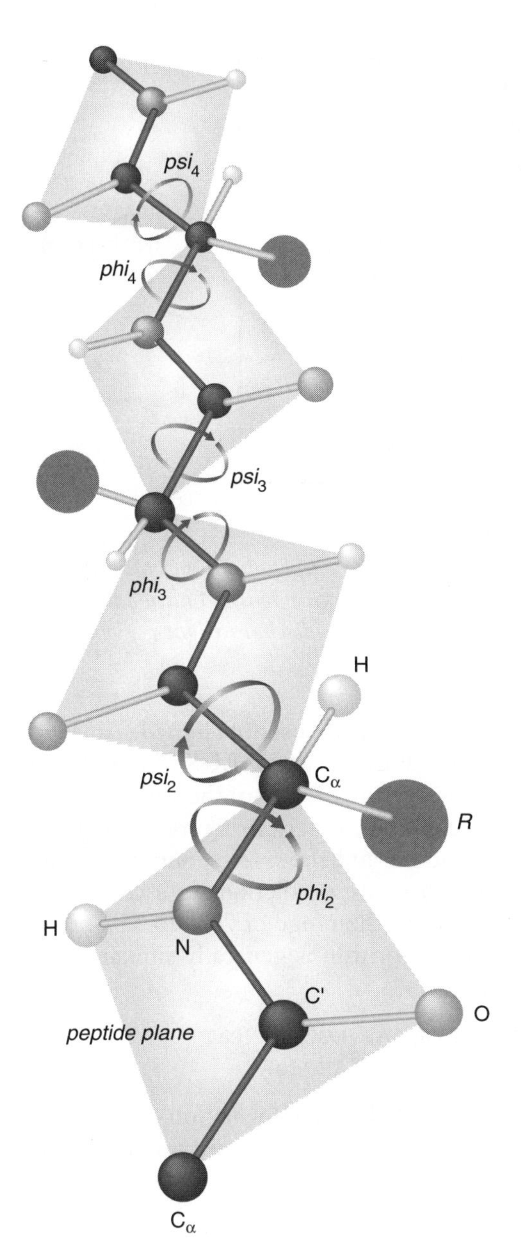

FIGURE 7.4 *Rotation about the C_α bonds are the only degrees of conformational freedom available to the protein backbone. The structure of the peptide bond forces the rest of the backbone atoms into a rigid planar configuration.*

U-turn-like structures formed when a beta strand reverses direction in an antiparallel beta sheet.

Accuracy of Predictions

Secondary structure prediction algorithms use a variety of computational techniques including neural networks, discrete-state models, hidden Markov models, nearest neighbor classification, and evolutionary computation. Most current

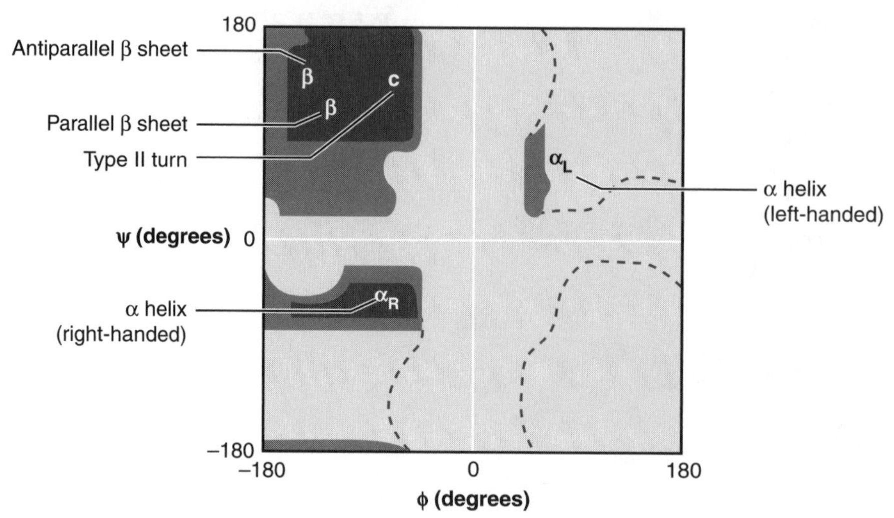

FIGURE 7.5 *The Ramachandran plot shows the values of phi and psi that are physically realizable without causing steric clash (dark regions). Additional conformations can be achieved by glycine due to its small side chain (light regions).*

secondary structure prediction algorithms begin with a set of aligned sequences produced with algorithms such as BLAST, FASTA, and CLUSTALW. Using these aligned sequences, the degree of conservation of each of the amino acids in the target sequence (the sequence for which the secondary structure is being predicted) is calculated. With a protein sequence and the corresponding conservation levels as input, modern secondary structure prediction algorithms, such as PHD and Predator, can obtain accuracies in the range of 70% to 75%. Output from a secondary structure prediction algorithm is generally similar to the following:

```
APAFSVSPASGASDGQSVSVSVAAAGETYYIAQCAPVGGQDACNPAT
---------HHHHHHH-HHHhhh---EEEEEeee---EEEEee----
```

In this case, H and h represent predictions of helical conformation (with strong and weak confidence, respectively), while E and e represent predictions of extended (beta-strand) conformation.

The Chou-Fasman and GOR Methods

The Chou-Fasman method takes a straightforward statistical approach to predicting secondary structure. Each amino acid is assigned several **conformational parameters,** *P*(a), *P*(b), and P(turn). These parameters, representing the propensity of each amino acid to participate in alpha helices, beta sheets, and beta turns, respectively, were determined based on observed frequencies in a set of sample proteins of known structure. In addition, each amino acid is assigned four turn parameters, $f(i)$, $f(i+1)$, $f(i+2)$, and $f(i+3)$, corresponding to the frequency with which the amino acid was observed in the first, second, third, or fourth position of a hairpin turn. The resulting table of **Chou-Fasman parameters** is shown in Table 7.1.

Using the Chou-Fasman parameters, the algorithm for assigning secondary structure proceeds as follows:

1. Identify alpha helices as follows:
   a. Find all regions where four of six contiguous amino acid residues have $P(a) > 100$.
   b. For each region identified in part (a), extend the region in both directions until a set of four contiguous residues with $P(a) < 100$ is encountered.
   c. For each region extended in part (b), compute $\Sigma P(a)$, the sum of $P(a)$ values for each residue in the region, and $\Sigma P(b)$. If the region is >5 residues in length, and $\Sigma P(a) > \Sigma P(b)$, then the region is predicted to be an alpha helix.
2. Identify beta sheets using the same algorithm as in step 1, but search for regions where four of six residues have $P(b) > 100$. Once the regions are extended (part b), a region is declared a beta strand if the average $P(b)$ over all residues in the region is greater than 100, and $\Sigma P(b) > \Sigma P(a)$.

TABLE 7.1 The Chou-Fasman parameters for the 20 common amino acids.

Amino acid	P(a)	P(b)	P(turn)	$f(i)$	$f(i+1)$	$f(i+2)$	$f(i+3)$
Alanine	142	83	66	0.06	0.076	0.035	0.058
Arginine	98	93	95	0.070	0.106	0.099	0.085
Asparagine	67	89	156	0.161	0.083	0.191	0.091
Aspartic acid	101	54	146	0.147	0.110	0.179	0.081
Cysteine	70	119	119	0.149	0.050	0.117	0.128
Glutamic acid	151	37	74	0.056	0.060	0.077	0.064
Glutamine	111	110	98	0.074	0.098	0.037	0.098
Glycine	57	75	156	0.102	0.085	0.190	0.152
Histidine	100	87	95	0.140	0.047	0.093	0.054
Isoleucine	108	160	47	0.043	0.034	0.013	0.056
Leucine	121	130	59	0.061	0.025	0.036	0.070
Lysine	114	74	101	0.055	0.115	0.072	0.095
Methionine	145	105	60	0.068	0.082	0.014	0.055
Phenylalanine	113	138	60	0.059	0.041	0.065	0.065
Proline	57	55	152	0.102	0.301	0.034	0.068
Serine	77	75	143	0.120	0.139	0.125	0.106
Threonine	83	119	96	0.086	0.108	0.065	0.079
Tryptophan	108	137	96	0.077	0.013	0.064	0.167
Tyrosine	69	147	114	0.082	0.065	0.114	0.125
Valine	106	170	50	0.062	0.048	0.028	0.053

3. If any of the helices assigned in step 1 overlap a beta strand assigned in step 2, then the overlapping region is predicted to be a helix if $\Sigma P(a) > \Sigma P(b)$, and a strand if $\Sigma P(b) > \Sigma P(a)$.
4. Finally, identify beta turns as follows:
   a. For each residue, i, calculate the turn propensity, $P(t)$, as follows: $P(t)$ = the $f(i)$ of the residue i + the $f(i+1)$ value of the following residue + the $f(i+2)$ value of the subsequent residue (position $i + 2$) + the $f(i+3)$ of the residue at position $i + 3$.
   b. Predict a hairpin turn starting at each position, i, that meets the following criteria:
      i. $P(t) > 0.000075$
      ii. The average $P(\text{turn})$ value for the four residues at positions i through $i + 3 > 100$
      iii. $\Sigma P(a) < \Sigma P(\text{turn}) > \Sigma P(b)$ over the four residues in positions i through $i + 3$.

Another statistical approach, the GOR (Garnier, Osguthorpe, and Robson) method, predicts secondary structure based on a window of 17 residues. For each residue in the sequence, 8 N-terminal and 8 C-terminal positions are considered along with the central residue. As with the Chou-Fasman method, a collection of sample proteins of known secondary structure was analyzed, and the frequencies with which each amino acid occupied each of the 17 window positions in helices, sheets, and turns was calculated, yielding a 17×20 scoring matrix. The values in this matrix are used to calculate the probability that each residue in a target sequence will be involved in a helix, sheet, or turn. The calculations used in the GOR method are based on information theory, and result in three-state prediction accuracies of around 65%.

Tertiary and Quaternary Structure

Predicting the secondary structure of a protein is only the first step in predicting the overall three-dimensional shape of a folded protein. The secondary structural elements of a protein pack together, along with less structured loop regions, to form a compact, globular native state. The overall three-dimensional shape of a folded polypeptide chain is referred to as its **tertiary structure.** Figure 7.6 illustrates some common tertiary structure motifs. For many proteins, the active form is a complex of two or more polypeptide units. **Quaternary structure** refers to the intermolecular interactions that occur when multiple polypeptides associate to form a functional protein, as well as the protein-to-protein contacts that can occur in multienzyme complexes.

The protein folding problem involves prediction of the secondary, tertiary, and quaternary structures of polypeptide chains based on their primary structure. Understanding the forces that drive a protein to fold is an active field of bio-

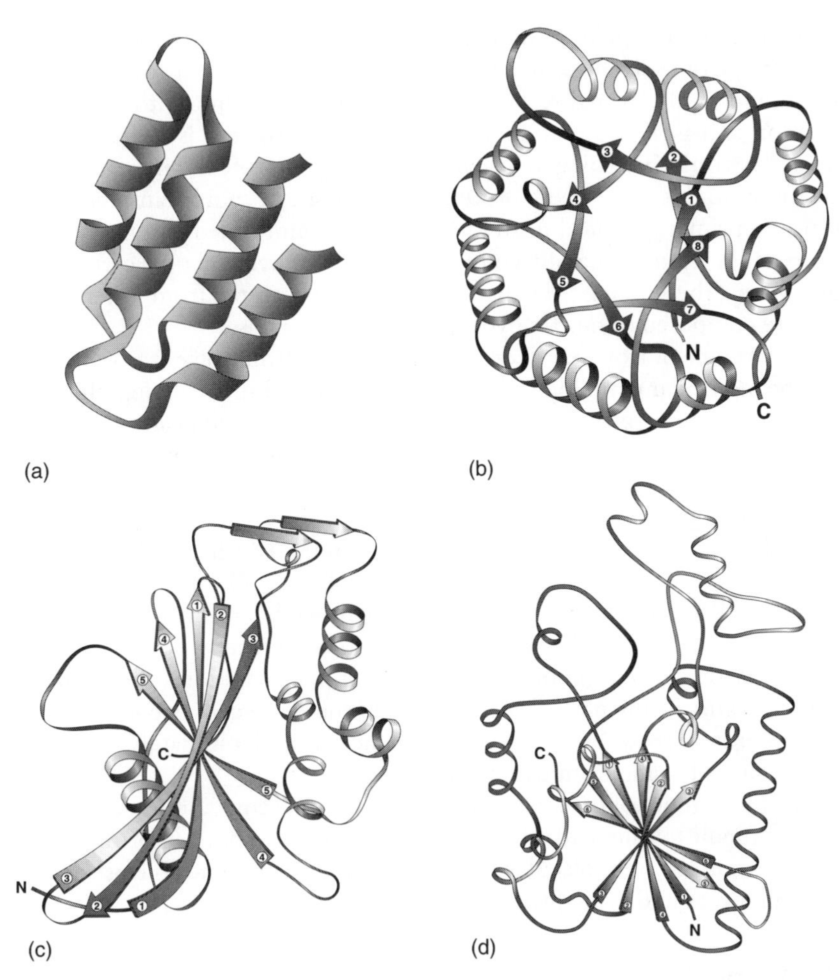

FIGURE 7.6 *Several common tertiary structure motifs: (a) a four helical bundle, (b) an alpha-beta barrel, (c) and (d) open twisted beta sheets.*

chemical research. A diverse array of forces, including electrostatic forces, hydrogen bonds, and van der Waals forces, all play important roles in protein folding. The formation of covalent bonds between cysteine residues can also play a key role in determining protein conformation. The problem is made even more complex by the action of a special class of proteins called chaperonins, which act by altering protein structures in as yet unpredictable but important ways.

Hydrophobicity

The hydrophobic effect is generally accepted as one of the central forces involved in driving proteins to adopt a compact globular form. The native structure of most proteins includes a hydrophobic core, where buried hydrophobic residues are sequestered from solvent, and a solvent-exposed surface consisting mostly or entirely of polar and charged residues. The process of folding into a compact conformation that isolates hydrophobic residues from solvent is sometimes referred to as **hydrophobic collapse.** Membrane-integral proteins constitute a notable exception to this rule. These proteins contain one or more transmembrane regions, often primarily helical in structure, that are embedded in a membrane. Because membranes are generally composed largely of hydrophobic carbon and hydrogen atoms, these "surface" helices are actually protected from water molecules and are composed of mostly hydrophobic amino acids.

The importance of isolating hydrophobic residues from solvent is well illustrated by the molecular pathology of sickle-cell anemia. Human hemoglobin, the protein responsible for oxygen transport in our blood, is biologically active as a tetramer consisting of two alpha-globin and two beta-globin chains. The mutation of a single surface residue of the beta-globin protein from a charged glutamic acid residue to a hydrophobic valine residue causes a hydrophobic patch to be present on the solvent-exposed surface of the protein. The hydrophobic effect drives the valine residues to avoid contact with solvent and causes beta-globin molecules to stick to each other. Long chains of hemoglobin result and distort red blood cells from their normal donut-like shape to a characteristic sickle shape. The effect is particularly pronounced when oxygen levels are low (as they are in our extremities and during physical exertion), and the sickle cells become tangled with each other in narrow blood vessels. Pain, anemia, and, ultimately, gangrene result all because of a single amino acid difference.

The exact energetics of the hydrophobic effect and its contribution to protein folding are difficult to calculate. However, most protein folding algorithms that base their calculations on molecular forces include hydrophobic collapse as one of the central forces in driving protein folding.

Disulfide Bonds

When the sulfhydryl (—SH) groups of cysteine residues come into proximity they can become oxidized to form covalent **disulfide bonds,** cross-linking residues that may be far removed from one another in the primary structure of the protein (Figure 7.7). These reduced cysteine residues are sometimes referred to as cystine, and their stabilizing effect on a folded protein structure can be significant. When experimental methods require that a protein be denatured, reducing agents such as β-mercaptoethanol are often used to break any disulfide bonds. As mentioned previously, Anfinsen's seminal work demonstrated that protein structure is specified by sequence. In these experiments, ribonuclease was denatured and then allowed to form disulfide bonds in the presence of a high concentration (8 *M*) of urea. The urea reduced the effect of hydrophobicity on the protein conformation, allowing the formation of disulfide bonds differing from those found in the native conformation. The "scrambled" ribonuclease, with its incorrectly cross-linked cystine residues, had only 1% of the enzymatic activity of ribonuclease in its native conformation.

Active Structures vs. Most Stable Structures

Because of the huge number of degrees of freedom in protein folding, it has remained impossible to evaluate, in general, whether the native state of a protein is actually the most stable (energetically favorable) conformation. It is clear, however, that natural selection favors proteins that are both active and robust. Mutations in a protein's primary structure that reduce structural stability are likely to produce an evolutionary disadvantage, and thus be selected against.

In his prominent 1968 paper, C. Levinthal noted that the number of possible folds for even a modestly sized polypeptide chain is so vast that a random search of all possible conformations would take many years. This observation, which has come to be known as the **Levinthal paradox,** suggests that proteins fold by proceeding from the unfolded structure along a pathway of successively more stable intermediates, until the native state is reached. Whether or not this pathway ends in a conformation of globally minimum energy remains a matter of some debate.

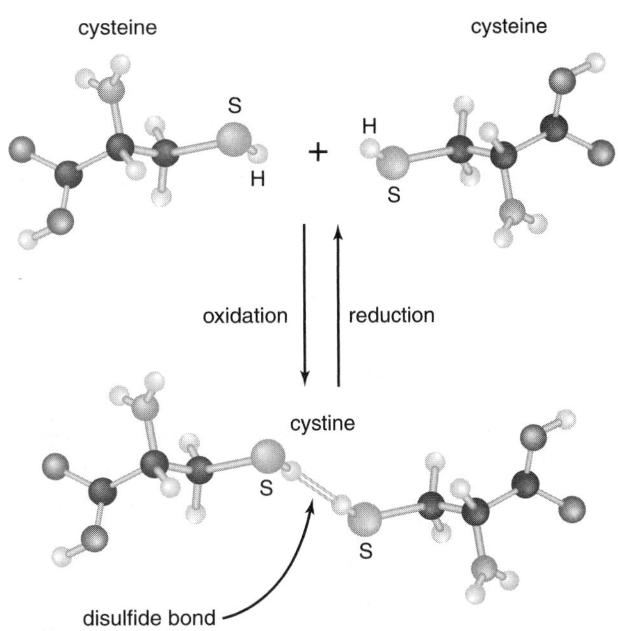

FIGURE 7.7 *The sulfhydryl groups of two cysteine residues are oxidized to form a disulfide bond. Disulfide bonds are generally the only nonbackbone covalent cross-linkages within a protein.*

Algorithms for Modeling Protein Folding

To better understand how the amino acid sequence of a protein determines its unique native conformation, numerous algorithms have been developed to simulate the protein folding process at various levels of abstraction. No algorithm developed to date can determine the native structure of a protein with the same accuracy as experimental methods, such as X-ray crystallography (described in more detail in Chapter 8). Nevertheless, computational models of protein folding have provided novel insights into the forces that determine protein structure and the process by which proteins fold. The accuracy and power of computational protein folding algorithms continue to improve as new optimization and machine learning algorithms are applied to the problem, and as experimental biochemistry learns more about the various forces that contribute to the stability of a folded protein.

Lattice Models

Even the rapidly advancing computing power of modern microprocessors is insufficient to model all of the interactions involved in a folding polypeptide chain for more than a few femtoseconds (1 femtosecond = 10^{-15} seconds). As a result, most computational protein folding methods simplify various aspects of the protein in order to make the computations more tractable.

One such approach is to limit the degrees of conformational freedom available to the protein backbone. Instead of allowing all physically possible protein conformations, the alpha carbons are restricted to positions lying on a two- or three-dimensional grid (or lattice). This simplification considerably reduces the number of conformations a protein can adopt. As a result an exhaustive search of the conformational space can be performed for modest-sized polypeptides, revealing the global minimum energy conformation. For larger polypeptides, simplified lattice models allow nonexhaustive methods to sample a much greater proportion of all possible conformations, resulting in more accurate estimates of the global minimum energy conformation.

The best studied simple lattice model is the **H-P** (for **hydrophobic-polar**) **model.** The H-P model further simplifies the protein by representing each amino acid residue as a single atom of fixed radius. Each atom in the representation has one of two types: hydrophobic or polar. Figure 7.8 shows two- and three-dimensional H-P model representations for a short polypeptide. By convention, the N-terminal amino acid is placed at the origin of the coordinate system, and the following residue is placed at coordinates (1,0).

Scoring in the H-P model is based on hydrophobic contacts. To evaluate a particular conformation in the H-P model, the number of H-to-H contacts in the grid is counted. Each H-to-H contact is assumed to provide an energy contribution of –1, except for those involving two residues that are contiguous in the primary structure of the polypeptide. (Since these H-to-H contacts are present in every possible conformation, they are excluded for the sake of simplicity.) The optimal conformation is the one with the most H-to-H contacts. As it turns out, maximizing the number of H-to-H contacts is generally achieved by forming a hydrophobic core containing as many H residues as possible, relegating the P residues to the surface of the polypeptide. The score for both the 2D and 3D conformations shown in Figure 7.8 would be –3.

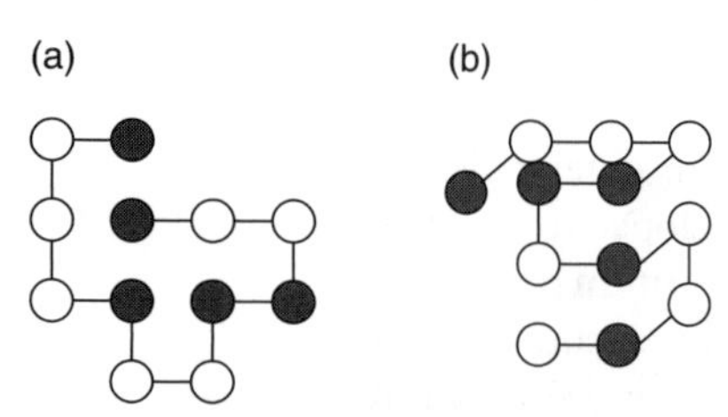

FIGURE 7.8 *(a) Two-dimensional and (b) three-dimensional H-P model representations for a 12-residue polypeptide. Hydrophobic residues are shown in black and polar residues are white.*

The assumption that hydrophobic collapse is the only significant factor in protein folding is obviously artificial, as are the conformational constraints imposed by a two- or three-dimensional lattice. Nevertheless, exhaustive searches of H-P models have provided some intriguing insights into the mechanisms of protein folding. For example, after examining optimal and near-optimal lattice structures for a variety of polypeptides, K. Dill suggested the **hydrophobic zipper** (Figure 7.9) as a possible mechanism for the formation of secondary structure.

As the size of a polypeptide chain grows, exhaustive searching of all possible conformations, even on a 90° lattice, becomes intractable. In this case, optimization (or search) algorithms must be used to find near-optimal conformations. Various advanced computational methods have been employed, including evolutionary computation, simulated annealing, Monte Carlo methods (described in Chapter 8), branch and bound search, and machine learning approaches.

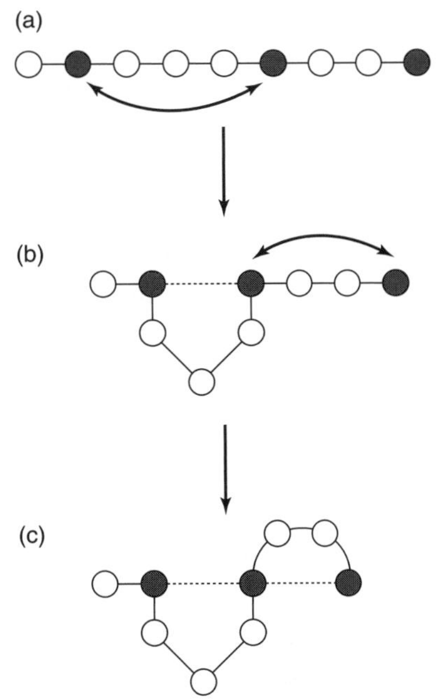

FIGURE 7.9 *The hydrophobic zipper mechanism. (a) Hydrophobic residues (shown as black filled circles) on the protein chain move together as a result of solvent interactions, causing intervening polar residues to form a loop as seen in (b). As the process is repeated (c), the basic structure of an anti-parallel beta sheet is formed.*

A primary consideration in implementing a search algorithm for lattice-based protein folding models is how to represent a particular configuration. The simplest approach is to place the first residue on the lattice at position (0,0), and then represent the direction moved for each subsequent residue. Using this **absolute direction representation** for a 2D model, the choices at each position are up, right, left, and down (U, R, L, D), while the 3D model includes up, right, left, down, back (increasing *Z*-axis values), and forward (decreasing *Z*-axis values) or (U, R, L, D, B, F). For the two-dimensional configuration in Figure 7.8, the representation using this approach would be (R, R, D, L, D, L, U, L, U, U, R), whereas the 3D configuration would be represented as (R, B, U, F, L, U, R, B, L, L, F).

We can reduce the number of possible choices at each position by using a **relative direction representation,** which represents the turns taken by the main chain for each residue. In the case of a two-dimensional square lattice model, each residue after the second has three options, left, right, and forward (often represented as L, R, and F); in a three-dimensional square lattice, the options are left, right, forward, up, and down (L, R, F, U, D). For this approach, we must not only keep track of the current position, but also which direction the current residue is "facing." For the 2D model, the first residue is placed at grid position (0,0) facing to the right. That is, if the first move is F, then the second residue will be placed at position (1,0). Thus, the 2D configuration shown in Figure 7.8 would have the representation (F, F, R, R, L, R, R, L, R, F, R). For the 3D model, the first residue is placed at position (0,0,0), facing to the right. As we move, we must keep track of which direction the current residue is facing, and which way it currently considers to be "up." Using this representation, the 3D configuration in Figure 7.8 would be represented as (F, L, U, U, R, U, U, L, L, F, L).

A key difficulty that arises in using either of these direction-based representations is that some of the configurations generated will place two residues in the same position. For example, for the relative (turn-based) representation, any 2D configuration that begins with (L, L, L, L) will result in two residues placed at the origin (0,0), resulting in a **bump,** or steric collision. Several approaches have been evaluated for dealing with bumps in conformational search. The simplest approach is to assign a very high energy to any configuration with bumps. Since the search algorithm is looking for low-energy configurations, configurations with bumps will be quickly eliminated from the search. Unfortunately, this draconian approach can hinder the search process by throwing away otherwise favorable conformations that might lead to a low-energy state if the bumps were resolved.

Alternate approaches include using local optimization strategies to resolve bumps before a configuration is scored, and using alternate representations that do not introduce configurations with bumps into the search. An example of such a representation is preference ordering. In this representation, each residue is assigned a permutation of all possible directions, rather than a single direction. For example, in a two-dimensional model, a single residue might be assigned the permutation {L, F, R}. This indicates that the most preferred direction is left; however, if a move to the left would result in a collision, forward is the next most preferred direction, followed by right. While this sort of representation can still introduce bumps (when a move in any direction results in a bump), they are introduced with much less frequency than in an absolute representation. Combinations of preference-ordered representations and local conformational search can be used to design representations that never introduce conformations containing bumps.

Off-Lattice Models

Moving off the lattice permits a protein model to adopt more realistic configurations. Using a complete backbone model and allowing phi and psi to adopt any values in the allowable regions of the Ramachandran plot, off-lattice folding models have produced configurations for small polypeptides that match experimentally observed configurations very closely. Error for off-lattice models is usually measured in terms of the C_α root mean squared deviation (RMSD) between the predicted and observed conformations. The C_α RMSD is simply the sum of the squared Euclidean distance between each C_α atom position in the predicted conformation and the observed conformation, when the two conformations are superimposed.

Improvements in the realism of protein models, however, come at the cost of added complexity. Off-lattice protein folding models may include alpha carbons only, all backbone atoms, or even all backbone and side-chain atoms. Backbone conformations are usually represented in terms of the phi and psi angles of each C_α atom. Side chains, if included, can be rigid, semiflexible, or fully flexible. For rigid side chains, the conformations of side chains in X-ray crystallographic structures are observed, and the most common conformation of each amino acid type is adopted as the only possible conformation for that type of amino acid.

For semiflexible side chains, a similar empirical method is used. Side-chain conformations in a set of X-ray structures are observed and the conformations are partitioned into similarly shaped groups. The average conformation of each group is called a **rotamer.** In a semiflexible model, each side chain is allowed to adopt any of its most commonly observed rotamers. This allows several possible conformations for each side chain, while reducing search complexity by disallowing conformations not commonly observed for a particular amino acid.

Energy Functions and Optimization

Off-lattice protein models, particularly less abstract models that include more backbone and side-chain atoms than the simple H-P model, require more sophisticated energy functions. In addition to hydrophobic contacts, energy functions for protein folding may consider hydrogen bonding, formation of disulfide bridges, electrostatic interactions, van der Waals forces, and solvent interactions. Since the relative contributions of each of these forces has remained difficult to calculate experimentally, the formulation of appropriate composite energy functions for protein folding remains an active field of study.

Two general approaches are used for formulating energy functions. The theoretical approach is based on the contributions of hydrogen bonds, electrostatic interactions, and other forces to the overall stability of a folded protein. The objective is to derive an approximated energy function, or force field, for which the crystallographically observed conformations of known proteins represent a minimum energy state. An example of such a force field might appear as follows:

$$\Delta G = \Delta G_{\text{van der Waals}} + \Delta G_{\text{H-bonds}} + \Delta G_{\text{solvent}} + \Delta G_{\text{Coulomb}}$$

The search for reliable energy functions is essentially a problem in molecular mechanics, and numerous competing functions have been devised.

While it makes sense to model proteins based on the physical forces and energies that actually drive their folding, these sorts of *ab initio* approaches have met with limited success for several reasons. First, the exact forces that drive the folding process and their various interactions are not yet well understood. In addition, these sorts of approaches, which attempt to model interactions among all atoms in a protein and the surrounding solvent, are often too computationally expensive to be feasible for realistic-sized polypeptides. An alternative to the *ab initio* approach is to devise an empirical pseudoenergy function based on the observed conformations of other proteins.

To formulate an empirical energy function, a set of X-ray crystallographic protein structures is selected (X-ray crystallography is discussed in detail in Chapter 8), and the three-dimensional neighbors of each amino acid are examined. A scoring matrix based on the relative positions of the various amino acids is thus formulated. For example, the number of times a serine residue and a threonine residue occur within 3.6 Å might be noted in the scoring matrix. To evaluate the empirical "energy" of a putative protein conformation, the neighbors of each residue in the protein are examined. Those local conformations that were found to be common in the example database receive low-energy scores. Uncommon local conformations receive higher, less favorable energy values. For example, if a particular serine residue is found to have three neighboring residues within 6 Å, an aspartate, a histidine, and a glutamate, and the scoring matrix indicated that Asp, His, and Glu were commonly found in proximity to Ser in the database of example protein structures, then the serine residue would receive a low-energy score. If, however, it was found that

Ser and Glu rarely occur in proximity, then the Ser residue might receive a higher empirical energy value. The local values are then summed over the entire protein to compute the global empirical energy. In effect, empirical energy functions favor conformations similar to the observed conformations of known proteins, while penalizing novel or unusual conformations.

In summary, formulation of an algorithm for protein folding consists of numerous steps. First, a protein model must be selected. Abstract models allow fast calculation and exhaustive searches, but cannot approximate actual protein conformations well. More exact models are realistic, but require expensive energy calculations to evaluate. Once a model is selected, a representation must be devised to represent each possible conformation. For lattice models, simple representations that encode which direction to move next are sufficient. For more exact models, the phi and psi angles of each C_α atom are often used. Next, a scoring function must be chosen to evaluate the favorability of a given conformation. Finally, an optimization method must be selected to search through all possible conformations for the structure that represents a global minimum for the energy function used.

An innovative approach to dealing with the computational complexity of *ab initio* methods has been devised by V. Pande. The Folding@Home program, which acts as a screensaver, uses idle CPU cycles to perform protein folding calculations. The calculations required to model a particular protein are divided into parts and distributed via the Internet to machines running the Folding@Home code. The results from each machine are sent back to the server, where they are combined and analyzed. By using the distributed computing power of the Internet, the Folding@Home algorithm can perform *ab initio* modeling of large polypeptides, a task that would be otherwise intractable for all but the largest supercomputers and networks. To participate in the Folding@Home project, you must be running a Windows, Linux, or Mac OS and be connected to the Internet. More information can be found at http://foldingathome.stanford.edu.

Structure Prediction

While protein folding models have helped molecular biologists and biochemists to understand the process of protein folding and the forces involved, no current *ab initio* protein folding algorithm is able to obtain very high accuracy (<3.0-Å backbone RMSD from the experimental structure) for large protein structures. For applications such as drug lead discovery and ligand design (discussed further in Chapter 8), a very accurate picture of the active site of a protein is required. Proteins often bind their ligands with such specificity that a deviation of less than an angstrom in the position of a key main-chain atom can result in a significant reduction in binding affinity. When the tertiary structure of one or more proteins similar in primary structure to a target protein is known, the target protein can be modeled, often with a high degree of accuracy, using comparative techniques.

Comparative Modeling

Comparative modeling, sometimes called homology modeling, seeks to predict the structure of a target protein via comparison with the structures of related proteins. The method relies on the robustness of the folding code. That is, small changes to the amino acid sequence of a protein usually result in minimal change to the tertiary structure of the protein. Numerous sequence changes must generally accumulate before a radical change in native conformation results. A generalized process for comparative modeling of a target protein proceeds as follows:

1. **Identify a set of protein structures related to the target protein.** Since the three-dimensional structure of the target protein is unknown, similarity is based on sequence. Sequence database search tools such as BLAST and FASTA are used to identify the related structures. Since these structures will serve as a template for the model, they are referred to as template structures.
2. **Align the sequence of the target with the sequences of the template proteins.** A multiple alignment tool, such as CLUSTALW, is used to generate the alignment. The multiple alignment allows us to identify regions of the target that are highly conserved across all template structures and those that are less conserved. When the sequence identity between the target and templates is less than ~30%, automated multiple alignment methods may not provide high-quality alignments. In this case, the alignments must be adjusted manually. For example, it is preferable to move gaps out of secondary structure elements and into surface loops, where they are less likely to disrupt the model structure.
3. **Construct the model.** Several methods may be employed to build the model structure. One of the most common is to superimpose the template structures and find the structurally conserved regions. The backbone of the template structure is then aligned to these conserved fragments of structure, forming a core for the model. When the structures of the template proteins diverge, methods such as secondary structure prediction for the target protein, sequence similarity, and manual evaluation must be used to select the correct structure for the model. Since the structures of the templates are far more likely to diverge in loop regions than in regions of secondary structure, loops are generally modeled separately after the core.
4. **Model the loops.** A number of computational methods are available for modeling loops. The two most common methods are (1) select the best loop from a database of known loop conformations, and (2) perform conformational search and evaluation. The second method is similar to a limited protein folding approach. While there are numerous methods to choose from in modeling loops, it is difficult to obtain an accurate conformation for loops longer than six residues.
5. **Model the side chains.** Once the backbone model has been constructed, the positions of the side-chain atoms must be determined. Again, a variety of

methods are commonly employed, ranging from rotamer library search to limited molecular dynamics approaches.

6. **Evaluate the model.** Various software packages are available to evaluate the quality of a protein structure. Examples include PROCHECK, WHAT CHECK, Verify-3D, and others. Validation algorithms usually check for anomalies such as phi/psi combinations that are not in the allowable regions of the Ramachandran plot, steric collisions, and unfavorable bond lengths and angles. Once structural anomalies are identified, the model is usually adjusted by hand, if possible, to correct the problems reported.

While comparative modeling involves numerous computational methods, it is nevertheless a manual process, involving expert intervention and decision making at each step. Automated methods are currently under development, but the need to produce very high-quality models for ligand design, active site analysis, and so on, suggests that the process will require a significant manual component for the near future. Nevertheless, comparative modeling remains the best available method for the construction of high-quality models for proteins of unknown structure.

Threading: Reverse Protein Folding

An approximate model of a protein's tertiary structure can be constructed quickly by finding another protein that assumes roughly the same conformation. If we assume a particular conformation, then we can use the same evaluation methods used in protein folding algorithms to determine whether the conformation is favorable. The process of assuming a particular conformation and evaluating its favorability is called **protein threading.** Threading is sometimes compared to protein folding in reverse, since you start by assuming a particular fold, and then evaluate the quality of the resulting structure.

A common application of threading algorithms is to determine which fold family a particular sequence is likely to belong to. Several hierarchical databases have been developed that identify groups of similarly folded proteins (e.g., CATH, SCOP, LPFC, Pclass, FSSP, and others) at various levels of abstraction. These databases group proteins of similar structure into categories such as fold families, superfamilies, and classes. To identify which fold family or superfamily a protein belongs to, you can take an average structure representing all of the structures in a family and evaluate the quality of the resulting structure if the target protein were to assume this conformation.

The evaluation is repeated for each fold family, and the conformation with the most favorable score is selected as the likely fold family for the target sequence. Assignment of a protein sequence to a particular fold family not only gives a rough approximation of the native structure of the protein, but also provides information about possible functions of the protein and its relationship to other proteins and biological pathways.

Predicting RNA Secondary Structures

Unlike DNA, which most frequently assumes its well-known double-helical conformation, the three-dimensional structure of single stranded RNA is determined by the sequence of nucleotides in much the same way the protein structure is determined by sequence. RNA structure, however, is less complex than protein structure and can be well characterized by identifying the locations of commonly occurring secondary structural elements.

For RNA, the elements of secondary structure are quite different than those for proteins. Like DNA, complementary base pairs of RNA will form hydrogen bonds, inducing a helical structure. However, in single stranded RNA, the base pairing occurs within a single RNA molecule, forming base-paired **stem** regions. Where the RNA chain reverses directions to allow this base pairing, a **hairpin turn** is formed. When a small number of bases along the RNA chain are not complementary, a small bulge or a larger loop region may form. Figure 7.10 illustrates the types of secondary structure that can be formed by an RNA molecule. The most difficult type of RNA structure to predict is the **pseudoknot,** where bases involved in a loop pair with other bases outside the loop. Because of the difficulty in pseudoknot prediction, many of the early algorithms for prediction of RNA secondary structure ignored them entirely, predicting the rest of the secondary structure elements as if no pseudoknot regions were present.

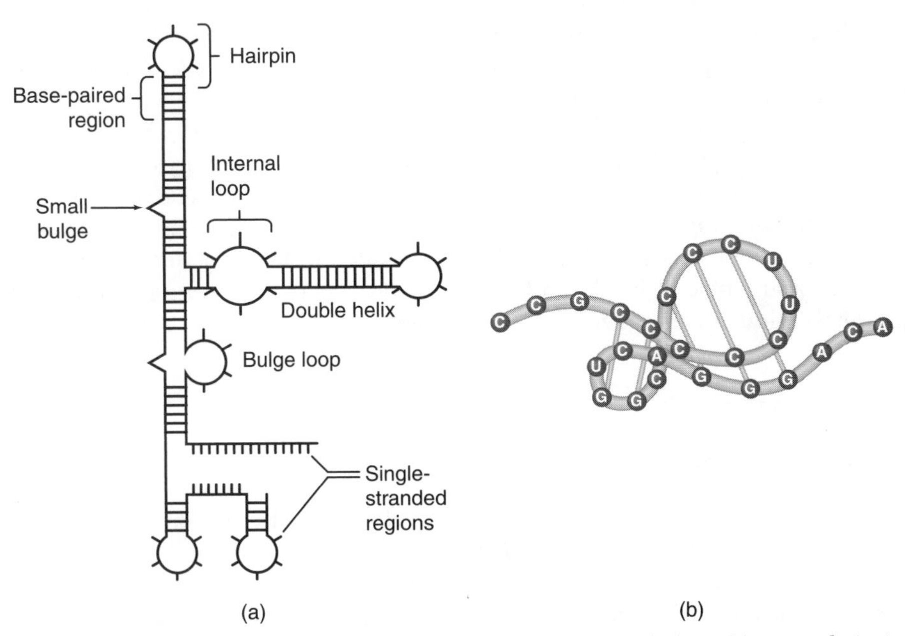

FIGURE 7.10 *RNA secondary structure elements, including (a) stems, hairpin turns, bulges/loops, and (b) pseudoknots.*

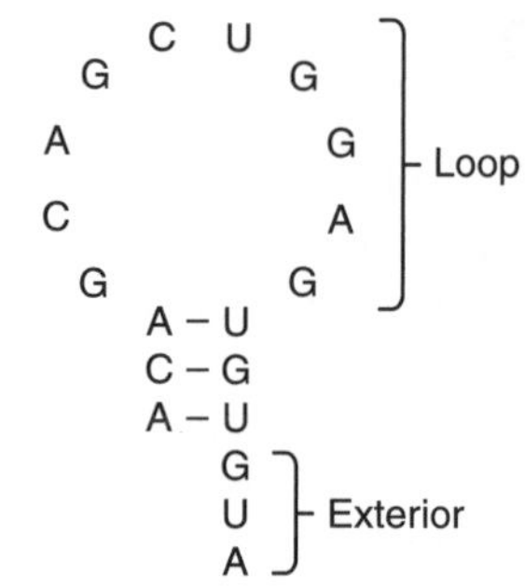

FIGURE 7.11 *Nearest neighbor energy rules: All of the nucleotides in the region labeled Loop can potentially interact, so all pairs within the loop must be considered when calculating the energy of this structure. The nucleotides in the region labeled Exterior, however, cannot pair with any of the nucleotides in the loop region without forming a pseudoknot. Thus, pairwise interactions between these bases and those in the loop are ignored by nearest neighbor energy rules.*

Various approaches have been devised for predicting the secondary structure of RNA molecules. Many of these methods attempt to minimize the free energy of the folded macromolecule, thus searching for the most stable structure. One of the most popular packages, Zuker's Mfold program, uses a set of **nearest neighbor energy rules** to calculate the energy of the structure. The rules are called "nearest neighbor" rules because the RNA structure being evaluated is broken into interacting regions and energy calculations are performed only on "neighbors"—base pairs that can potentially interact with one another. Thus, pairwise interactions that cross base-paired stem regions do not have to be considered (see Figure 7.11), speeding up the calculations considerably.

Because RNA is the intermediary language between DNA and proteins, accurate prediction of RNA secondary structure is important for understanding gene regulation and expression of protein products. Furthermore, many RNAs have been found to bear catalytic properties of their own. These RNAs, now referred to as **ribozymes,** have been implicated in the splicing of tRNA molecules, the activity of ribosomes, the processing of eukaryotic hnRNAs, and many other functions. While they usually occur in the context of a protein–RNA complex, it has been demonstrated that ribozymes can exhibit catalytic activity even without their protein partners under some circumstances. Often RNA acts as the structural scaffolding for DNA, RNA, and polypeptide processing reactions. In addition, since some viruses (such as HIV) are encoded in RNA, understanding the secondary structure of RNA can help in the process of discovering and testing pharmaceutical agents against these pathogens.

Chapter Summary

Proteins are complex macromolecules that display a diverse range of three-dimensional structures. Scientists are just beginning to understand how it is that the sequence of amino acids that makes up a protein chain encodes a unique three-dimensional structure. The secondary structure of a protein can be predicted with reasonably high accuracy (>75%) using neural networks, hidden Markov models, and other advanced computational techniques. Tertiary and quaternary structures are far more difficult to predict, and a variety of algorithms have been devised to solve the problem of protein fold prediction at various levels of abstraction. Protein folding algorithms, comparative modeling methods, and threading are advancing our knowledge and understanding of the process of protein folding, while producing increasingly accurate structural predictions. No less important is the prediction of RNA secondary structure, which, like protein secondary structure prediction, can be achieved with good accuracy using neural networks, hidden Markov models, and other computational methods.

Readings for Greater Depth

Numerous methods are available for prediction of protein secondary structure, including PhD, Jpred, GOR, NNSP, Predator, DSC, MULPRED, and many others. The web page for Jpred provides a brief summary and link to these algorithms and others: http://jura.ebi.ac.uk:8888/refs.html.

Chou and Fasman's well-known technique for the prediction of secondary structure is described in P. Prevelige, Jr., and G. D. Fasman, 1989, Prediction of secondary structure, in G. Fasman (ed.), *Prediction of Protein Structure and the Principles of Protein Conformation*, pp. 1–91, Plenum Press, New York.

The GOR method for secondary structure prediction is described in J. Garnier, D. J. Osguthorpe, and B. Robson, 1978, *J. Mol. Biol.* **120:** 97–120, as well as in J. Garnier and B. Robson, 1989, The GOR method for predicting secondary structures in proteins, in G. Fasman (ed.), *Prediction of Protein Structure and the Principles of Protein Conformation*, pp. 417–465, Plenum Press, New York.

The famous Levinthal paradox was first described in C. Levinthal, 1968, Are there pathways for protein folding?, *J. Chem. Phys.* **65:** 44–45; which can be obtained online at http://brian.ch.cam.ac.uk/~mark/levinthal/levinthal.html.

Anfinsen's experiments on protein renaturation are described in C. B. Anfinsen, 1973, Principles that govern the folding of protein chains, *Science* **181:** 223–230.

John Moult provides a detailed comparison of the two basic methods for formulating energy functions for protein folding in J. Moult, 1997, Comparison of database potentials and molecular mechanics force fields, *Curr. Opin. Structural Biol.* **7:** 194–199.

Worthwhile summaries of secondary structure prediction methods can be found at http://cubic.bioc.columbia.edu/predictprotein and http://genamics.com/expression/strucpred.htm.

A step-by-step summary of the process of comparative structure prediction is provided at http://guitar.rockefeller.edu/~andras/watanabe.

Details of Zuker's Mfold algorithm can be found in M. Zuker, D. H. Mathews, and D. H. Turner, 1999, Algorithms and thermodynamics for RNA secondary structure prediction: A practical guide, in J. Barciszewski and B. F. C. Clark (eds.), *RNA Biochemistry and Biotechnology*, NATO ASI Series, Kluwer Academic Publishers, Boston, and at http://bioinfo.math.rpi.edu/~zukerm/seqanal.

Questions and Problems

* **7.1** The charges associated with proteins often play important roles in their biochemical separation from other proteins also found in cells. Consider a polypeptide that is found to bind tightly to a column packed with a cation-exchange resin at pH 3.5. Explain why it might be possible to wash this protein off the column by passing a buffer of pH 8 through the column.

7.2 Using the Chou-Fasman algorithm and the parameters in Table 7.1, predict the regions of alpha helices and beta strands for the following sequence:

```
CAENKLDHVADCCILFMTWYNDGPCIFIYDNGP
```

* **7.3** For the sequence in Question 7.2, use the Chou-Fasman method to predict the hairpin turns.

7.4 Assuming an energy contribution of –1 for each nonbackbone hydrophobic contact, determine the energy score for the following 2D H-P model configuration:

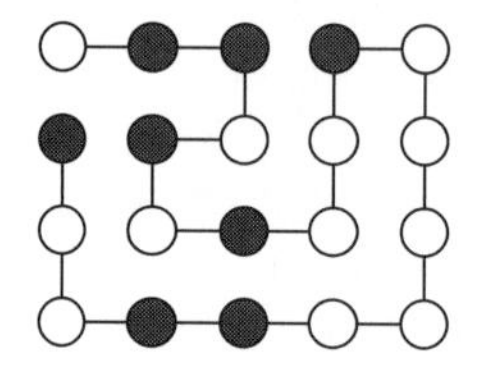

* **7.5** Draw a 2D H-P configuration for a peptide with an energy score of –7.

7.6 Assuming a relative direction representation using L = left, R = right, and F = forward, what is the representation for the 2D configuration in Question 7.4?

* **7.7** A preference ordering representation using L = left, R = right, and F = forward is used along with a 2D H-P model. Draw the conformation corresponding to the following representation (assume all residues are polar): (FLR), (RFL), (LFR), (LRF), (LFR), (LFR), (LFR), (LFR), (LRF), (LFR), (LRF), (LFR).

CHAPTER

8 Proteomics

If a little knowledge is dangerous, where is a man who has so much as to be out of danger?

Thomas Henry Huxley (1825–1895)

While a genome is the sum total of an organism's genetic material, a **proteome** is the sum total of its proteins. The very nature of genes (their simple chemical makeup and their ability to serve as templates for exact copies of themselves) has made them comparatively easy to study and amenable to automated analysis. The nature of proteins (with their 20 building blocks, complex chemical modifications, and barriers to duplication) is very different. Many detractors of the numerous completed and ongoing genome projects have questioned the wisdom of investing so much in those DNA-level efforts when they actually give us very little knowledge about how an organism actually works. Proteins are what get things done for cells, and now that biology is beginning to enter into what many are already calling the "post-genome era" the larger task of characterizing proteomes looms on the horizon.

The previous chapter described the still imprecise process of inferring a protein's three-dimensional structure from just the nucleotide sequence of the gene that codes for it. This chapter highlights the difficult task of simply cataloging a cell's set of proteins as well as the problems of determining their relative abundance, modes of interaction, and roles that they play. Some laboratory and computational tools do exist, but much still needs to be learned before proteomes can be reverse engineered back to the level of genomes and the bioinformatics story is made complete.

From Genomes to Proteomes

Despite their exquisite sensitivity and capacity to generate staggering amounts of data, state-of-the-art gene expression analyses such as those described in Chapter 6 give disappointingly little information about what proteins are present in cells, let alone what those proteins do and how they function. First, the longevity of an mRNA and the protein it codes for within a cell are usually very different—the correlation between the relative abundance of an mRNA and the relative abundance of its corresponding protein within any given cell is routinely less than 0.5. Second, many proteins are extensively modified after translation in a wide variety of ways including proteolytic cleavage, glycosylation, and phosphorylation (as described later in this chapter). These modifications almost invariably alter the activity of proteins and clearly happen differently in different tissues and circumstances, yet the amino acid sequence of a protein explains only some of the molecular features that are scrutinized prior to modification. Third, many proteins are not functionally relevant until they are assembled into larger complexes or delivered to an appropriate location either inside or outside of a cell. Amino acid sequences alone only hint at what those interactions and final destinations might be.

These difficulties in inferring the population of proteins in a cell and the roles that those proteins play are compounded by the fact that proteins themselves are much less amenable to direct analysis than are nucleic acids. Proteins require much more careful handling than DNA because their functionally important tertiary structures can be easily altered when they come in contact with an inappropriate surface or environment. Also, as described in Chapter 1, the ability of nucleic acids to specifically base pair with other strings of nucleotides makes DNA identification a fairly simple task. Protein identification is much more difficult and requires complicated mass spectrometric analyses in conjunction with sophisticated software (described in more detail below) or the generation of specific antibodies. Finally, most analyses of both nucleic acids and proteins rely on the ability to manipulate large numbers (typically billions) of identical molecules at some point in the process. Generating large numbers of copies of any given gene is fairly easy due to their ability to serve as a template for their own amplification (usually PCR in an efficient, controlled, and cell-free environment). Obtaining large numbers of protein molecules for analysis almost always requires inefficient and laborious chemical isolation from large numbers of living cells.

In the final analysis, it is fair to say that "genes were easy" when considering the challenges facing proteomics researchers. None of the obstacles to proteome analysis are insurmountable and workable solutions are already in place for many. The power that should come from the knowledge of an organism's proteome promises to make the remaining challenges worth facing. Proteomic insights have already played important roles in understanding the molecular basis of diseases such as cystic fibrosis, converting cells to molecular factories, improving the efficiency of genetically engineered organisms, and designing new drugs. The real work of bioinformatics has just begun.

Protein Classification

Indexing and cataloging proteomic data are challenging tasks due to the wide variety of different proteins cells need to manipulate their internal environments. Several alternative methods of classifying proteins have been suggested to help with that process and to facilitate comparisons between different organisms and cell types. The oldest systematic method, that of the International Enzyme Commission, is based on protein function and assigns proteins to one of six different categories. Alternative means of classification include ones based on evolutionary history (dividing all proteins into 1 of about 1,000 different homologous families) and structural similarity.

Enzyme Nomenclature

Rapid growth in the number of known enzymes during the 1950s made it necessary for a set of widely accepted naming conventions to be put in place. Prior to the formation of the International Enzyme Commission in 1955 it was not uncommon for a single enzyme to be known by several different names, while in other instances the same name was given to distinctly different enzymes. Further, many of the names conveyed little or no idea of the nature of the reactions catalyzed. By 1965 a systematic approach to naming enzymes was suggested that

TABLE 8.1 The main enzyme classes according to the International Enzyme Commission.

Main Class	Type of Reaction Catalyzed
1. Oxidoreductases	Oxidation-reduction reactions of all types
2. Transferases	Transferring an intact group of atoms from a donor to an acceptor molecule
3. Hydrolases	Cleaving bonds hydrolytically
4. Lyases	Cleaving bonds by means other than hydrolysis or oxidation
5. Isomerases	Interconversion of various isomers
6. Ligases	Bond formation due to the condensation of two different substances with energy provided by ATP

divided all enzymes into six main classes on the basis of the general type of reaction they catalyze (Table 8.1).

By using a numbering system throughout the scheme, each enzyme (i.e., chitinase) can be assigned a numerical code, such as 3.2.1.14, where the first number specifies the main class, the second and third numbers correspond to specific subclasses, and the final number represents the serial listing of the enzyme in its subclass. For example, the enzyme traditionally known as alcohol dehydrogenase is identified as 1.1.1.1 (main class: oxidoreductase; class: acting on the CH—OH group of donors; subclass: with NAD or NADP as an acceptor; alcohol dehydrogenase: the first of 269 enzymes entries in this category). By the same token, DNA-directed RNA polymerases are identified as 2.7.7.6 (main class: transferases; class: transferring phosphorus-containing groups; subclass: nucleotidyltransferases; DNA-directed RNA polymerases: the sixth of 60 entries in this category). New enzymes are constantly being added to the list of thousands assigned these identifiers, and updates are routinely available at the International Enzyme Commission's web site (http://www.chem.qmw.ac.uk/iubmb/enzyme).

Families and Superfamilies

Sequence similarities among the many thousands of proteins for which an amino acid sequence is available suggest that all modern-day proteins may be derived from as few as 1,000 original proteins. It is unclear if this fairly small number is dictated more by the physical restraints on the folding of polypeptide chains into three-dimensional structures (Chapter 7), by the fact that these different proteins provided sufficiently varied structural and chemical properties such that no others have been needed during the course of evolution, or by some combination of these two possible explanations. One of the strongest arguments for an evolutionary explanation came in a study published in 1991 by Dorit, Schoenbach, and Gilbert (see the suggested readings at the end of this chapter). They suggested that exons themselves correspond closely to the functional domains of proteins and that all proteins are actually derived from various arrangements of as few as

1,000 to 7,000 exons. Regardless of the basis for the similarities, though, sequence alignments and database similarity searches such as those described in Chapter 2 are often employed to discover familial relationships between different proteins. These relationships have subsequently been extremely helpful in attempts to predict protein structures since the shapes of proteins seem to be under tighter evolutionary constraint than their specific amino acid sequences.

By definition, proteins that are more than 50% identical in amino acid sequence across their entire length are said to be members of a single **family.** By the same token, **superfamilies** are groups of protein families that are related by lower but still detectable levels of sequence similarity (and therefore have a common but more ancient evolutionary origin). All proteins can be further categorized on the basis of their predominant secondary structural features as shown in Table 8.2. Features include membrane/cell surface proteins; mainly alpha; mainly beta; and both alternating alpha/beta structures and alpha+beta structures.

Several databases that group proteins in this way have been assembled with varying combinations of manual inspection and automated methods. They include SCOP (Structural Classification of Proteins; http://pdb.wehi.edu.au/scop), CATH (Class, Architecture, Topology and Homologous superfamily; http://www.biochem.ucl.ac.uk/bsm/cath_new/index.html), and FSSP (Fold classification based on Structure-Structure alignment of Proteins; http://www.hgmp.mrc.ac.uk/Databases/fssp). All intend to provide a detailed and comprehensive description of the structural and evolutionary relationships between all proteins whose structures are known through X-ray crystal analysis (described later).

Folds

Whereas protein families have clear evolutionary relationships, and protein superfamilies have probable evolutionary relationships, proteins are said to have a common **fold** if they have the same major secondary structures in the same arrangement and with the same topological connections. Different proteins with

TABLE 8.2 Structural classification of proteins.

Class	Folds	Superfamilies	Families
All alpha proteins	144	231	363
All beta proteins	104	190	303
Alpha and beta proteins (α/β)	107	180	409
Alpha and beta proteins ($\alpha+\beta$)	194	276	428
Multidomain proteins	32	32	45
Membrane and cell surface proteins	11	17	28
Small proteins	56	81	123
Total	648	1,007	1,699

Note: According to the October 2001 release of the SCOP Database (v. 1.57). A total of 14,729 protein database entries (http://www.rcsb.org/pdb) were analyzed and divided into 35,685 domains.

the same fold often have peripheral elements of secondary structure and turn regions that differ in size and conformation. In many respects, the term *fold* is used synonymously with *structural motif* but generally refers to larger combinations of secondary structures—in some cases, a fold comprises half of a protein's total structure.

Proteins placed together in the same fold category may not have a common evolutionary origin though they may be the result of **exon shuffling,** in which proteins with new functions are created by the process of recombining exons corresponding to functional domains of existing genes at the level of DNA. Alternatively, the structural similarities could arise just from the physics and chemistry of proteins favoring certain packing arrangements and chain topologies.

Experimental Techniques

As was the case with genome analysis, much of proteome analysis is limited by the experimental techniques that are currently available. Unfortunately, from the perspective of proteomics, the nature of proteins makes laboratory analysis especially difficult and much less precise than what is available for genome analysis. Since it plays such an important role in the field's progress, it is especially important that bioinformaticians appreciate the few strengths and significant weaknesses of proteomic methods.

2D Electrophoresis

The workhorse laboratory technique of proteomics is two-dimensional (2D) gel electrophoresis. As the name implies, the approach involves two different gel electrophoretic separations that give rise to a pattern of protein dots with distinctive x and y coordinates such as those shown in Figure 8.1. The first separation involves **isoelectric focusing** (IF) of proteins. This step exploits differences in the pI value of proteins (the pH at which the net molecular charge is zero; see Chapter 7) and is capable of separating proteins with extremely small differences in pI (as little as 0.0025 pH units). The relatively recent availability of immobilized pH gradient (IPG) strips has significantly standardized the first dimension of separation, which had been fraught with artifacts and variability when the gradients had been made by hand for each 2D gel. After the first dimension is run, the strips are laid on a sodium dodecyl sulfate (SDS) polyacrylamide gel to separate the molecules by molecular weight in much the same way that nucleic acids can be size fractionated (see Chapter 1). The SDS in the gel is actually a detergent that binds uniformly to all proteins in the gel and confers on them a negative charge that allows the electric current to pull them through the gel's molecular sieve.

The combined resolution of this IF-SDS procedure is outstanding and routinely yields autoradiograms like the ones shown in Figure 8.1. Clear differences in the protein populations of different cell types such as human kidney

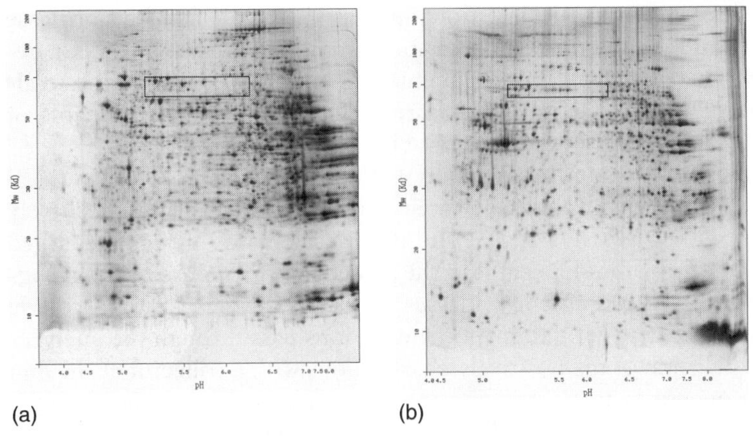

F I G U R E 8.1 *2D gels of proteins in two different types of human cells. Each spot corresponds to a single different protein present in cell extracts from each type of tissue. (a) A 2D gel generated from human liver cells and (b) 2D gel generated from human kidney cells.*

(Figure 8.1a) and human liver (Figure 8.1b) are readily apparent, and relative amounts of each protein can be determined by quantitating the intensity of each spot. Those differences can be readily correlated with a variety of disease states as well as different stages of development, and a large collection of heavily analyzed 2D gels is available for perusal on the Internet (http://us.expasy.org/ch2d/). Differential labeling of proteins (i.e., radioactively labeling one preparation with ^{15}N and another with ^{14}C) can even allow direct comparisons to be made on a single 2D gel. Several companies have risen to the challenge of generating sophisticated software packages that automate the process of aligning spots between gels and integrating the intensities of each spot. They include Amersham Pharmacia Biotech's ImageMaster and Compugen's Z3.

At first glance 2D electrophoresis might appear to be a method perfectly suited for annotating and tracking the proteome. Literally thousands of proteins can be reproducibly separated in a single experiment. However, the method has several important limitations including the fact that the human genome codes for many tens of thousands of proteins, underrepresentation of membrane-bound proteins due to poor solubility in sample preparations and gels, and relatively insensitive detection methods. The single greatest difficulty associated with 2D gel electrophoresis of proteomes, however, is that of determining exactly which protein is represented by each spot. Those determinations are typically only made after extensive computational analysis of mass spectra data as described below.

Mass Spectrometry

The ability of mass spectrometry to uniquely identify the proteins associated with each spot in a 2D gel has been the driving force behind most recent progress in proteomics. After proteins have been separated on 2D gels, identification usually requires spot excision, enzymatic digestion into peptide fragments with a protease such as trypsin, and then deposition on a substrate for matrix assisted laser desorption ionization (**MALDI**) for subsequent mass spectrometric analysis. An alternative to MALDI is electrospray ionization (**ESI**), and both are capable of identifying and quantifying a wide variety of large biological molecules.

Mass spectrometry itself is a technique that is commonly used by analytical chemists. For proteomic applications, protein fragments are made into positively charged ions by either MALDI or ESI and then accelerated through a charged array into an analyzing tube. The paths of the protein fragments are bent by magnetic fields such that those with low masses are deflected the most and collide with the walls of the analyzer. High-momentum ions pass through the analyzer more quickly. Collision of the positively charged ions with a collector at the end of the analyzer generates an electric current that can be amplified and detected as a series of signal peaks corresponding to a **peptide mass fingerprint.**

Identification of proteins with mass spectrometry requires a marriage between instrumentation and computation, in which a peptide mass fingerprint must be matched up with the theoretical mass spectrum of any of a large number of proteins derived from a genomic database. The problem gets even more complicated when trying to identify a protein from an organism whose genome sequence is not yet complete. Innovative approaches to these complex search problems have been essential, and examples of mass spectrum analysis software include ProFound (http://prowl.rockefeller.edu/cgi-bin/ProFound), which correctly identified the protein in a 2D gel spot after the MALDI mass spectrometric analysis shown in Figure 8.2.

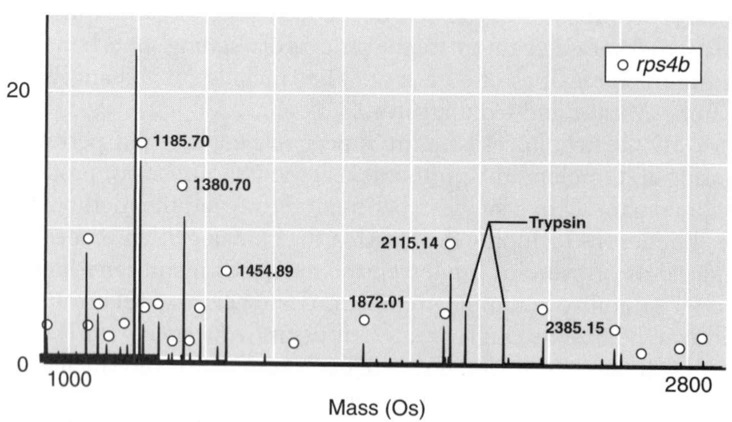

FIGURE 8.2 *MALDI spectrum of the proteolytic products from a tryptic digest of a 30-kDa 2D gel protein spot. Open circles indicate peaks that match with masses of the theoretical tryptic fragments of* rps4b. *Trypsin (the proteolytic enzyme that was used to digest the original spot) self-digestion products are labeled "Trypsin." (Derived from Figure 2 in Zhang and Chait, 2000.)*

Protein Microarrays

Protein microarrays are beginning to get widespread attention because they have the potential to allow large-scale analyses of proteins in much the same way that gene chips have revolutionized transcriptome analyses (see Chapter 6). The concept behind protein chips is very similar to the one that underlies gene chips: Tiny amounts of individual probes are covalently attached to the surface of silica (glass) chips in high-density arrays. Protein extracts from cells are then fluorescently labeled and washed over these chips. Just as with gene chips, amounts of material (now proteins) bound to the probes are determined by laser excitation of the fluorescent labels. In addition to arrays of proteins for detecting protein–protein interactions, protein–compound interactions, and so on, researchers may use arrays of capture probes (i.e., antibodies) that bind proteins in a sample such that their relative expression levels can be detected.

Several hurdles still need to be overcome before these protein chips have the same impact brought about by gene chips. First, unlike DNA sequences with their unique binding characteristics determined exclusively by base pairing rules, it is reasonable to expect that single proteins will interact with multiple, different probes. Also, the binding kinetics (see Appendix 2) of each probe are likely to be subtly different, and differences in signal intensity might be due to differences in binding affinities. Finally, proteins are notoriously sensitive to the chemistry of their environment and the surfaces they encounter, and both the cell extracts and the probes themselves may behave differently than expected when exposed to the testing procedure.

These problems and others should be solvable through the application of computer analyses of the resulting data. In the meantime, researchers are likely to use gene chip arrays first to help them home in on proteins of interest. Once they have narrowed the field, proteomic analyses of small subsets of proteins will be analyzed on custom-made protein chips. In one study, researchers developed methods to array more than 10,000 proteins on a single glass microscope slide and then showed that their arrays could be used to study the enzymatic activity of proteins (*Science*, 2000; **289**: 1760–1763). Notably, they were able to detect the binding of small, drug-like molecules to particular proteins on microarrays.

Inhibitors and Drug Design

One of the most important applications of bioinformatics is the search for effective pharmaceutical agents to prevent and cure human disease. The development and testing of a new drug is a costly and time-consuming undertaking, often spanning as many as 15 years, at a cost of up to $700 million. Functional genomics, structural bioinformatics, and proteomics promise to reduce the labor associated with this process, allowing drugs to be developed faster and at a lower cost.

While the exact steps in the development of a drug vary, the overall process can be broken into two major components: discovery and testing. The testing process, which involves preclinical and clinical trials, is generally not subject to significant enhancement using computational methods. The discovery process, however, is labor intensive and expensive and has provided a fertile ground for bioinformatics research. The discovery process itself can be broken into several components, including target identification, lead discovery and optimization, toxicology, and pharmacokinetics.

The objective of **target identification** is to identify a biological molecule that is essential for the survival or proliferation of a particular disease-causing agent, or **pathogen.** Once a target is identified, the objective of drug design is to develop a molecule that will bind to and inhibit the drug target. Since the function of the target is essential to the life processes of the pathogen, inhibition of the target either stops the proliferation of the pathogen or destroys it. An understanding of the structure and function of proteins is an important component of drug development because proteins are the most common drug targets. For example, HIV protease is a protein produced by the human immunodeficiency virus (HIV)—the pathogen that causes AIDS—in the context of a human host cell. HIV protease is essential to the proliferation of the virus. If this protein can be effectively inhibited, then the virus cannot infect any additional cells, and the advance of the disease is stopped.

How might a molecule act to inhibit an enzyme like HIV protease? Proteases are proteins that digest other proteins, much like restriction enzymes are used to specifically cut DNA molecules (Chapter 1). Many of the proteins that HIV needs to survive and proliferate in a human host are produced as a single, large polypeptide chain. This polypeptide must then be cut into the functional component proteins by HIV protease. Like many enzymes, HIV protease has an **active site** where it binds to and operates on other molecules. A drug design objective in this case, then, might be to discover and refine a molecule that will bind in the active site of HIV protease in a way that prevents it from functioning normally.

Ligand Screening

The first step in the discovery of an inhibitor for a particular protein is usually to identify one or more **lead compounds** that might bind to the active site of the target protein. Traditionally, the search for lead compounds has been a trial-and-error process in which numerous compounds are tested using various biochemical assays until a sufficient number of compounds with some inhibitory effect is found. Recently, high-throughput screening (HTS) methods have made the procedure much more efficient, but the underlying process remains essentially an exhaustive search of as many potential leads as possible. Ligand docking and screening algorithms strive to streamline the lead discovery process by moving as much of the *in vitro* experimentation as possible into the realm of computation. Since lead discovery is an expensive and time-consuming process, some of the most dramatic contributions of computational methods to drug development will likely involve reducing the human effort and time associated with finding lead compounds.

Ligand Docking

In many cases, the three-dimensional structure of a protein and its ligand are known (see X-Ray Crystal Structures section below), but the structure of the complex they form is unknown. The objective of computational docking is to determine how two molecules of known structure will interact. In the case of drug design, molecular docking is most often employed to aid in determining how a particular drug lead will bind to the target, or how two proteins will interact to form a binding pocket.

Molecular docking algorithms share a great deal in common with algorithms for protein folding. Both problems involve identifying the energy of a particular molecular conformation, and then searching for the conformation that minimizes the free energy of the system. As with protein folding, there are far too many degrees of freedom to exhaustively search all possible binding modes or conformations, so **heuristics**—general rules that tend to lead to good solutions, but cannot be guaranteed to lead to optimal solutions—and search algorithms must be used to find suboptimal solutions that approach the global energy minimum.

As in protein folding, there are two primary considerations in designing a molecular docking algorithm. First, an energy function must be formulated to evaluate the quality of a particular complex, and then a search algorithm must be employed to explore the space of all possible binding modes and conformations in search of a minimal energy structure. Another important consideration in the design of a docking algorithm is how to deal with flexibility in both the protein and the putative ligand. **Lock and key approaches** to docking assume a rigid protein structure, to which a flexible ligand structure is docked. This approach is attractive because it is far less computationally expensive than **induced fit docking,** which allows for flexibility in both the protein and the ligand. Induced fit approaches vary in the amount of flexibility afforded to proteins. A common compromise is to assume a rigid protein backbone, while allowing for flexibility in the side chains near the ligand binding site.

AutoDock (http://www.scripps.edu/pub/olson-web/doc/autodock), a well-known method for docking of rigid or flexible ligands, uses a grid-based force field for evaluating a particular complex. The force field is used to score a complex based on formation of favorable electrostatic interactions, the number of hydrogen bonds formed, van der Waals interactions, and several other factors.

To find a docking that minimizes the free-energy function, the original version of AutoDock employed a Monte Carlo/simulated annealing approach. The **Monte Carlo algorithm** is a commonly used method of energy minimization that essentially makes random changes to the current position and conformation of the ligand, keeping those that result in a lower energy (that is, a more stable configuration) than the current configuration. When a move results in a higher energy, it is usually not accepted. However, in order to allow the algorithm to find low-energy states that may have an energy barrier (Figure 8.3), moves that result

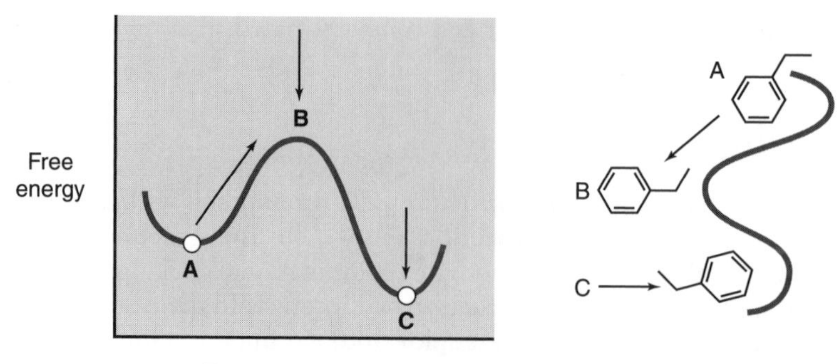

FIGURE 8.3 *Sometimes a docking algorithm must accept a move that results in an unfavorable energy state in order to find a global minimum.*

in higher energies are sometimes accepted. The frequency with which higher energy states are accepted is set relatively high at the beginning of the algorithm, and then slowly decreased as the algorithm iterates. This probability is often thought of as a temperature factor, and the gradual reduction of the temperature factor during a run is thought of as a cooling process. Thus the analogy with the process of annealing metals and the name *simulated annealing*. More recent versions of AutoDock use variations of genetic algorithms—optimization programs that emulate the dynamics of natural selection on a population of competing solutions—to accomplish the search for a minimal energy configuration.

AutoDock is, of course, only one example of the numerous algorithms for docking biological molecules. FTDock (http://www.bmm.icnet.uk/docking) is often used for protein–protein docking. Binding modes determined by FTDock can be further refined by its companion program, MultiDock. The HEX algorithm is specialized for docking proteins to other proteins and does so with surprising speed. Other algorithms, including DOCK, Hammerhead, GOLD, FLEXX (http://cartan.gmd.de/flexx), and many more, use a variety of protein representations, force fields, and search algorithms.

Database Screening

A primary consideration in designing ligand docking algorithms is the balance between the need for a complete and accurate search of the possible ligand conformations and binding modes, and the necessity of designing an algorithm with a reasonable computational complexity. A docking algorithm that always produces perfect results is not useful if it takes years to complete. For screening of databases containing possible drugs, algorithms need to essentially dock thousands of ligands to a protein active site and, therefore, need even greater efficiency.

A common approach for screening ligand databases is to use a fast docking algorithm, such as Hammerhead, FLEXX, or HEX, to dock a variety of ligands to the target protein, using the energy score from the docking algorithm as a predictor of binding affinity. For large databases containing hundreds of thousands of lead compounds, however, such an approach is impractical, requiring hundreds of days to screen a database for a single drug target.

Methods that are designed specifically for database screening, such as the SLIDE algorithm, often reduce the number of compounds that must be docked using database indexing techniques to rule out lead compounds that are highly unlikely to bind to the target active site prior to docking. SLIDE characterizes the target active site according to the positions of potential hydrogen bond donors and acceptors, and hydrophobic interaction points, forming a template. Each potential ligand in the database is characterized in a similar manner, and a set of indices is constructed. These indices allow SLIDE to quickly rule out ligands that are, for example, too large or too small to fit the template (Figure 8.4 on page 192). By reducing the number of ligands that undergo the computationally expensive docking procedure, SLIDE and similar algorithms can screen large databases of potential ligands in days (and sometimes even hours) rather than months.

X-Ray Crystal Structures

Many of the bioinformatics techniques that we have discussed in this and the previous chapter, including computational docking, ligand database screening, and the formulation of empirical energy functions for protein folding, depend on experimentally observed three-dimensional protein structures. While even the most powerful microscopy techniques are insufficient to determine the molecular coordinates of each atom in a protein, the discovery of X-rays by W. C. Roentgen in 1895 has led to the development of a powerful tool for analyzing protein structure: X-ray crystallography. In 1912, M. von Laue discovered that **crystals,** solid structures formed by a regular array of molecules, diffract X-rays in regular and predictable patterns. In the early 1950s, pioneering scientists such as D. Hodgkin were able to crystallize complex organic molecules and determine their structure by looking at how they diffracted a beam of X-rays. Today, X-ray crystallography has been employed to determine ~15,000 protein structures at very high levels of resolution.

The first step in crystallographic determination of a protein's structure is to grow a crystal of the protein. While this can often be a sensitive and demanding process, the basic idea is simple. Just as sugar crystals can be grown by slow evaporation of a solution of sugar and water, protein crystals are often grown by evaporation of a pure protein solution. Protein crystals, however, are generally very small (~0.3 to 1.5 mm in each dimension), and consist of as much as 70% water, having a consistency more like gelatin than sugar crystals (Figure 8.5 on page 193). Growing protein crystals generally requires carefully controlled conditions and a great deal of time; it sometimes takes months or even years of experimentation to find the appropriate crystallization conditions for a single protein.

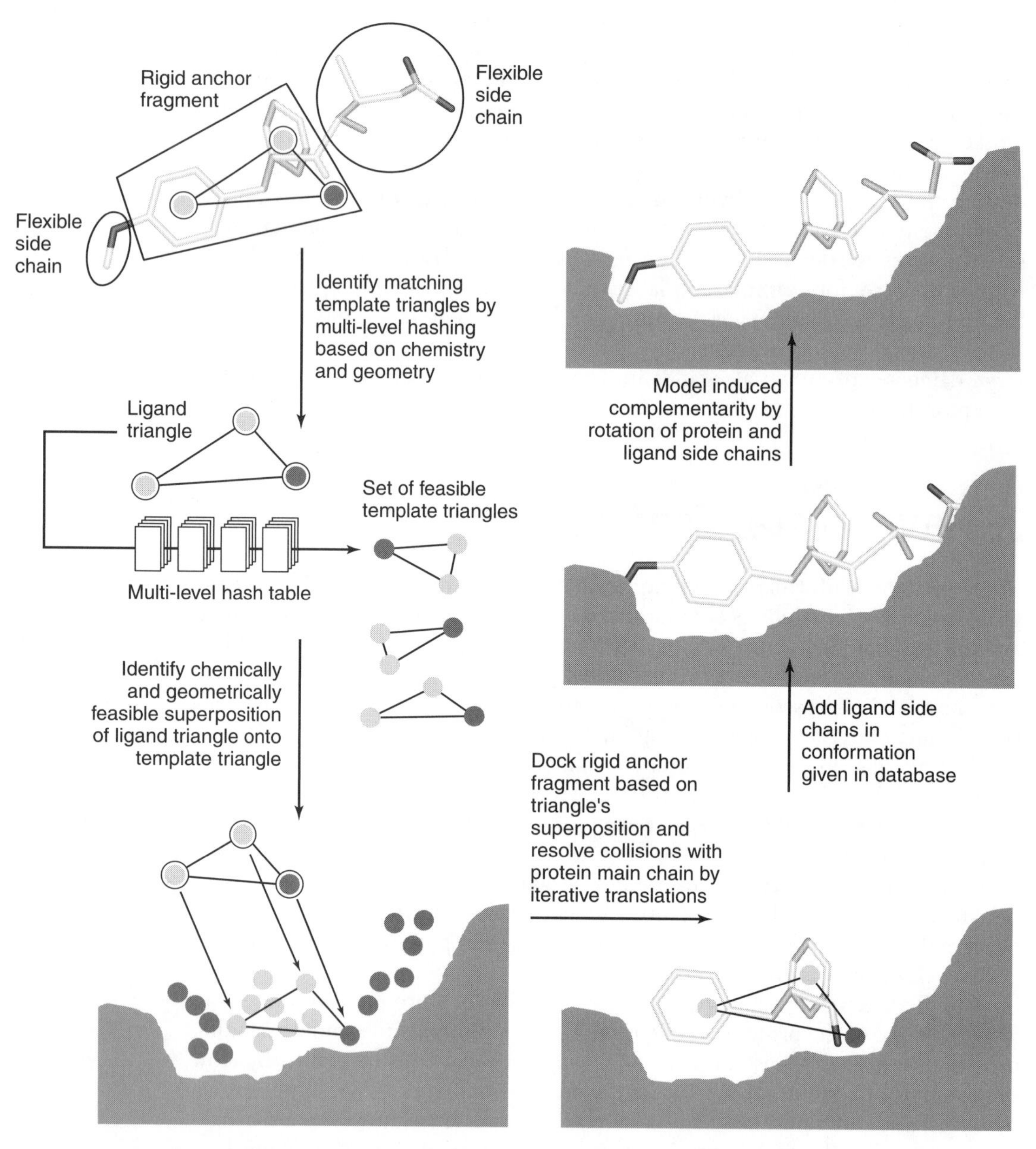

FIGURE 8.4 *The steps involved in finding potential ligands in SLIDE. Ligands are characterized by hydrogen bond donor and acceptor positions and by hydrophobic interaction points. Each triangle composed of three such points is indexed in a set of hash tables according to the length of the longest side, the perimeter distance, and other features. The hash tables allow ligands that will not fit into the active site to be eliminated before the computationally expensive docking steps begin.*

FIGURE 8.5 *A protein crystal.*

Once protein crystals are obtained, they are loaded into a capillary tube and exposed to a beam of X-ray radiation, which is then diffracted by the protein crystal (Figure 8.6 on page 194). In early crystallography, the diffraction pattern was captured on X-ray film. Modern crystallography equipment uses X-ray detectors that transfer the diffraction pattern directly to a computer for analysis. Figure 8.7 on page 194 shows an example of an X-ray diffraction pattern.

Once the diffraction data are obtained, a crystallographer uses a very complex mix of reverse Fourier transformations, crystallographic software tools, and protein modeling skills to determine the three-dimensional structure of the protein. The Protein Data Bank (PDB) at http://www.rcsb.org serves as the primary repository for protein structures derived from X-ray crystallography. PDB also serves as a repository for protein structures derived from nuclear magnetic resonance (**NMR**), another technique for resolving protein structures. The structures stored in the PDB are kept in two distinct formats. The original PDB format is text based and remains the most commonly used method for storing and processing crystallographic structures. The format of text-based PDB files is convenient for parsing and analysis using string processing languages such as Perl. For each atom in the protein, a line in the PDB file contains the X, Y, and Z coordinates in angstroms (0.1 nm or 10^{-10} m), along with a few other pieces of information relevant to that atom. Example number 8 in Appendix 3 demonstrates how to read and parse the molecular coordinates in a text-based PDB file.

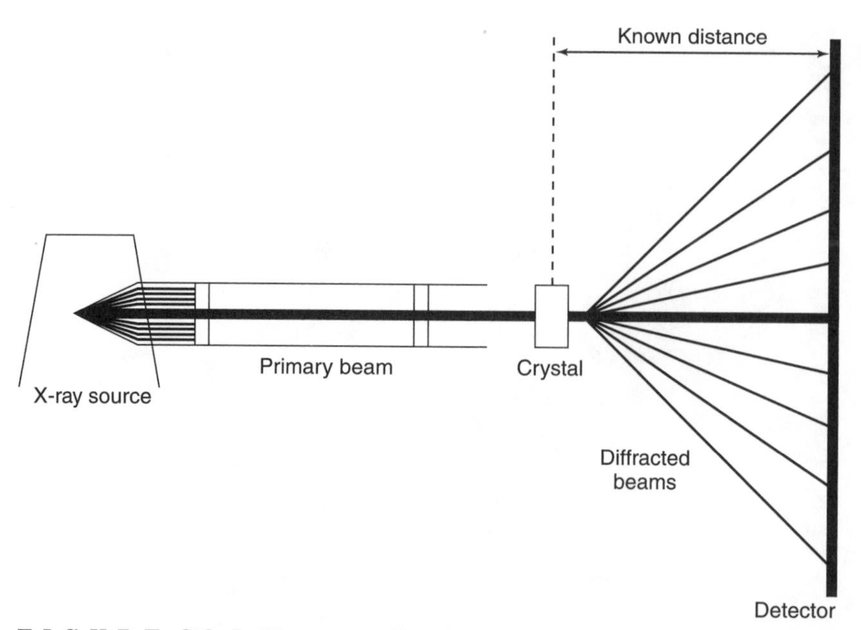

FIGURE 8.6 *In X-ray crystallography, X-ray radiation is directed at a protein crystal. The electrons in the protein diffract the X-rays, producing a diffraction pattern that is captured by the detector.*

FIGURE 8.7 *An X-ray diffraction pattern.*

Each structure in the PDB is assigned a four-character PDB code. For example, the PDB entry 2APR contains the molecular coordinates of rhizopuspepsin, an aspartic protease. PDB formatted text files are generally named either XXXX.pdb, or pdbXXXX.ent, where XXXX is the four-letter PDB code for the structure.

In 1990 the International Union of Crystallographers (IUCr) formalized a new standard, the mmCIF (macromolecular crystallographic information file) format. This format is designed to provide additional flexibility and richness over the original PDB file format. However, PDB text files currently remain the most common format for distribution of protein structural data. Further details on the PDB and mmCIF formats can be found at http://www.rcsb.org.

The Protein Information Resource (PIR) is another rich source of protein sequence and structure data. The PIR web site (http://pir.georgetown.edu) is the home of the PIR-PSD, a richly annotated protein sequence database, the *i*ProClass protein classification database, various sequence search and prediction algorithms, and numerous other tools and data sources related to protein structure and function. Additionally, a number of proteomics tools and databases can be found at the ExPASy server (http://www.expasy.ch), including the SwissProt database, a richly annotated protein sequence database; the Prosite database of protein families and domains; and numerous proteomics tools, including software tools for some of the techniques mentioned earlier in this chapter, such as peptide mass fingerprinting.

While the three-dimensional coordinates found in a PDB entry are not particularly well suited for human viewing, numerous molecular graphics tools are available for visualizing protein structures from X-ray coordinates. RasMol, along with the Microsoft Windows® version, RasWin (http://www.bernstein-plus-sons.com/software/rasmol), is among the easiest to install and learn, and provides various views of a protein that can be rotated and manipulated on even a modestly powered Windows, Mac, or UNIX-based system.

Other useful and easy to learn visualization tools include the Swiss PDB viewer (http://www.expasy.ch/spdbv), VMD (http://www.ks.uiuc.edu/Research/vmd), Spock (http://quorum.tamu.edu/jon/spock), and Protein explorer (http://www.umass.edu/microbio/chime/explorer). Protein explorer provides numerous molecular graphics and protein analysis tools and runs within the Netscape web browser. Protein explorer can download PDB coordinates directly from the PDB site, obviating the need to download PDB files to your own hard disk. One of the most useful tools in Protein explorer is the comparator, which visualizes two structures side by side, allowing simultaneous manipulation and comparison of both structures. DINO (http://www.bioz.unibas.ch/~xray/dino) provides a wide array of visualization methods, at the expense of a slightly steeper learning curve.

Numerous other molecular graphics packages are available, representing a wide range of capabilities and complexity. Many of them are listed in the Links section of the PDB web site: http://www.rcsb.org/pdb/software-list.html#Graphics. Figure 8.8 shows some of the protein representations that can be generated by some of the molecular graphics tools mentioned above.

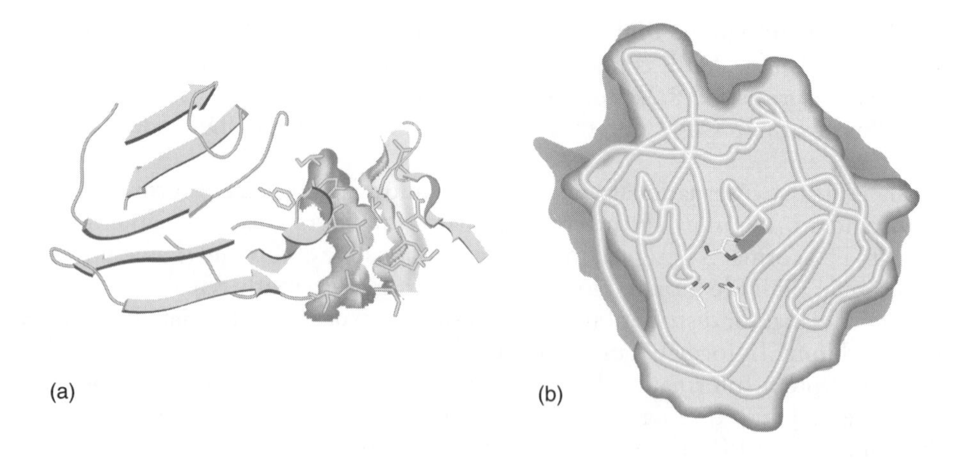

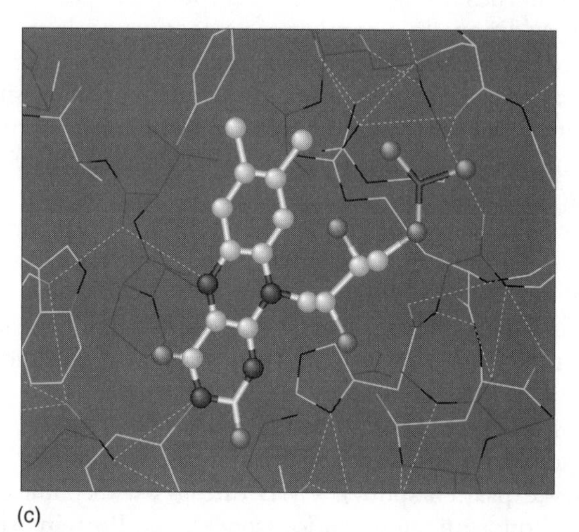

FIGURE 8.8 *Several molecular graphics representations of proteins. (a) The cartoon method highlights regions of secondary structure. (b) Molecular surface representations reveal the overall shape of the protein. (c) The wireframe and ball-and-stick representations illustrate specific molecular interactions.*

Finally, note that crystallographic structures are essentially averaged over the many copies of the protein in a single crystal and also over the time that the crystal is exposed to the X-ray. Proteins in a crystal are not completely rigid, and the mobility of specific atoms in a protein will "blur" the crystallographic signal. The positions of water molecules in the crystal, which are often included in a PDB entry, are particularly difficult to resolve and subject to noise. Nevertheless, protein crystallography remains the primary method for obtaining a glimpse of the structure of a protein at an atomic resolution.

NMR Structures

Nuclear magnetic resonance spectroscopy provides an alternative method for determining macromolecular structures. At the heart of the NMR technique is the fact that the nuclei of the atoms of some elements (such as hydrogen and radioactive isotopes of carbon and nitrogen) vibrate or resonate when the molecules they are a part of are placed in a static magnetic field and exposed to a second oscillating magnetic field. The magnetic fields of these atomic nuclei vibrate as they try to become aligned with the external magnetic fields, and those resonances can be detected by external sensors as NMR spectra. The behavior of any atom is influenced by neighboring atoms in adjacent residues such that closely spaced residues are more perturbed than distant residues. Interpretation typically requires extensive and complex reverse Fourier transformations, much like X-ray crystallography. The complexity of the data analysis alone usually limits the utility of the approach for any proteins or domains larger than 200 amino acids.

A key advantage of NMR methods is that they do not require crystallization of the protein. Because not all proteins can be effectively crystallized (many membrane integral proteins, in particular, have proven resistant to crystallization), NMR is an important source of 3D structural information. The result of an NMR experiment is a set of constraints on the interatomic distances in a macromolecular structure. These constraints can then be used, along with a protein's known sequence, to provide a model of the protein's structure. It is, however, generally the case that more than one model can effectively satisfy the constraints obtained by NMR, so NMR structures usually contain several models of a protein (that is, several sets of coordinates), while crystallographic structures usually contain only a single model.

Empirical Methods and Prediction Techniques

The wealth of protein structural data provided by crystallography has opened up the way for numerous empirical studies of protein structure and function. Computational techniques from statistics, data mining, computer vision and pattern recognition, evolutionary computation, and many other fields have all been applied to glean a deeper understanding of how protein molecules behave and interact with other biological molecules. While far too many studies of this kind exist to provide a comprehensive treatment of the subject here, many empirical algorithms follow a pattern recognition approach in their general design. Thus, a simple example can be employed to demonstrate some of the features common to many empirical methods.

Suppose we are given the task of designing an algorithm that, given the three-dimensional structure of a protein, predicts which residues in the protein are likely to be involved in protein–protein contacts. Since many proteins are only active when associated with other proteins in multienzyme complexes, this might be an interesting question to pursue. How would one go about designing an algorithm to make such predictions about specific residues, given only the three-dimensional coordinates of a protein? For an empirical algorithm, the answer is "by observing the properties of residues that are known to be involved in protein–protein contacts (we'll call them *interfacial* residues) and those that are not."

From the structures in the PDB, we might begin by selecting a set of example structures that show two or more proteins complexed together. In each of these structures, there will be some interfacial residues involved in the contact surface between two proteins, and some noninterfacial residues that are not near the protein–protein interface. For each residue, we can select a set of features to measure. For example, we might choose to measure the number of other residues within a certain radius of each residue; this would give us an idea of how crowded the area around the residue is. Other features we choose to measure might include the net charge of the residue and surrounding residues, the level of hydrophobicity of the residue, the hydrogen bonding potential of the residue, and any other feature we think might be relevant to whether or not the residue is part of a protein–protein interface.

Once the features are measured, each residue in our database is represented as a **labeled feature vector.** The label represents whether the residue is an interfacial residue or not, and the feature vector is just the list of feature values for that residue. These labeled feature vectors serve as the training data for a prediction algorithm, as illustrated in Figure 8.9. The prediction algorithm can be a statistical method, a neural network, an evolutionary algorithm, or one of any number of other computational techniques for learning from example data. The goal of the prediction algorithm is to learn, by observing the labeled training data, which feature values are associated with interfacial residues and which values indicate noninterface residues. Ideally, once the predictor is trained, we can give the algorithm a new residue without the label and it will predict whether the new residue is an interface or noninterface residue.

This general method has been applied to develop algorithms that predict a diverse array of protein properties, including secondary and tertiary structure; binding to ligands, other proteins, and water molecules; active site location; and even general biological function.

Post-Translational Modification Prediction

The wide variety of protein structure and functions is due, in part, to the fact that proteins undergo a wide variety of modifications after being transcribed. These modifications may remove segments of a protein; covalently attach sugars, phosphates, or sulfate groups to surface residues; or cross-link residues within a pro-

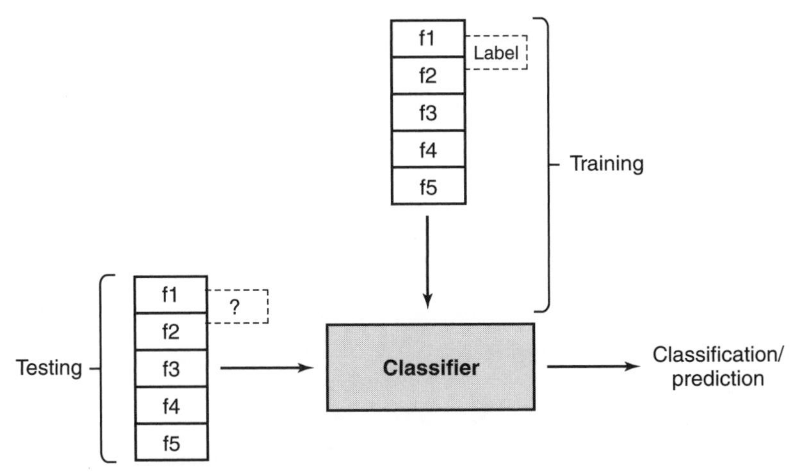

FIGURE 8.9 *During training, labeled feature vectors are provided to the classification algorithm. After it has learned the feature values that distinguish each class, the algorithm is presented with an unlabeled feature vector to classify.*

tein (disulfide bond formation). Many of these modifications are themselves performed by other proteins that must recognize specific surface residues as appropriate for modification. The question of what signals a protein-modifying enzyme to operate on a particular residue is an important question to understanding protein function, regulation, and localization.

Protein Sorting

The presence of internal, membrane-bound compartments is one of the single most distinguishing features of eukaryotic cells. The chemical environment within those different compartments can differ dramatically as can the population of proteins found within them. It is both functionally and energetically imperative that eukaryotes deliver proteins to the appropriate compartment. For example, histones (DNA binding proteins associated with chromatin; see Chapter 6) are only functionally useful within a eukaryotic cell's nucleus where its chromosomes are sequestered. Other proteins such as the proteases found within peroxisomes are quite dangerous to a cell anywhere but within their proper compartment. Eukaryotic cells appear to consider proteins as belonging to one of only two distinctly different classes with regard to their localization (Figure 8.10 on page 200): those that are not associated with membranes, and those that are associated with membranes.

The first set of proteins is translated exclusively by ribosomes that remain suspended (sometimes called "free-floating") within the cytoplasm. mRNAs translated by free-floating ribosomes can ultimately be delivered to some of the following important destinations: the cytoplasm itself, the nucleus, the mitochondria, chloroplasts, and peroxisomes. Residence within the cytoplasm seems to be the default state for proteins. In contrast, delivery to the nucleus, mitochondria,

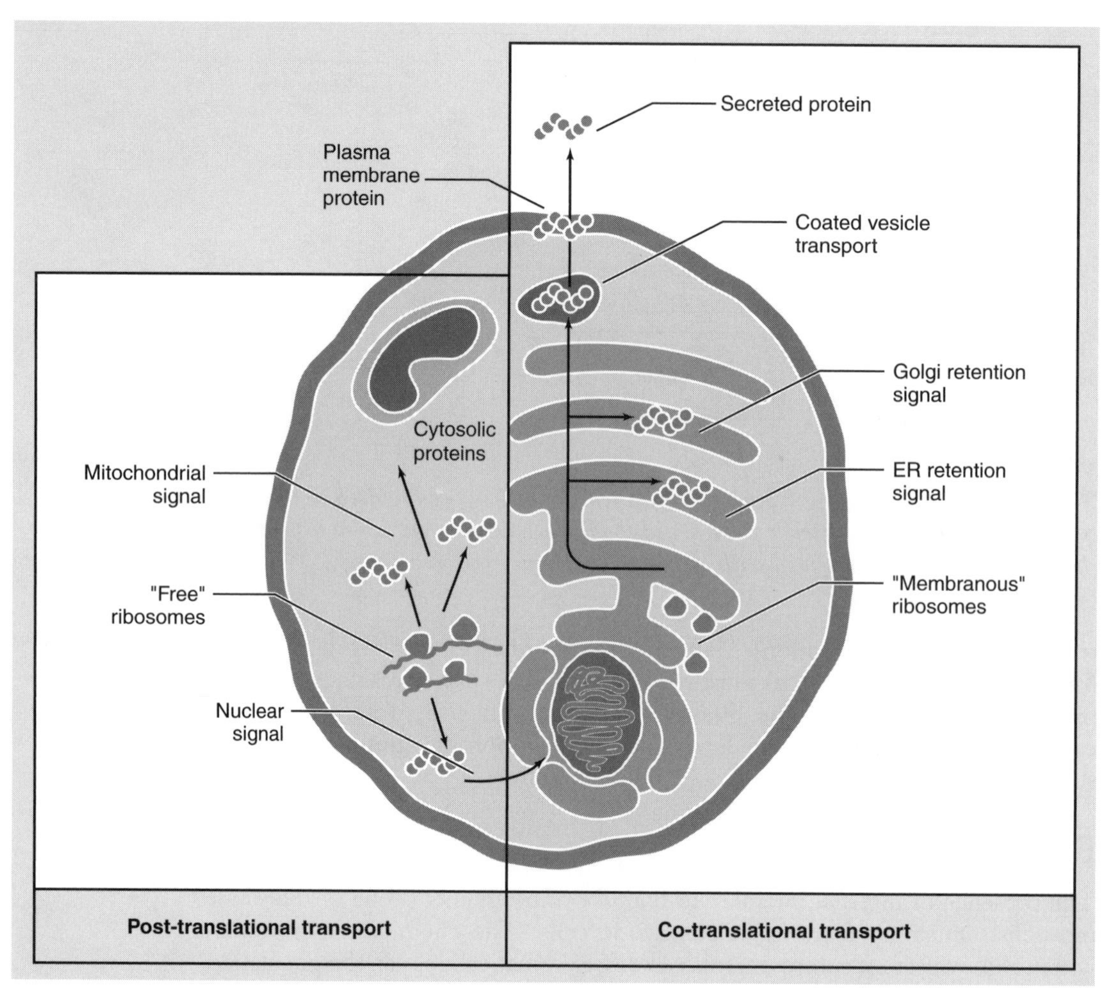

FIGURE 8.10 *Proteins released into the cytosol by free ribosomes may depend on various protein sorting signals to determine their cellular destination, or they may remain in the cytoplasm. Proteins synthesized by membrane-bound ribosomes in the ER move to the Golgi apparatus and then through the plasma membrane unless they contain specific localization signals to determine their final destination.*

chloroplasts, or peroxisomes requires the presence and recognition of specific localization signals (Table 8.3).

Nuclear proteins (such as histones and transcription factors) are targeted to the nucleus by their possessing a **nuclear localization sequence:** a stretch of 7 to 41 amino acids that is rich in lysine and/or arginine. Mitochondrial proteins all possess an amphipathic helix 12 to 30 amino acids long at their N terminus. This **mitochondrial signal sequence** is ultimately recognized by a receptor on the surface of mitochondria and then often removed to activate the protein as it is transported into the mitochondria. Chloroplast proteins encoded by nuclear

TABLE 8.3 Several protein localization signals used by all eukaryotic cells.

Organelle	Signal location	Type	Signal length
Mitochondria	N terminus	Amphipathic helix	12–30
Chloroplasts	N terminus	Charged	Roughly 25
Nucleus	Internal	Basic	7–41
Peroxisome	C terminus	SKL	3

genes all possess a **chloroplast transit sequence** (a string of roughly 25 charged amino acids at their N terminus) that is similarly recognized by receptor proteins on the surface of chloroplasts. By the same token, proteins destined for peroxisomes all have one of two **peroxisomal targeting signals** that are recognized by receptors that ensure their delivery to the correct destination.

The second set of proteins is translated by membrane-bound ribosomes that are associated with the **endoplasmic reticulum (ER).** The endoplasmic reticulum itself is a web-like network of membranes that is intimately associated with the Golgi apparatus in which a great deal of additional protein processing (such as glycosylation and acetylation) also takes place, as described below. All proteins translated by ER ribosomes actually begin to be translated by free-floating ribosomes in the cytoplasm. When the first 15 to 30 amino acids to be translated correspond to a special **signal sequence,** a signal recognition particle binds to it and stops further translation until the ribosome and its mRNA are delivered to the ER. While no particular consensus sequence for the signal sequence is evident, a hydrophobic stretch of 10 to 15 residues, ending with one or more positively charged residues, is found in almost all of these targeting motifs. When translation resumes, the new polypeptide is extruded through a pore in the membrane of the ER into the ER's lumen (inner space). Once a protein is transported across the ER membrane, a **signal peptidase** cleaves the N-terminal targeting sequence from the protein unless it is to be retained permanently as a membrane-bound protein itself.

Various computational methods have been employed for the identification of targeting signals and predicting protein localization. PSORT (http://psort.nibb.ac.jp) uses a **nearest neighbor classifier.** The nearest neighbor classifier is a statistical method that classifies according to similarity. A database of examples of signal peptides and nonsignal peptides is consulted. To predict if a particular sequence is a signal peptide, it is compared with the sequences in the database. The sequences that are very similar to the sequence being predicted are identified and tallied. If the majority of these "near-neighbor" sequences are signal peptides for a particular cellular location, then the sequence in question is also predicted to be a signal peptide for that location. Another localization signaling prediction tool, SignalP (http://www.cbs.dtu.dk/services/SignalP), uses artificial neural networks, a computational method that has been successfully employed in the analysis of a wide range of bioinformatics problems and data sets as described previously in the context of gene recognition (Chapter 6).

Proteolytic Cleavage

Both prokaryotic and eukaryotic organisms have numerous enzymes responsible for the cleavage and degradation of proteins and peptides. Examples of proteolytic cleavage are abundant, ranging from the removal of the initial methionine residue present at the start of every polypeptide (because start codons also code for methionine; see Chapter 1), to the removal of the signal peptides discussed in the previous section. Many of these enzymes recognize specific sequence motifs as indicators of cleavage sites. Sometimes, the cleavage signal can be as short as a single residue. Examples include chymotrypsin, which cleaves polypeptides on the C-terminal side of bulky and **aromatic** (ring-containing) residues such as phenylalanine; trypsin, which cleaves the peptide bond on the carboxyl side of arginine and lysine residues; and elastase, which cleaves the peptide bond on the C-terminal side of small residues such as glycine and alanine. In many cases, however, the sequence motif is larger and more ambiguous, much like the localization motifs previously discussed. Protein cleavage by **proteasomes** (multienzyme structures involved in protein degradation and immune response) can be predicted with high levels of accuracy (>98%) using a neural network (http://www.paproc.de).

Glycosylation

Glycosylation is the process of covalently linking an **oligosaccharide** (a short chain of sugars) to the side chain of a protein surface residue. The presence of glycosylated residues can have a significant effect on protein folding, localization, biological activity, and interactions with other proteins. There are several types of glycosylation in eukaryotes, the most important of which are N-linked and O-linked glycosylation.

N-linked glycosylation is the addition of an oligosaccharide to asparagine residues during protein translation. The primary signal that an asparagine (Asn) residue should be glycosylated is the local amino acid sequence. The consensus sequence Asn-X-Ser or Asn-X-Thr, where X is any residue other than proline, is nearly always found at glycosylated Asn residues. However, the sequence alone is insufficient to determine glycosylation, because some Asn-X-Ser/Thr triplets have nonglycosylated asparagine residues. The most common computational technique for identifying glycosylation sites involves the use of neural networks trained to identify glycosylated Asn residues using a procedure much like that described in the previous section on empirical methods and prediction techniques. Modern neural network-based approaches obtain accuracies around 75% for predicting N-linked glycosylation sites.

O-linked glycosylation is a post-translational process in which the enzyme *N*-acetylglucosaminyl transferase attaches an oligosaccharide to the oxygen atom of a serine or threonine residue. Unlike N-linked glycosylation, there is no known sequence motif that signals an O-glycosylation site, other than the presence of proline and valine residues in the vicinity of the Ser or Thr to be glycosylated. Again, a common method for identification of these sites is to use prediction algorithms based on neural networks, trained using local sequence context and surface accessibility as features. The resulting algorithms have >85% accuracy in predicting glycosylated residues.

Phosphorylation

Phosphorylation (attachment of a phosphate group) of surface residues is probably the most common post-translational modification in animal proteins. **Kinases,** the enzymes responsible for phosphorylation, are involved in a wide variety of signaling and regulatory pathways. Since phosphorylation frequently serves as an activation signal for an enzyme, it is often an ephemeral condition. **Phosphatases** are the enzymes responsible for removing phosphate groups from phosphorylated residues.

Often, kinases themselves are activated by phosphorylation, so that activation of a particular kinase can result in a cascade of successive phosphorylation events. One of the most well studied of these **signal cascades** is the cyclic AMP (cAMP) cascade, which controls glycogen synthesis and breakdown. Glycogen is a glucose polymer that serves as an energy-storage molecule. The synthesis and breakdown of glycogen are controlled by a number of hormones including glucagon, insulin, and epinephrine. When epinephrine is released by the adrenal medulla it leads to an increase in the cytosolic level of cAMP, which activates protein kinase A. Protein kinase A (PKA) then phosphorylates and activates phosphorylase kinase, which, in turn, phosphorylates and activates phosphorylase, the enzyme responsible for glycogen breakdown. PKA simultaneously phosphorylates and *deactivates* glycogen synthase, the enzyme responsible for glycogen synthesis. The multiple phosphorylation events in this cascade facilitate a very rapid response to epinephrine release.

Because phosphorylation of key tyrosine, serine, and threonine residues serves as a regulation mechanism for a wide variety of molecular processes, the various types of kinases involved in each process must have high specificity in recognizing particular enzymes. As a result, no single consensus sequence identifies a residue as a phosphorylation target. However, some patterns do exist, allowing computational prediction of phosphorylation sites. Neural network-based approaches, such as NetPhos, have obtained sensitivity greater than 70% in predicting phosphorylation targets based on sequence and structural information.

Chapter Summary

While genomics is rapidly becoming a mature research area, proteomics techniques are only now beginning to identify the proteins encoded within the genome and their various interactions. Characterization of an organism's proteome promises to bridge the gap between our understanding of the genome and the physiological and morphological effects of the genes therein. Various taxonomies have been developed to classify and organize proteins according to enzymatic function, sequence similarity, and three-dimensional shape. Armed with these databases of protein families, superfamilies, and folds, along with experimental techniques such as 2D electrophoresis, peptide mass fingerprints, and mass spectrometry, investigators can separate, purify, and identify the various proteins expressed by a cell at a particular time. One important application of proteomic information is in pharmaceutical drug design. Advances in knowledge of protein structure and X-ray crystallography have allowed computational methods for protein ligand screening and docking to contribute to the process of drug discovery. While the three-dimensional structure of a protein is the keystone for understanding a protein's function and interactions with other proteins, some important information can be obtained from sequence alone. Protein localization and various post-translational modifications are signaled by sequence motifs in the protein's primary structure.

Readings for Greater Depth

A summary of the means by which proteins can be grouped into families and superfamilies on the basis of their evolutionary history can be found in C. Chothia, 1992, Proteins, one thousand families for the molecular biologist, *Nature* **357**: 543–544.

The idea of exons being regions that code for portions of proteins with distinct functionalities is well described in R. L. Dorit, L. Schoenback, and W. Gilbert, 1991, How big is the universe of exons? *Science*, **253**: 677–680.

The integration of fast database searching techniques and Bayesian statistics for protein identification from mass spectra data is spelled out in W. Zhang and B. T. Chait, 2000, ProFound—An expert system for protein identification using mass spectrometric peptide mapping information, *Analyt. Chem.* **72**: 2482–2489.

The AutoDock algorithm, one of the first methods for completely automated molecular docking, was first described in D. S. Goodsell and A. J. Olson, 1990, Automated docking of substrates to proteins by simulated annealing, *Proteins: Structure, Function, and Genetics*, **8**: 195–202. The most recent version of AutoDock, version 3.0, is described in G. M. Morris, D. S. Goodsell, R. S. Halliday, R. Huey, W. E. Hart, R. K. Belew, and A. J. Olson, 1998, Automated docking using a lamarckian genetic algorithm and an empirical binding free energy function, *J. Computational Chem.* **19**: 1639–1662.

The Fourier transform-based methods of FTDock are described in E. Katchalski-Katzir, I. Shariv, M. Eisenstein, A. A. Friesem, C. Aflalo, and I. A. Vakser, 1992. Molecular surface recognition: Determination of geometric fit between proteins and their ligands by correlation techniques, *Proc. Nat. Acad. Sci. U.S.A.* **89**: 2195–2199.

The speedy HEX docking program is detailed in D. W Ritchie and G. J. L. Kemp, 2000, Protein docking using spherical polar Fourier correlations, *Proteins: Structure, Function, and Genetics*, **39**: 178–194.

SLIDE, an algorithm for screening large databases for protein ligands, is described in V. Schnecke and L. A. Kuhn, 1999, Database screening for HIV protease ligands: The influence of binding-site conformation and representation on ligand selectivity, *Proc. Seventh Int. Conf. Intelligent Systems for Molecular Biology*, pp. 242–251.

NetPHOS, a neural network-based method for predicting phosphorylation sites, is discussed in N. Blom, S. Gammeltoft, and Sùren Brunak, 1999, Sequence and structure-based prediction of eukaryotic protein phosphorylation sites, *J. Molecular Biol.* **94**: 1351–1362.

More information on neural networks, nearest neighbor classification, and other computational pattern recognition methods can be found in R. Duda and P. Hart, 1973, *Pattern Classification and Scene Analysis*, John Wiley & Sons, New York.

Many of the computational prediction algorithms referenced in this chapter can be found at the Center for Biological Sequence Analysis, Technical University of Denmark's web page: http://www.cbs.dtu.dk, and at the ExPASy tools page: http://www.expasy.ch/tools.

Questions and Problems

* **8.1** Briefly describe how the cAMP cascade allows a rapid response to the release of epinephrine.

8.2 Modify the Perl code in Example 8 of Appendix 3 to find the centroid of a protein by finding the average X, Y, and Z values of all the ATOM records in a PDB file.

* **8.3** Starting with your solution for Question 8.2, add some Perl code to find and print the atoms closest to and furthest from the centroid.

8.4 Briefly describe the key difference between ligand docking and ligand screening algorithms. What trade-offs must be considered when designing a ligand screening algorithm?

* **8.5** Describe in detail how one might go about designing an empirical algorithm to predict whether or not a particular cysteine residue will form a disulfide bond, based on crystallographic structures from the PDB.

8.6 Given that glycosylation typically takes place within the Golgi apparatus, how likely is it that nuclear proteins like histones and transcription factors will be normally glycosylated within eukaryotic cells? Why?

* **8.7** Assuming that all amino acids are utilized with the same frequency and that there are a total of approximately 1,000 unique destinations for a protein within a eukaryotic cell, what is the minimum length that a signal sequence must be to ensure that its associated protein will be delivered appropriately?

8.8 Why might it be advantageous for receptor proteins such as those on the surface of the mitochondria and chloroplasts to cleave off signal sequences as proteins are delivered to their final destination?

APPENDIX 1

A Gentle Introduction to Computer Programming and Data Structures

Anyone who uses a computer regularly will quickly learn to appreciate the advantages of knowing how to write programs. Nowhere is this more true than in the field of bioinformatics, where new algorithms and techniques are constantly being developed, and the standards for file and data formats are diverse and rapidly evolving. Learning the basics of computer programming will allow you to read, reformat, and write data files; perform basic calculations on your data; and even make additions and refinements to the bioinformatics tools that you use in your research, since the source code for many bioinformatics algorithms is freely available on the World Wide Web.

The objective of this chapter is to teach the basic skills and concepts involved in writing and debugging computer programs. We have chosen Perl as an example language because it is commonly used in bioinformatics software, and because it is a simple enough language to be examined in the scope of this appendix. Once you have learned the programming concepts presented here, you will find that you can not only write simple programs in Perl, but also quickly learn and use other programming languages. Most commonly used languages are based on the underlying concepts discussed in this chapter. Once you understand the basics of programming in Perl, you will find it much easier to learn to use other languages including C, C++, Java, Python, and many more.

This chapter is not intended to be a complete Perl language reference, but rather an introduction to the fundamental concepts of computer programming, using Perl as an example language. For a more detailed discussion of the Perl programming language, refer to the additional resources listed at the end of the chapter.

Creating and Executing Computer Programs

A computer program is actually nothing more than a text file (or, more often, a collection of text files) that contains instructions for accomplishing a certain task. Writing a computer program is a lot like writing a recipe or a laboratory protocol—the first step is to outline exactly what needs to be done. The main difference between writing a computer program and writing a recipe is that the author of a computer program is limited to a specific list of legal instructions. While there may be many ways to tell a cook to include a dash of salt in the soup, there is generally only one instruction that accomplishes a specific task in a computer programming language.

Many programming languages exist, but computers actually speak only one language, **machine language.** Most programs are not written in machine language directly because machine language is not very easy for humans to read and write. Instead, we use a **compiler** to translate a list of instructions (i.e., a computer program) from a programming language (such as C++) into machine language. The resulting program is then ready to run on your computer. In the Windows operating system, programs that have been compiled and are ready to run have file names that end in ".exe" for "executable." Perl, however, is a special type of programming language that does not require a compiler. To write a Perl program, we simply create a text file containing a list of Perl commands. This text file is called a **Perl script,** or sometimes a Perl program. To run the program, we must start the **Perl interpreter,** and tell it to run our newly written Perl script. The way in which we start the Perl interpreter depends on what type of computer we are using. On UNIX and most Windows systems, we can usually type **`perl`** ***`scriptname`*** where *scriptname* is replaced with the name of our Perl script. Perl programmers often use file names ending in ".pl" for their scripts. On properly configured Mac and Windows systems, a Perl script can be executed by double-clicking on the script file.

In summary, writing a computer program in Perl includes the following steps:

- First, we write the program as a text file. As a general rule, we give the file a name ending in ".pl", such as "myscript.pl".
- Next, we run the interpreter, which executes the script. The way in which we start the interpreter depends on the type of computer we are using.

Variables and Values

One of the most important concepts in computer programming is the idea of a variable. In general, variables in computer programs work in much the same way as variables in algebra. If we have the following two algebraic equations:

$$x = 7$$
$$y = x + 5$$

we can conclude that $y = 12$. Each variable acts as a named storage location where a value is stored. In algebra, variables are usually named using single letters, such as x and y in the above example. In computer programs, longer names are often used to make the program easier to read. The legal names for variables depend on the language you are using. In Perl, most combinations of letters, numbers, and a few other characters including the hyphen (-) and underscore (_) characters, are legal variable names. For now, all of our variable names must begin with the character $. In Perl, a dollar sign indicates that a variable is a **scalar variable,** or a variable that can hold only one value. We will discuss scalar and other types of variables in more detail shortly. The value stored in a variable can be retrieved or changed at any time, and we can perform operations on it. The Perl language includes most of the basic arithmetic operations. Consider the following bit of Perl code:

```
$var1 = 9;
$var2 = 7;
$product = $var1 * $var2;
$sum = $var1 + $var2;
$remainder = $var1 % $var2;
```

In this example, `$product` is a variable containing the value 63, and `$sum` is a variable containing the value 16. In Perl, the symbol % is used to compute the remainder when the first operand is divided by the second, so in the above example the variable `$remainder` contains the value 2.

Data Typing

Unlike algebra, the variables in our computer programs can contain more than just numbers. In Perl, a variable might contain an integer number, such as 7, a real number such as 3.1417, or a character string such as "hello world," among other things. In many languages, you must declare what kinds of values a variable can contain before using it. In C and C++, for example, if you decide that a certain variable can contain integers, then that variable cannot contain text, real numbers, or any other kind of data. In Perl, however, a variable can contain any kind of data at any time. The only decision we must make when creating a variable in Perl is whether the variable will hold only one value or many values. A variable that can hold only a single value is called a scalar variable, as described previously, and begins with the character $. Two other types of variables, arrays and hashes, can hold many values. Arrays and hashes will be discussed in more detail in the Data Structures section.

In most languages, you must list all of the variables you wish to use at the start of your program. This is called **declaring** the variables. In Perl, declaring variables is not necessary. Variables are created the first time you refer to them. The following Perl script uses a single variable to store several types of data:

```
$my_variable = 7;
print("my variable contains the value: $my_variable \n");
$my_variable = 7.3217;
print("my variable contains the value: $my_variable \n");
$my_variable ="apple";
print("my variable contains the value: $my_variable \n");
```

If you type the lines above into a text file and execute it in Perl, you will see the following output:

```
my variable contains the value: 7
my variable contains the value: 7.3217
my variable contains the value: apple
```

In this short script, the variable `$my_variable` contained first an integer value, then a real value, and finally a text string. This flexibility in handling variables makes Perl one of the easiest programming languages to learn and to write short programs in quickly.

Before we go on, a few features of the preceding Perl script merit further discussion. First, note that each line of Perl code ends in a semicolon. The reason for this is discussed further in the Program Control section. Also note that text strings are enclosed in double quotes. To store the word *apple* in the variable `$my_variable`, we enclose it in double quotes. Text strings can consist of more than one word, so the following instruction would also be legal in Perl:

```
$my_variable ="a ripe red apple";
```

Finally, note that we can use the `print` instruction to send output to the computer's display. The "\n" at the end of each print instruction is a special code to move to the next line of the display screen. Without the "\n" in each print instruction, all the output would have printed on the same line of the display screen, which would have been difficult to read. Text strings and special characters, such as \n, will be discussed further in the Data Structures section.

Basic Operations

As mentioned previously, Perl contains symbols for most of the basic algebraic and math operations. As a convenience for programmers, Perl also contains some shortcuts for common combinations of operations. For example, it is often necessary to add or subtract a number from a variable, as in the following example:

```
$var2 = $var2 + 12;
```

Perl provides the += operator to make this operation easier. The following line of Perl code produces exactly the same results as the previous line:

```
$var2 += 12;
```

Table A1.1 shows some common operators used in Perl and their effects.

Program Control

At this point we have discussed enough of the Perl language to do simple numeric calculations, but we haven't yet seen anything that could not be done faster and more easily on a pocket calculator. The real power of a computer lies largely in its ability to make decisions. To achieve this ability, all programming languages

TABLE A1.1 A few commonly used Perl operators.

Operator	Effect
*, /, +, -	Algebraic multiplication, division, addition, and subtraction, respectively
%	Modulus (remainder)
++, - -	Increment or decrement (`$var++` adds 1 to the numeric value in $var, `$var--` subtracts 1 from $var)
+=, -=, *=, /=	Shortcuts for algebraic operations (`$var1 += 12` is the same as `$var1 = $var1 + 12`)
**	Exponentiation: `$var**3` is `$var` cubed
.	String concatenation: If `$var1` = "hi" and `$var2` = "there", `$var1 . $var2` = "hithere"
×	String repetition: `$string = "hi" × 3` sets the value of $string to "hihihi"

provide some facility for **program control.** That is, there must be a way to tell the computer to execute one set of instructions if a certain condition is true, and another set of instructions otherwise. In Perl, program control is mostly realized by **conditional** or **if statements,** and by loops. Before we can discuss these two structures, however, it is necessary to have an idea of the basic units of structured code in Perl: the statement and the block.

Statements and Blocks

The basic unit of execution in Perl is the statement. A **statement** represents one instruction, and is generally terminated with a semicolon. In each of the previous examples there is a semicolon at the end of each statement. Sometimes a statement needs to be longer than a single line of code. Because each Perl statement ends in a semicolon, there is no difficulty in determining where a statement ends, even when it spans multiple lines, as in the following example:

```
$total_annual_sales = $first_quarter_total +
  $second_quarter_total + $third_quarter_total +
  $fourth_quarter_total;
```

The second and third lines of the statement are indented to make the line more readable. The indentation, however, is only used to help human readers see where the statement ends, and it is optional. The Perl interpreter knows where the statement ends by looking for the semicolon.

A **block** is a set of statements enclosed in curly braces. We will discuss code blocks further shortly. For now, simply remember that each block of code starts with an opening curly brace ({) and ends with a closing curly brace (}).

Conditional Execution

As you begin to write Perl programs, you will find that you often want to include some statements that are only executed under certain conditions. If these conditions are not met, then the code is skipped. In Perl, we can easily write this sort of code using the `if` statement. The `if` statement in Perl is the most commonly used mechanism for **conditional execution.** In the following example, we wish to take the absolute value of the variable `$x` and place it in the variable `$abs_x`. The actions we need to take depend on whether the value of `x` is positive or negative:

```
$abs_x = $x;
if ($abs_x < 0)
{
  $abs_x *= -1;
}
print("The absolute value of x is $abs_x \n");
```

Let's examine how this code will execute in detail. When the first line is executed, the value of `$x` is copied into `$abs_x`. Next, the `if` statement tests whether the value of `$abs_x` is less than zero. If so, the block of code following the `if` statement is executed. If not, the block of code is skipped and the program continues with the `print` statement, which will print the value of `$abs_x`. For example, if `$x = -7`, then the `print` statement above will print the following line to the computer screen:

```
The absolute value of x is: 7
```

There are several important things to note about this seemingly simple piece of code. First, notice that after the `if` statement we include a test inside a pair of parentheses. If the result of this test is true, then the block of code following the `if` statement is executed, otherwise it is skipped. This is the one of the most important uses of blocks in Perl: After an `if` statement, the next block of code is the **conditional code**—the code that will only be executed if the test is true. Any code that follows the closing brace of the block, such as the `print` statement in the above example, will be executed regardless of whether the test in the `if` statement is true or false.

Sometimes, when the test condition is not true, you may wish to execute an alternate block of code, rather than simply skipping the conditional block. For this purpose most languages that provide the `if` statement also provide the capability for an `else` block, which is executed when the test in the `if` statement is false. In Perl, the use of an `else` block looks like this:

```
if ($x > 0)
{
  print("$x is positive\n");
}
else
{
  print("$x is either zero or negative\n");
}
```

In fact, Perl and many other languages take the concept a step further: By using the `elsif` statement, we can include additional conditions, as in the following example:

```
if ($x > 1000)
{
  print("x is large and positive\n");
}
elsif ($x > 0)
{
  print("x is small and positive\n");
}
elsif ($x > -1000)
{
  print("x is negative\n");
}
```

```
else
{
  print("x is VERY negative\n");
}
print("done!\n");
```

In this example, the first line will test whether the value stored in `$x` is greater than 1,000; if so, the first conditional block is executed, printing `"x is large and positive"`, then the program will jump to the end of the `else` block, and the last `print` will be executed, printing `"done!"`. If, however, `$x` is not greater than 1,000, then the first `elsif` test will be performed: If `$x` is greater than 100, then the conditional block following this test will be executed, printing `"x is small and positive"`, and as before we will jump to the end of the `if/else` code and execute the last `print` statement. If `$x` is not greater than 100, we move on to the next `elsif`, and so on until we find one for which the test is true. If we reach the final `else` statement, and none of the `elsif` tests have passed, then we execute the final conditional block and print out "`x is VERY negative`".

This `if/else` construct ensures that one, and only one, of the conditional blocks will be executed, depending on the value of `$x`. Regardless of which conditional block is executed, the program will always move on to the final `print` statement and print `"done!"` afterward.

Most languages provide some sort of conditional execution mechanism, and most of them look very similar to the `if`, `elsif`, and `else` statements provided by Perl. In all of these examples, we are using the simple relational operator '>' to test if the value of a variable is greater than a certain number. Perl includes numerous other tests that we can include inside the parentheses of a conditional statement. Table A1.2 shows some of the operators that we can use to test conditions for `if` statements and other program control statements.

TABLE A1.2 Common relational and logical operators in Perl.

Operator	Description
<, >, <=, >=	Numerical comparisons. `($x < $y)` returns TRUE when the value of $x is less than $y. <= and >= are "less than or equal to" and "greater than or equal to," respectively.
lt, gt, le, ge	String comparisons. `($x lt $y)` returns TRUE when the string $x would come before $y if the two strings were sorted alphabetically.
= =, !=	Numerical equality. = = returns TRUE when the two arguments are numerically equal. != returns TRUE when the arguments are not equal.
eq, ne	String equality. Similar to = = and !=, but for string comparisons.
&&, ||	Combination operators. `(test) && (test)` returns TRUE only if both of the tests in parentheses return TRUE. `(test) || (test)` returns TRUE if either of the tests in parentheses returns TRUE.
!	Logical "not" operator. `!(test)` returns TRUE when the test in parentheses returns FALSE.

Note that each of the comparison and equality operators has both a numerical (<, >, <=, etc.) and a string version (lt, gt, le, etc.). The difference between these operators is most clearly illustrated by a simple example. Suppose we assign values to a pair of variables as follows:

```
$var1 ="0012.0";
$var2 ="12";
```

Numerically, the two strings have equal values (12), so the test `($var1 == $var2)` would return TRUE. The two strings, however, are not the same, so the test `($var1 eq $var2)` would return FALSE.

It is important to note that the numerical comparison operator uses two equals signs. A common mistake for beginning programmers is to write code similar to the following:

```
if ($var1 = $var2) {
  print("the two variables are equal\n");
}
```

The test in this `if` statement should be `($var1 == $var2)`, the numerical test for equality. Instead, the `if` statement contains the assignment `$var1 = $var2`. This will copy the value of $var2 into $var1, and will also return TRUE (indicating that the value was copied successfully). Not only will the value of $var1 be unexpectedly changed, but the conditional code will always be executed. This sort of error can be very difficult to find, so it is important to exercise care to always use two equals signs when testing for numerical equality.

We can use the operators `&&` and `||` to combine conditions. When `&&` is used, the conditional block is only executed if both conditions are true. When `||` is used, the conditional block is executed if either one of the conditions is true. Parentheses placed around each of the conditions helps to keep the meaning clear. In the following example, the conditional code would be executed only if both of the conditions, `$x > 5` and `$y <= 10`, are true:

```
if (($x > 5) && ($y <= 10)) {
  conditional code;
}
```

Another useful operator is `!`, which tests for the opposite of a condition. The `!` operator is sometimes read as "not." For example, if I wish to execute some code only if the value of `$x` is not 3, I can either use the `!=` operator, as follows:

```
if ($x != 3)
{
  do something . . .
}
```

or I can use the `!` operator to test for the opposite of the condition `($x == 3)`, as shown below:

```
if (!($x == 3))
{
  do something . . .
}
```

Loops

Another common need in programming is the ability to repeat a block of code either a fixed number of times or until a certain condition is met. For that purpose, most languages provide **loop statements.** Two of the most commonly used loop statements in Perl, as well as many other languages, are the `while` loop and the `for` loop.

A `while` statement is structured very similar to an `if` statement:

```
$x=0;
while ($x < 5) {
  print("The value of x is $x\n");
  $x++;
}
```

Unlike an `if` statement, however, the block of code following the `while` statement is repeatedly executed for as long as the condition between the parentheses following the keyword `while` is true. In the example above, the block following the `while` statement will be executed five times (recall that the last line of code in the conditional block is shorthand for the statement `$x = $x + 1;`). Execution of this code will produce the following output on the screen:

```
The value of x is: 0
The value of x is: 1
The value of x is: 2
The value of x is: 3
The value of x is: 4
```

Note that the condition in the `while` statement is tested before executing the following block of code, so it is possible that the conditional block will not be executed at all. In the previous example, if the first line had read `$x = 7;` then the block of code following the `while` statement would not have been executed at all.

Another common loop structure is the `for` loop. A `for` loop is very similar to a `while` loop, but is slightly more limited in use. A `for` loop is used when we wish to execute a block of code a fixed number of times. The previous example can be rewritten using a `for` loop as follows:

```
for($x = 0; $x < 5; $x++)
{
  print("The value of x is: $x\n");
}
```

Note that the expression in parentheses in the `for` statement consists of three parts, separated by semicolons. The execution of a `for` loop proceeds as follows:

1. The first part of the `for` statement (`$x = 0`) is executed immediately when the `for` statement is reached.
2. Next, the second part of the `for` statement (`$x < 5`) is tested; if the condition is not met, then the rest of the `for` statement is skipped and we proceed with the code following the conditional block.

3. If the condition in the second part of the `for` statement is true, then the conditional code is executed.
4. The last part of the `for` statement (`$x++`) is executed.
5. Go to step 2 and continue. As long as the test in the second part of the `for` statement remains true, the conditional block and the last part of the `for` statement will continue to be executed. As soon as the test becomes false, the rest of the `for` statement is skipped and execution continues with the line following the conditional code.

You might have noticed at this point that using `for` loops in Perl is not strictly necessary—anything that can be done using a `for` loop can also be done with a `while` loop and a few more lines of code. The `for` loop is provided as a convenience to make it easier to write and read code for the very common situation in which we wish to have a block of code executed a fixed number of times. Examine the example shown earlier until you are convinced that it will produce the same results as the previous example using a `while` loop.

A final type of loop provided by Perl is the `foreach` loop. This loop is very similar in function to a `for` loop, but it executes the code in the loop once for each value in a list of values. The concept is best illustrated by an example:

```
foreach $value (7, 3, -3, 5, 2)
{
  print("The value is: $value\n");
}
```

This loop will execute five times, once for each value in the list in parentheses. For each iteration, the next value in the list will be assigned to the variable `$value`.

Readability

Structured Programming

As you begin to write longer and more complex programs, it becomes increasingly important to write code that is easy to read and understand. Although the examples we have looked at so far have only been a few lines long, real-world programs can range from a few to tens of thousands of lines of code. For such large and complex programs, it is essential to write well-organized and readable code. Furthermore, it is not uncommon for one programmer to pass along code to another to be modified for use on a slightly different problem. When sharing code with other programmers, writing clear and readable programs becomes even more important.

There are several habits that you can develop that will help you to produce easily readable code. The first is to use a **structured programming** style such as the one demonstrated in the previous examples. The basic elements of a well-structured program include appropriate use of new lines and consistent indenting of code blocks. The following example shows a well-structured program fragment. Note how each **nested block** of code is indented further than the previous block. It an easy matter to identify which opening and closing braces are paired because they are indented to the same level. Using correct program structure not only makes your programs easier to read, but also helps you to avoid mismatched parentheses and braces.

```
if ($x < 0)
{
  print("$x is negative, counting up . . . \n");
  while ($x < 0)
  {
    print("the value of x is $x\n");
    $x++;
  }
}
else
{
  print("x is positive, counting down . . . \n");
  while ($x > 0)
  {
    print("the value of x is $x\n");
    $x--;
  }
}
```

Comments

As your programs become longer and more complex, they will become more difficult for others to read and understand. Even when looking at your own code, it can be difficult to recall the function of a particular loop or block of code. Fortunately, Perl and most other commonly used programming languages provide a mechanism for jotting down notes, or **comments,** in your code that can help yourself and others to understand what the code does. You should employ comments liberally in your code to describe the function of complex blocks of code, variables, and so on. In Perl, comments are initiated using the # character. Any text following a # character is considered a comment. The following example shows a complete Perl script, with appropriate comments:

```
###############################################
# Factorial program                           #
#    John C. Programmer; July 2001            #
# This program prints the integers from       #
# 1 to 10 and the factorial of each to the    #
# screen.                                     #
###############################################
```

```
# For each number from 1 to 10 . . .
for ($number = 1; $number <= 10; $number++)
{
  # Print the number to the screen:
  print("Number: $number\t");

  # Initialize factorial to 1
  $factorial = 1;

  # Compute the factorial by multiplying the
  # current value by 2, 3, 4, . . . $number
  for ($i = 2, $i <= number, $i++)
  {
    $factorial *= $i;
  }

  # Now the factorial is computed; print it
  # and go on to the next number:
  print("Factorial: $factorial\n");
}
```

The control character \t, used in the first print statement, prints a tab character to the screen. Using the \t instead of \n in the first print statement allows each number and its factorial to be printed on the same line. You should examine this code until you are convinced that it will print the factorial of every integer from 1 to 10. Note how the consistent indentation makes it easier to see where the two nested `for` loops in this program start and end. There is no doubt that the final closing brace in this script matches the opening brace from the first `for` loop, because they are vertically aligned. Likewise, the indention of the second `for` loop makes it clear where this inner loop starts and ends. The liberal use of comments in the code clarifies the function of each block of code and makes the entire script easier to read and understand.

Descriptive Variable Names

Note the variable names in the previous example. Using descriptive variable names, like `$number` and `$factorial`, instead of nondescriptive names such as `$x` and `$y` helps to make your code easier to understand. One exception to this rule is simple loop counters, like `$i` in the example above, which are often reused several times and are quite often named using single letters. In general, the more descriptive the variable names, the better. Don't be afraid of using long variable names such as `$velocity_squared`—you will thank yourself later when you don't have to keep track of your variables on a piece of scratch paper while you are programming.

Data Structures

So far we have dealt only with simple scalar variables. Each scalar variable can hold only a single value. Sometimes, however, it is necessary to store and manipulate very large amounts of data. Data structures are variables that can store more than one value. Correct use of data structures is a key element of writing efficient and readable code.

Arrays

Suppose you are asked to write a program that stores 1,000 numbers entered by the user, and then sorts them by value and prints them to the screen. Using only the simple variable types we have discussed so far, it would be necessary to use 1,000 variables to store all of the numbers until they are sorted. Such a cumbersome solution is not necessary, however, because Perl provides us with **arrays**—variables that can store multiple values. Using an array, we can use a single variable to store all 1,000 numbers. An array variable in Perl begins with the @ character. For our array of 1,000 numbers, we might use an array named `@numbers`. However, we will only need to use the name `@numbers` when we wish to refer to the entire array at once, which will not occur very often. Instead, we will reference individual elements of the array, which are accessed using square braces as follows:

```
$numbers[20] = 20;
$numbers[21] = 15;
```

Note that each element of an array is an ordinary scalar value, so it starts with a $ character. It is not necessary to include any special code in our Perl script to create the array `@numbers`. The first time we access an element of the array (e.g., `@numbers[3] = 5;`) the array `@numbers` will be created. It is important to distinguish between the different types of variables that might have similar names:

- `$numbers`—a scalar variable containing one value,
- `@numbers`—an array variable containing many values, and
- `$numbers[3]`—an element of the array `@numbers`.

In Perl, the elements of an array are usually numbered starting at 0. So the 1,000 elements of our `@numbers` array are named `$numbers[0]` through `$numbers[999]`. The number of a particular array element (in square brackets) is called the **index.** Once we have declared our array, we can even use another variable to identify which element to access. So the code to print 1,000 values stored in our `@numbers` array can be as simple as this:

```
for ($i = 0; $i < 1000; $i++)
{
  print("The next number is $numbers[$i]\n");
}
```

Note that this `for` loop will result in values of `$i` ranging from 0 to 999. The following example program computes the average of 10 numbers entered by the user. The `<STDIN>` operation, used in this example to get a number typed by the user, will be discussed in detail in the Input and Output section. Note in this example how an array is used to store the numbers entered by the user and how a `for` loop is used to step through the array and sum the values.

```
####################################################
# Averaging program; John C. Programmer, 2001      #
#                                                  #
# This program prompts the user for 10 numbers     #
# and then prints their average.                   #
####################################################

# prompt the user for all 10 numbers and store
# them in the array 'numbers'
for($i = 0; $i < 10; $i++)
{
  print("Enter another number\n");
  $numbers[$i] = <STDIN>;
}

# Sum up all the values in the array 'numbers' */
$sum = 0;
for($i = 0; $i < 10; $i++)
{
  $sum += $numbers[$i];
}

# Divide sum by 10 to get the average
$average = $sum / 10.0;

# Print the average
print("The average of all 10 numbers is $average\n");
```

Hashes

One of the most useful data structures provided by Perl, particularly for bioinformatics applications, is the **hash.** A hash is similar to an array, but its indexes are not limited to integers. Instead, a hash can use any numeric or string value to select or index its elements. For example, suppose we wish to store the number of heavy (nonhydrogen) atoms in the side chain of each of the 20 amino acids. Using an array, we would have to number the amino acids from 0 to 19, and then store the counts as follows:

```
$atom_count[0] = 1; # Alanine (A) is amino acid 0
$atom_count[1] = 2; # Cysteine (C) is amino acid 1
$atom_count[2] = 4; # Aspartate (D) is amino acid 2
. . .
```

The primary difficulty with this method is that we have to remember the number for each amino acid. Using a hash instead, we can use a string, such as the one- or three-letter abbreviation for the amino acid, as the index for the hash entries, as in the following example. Note that the index for a hash is enclosed by braces, rather than the square brackets used for array indices.

```
$atom_count{"ALA"} = 1;
$atom_count{"CYS"} = 2;
$atom_count{"ASP"} = 4;
 . . .
```

Just as with arrays, the individual elements of a hash are ordinary scalar variables, and so we refer to them using a dollar sign. To refer to the entire hash at once, the hash name is prefixed with a % sign. The hash created in the previous example would be named `%atom_count`. We will examine some examples where we need to refer to an entire array or an entire hash at one time in the Functions section.

Working with Strings

Because Perl uses scalar variables to hold strings, and because Perl makes it very easy to create and work with strings, it may seem that strings are not really much of a data structure at all. However, note that in many languages string variables are more tedious to create and manipulate. One of the great strengths of Perl for bioinformatics is the ease with which it allows us to read, write, and manipulate strings of text. Nevertheless, there are a few important things to keep in mind when dealing with scalar variables that contain text strings. One of the most important concepts in working with strings is knowing when variables will be **evaluated,** or replaced with their contents, and when they will not. Most of our examples so far have contained statements similar to the following:

```
print("The value of x is $x\n");
```

Assuming that the variable `$x` contains the value 5, this will print the sentence "The value of x is 5" to the screen. When the string is printed, the variable `$x` is replaced with the value 5. Any time we create a string using double quotes, any variables inside the string are replaced with their contents. Consider the following example:

```
$first_name = "John";
$middle_initial = "Q.";
$last_name = "Programmer";

$full_name = "$first_name $middle_initial $last_name";
```

After this code is executed, the variable `$full_name` will contain the string "John Q. Programmer". When we use double quotes to create the string in the last line, the variables `$first_name`, `$middle_initial`, and `$last_name` are replaced with their values.

What if we want to create a string in which the variables are not evaluated (replaced with their values)? We can do this in Perl by enclosing the string in single quotes. If the last line of the previous example appeared as follows:

```
$full_name = '$first_name $middle_initial $last_name';
```

Then the value of the variable `$full_name` would be the string `"$first_name $middle_initial $last_name"`. No variable evaluation would occur.

Occasionally, when working with variables and strings, it is not clear where a variable name should end. For example, suppose I wish to append the letter "y" to a string variable. If I use the following code:

```
$name = "Mike";
$nickname = "$name y";
```

I would get the string "Mike y" in `$nickname`, not "Mikey". However, if I do this:

```
$name = "Mike";
$nickname = "$namey";
```

then Perl will look for the variable `$namey`, which doesn't exist. The solution is to use curly braces around the variable name. In Perl, we can use braces after a dollar sign to surround the name of the variable. So `${name}` is the same as `$name`, but the former method makes it clear where the name of the variable starts and ends. Now we can create our nickname variable as follows:

```
$name = "Mike";
$nickname = "${name}y";
```

As we hoped, the variable `$nickname` now contains the string "Mikey".

Another difficulty in dealing with strings occurs when we wish to include a double or single quote within a string. For example, what if we wish to print the following string to the screen, exactly as it appears below?

```
His name is Mike, but people call him "Mikey".
```

How can we include the double quotes in the string above, without prematurely ending our string? The answer is to use **escape sequences,** special character sequences that, in Perl, are initiated with a backslash. We already know two escape sequences: \n, which moves the cursor to the beginning of the next line on the screen, and \t, which moves the cursor right to the next tab stop position (on most screens, the tab stop positions occur every eight characters). The escape sequences \" and \' allow us to insert a double or single quote into a string, without ending the string earlier than we intend. To print the string from the previous example, we can use the following code:

```
print("His name is Mike, but people call him \"Mikey\".");
```

To place an ordinary backslash character into a string, we can use the escape sequence \\.

Subroutines and Functions

As you write more programs in Perl, you will often find that you need to perform the same calculations repeatedly in the same program. For example, if you are writing a program that deals with protein atom coordinates, you might find that you need to compute the Euclidean distance between two atoms several times in one program. One way to deal with this would be to write some code to compute the distance, and then cut and paste this code into each spot where you need it. This approach, however, has several drawbacks. First, you will end up with a very long program consisting largely of repeated code. Even worse, if you should discover an error in your Euclidean distance calculation, you have to find every place in your program where you have pasted your code and fix the error in every one of them. Finally, your program won't be particularly efficient. That is, it will use more computer memory than it really needs because each copy of your distance calculation code takes up some space in memory.

Fortunately, most programming languages provide a way to avoid repeated code by writing a single block of code that can be used repeatedly in your program without cutting and pasting. In Perl, a reusable block of code is called a **subroutine.** The following simple example shows a subroutine that prints the coordinates of an atom:

```
#################################################
# Program to print the coordinates of two atoms #
#################################################

# Create atom 1, use an array with three elements
# for the three dimensions
$atom1[0] = 3.12;     # X-Coordinate
$atom1[1] = 22.5;     # Y-Coordinate
$atom1[2] = 112.34;   # Z-Coordinate
# Call the subroutine to print the coordinates
print("Atom 1:\n");
print_coords($atom1[0], $atom1[1], $atom1[2]);

# Create another atom
$atom2[0] = 121.1;      # X-Coordinate
$atom2[1] = 62.25;      # Y-Coordinate
$atom2[2] = 23.12;      # Z-Coordinate

# Call the subroutine to print the coordinates
print("Atom 2:\n");
print_coords($atom2[0], $atom2[1], $atom2[2]);
print("\n");

# Exit the program
exit;
```

```
##########################################################
# print_coords -- print three-dimensional coordinates to
#                 the screen
#
# Inputs: Three coordinates: x, y, and z.
#
# Outputs: None
#
# This subroutine expects three arguments, which are
# printed to the screen.
##########################################################
sub print_coords
{
  # Copy the three arguments to the subroutine into
  # $x, $y, and $z
  my $x = $_[0];
  my $y = $_[1];
  my $z = $_[2];

  print("X Coordinate: $x\n");
  print("Y Coordinate: $y\n");
  print("Z Coordinate: $z\n");
}
```

This program would produce the output:

```
Atom1:
X Coordinate: 3.12
Y Coordinate: 22.5
Z Coordinate: 112.34

Atom2:
X Coordinate: 121.1
Y Coordinate: 62.25
Z Coordinate: 23.12
```

The main program in this example simply declares two arrays named `atom1` and `atom2`, each of which contain three elements: the `x, y,` and `z` coordinates of the atom. After the array is created, the three elements are passed to the subroutine `print_coords`, which prints them to the screen. The line `print_coords($atom1[0], $atom1[1], $atom1[2]);` calls the subroutine, and passes it three values, the `x, y,` and `z` coordinates of `atom1.` To call a subroutine, we simply use the subroutine name, followed by any **parameters** to the subroutine, enclosed in parentheses. Parameters are the values on which we want the subroutine to operate. The benefit of using a subroutine is that we can call it many times with different parameters. Sometimes the parameters to a subroutine are also referred to as its **arguments.**

The definition of the subroutine begins with the line `sub print_coords`. This defines the name of the subroutine to be `print_coords`. Everything enclosed in the pair of braces following this line is part of the subroutine. The first few lines of the subroutine define some **local variables.** These are variables that exist only inside this subroutine, and are declared using the keyword `my`. If there is a variable called `$x` in our main program, or in another subroutine, it is different from the `$x` used in `print_coords`, because we declared `$x, $y,` and `$z` preceded by the keyword `my`. To put it another way, changes to `$x` in `print_coords` will not change the values of any other variables named `$x` elsewhere in this program. If we don't use `my` when creating variables in subroutines, then they are **global variables,** meaning they are visible elsewhere in the program, and any changes we make will affect any other variables elsewhere that have the same name. Using local variables inside your subroutines is a good programming practice to follow, because it eliminates the need to check the variable names in all the other subroutines in a program to make sure they aren't the same as the variables in the current subroutine.

In addition to declaring `$x, $y,` and `$z` as local variables, the first few lines of the subroutine copy the three parameters to the subroutine into these variables. In Perl, the parameters to a subroutine are stored in a special array whose name is simply the underscore (_) character. This array has as many elements as there were parameters passed to the subroutine when it was called. Since we called `print_coords` with three arguments, this array has three elements in the subroutine. Because array numbering in Perl starts at 0, the elements are named `$_[0], $_[1], $_[2]`. The rest of the subroutine is fairly straightforward; it simply prints the values of the three arguments to the subroutine using the now-familiar `print` statement.

When an entire array is passed as an argument, Perl passes all of the elements of the array as if they were typed out in the parentheses following the subroutine name. So, instead of listing all three elements of the array `atom1`, we could have called the `print_coords` subroutine as follows, with the same results:

```
print_coords(@atom1);
```

Note that we use the `@` character to refer to the entire array `atom1` at once.

This example subroutine captures the first three variables in the `@_` array into `$x`, `$y`, and `$z`. If we call the subroutine with more than three parameters, the extras are simply ignored. Sometimes it is useful to write a subroutine that can accept any number of parameters. Using the `foreach` control structure, this is very easy to do in Perl, as in the following example:

```
####################################################
# Example program using foreach
####################################################

# Declare array1 with 5 elements
@array1 = (7, 5, 9, 12, -3);

# Call the print_array subroutine
print_array(@array1);
```

```
Declare array2 with 7 elements
@array2 = (3, 5, 9, 12, 3, -3, 5);

# Call the print_array subroutine
print_array(@array2);

##################################################
# print_array -- print the contents of an array
##################################################
sub print_array
{
  $count = $#_+1;
  print("The array contains $count elements.\n");
  print("The elements of the array are:\n");
  foreach $element (@_) {
    print("$element\n");
  }
}
```

Note that this example introduces a new way to create an array. Rather than having to tediously create an array element by element (`$array1[0] = 7; $array[1] = 5;` etc.), we can create an array all at once using a list of values enclosed in parentheses. Secondly, note that the variable `$#arrayname` contains the index of the last element in the array `@arrayname`. Since array numbering starts at zero, the number of parameters passed to the subroutine will equal `$#_+1`. By using an array in place of the usual list of values in a `foreach` statement, the loop iterates with each value of the array assigned to the variable `$element`. This sort of loop is very convenient for creating subroutines that act on each input parameter, no matter how many parameters are passed to the subroutine.

It is often useful to create a subroutine that performs some computation on the parameter values and returns the result to the main routine. A subroutine that returns a value is often referred to as a **function.** In Perl, there is not much difference between a function and a subroutine, except that functions use the `return` statement to return a value to the calling program. When you call a function, you can assign the returned value to a variable. For example, if I wrote a function called `average` that returned the mean of the input parameters, I would assign the result to a variable as follows:

```
@values = (23, 17, 83, 21, 54, 23, 87);
$mean = average(@values);
```

The following example program demonstrates the use of a function to compute the Euclidean distance between two points.

```
#########################################################
# This program example computes the Euclidean distance
# between two atoms.
#########################################################

# Store the coordinates for atom1 in the array @atom1
@atom1 = (1.212, 19.215, 102.23);
```

```
# Store the coordinates for atom2 in the array @atom2
@atom2 = (32.202, 220.21, 23.022);

# Compute the distance between the two atoms
$distance = euclid_dist(@atom1, @atom2);
print("The distance from atom1 to atom2 is $distance\n");

#####################################################
# euclid_dist - compute the Euclidian distance      #
#               between two points                  #
#                                                   #
# Inputs: Two points in 3D space, specified by      #
#         their x, y, and z coordinates. Each       #
#         point's coordinates are stored in a       #
#         hash with elements x, y, and z.           #
#                                                   #
# Returns: The distance between the two points      #
#####################################################
sub euclid_dist
{
  # Call the input parameters point1 and point2
  my @point1 = ($_[0], $_[1], $_[2]);
  my @point2 = ($_[3], $_[4], $_[5]);

  # compute the distance along each dimension
  my $xdist = $point1[0]-$point2[0];
  my $ydist = $point1[1]-$point2[1];
  my $zdist = $point1[2]-$point2[2];

  # compute the sum of the squared distances
  my $edist = $xdist * $xdist;
  $edist += $ydist * $ydist;
  $edist += $zdist * $zdist;

  # take the square root for euclidean distance
  $edist = sqrt($edist);

  return($edist);
}
```

When the two arrays are passed to the subroutine `euclid_dist`, the arrays are expanded as though all elements of each array were parameters to the subroutine, as previously discussed. Thus, the line

```
$distance = euclid_dist(@atom1, @atom2);
```

is equivalent to

```
$distance = euclid_dist($atom1[0], $atom1[1], $atom1[2],
  $atom2[0], $atom2[1], $atom2[2]);
```

The subroutine assigns the first three parameters to the array `@point1` and the second three to the array `@point2`. It then computes the distance in each dimension, squares them, and adds them to `$edist`. Finally, the square root of this sum of squared distances in `$edist` is computed and returned to the calling program.

Input and Output

One of the reasons why Perl is a popular language for bioinformatics applications is the ease with which it allows manipulation of text files. Many bioinformatics tools read text files containing sequences, sequence alignments, or protein structures as their input, and produce text files of various formats as output. As you develop your Perl programming skills, you will find that Perl is a powerful tool that you can use to convert data between formats, summarize your results, and even perform sophisticated analyses of sequence and structural data. To do all of these things, it is necessary to know how to read and write information to and from text files in Perl.

Before a file can be read from or written to in Perl, it must first be *opened.* When you open a file, you create a special variable associated with the file. Once the file is opened, we use the variable to refer to the file whenever we want to perform a read or a write operation. The Perl function `open` is used to open a file, as follows:

```
# This program will echo the contents of the file
# poem.txt to the screen

open(POEM,"<poem.txt");

while(<POEM>) {
  $line = $_;
  print($line);
}
```

This fairly short script, which prints the contents of the file "poem.txt" to the screen, presents a number of new concepts. First, take a look at the `open` function. The first argument to the function is the special variable to be associated with the file we are about to open. This variable is called the **file pointer,** and it is neither a scalar variable, a hash, nor an array, so it doesn't start with any of the special characters ($, %, @) we usually associate with Perl variables. To make it easier to remember that this is a file pointer variable, it is common Perl programming practice to use a name consisting of all capital letters (in this case, POEM). The second argument to `open` is the name of the file to open. In this example, we want to open the file "poem.txt". By placing the character '<' before the file name, we have indicated that we wish to open the file only for reading. Perl looks at the first character of the file name in an open statement to determine whether the file should be opened for reading, writing, or both. For our file "poem.txt" some of the file names we might use include:

- "<poem.txt"—open poem.txt for reading only.
- ">poem.txt"—open poem.txt for writing only. Erase the current contents of the file, and start writing at the beginning.
- ">>poem.txt"—open poem.txt for append. Leave the current contents intact and begin writing/appending at the end of the file.
- "+>poem.txt"—open the file for both reading and writing.

To read a line from a file that has been opened for reading, we use the operator <> with the file pointer variable. In the example code, the command `<POEM>` reads the next line of the file. The file pointer variable `POEM` keeps track of the last line read, so each time we execute `<POEM>` we get the next line of the file. As the line is read, it is placed in the special variable `$_`. The first statement inside the `while` loop copies the line into the variable `$line`, and the statement prints the line to the screen.

In this example, the `<POEM>` command is also used as a test. Each time the command executes, it reads the next line of the file poem.txt and places the line into the variable `$_`. If this process succeeds (that is, there was another line in the file to read), then `<POEM>` also returns the value `TRUE`, just like the test `($x < 3)` returns `TRUE` when the value of `$x` is less than 3. When there are no more lines to read in poem.txt, `<POEM>` returns `FALSE`. Thus, we can use it as a test in the while loop. Thus, the `while` loop in this example proceeds as follows:

```
while (there are more lines to read in poem.txt) {
  get the next line into the variable $_;
  copy the line into $line;
  print($line);
}
```

You may notice that copying the value into the variable `$line` is not strictly necessary. In the example, this was done only to clarify the operation of the `<POEM>` operation. We could just as well rewrite the loop as follows:

```
while(<POEM>) {
  print($_);
}
```

Believe it or not, we can shorten the loop even a bit further. When the `print` function is called with no arguments, it prints the special variable `$_` by default. So we could actually write the `while` loop as:

```
while(<POEM>) {
  print;
}
```

Perl has a number of shortcuts, similar to this one, that can make your programs more concise. However, these shortcuts generally come at the cost of making your program less readable as well. Given the choice between the original `while` loop given in the example and the loop above, the former provides the same function while being easy to read, even for a nonexpert in Perl.

Regular Expressions

One of the most powerful components of Perl is the regular expression, but unfortunately it is also among the easiest ways to write short, cryptic sections of code. There is nothing that a regular expression does that cannot also be performed by a larger amount of code, but depending on the regular expression, this code could turn out to be rather large and difficult to write. This section is intended to provide an introduction to regular expressions so that you will understand what you're seeing if you encounter them in Perl code written by someone else, but it is not intended to be a comprehensive guide.

In short, a regular expression is a shorthand method for matching portions of strings, by stating a template against which the string is compared. Two of the most common uses of regular expressions are in `if` statements and transliteration, or `tr` statements, which are covered later in this section.

Now that the generalities have confused you, let's look at an example that will hopefully make things clearer. Suppose we wanted to look at a string and see if it contained the word *alpha* somewhere within it. Assuming that the string we want to compare is contained within the variable `$string`:

```
if ( $string =~ /alpha/)
{
  printf( "Match!\n");
}
else
{
  printf( "No match.\n");
}
```

Several new elements are introduced here, so let's consider each one separately. First off, there's the shorthand notation `/alpha/`—this is the string that we wish to match. To show that matching is what we want to do, we use the `=~` or **binding operator,** which in this context you can think of as being the same as `==`, but for regular expressions.

Suppose we want to match not just a particular string, but several variations of that string, such as CAU or CAC, representing histidine. We could do an `if` containing an or clause, but there is an easier way using regular expressions. We can use a matching string that contains a **character class,** as `/CA[CU]/`. This says that the string must match CA, followed by either C or U. Although not appropriate to this example, we can also use the hyphen character to indicate a sequence of characters; instead of saying `[ABCDEFabcdef]`, we can say `[A-Fa-f]`.

After specifying a character class, suppose that we want to indicate that it can appear multiple times. This requires the special **quantifiers** `*` and `+`. The `*` quantifier means that the class can appear 0 or more times, while `+` means that it can appear 1 or more times but must occur at least once. For instance, if we want to indicate that a number must occur, we might say something like `[0-9]+`, or if we want to say that it cannot start with a zero, `[1-9][0-9]*`.

There is also a special wildcard character, indicated by a period. For instance, if we wanted to indicate that we're looking for the words *alpha* and *helix*, with anything at all in between them (or nothing at all), we would use the regular expression `/alpha.*helix/`. If we wanted to indicate that something had to occur between them, we would simply change the `*` to a `+`.

You can anchor an expression by using the special characters `^` or `$` in the regular expression. These mean that the given pattern must occur at either the beginning or end of the string, respectively. For instance, `/^alpha/` would only match if the word *alpha* occurs at the beginning of the string, while `/helix$/` only matches if the word *helix* occurs at the end.

As mentioned previously, besides matching, regular expressions can also be used to perform **transliteration.** This means that the matching characters are changed, or transliterated, into other characters. The operator used to do this is `tr` and, as a simple example, suppose that we wanted to take an RNA sequence and convert it to DNA, changing all the U's to T's. Using `tr`, we can do this simply as:

```
$sequence ="GCUACGUUCGAAGCU";
$sequence =~ tr/U/T/;
printf("sequence is $sequence\n");
```

This would result in the output: `sequence is GCTACGTTCGAAGCT`. Note that the `tr` operator can be destructive—the contents of `$sequence` have been changed.

Additionally, the output of the `tr` operator is a count of the number of characters transliterated. This can be used to our advantage, allowing us to not only count the number of changes, but to count changes without actually making any! We do this by simply not stating what the matching characters should be changed to. For example, if we want to count the number of U's in a sequence without changing them:

```
$sequence ="GCUACGUUCGAAGCU";
$Ucount = ( $sequence =~ tr/U//);
printf( "There are $Ucount uracil in the sequence\n");
```

Where to Go from Here

The tools you have learned about in this chapter will allow you to write simple Perl scripts to perform a variety of tasks. Many of the concepts presented are common to other languages including C, C++, Java, and Python, so you should now have the tools you need to learn and use most of the programming languages commonly employed for bioinformatics applications. For all but the most demanding or specialized bioinformatics programs, Perl is more than powerful enough to get the job done. However, this appendix has only scratched the surface of the many functions, operators, and other facilities available in Perl. For bioinformatics applications, the string parsing and pattern matching features of Perl, in particular, are indispensable. Other aspects of Perl that are of interest for bioinformatics programming include:

- Regular expressions and string substitution operators,
- Command-line arguments: @ARGV,
- Perl modules and the Perl module library,
- References, typeglobs, and parameter passing, and
- Lists of lists and multidimensional arrays in Perl.

Numerous references are available to help you learn to use these operators. With what you now know about programming in general, and Perl in particular, these references should be accessible and useful for creating powerful bioinformatics applications in Perl.

Readings for Greater Depth

Numerous introductory materials are available on the World Wide Web for learning basic and advanced Perl programming. Since these resources change at a rapid pace, the best strategy is to use a web search engine with keywords such as "Perl tutorial" and "Perl introduction" to find the most recent material.

Several books from O'Reilly are available that present various aspects of the Perl language, including *Learning Perl*, *Learning Perl on Win32 Systems*, *Perl in a Nutshell*, *Perl Cookbook*, and *Advanced Perl Programming*.

A good next step for bioinformaticians wishing to learn additional programming skills in Perl is J. Tisdall, 2001, *Beginning Perl for Bioinformatics*, O'Reilly & Associates, Sebastopol, CA.

For further exploration of fundamental concepts in programming, a college-level textbook on data structures and algorithms is recommended, such as T. H Corman, C. E. Leiserson, and R. L. Rivest, 2001, *Introduction to Algorithms*, 2nd edition McGraw Hill, New York, NY.

If you are using Perl on a UNIX system, *man* pages are generally available that can serve as useful references and reminders of the format for the various Perl functions and operators. From a UNIX command prompt, type `man perl` for more information. Two of the most useful are the page on Perl operators (`man perlop`) and the page describing Perl functions (`man perlfunc`).

Questions and Problems

* **A1.1** Write a short Perl script to print the integers from 1 to 100 to the screen.

A1.2 Write a Perl script to print the integers from 1 to 100 that are divisible by 7 to the screen. Use the modulus function (%) to determine if a number is divisible by 7.

* **A1.3** Write a Perl script to read a file named “values.txt” containing real numbers, one number per line, into an array. Once all of the numbers have been stored in the array, print the average (mean) of the numbers to the screen.

A1.4 Rewrite the script in Question A1.3 using a function to compute the mean. Pass the array as an argument to the function.

* **A1.5** Create a text file containing the following triplets of *X*, *Y*, and *Z* coordinates:

```
45.010    48.193    104.291
60.160    55.939    117.081
39.849    48.051    135.613
46.676    46.998     66.327
31.434    85.158     97.469
32.996    37.720    111.954
47.852    44.686    100.663
29.227    37.351    103.290
54.642    90.459    126.119
```

Assuming that each line represents the *X*, *Y*, and *Z* coordinates of an atom, in angstroms, write a Perl script to count the number of atoms within 20 angstroms of the point 45.0, 45.0, 100.0. [*Hint:* You can use the `split()` function to split a line into several variables. For example, assuming you have read the first line of the file into `$line`, you can use `($x, $y, $z) = split(″, $line)`, which will set `$x` = 45.010, `$y` = 48.193, and `$z` = 104.291. There is a single space between the two single quotes, which tells `split` to split the line wherever there are spaces between characters.]

A1.6 Write a subroutine that checks a string of characters and returns “true” if it is a DNA sequence. Write another that checks to see if it is a protein sequence written in the one-letter code for amino acids.

APPENDIX 2

Enzyme Kinetics

An important emerging area of bioinformatics, systems biology, strives to model biological systems (including biochemical pathways, genetic regulation and even interactions between cells, tissues, and whole organisms). The remarkable integration and interdependency of metabolic processes pose daunting challenges to systems biologists. Learning which molecules are in a particular metabolic pathway is a good starting point, but knowing how each protein functions and interacts with other molecules is essential for useful understanding. Despite the underlying complexity of protein structure and almost unique catalytic strategy of every enzyme, much of enzyme kinetics can be reduced to a few fairly simple equations and essential parameters. Determination of an enzyme's V_{max} and K_m, although considered to be "old-fashioned biochemistry" by some, gives remarkably clear insight into the normal physiology of an organism and is a fundamentally important starting point for any attempts at higher order modeling.

Enzymes as Biological Catalysts

From an energetics perspective, chemical reactions are substantially more than the conversion of reactants to products (Figure A2.1). **Transition state theory** states that products are formed only after reactants have (1) collided in an appropriate spatial orientation and (2) acquired enough **activation energy (E_{act})** to reach a transition state. Transition states represent something of a halfway point in a chemical reaction when the chemical bonds of the reactant(s) are distorted in a way that allows conversion to product(s). The lower a reaction's E_{act}, the more easily the transition state will be reached and faster the reaction will occur. Catalysts, including enzymes, increase rates of chemical reactions by selectively lowering E_{act}. This property of enzymes can have extremely dramatic effects. For instance, carbonic anhydrase is an enzyme within mammalian blood that converts carbon dioxide (CO_2) and water (H_2O) to carbonic acid (H_2CO_3). Even when no catalyst is present, the reaction takes place at a rate of about one product being produced every 2 minutes at normal concentrations and temperature in blood. A single molecule of carbonic anhydrase in the same circumstances increases the reaction rate by over one million fold—100,000 molecules of CO_2 are converted to carbonic acid every second. A surprisingly small reduction in activation energy (2×10^4 calories/mole versus 1×10^4 calories/mole in the presence of carbonic anhydrase) is responsible for this dramatic effect. While some enzymes enhance reaction rates by a factor of up to 10^{15}, most enzymes cause specific chemical reactions to proceed 1,000 to 10,000 times faster than they would have occurred in their absence and have fairly small effects on activation energies.

Notice in Figure A2.1 that the chemical reaction proceeds from a state in which the reactants have a high energy to a product that has a lower amount of energy associated with it. The energy associated with a closed system is determined solely by the energy stored in its chemical bonds (H, usually described in terms of joules/mole); its temperature (T, described in terms of degrees Kelvin, K); and the extent to which it is disordered (S, also usually described in terms of

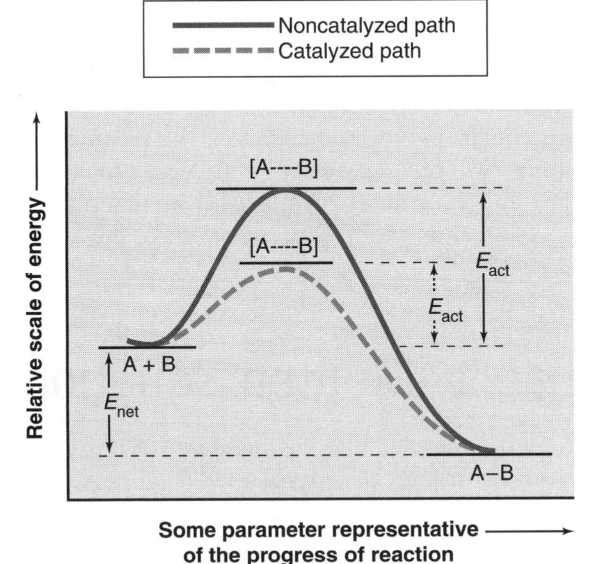

FIGURE A2.1 *Energy profile diagrams of a chemical reaction with and without a catalyst. In the hypothetical reaction A + B → B – A, the transition state energy* (E_{act}) *is greater than either the energy levels of the reactants or the product. In the presence of a catalyst for this reaction (dashed line),* E_{act} *is lower than in its absence (solid line) and the reaction proceeds more quickly. The total amount of energy released during the course of the reaction* (E_{net}) *remains the same both with and without a catalyst.*

joules/K/mole). The total energy of a system is usually described in terms of its *Gibbs free energy* (G), and changes in G (ΔG) can also be described in terms of chemical energy, temperature, and disorder ($\Delta G = \Delta H - T\Delta S$). Chemical reactions that release energy are **exergonic** and have negative Gibbs free energies ($-\Delta G$; $G_0 > G_1$). The energy released in reactions like these can be used to do work if it is not lost as heat. In contrast, reactions that require a net input of energy to convert reactants to products are **endergonic** and have positive ΔG's ($G_0 < G_1$). Many of the chemical reactions that are necessary for life are very endergonic and can only occur when they are coupled by enzymes to other chemical reactions that are even more exergonic. The large negative values for the ΔG of the exergonic reactions offset the positive ΔG of the desired reaction. While chemical reactions with large negative values for their ΔG are more **thermodynamically favorable** than those with positive or smaller negative values for their ΔG, they do not necessarily take place any more quickly. Reaction rates are determined entirely by the amount of energy that must be acquired to reach the reaction's transition state (and not necessarily the energy difference between reactants and products).

Appreciate also that chemical reactions are usually **reversible,** meaning that products can be converted back to reactants. In Figure A2.1 the activation energy for the forward reaction (A + B → A – B) is lower than that of the reverse reaction (A – B → A + B). That difference means that if reactants and products are present at the same concentrations the rate k_1, at which the forward reaction occurs, will be greater than the rate k_{-1}, at which the reverse reaction occurs. In time, though, more product (A – B) will be available for conversion to reactants (A + B), and the net number of conversions between the two states will be the same. The point at which the relative concentration of products and reactants no longer changes occurs when an **equilibrium constant, K_{eq},** is reached. If the forward reaction occurs 100 times more rapidly than the reverse reaction, then $K_{eq} = k_1/k_{-1} = 100$ and there will be 100 times more product than reactants when the reaction is in equilibrium. For any given reaction, enzymes always enhance both k_1 and k_{-1} to the same degree—as a result, they have no effect on K_{eq}.

Among the most powerful insights from thermodynamics is the relationship between K_{eq} and ΔG: $\Delta G = -RT\ln K_{eq}$ (where $R = 2.0 \times 10^{-3}$ kcal/degree/mole and T = temperature in degrees Kelvin). Values for K_{eq} can usually be determined easily in the laboratory and, as a result, allow ready determinations of the ΔG of reactions as well.

The Henri–Michaelis–Menten Equation

In the late 1800s studies of enzymes had revealed that the initial rate of a reaction was directly proportional to the concentration of enzyme used (Figure A2.2a). However, reaction rates were also observed to increase in a nonlinear manner and to approach a limiting maximum rate as reactant concentration was increased (Figure A2.2b). By 1903, V. Henri was able to use those observations to theoretically derive a general mathematical equation that related v_0 (the rate at which an enzyme catalyzed a reaction when substrate concentrations were low, as in the left-hand part of the curve in Figure A2.2b) to [E] (enzyme concentration) and to [S] (a specific concentration of reactants, or substrates).

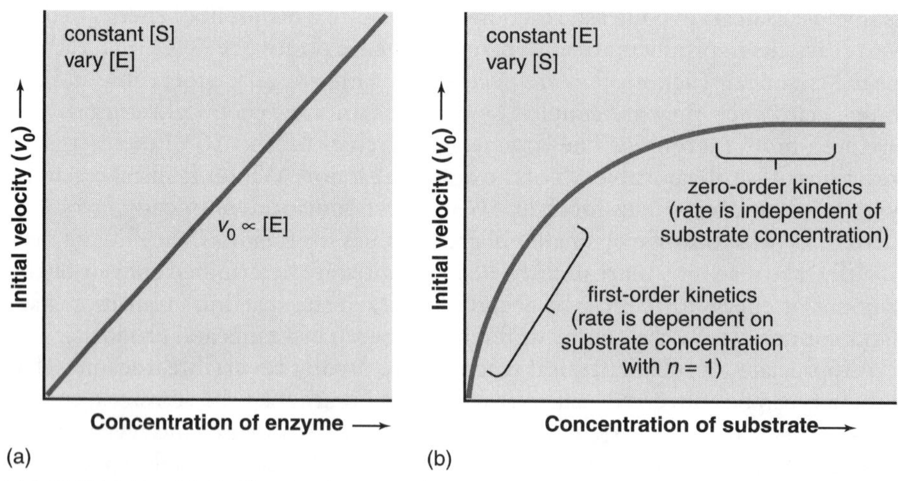

FIGURE A2.2 *Kinetic features of an enzyme-catalyzed reaction. (a) Reaction rate is linearly related to enzyme concentration. (b) Reaction rate and substrate concentration have a hyperbolic relationship. When substrate concentrations are low, the reaction rate increases in a way that is directly proportional to substrate concentration. However, when substrate concentrations are high, reaction rates change very little and asymptotically approach a maximum velocity (V_{max}).*

At the heart of Henri's model was the idea that enzyme-catalyzed reactions occurred in two steps: (1) substrates (S) bound to the enzyme (E), and (2) products (P) were released from the enzyme. Each of those steps appeared to have their own characteristic equilibrium constants so that the overall reaction could be written as:

$$E + S \underset{k_{-1}}{\overset{k_1}{\longleftrightarrow}} ES \underset{k_{-2}}{\overset{k_2}{\longleftrightarrow}} P + E.$$

Over the course of the next 10 years, L. Michaelis and M. Menten used carefully designed and controlled laboratory experiments to test that model and presented a slightly modified version of Henri's rate equation that is now commonly known as the **Henri–Michaelis–Menten equation:**

$$v_0 = (V_{max}\,[S])/(K_m + [S])$$

where V_{max} is the maximum rate at which an enzyme catalyzes a reaction when substrate concentrations are high (as in the right-hand part of the curve in Figure A2.2b); and K_m is the Michaelis–Menten constant {$K_m = (k_{-1} + k_2)/k_1$; a grouping of the rate constants that have the biggest effect on [ES]}. Simple inspection of the Henri–Michaelis–Menten equation reveals that K_m corresponds to the substrate concentration that allows an enzyme to generate products at half their maximum velocity ($1/2V_{max}$).

V_{max} and K_m

Experimentally determined values for V_{max} and K_m have proven to be fundamentally important descriptors of any given enzyme's kinetic activity as well as the intracellular environment in which natural selection has honed the enzyme to work best. For instance, V_{max} describes what is known as the **turnover number** of an enzyme [the moles of substrate converted to product per mole of enzyme per unit time (usually, 1 minute)]. An enzyme's K_m gives invaluable insights into the normal physiological state inside cells in at least four different ways. First, K_m generally corresponds to the substrate concentration that an enzyme normally encounters within a cell. (Substrate concentrations much below K_m would generate wide variations in reaction velocity and those velocities would be far below V_{max}; substrate concentrations much above K_m would result in only marginally faster reaction rates—a 10,000-fold increase in substrate concentration above K_m results in only a doubling of the reaction rate.) Second, since K_m is a constant for any given enzyme, it is possible to directly compare the efficiency of related (or altered/mutated) enzymes from different tissues or organisms simply by comparing K_m values. Third, for enzymes that utilize a variety of substrates, it is possible to compare the relative efficiency with which they are utilized by comparing the K_m for each substrate. (Lower K_m values correspond to greater enzymatic specificity and imply the substrate is preferred.) Fourth, measurements of the effects of different compounds on an enzyme's K_m provide an objective way of determining their role as activators or inhibitors of the reaction—an issue of particular importance in the field of drug discovery.

Direct Plot

Estimated values for V_{max} and K_m can be determined fairly easily from experimental data. For instance, a direct plot of initial catalytic velocity, v_0, determined at a range of different substrate concentrations, [S], when enzyme concentration is held constant allows a visual estimation of both V_{max} and K_m. Consider the reaction catalyzed by hexokinase, which yields the data provided in Table A2.1. Hexokinases play a central role in metabolism by transferring a phosphate group to six-carbon sugars like glucose, as shown in Figure A2.3. A direct plot of the data (y = [S], $x = v_0$) from Table A2.1 is shown in Figure A2.4 on page 242. A value for V_{max} can be estimated from even a small number of data points by determining the value being asymptotically approached by the curve. One-half the value for V_{max} ($V_{max}/2$) corresponds to the initial velocity observed when the substrate concentration is equal to K_m.

Lineweaver–Burk Reciprocal Plot

Despite the relative simplicity of direct plots such as the one shown in Figure A2.4, the asymptotic nature of the curve prevents precise measurement of values for V_{max}. Algebraic rearrangement of the Henri–Michaelis–Menten equation gives an alternative graphical representation from what is known as the Lineweaver–Burk equation:

$$1/v_0 = (K_m/V_{max})(1/[S]) + 1/V_{max}$$

This equation still conforms to the model originally put forward by Henri, but has the straight-line form of an equation in the format of $y = mx + b$. Here, the two variables (y and x; $1/v_0$ and $1/[S]$, respectively) are described in terms of each

TABLE A2.1 Biochemical data for human hexokinase.

Glucose (mM)	Initial velocity, v_0, (μmol/min)	1/[S], (mM^{-1})	$1/v_0$, (min/μmol)	v_0/[S]
0 (blank)	0	—	—	—
0.05	25	20	0.040	500
0.10	40	10	0.025	400
0.15	50	6.7	0.020	333
0.20	57	5.0	0.018	285
0.25	63	4.0	0.016	252
0.30	67	3.3	0.015	223
0.35	70	2.9	0.014	200
0.40	73	2.5	0.014	183

Note: Substrate (glucose) concentrations and corresponding initial velocities of conversion of glucose to glucose-6-phosphate (G6P) are shown. Reciprocal values (for use in a Lineweaver–Burk reciprocal plot) are also shown, as are values for v_0/[S] (for use in an Eadie–Hofstee plot).

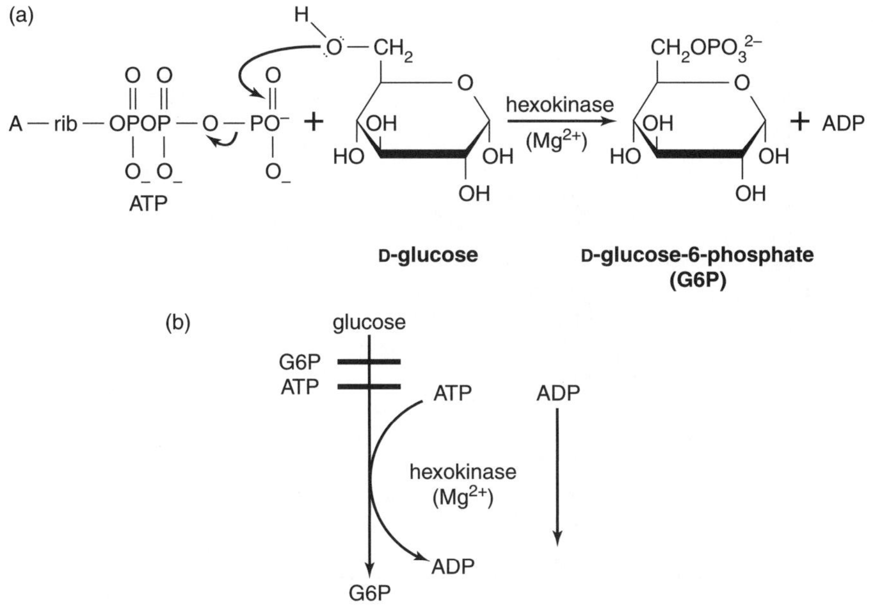

FIGURE A2.3 *The reaction catalyzed by hexokinase. (a) Hexokinase transfers a phosphate group from ATP to glucose at the very start of glycolysis to produce glucose-6-phosphate (G6P). Magnesium is a required co-factor (as it is for most kinase reactions). (b) Hexokinase is sensitive to two inhibitors (G6P and ATP) and one activator (ADP). Bars across reaction arrows are often used to signify inhibitors, while parallel arrows are used for activators.*

other, a slope ($m = K_m/V_{max}$) and the intercept of the y axis ($b = 1/V_{max}$) (see Figure A2.5 on page 243). A precise value for K_m can be determined by multiplying the slope by the value for V_{max} or by extrapolating the line and determining the negative value of the x intercept.

Eadie-Hofstee Plot

A more recently developed alternative to direct plots, Eadie–Hofstee plots, avoids some potential extrapolation errors introduced by the tendency for data points to be unevenly distributed in Lineweaver–Burk plots. Algebraic manipulations of the Henri–Michaelis–Menten equation also yield the Eadie–Hofstee equation:

$$v_0 = -K_m\,(v_0/[S]) + V_{max}$$

Here, the two variables (y and x; v_0 and $v_0/[S]$, respectively) are described in terms of each other. Just as before, this equation corresponds to a straight line but now $m = -K_m$ and $b = V_{max}$, as shown in Figure A2.6 on page 243.

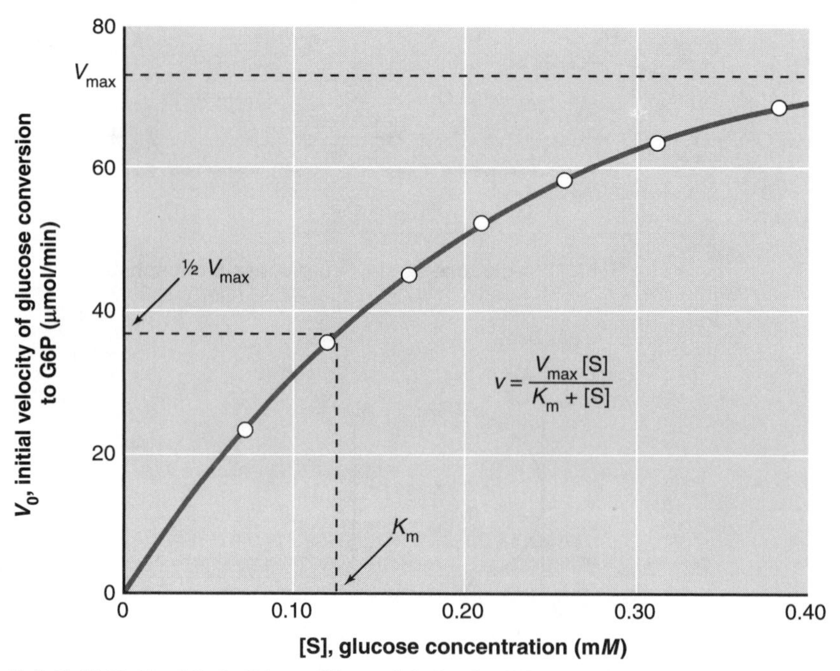

FIGURE A2.4 *Direct Henri–Michaelis–Menten plot of experimental enzyme kinetic data for human hexokinase. Data are taken from Table A2.1.* V_{max} *is determined by estimating the asymptotic value of the curve. Uncertainties in estimating* V_{max} *lead directly to errors in the estimation of* $K_m(V_{max}/2)$ *by this approach.*

Simple Inhibition Systems

Any substance that increases the velocity of an enzyme-catalyzed reaction is called an **activator.** By the same token, substances that decrease reaction velocities are called **inhibitors.** Regulation of enzyme activity, particularly by inhibition, is one of the principal ways that cells control the chemical reactions that distinguish them from nonliving materials. In the case of hexokinase, enzyme activity is suppressed by the presence of glucose-6-phosphate and ATP and enhanced by ADP; the regulators are in fact one of the substrates and two of the products of the reaction it catalyzes (Figure A2.3). Practical and economically important applications for these regulatory substances come from their common use as drugs, poisons, antibiotics, and preservatives. It has been argued that the entire history of pharmacology has been built around artificially regulating the activity of approximately 400 enzymes—almost always through the action of inhibitors. The remainder of this appendix describes the way in which studies of enzyme kinetics can distinguish between three different kinds of simple inhibition and other factors that can be used to alter the activity of biological catalysts.

Competitive Inhibition

As their name suggests, competitive inhibitors (I) are molecules that limit access of substrates (S) to an enzyme's (E) active site (i.e., I and S have similar chemical

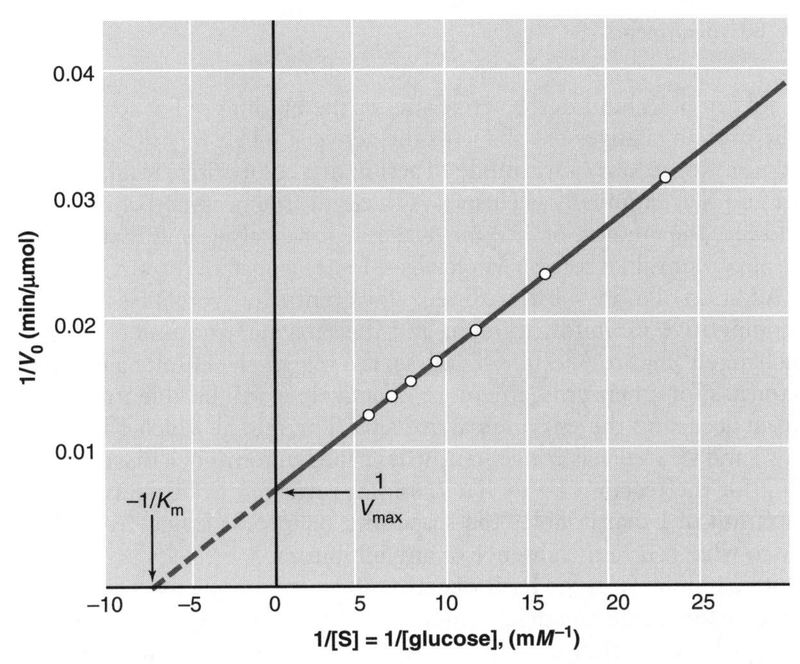

FIGURE A2.5 *Reciprocal Lineweaver–Burk plot of experimental enzyme kinetic data for carbonic anhydrase. Data are once again taken from Table A2.1, but here the reciprocal of each value is used. Portions of the line determined by extrapolation are dashed.*

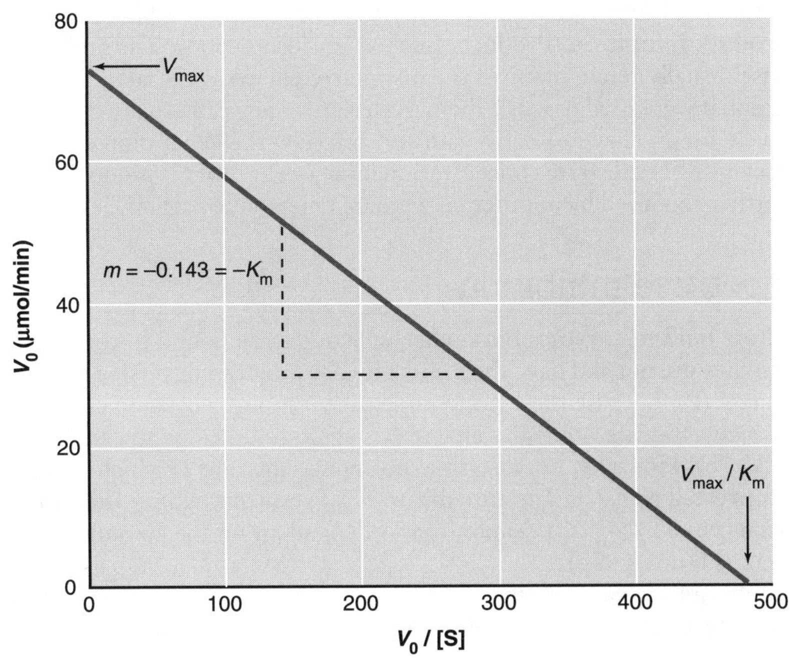

FIGURE A2.6 *Eadie–Hofstee plot of the experimental enzyme kinetic data for carbonic anhydrase. Data points are once again derived from the values in Table A2.1.*

properties and can both bind to the active site, or the binding of I at some other region of the protein changes the shape of the active site in a way that prevents S from binding). Regardless of the mode of action of a competitive inhibitor, the binding of I and S are mutually exclusive. As a result, the presence of inhibitor effectively lowers the amount of enzyme available for catalysis and that in turn changes the apparent values for the Michaelis–Menten constant (now K'_m) (Figure A2.7a). Addition of more substrate or enzyme eventually overcomes the presence of a competitive inhibitor, though, and the enzyme's maximum velocity remains unchanged (Figure A2.7b). Of course, the higher the affinity a competitive inhibitor has for an enzyme, the more effectively it will be able to prevent substrate from accessing the enzyme's active site. The rate at which EI is converted to E + I and vice versa is conveniently described in terms of a **dissociation constant,** K_I, for the reaction EI $\leftrightarrow$ E + I, and K_I turns out to be equivalent to the concentration of I that doubles the slope of a reciprocal Lineweaver–Burk plot relative to what it is in the absence of any inhibitor.

Competitive inhibition is the basis of action for many widely used drugs and a natural strategy for bioinformaticians interested in rational drug design. For example, benazepril and a variety of other high blood pressure medications are synthetic compounds that competitively inhibit the action of angiotensin-converting enzyme (ACE) in mammals. These drugs are derivatives of the amino acid proline and have very similar chemical features to a proline-rich region near a Phe-His peptide bond in angiotensin I that is cleaved by ACE. Commonly referred to as *ACE inhibitors*, these ACE-competitive inhibitors effectively lower the efficiency of the enzyme while still allowing appropriate physiological responses when high levels of angiotensin I are present.

Competitive inhibition is also commonly used in nature in a phenomenon known as **product inhibition.** Products (such as glucose-6-phosphate and ATP for hexokinase) usually retain many of the distinctive chemical characteristics of the substrates that were used to make them. As a result, they can usually compete quite effectively for an enzyme's active site and can slow the generation of additional product until (1) substrate concentration increases and/or (2) product molecules are further modified by another enzyme in a metabolic pathway.

Noncompetitive Inhibition

Some inhibitors bind to enzymes both when substrates are bound to the active site and when they are not and have their effect by limiting the conversion of substrate to product (P). Because these noncompetitive inhibitors have no effect on the rate at which substrate binds the enzyme, they do not change the enzyme's K_m (Figure A2.7b). However, because they do reduce the rate at which the ES complex is converted to EP and, eventually, to E + P, noncompetitive inhibitors do reduce an enzyme's V_{max} in a fashion that is dependent on the amount of inhibitor present (Figure A2.7b).

Some inhibitors display aspects of both noncompetitive and competitive inhibition and are known as **mixed inhibitors.** Mixed inhibitors usually have a

(a)

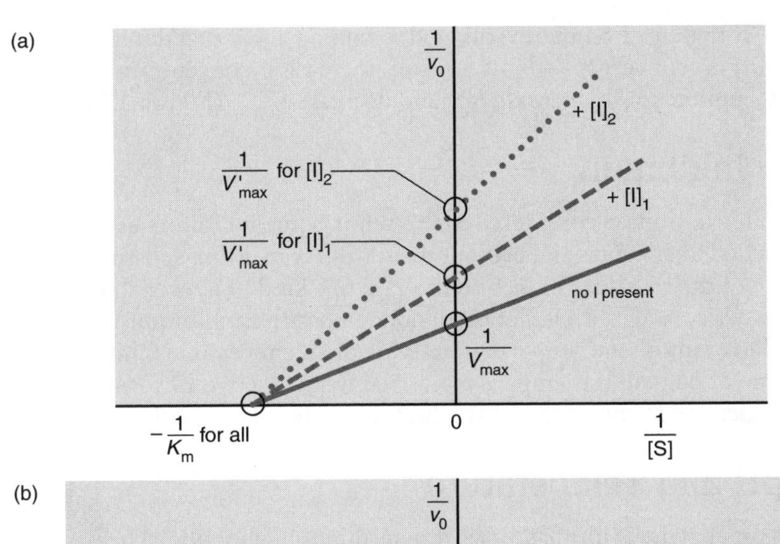

(b)

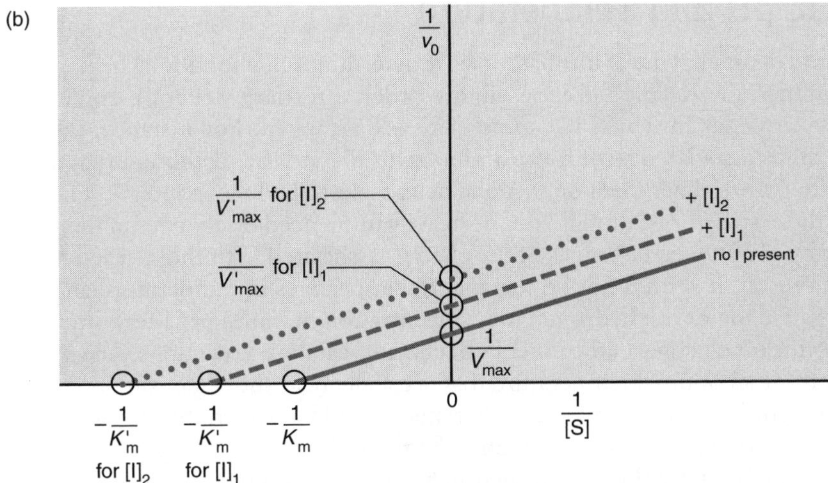

(c)

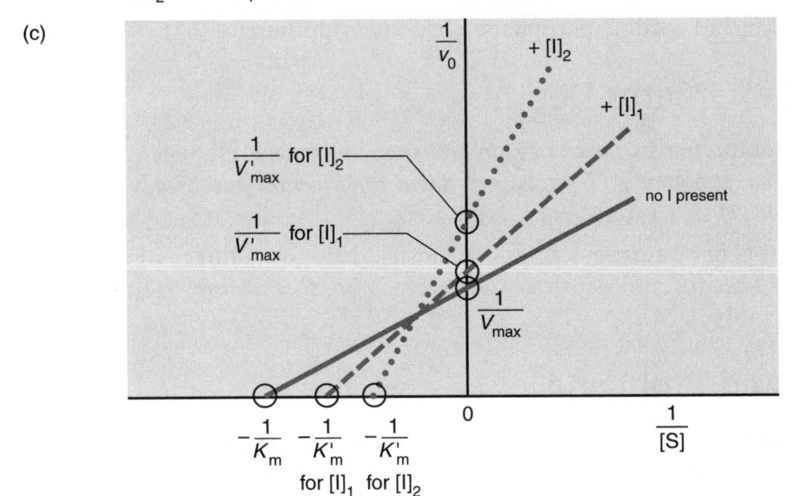

FIGURE A2.7 *Enzyme kinetic effects of inhibitors. (a) Lineweaver–Burk plots obtained when different amounts of competitive inhibitor are present. The solid line indicates the plot obtained when no competitive inhibitor is present; the dashed line is obtained with a low concentration of competitive inhibitor; and the dotted line is obtained with higher concentrations of competitive inhibitor. (b) Lineweaver–Burk plots obtained when different amounts of noncompetitive inhibitor are present. The solid line indicates the plot obtained when no noncompetitive inhibitor is present; the dashed line is obtained with a low concentration of noncompetitive inhibitor; and the dotted line is obtained with high concentrations of noncompetitive inhibitor. (c) Lineweaver–Burk plots obtained for mixed inhibition. The solid line indicates the plot obtained when no mixed inhibitor is present; the dashed line is obtained with a low concentration of mixed inhibitor; and the dotted line is obtained with high concentrations of mixed inhibitor.*

higher affinity for E than for ES but can still bind to both in a way that diminishes the enzyme's ability to convert S → P. As a result, increasing concentrations of mixed inhibitors simultaneously increase K'_m and decrease V'_{max} (Figure A2.7c).

Irreversible Inhibition

Unlike the reversible inhibitors considered to this point, some inhibitors become permanently attached to enzymes and prevent them from converting substrate to products. Because of the resulting reduction in V_{max}, this kind of irreversible inhibition is sometimes considered a variation of noncompetitive inhibition. Penicillin is a particularly famous and important example of an irreversible inhibitor. The target enzyme of penicillin therapy is converted to an inactive EI form that can no longer participate in the chemical synthesis of bacterial cell walls.

Effects of pH and Temperature

The structure of an enzyme is ultimately what determines its function. Much of what determines an enzyme's precise, highly ordered tertiary structure comes from the linear order in which its amino acids are arranged. For instance, the chemical groups associated with those amino acids dictate the likelihood that a region will fold into a beta sheet or an alpha helix as described in Chapter 7. The stability of those structures, though, can be heavily influenced by changes in temperature or by pH-induced changes of the charges associated with those reactive groups. As a result, it should not be surprising that changes in temperature and pH (the negative log of the hydrogen ion concentration in moles per liter) often result in significant changes in the kinetics of enzyme-catalyzed reactions. All enzymes operate at an optimal temperature that is usually near the organism's body temperature. Similar optimum pHs usually reflect the pH at which the enzyme is typically found within an organism and can differ radically from one enzyme to another (i.e., within humans: the enzyme pepsin has a pH optimum of 2; trypsin has a pH optimum of 8; and alkaline phosphatase has a pH optimum of 9.5).

Readings for Greater Depth

The classic and comprehensive textbook on enzyme kinetics is I. H. Segel, 1975, *Enzyme Kinetics: Behavior and Analysis of Rapid Equilibrium and Steady-State Enzyme Systems.* Wiley-Interscience, New York.

A commonly used general textbook on biochemistry that covers this material in much greater breadth and depth is L. Stryer, 2002, *Biochemistry* (5th ed.), W. H. Freeman, New York.

Questions and Problems

* **A2.1** Assume that the rate at which carbonic acid is converted to carbon dioxide and water by carbonic anhydrase is 1 molecule/second. What would the relative

concentrations of carbonic acid and carbon dioxide be at equilibrium if the enzyme catalyzes the reverse reaction at a rate of 1 molecule per minute?

A2.2 A reaction is found to have a K_{eq} of 1.3×10^{-4} at room temperature (conventionally 25°C or 298.15 K). What is its ΔG?

* **A2.3** What are the advantages of using a Lineweaver–Burk plot relative to a direct plot of enzyme kinetic data? What are the advantages of an Eadie–Hofstee plot relative to a Lineweaver–Burk plot?

A2.4 If the rate at which a forward reaction occurs is 3.5×10^5 reactions per minute and the K_{eq} for substrates and products is 1.0×10^2, then what is the rate at which the reverse reaction occurs?

* **A2.5** Just as inhibitors decrease an enzyme's activity, activators increase it. Most activators are noncompetitive. Would such an activator primarily affect an enzyme's K_m or its V_{max}?

APPENDIX 3

Sample Programs in Perl

Example 1: Conceptual Translation

Example 2: Dot Plot

Example 3: Relative Rate Test

Example 4: UPGMA

Example 5: Common Ancestor

Example 6: Splice Junction Identification

Example 7: Hydrophobicity Calculator—The 2D-HP Model

Example 8: Atomic Density Calculation

Example 9: Enzyme Kinetics—Linear Regression

The example programs contained herein were selected, one from each chapter plus Appendix 2, to present algorithms that are reasonably straightforward to implement, yet provide a range of challenges for students of varying computational backgrounds. We have endeavored to select programs that can be implemented in two to four pages of code and to provide meaningful comments within the code. Before each program we present a discussion of the overall approach used, along with samples of input and output to make it easier for readers to test their own implementations, and also provide a stepping stone to further experimentation.

Readers with a limited computational background might wish to start with examples 3, 1, 2, and 8. Those with a moderate background might find examples 4, 6, and 9 to be the most interesting, while examples 5 and 7 present the most computational challenge.

Example 1: Conceptual Translation

This program translates from nucleotide strings to amino acids. It prints all three reading frames in the forward direction, skipping nucleotides at the beginning to produce the additional frames, and ignoring nucleotides at the end that do not form a group of 3.

Sample input:

```
UAAUGCAUAGGCUACUCUAG
```

Corresponding output:

```
Nucleotide sequence: UAAUGCAUAGGCUACUCUAG

Reading frame 0: STP Cys Ile Gly Tyr Ser
Reading frame 1: Asn Ala STP Ala Thr Leu
Reading frame 2: Met His Arg Leu Leu STP
```

Code:

```
#!/usr/local/bin/perl

# Example program 1. Perform conceptual translation from
# nucleotides to amino acids. Do this for three reading
# frames, skipping the first 0-2 nucleotides to produce the
# reading frames.

# Define some constants that we'll need later.

$minlength      = 3;
$readingframes  = 3;
$unknown        = "UNK"; # If a nucleotide is unknown,
                         # print this.
```

```
# Define a hash to do matching/printing. This allows us to
# say things like $nucleohash{"UUU"} and receive "Phe".

 %nucleohash =
 ( "UUU", "Phe", "UUC", "Phe", "UUA", "Leu", "UUG", "Leu",
   "UCU", "Ser", "UCC", "Ser", "UCA", "Ser", "UCG", "Ser",
   "UAU", "Tyr", "UAC", "Tyr", "UAA", "STP", "UAG", "STP",
   "UGU", "Cys", "UGC", "Cys", "UGA", "STP", "UGG", "Trp",
   "CUU", "Leu", "CUC", "Leu", "CUA", "Leu", "CUG", "Leu",
   "CCU", "Pro", "CCC", "Pro", "CCA", "Pro", "CCG", "Pro",
   "CAU", "His", "CAC", "His", "CAA", "Gln", "CAG", "Gln",
   "CGU", "Arg", "CGC", "Arg", "CGA", "Arg", "CGG", "Arg",
   "AUU", "Ile", "AUC", "Ile", "AUA", "Ile", "AUG", "Met",
   "ACU", "Thr", "ACC", "Thr", "ACA", "Thr", "ACG", "Thr",
   "AAU", "Asn", "AAC", "Asn", "AAA", "Lys", "AAG", "Lys",
   "AGU", "Ser", "AGC", "Ser", "AGA", "Arg", "AGG", "Arg",
   "GUU", "Val", "GUC", "Val", "GUA", "Val", "GUG", "Val",
   "GCU", "Ala", "GCC", "Ala", "GCA", "Ala", "GCG", "Ala",
   "GAU", "Asp", "GAC", "Asp", "GAA", "Glu", "GAG", "Glu",
   "GGU", "Gly", "GGC", "Gly", "GGA", "Gly", "GGG", "Gly"
 );

# Retreive and check the command line parameter.

$input = @ARGV[0];

if ( length( $input) < $minlength )
{
  die( "$0: Place the nucleotide string on the commmand"
      . " line.\n\n" );
} # if

printf( "Nucleotide sequence: $input\n\n" );

# Run through all 3 possible reading frames, skipping the
# first letter or two for frames 1 and 2.

for ( $i = 0; $i < $readingframes; $i++ )
{
  printf( "Reading frame $i:" );

  # Find out how many 3-letter sequences remain, after
  # skipping 0-2 for the reading frame, and loop through
  # all of these sequences.

  $len = int( length( substr( $input, $i ) ) / 3 );

  for ( $j = 0; $j < $len; $j++ )
  {
    # Take the current 3-letter sequence, look up the
    # corresponding amino acid. If it isn't in the hash
```

```
      # table, it is unknown.
      $nuc = substr( $input, $i + $j * 3, 3 );

      if ( defined( $nucleohash{ $nuc } ) )
      {
        $aa = $nucleohash{ $nuc };
      } # if
      else
      {
        $aa = $unknown;
      } # else

      printf( "$aa " );
    } # for j

    printf( "\n" );
  } # for i

  printf( "\n" );
```

Example 2: Dot Plot

Given two nucleotide sequences, the goal is to produce the dot plot of the matching substrings, allowing the user to select the values for the window length and match criteria. The dot plot produced shows the second sequence on the *y* axis and the first sequence on the *x* axis.

Sample input:

```
5 3 ATAAAAATTTT TAATAAA
```

Corresponding output:

```
 Window length: 5
Match criteria: 3
    Sequence 1: ATAAAAATTTT
    Sequence 2: TAATAAA

 --------
|******  |
|******* |
|*****  *|
 --------
```

Code:

```
#!/usr/local/bin/perl

# Example program 2. Compute the dot plot for two
# sequences, based only on exact matches. Thus this
```

```
# program is equally useful for both nucleotide and amino
# acid sequences. The input should be of the form
# <window length> <match criteria> <sequence1> <sequence2>.

# Define the output characters for easy modification.

$match_char    = "*";
$mismatch_char = " ";
# Make sure there are 4 inputs, store them for future use,
# and print them out.

if ( @ARGV != 4 )
{
  die( "$0: <window length> <match criteria>"
      . " <sequence 1> <sequence 2>\n\n" );
} # if

$window_length  = $ARGV[0];
$match_criteria = $ARGV[1];
$sequence_1     = $ARGV[2];
$sequence_2     = $ARGV[3];

printf( " Window length: $window_length\n" );
printf( "Match criteria: $match_criteria\n" );
printf( "    Sequence 1: $sequence_1\n" );
printf( "    Sequence 2: $sequence_2\n\n" );

# Set the size of the plot dimensions, based on the
# length of the strings and the window length.

$sizex = length( $sequence_1) - $window_length + 1;
$sizey = length( $sequence_2) - $window_length + 1;

# Do some error checking on the inputs to be sure that they
# make sense.

if ( $match_criteria > $window_length )
{
  die( "$0: Window length ($window_length) must be >="
      . " Match criteria ($match_criteria).\n\n" );
} # if

if (     ( $sizex < 0 )
      || ( $sizey < 0 )
    )
{
  die( "$0: Each input sequence must be of length >="
      . " Window length ($window_length).\n\n" );
} # if
```

```
# Draw a line of dashes at the top of the plot.
draw_line( $sizex + 2 );

# Run through all possible plot positions, with sequence 2
# on the y axis, and sequence 1 on the x axis.

for ( $y = 0; $y < $sizey; $y++ )
{
  # Extract the section of sequence 2 to be compared
  # against sequence 1.
  $substr2 = substr( $sequence_2, $y, $window_length );
  printf( "|" );

  for ( $x = 0; $x < $sizex; $x++ )
  {
    # Extract the section of sequence 1 to be compared
    # against $substr2.
    $substr1 = substr( $sequence_1, $x, $window_length );

    # If the number of matches is >= the match criteria, a
    # match has occurred. Mark the appropriate character
    # in the dot plot.
    if (     num_matches( $substr1, $substr2 )
         > = $match_criteria )
    {
      $dotplot[$x][$y] = $match_char;
    } # if
    else
    {
      $dotplot[$x][$y] = $mismatch_char;
    } # else

    printf( "$dotplot[$x][$y]" );
  } # for x

  printf( "|\n" );
} # for y

# Draw a line of dashes at the bottom of the plot.
draw_line( $sizex + 2 );
printf( "\n" );

# Function to compute the number of character positions at
# which two strings of equal length match.

sub num_matches
{
  my( $a, $b) = @_;
  my( $i, $matches );
  for ( $i = 0; $i < length( $a); $i++ )
```

```
  {
    if ( substr( $a, $i, 1 ) eq substr( $b, $i, 1 ) )
    {
      $matches++;
    } # if
  } # for i

  return $matches;
} # num_matches

# Function to draw a line of dashes of a given length.

sub draw_line
{
  my( $len ) = @_;
  my( $i );

  for ( $i = 0; $i < $len; $i++ )
  {
    printf( "-" );
  } # for i

  printf( "\n" );
} # draw_line
```

Example 3: Relative Rate Test

This is the simplest of the algorithms presented, and as such is an excellent starting point for the reader with a limited computational background. Given the similarity measures between three organisms, the program computes the amount of divergence that has occurred between the two members of the ingroup and their common ancestor 'A'.

Sample input:

```
95 70 75
```

Corresponding output:

```
d12: 95
d13: 70
d23: 75

dA1: 45
dA2: 50
```

Code:

```
#!/usr/local/bin/perl
```

```
# Example program 3. Perform a relative rate test for
# three related sequences. The program accepts three
# inputs from a dissimilarity matrix. As in Chapter 3,
# elements 1 and 2 represent the ingroup, element 3 the
# outgroup. The inputs to this program are d12, d13, d23.

# Make sure there are 3 inputs, store them for future use,
# and print them out.

if ( @ARGV != 3 )
{
  die( "$0: <d12> <d13> <d23> \n\n" );
} # if

$d12 = $ARGV[0];
$d13 = $ARGV[1];
$d23 = $ARGV[2];

printf( "d12: $d12\n" );
printf( "d13: $d13\n" );
printf( "d23: $d23\n\n" );

# Make sure that the inputs make some sense: that
# element 3 seems to represent the outgroup.

if (     ( ( 2 * $d12 ) >= $d13 )
      && ( ( 2 * $d12 ) >= $d23 )
   )
{
  die( "$0: Element 3 does not appear to represent"
     . " the outgroup.\n\n" );
} # if

# Compute and print dA1 and dA2.

$dA1 = ( $d12 + $d13 - $d23 ) / 2;
$dA2 = ( $d12 + $d23 - $d13 ) / 2;

printf( "dA1: $dA1\n" );
printf( "dA2: $dA2\n\n" );
```

Example 4: UPGMA

Grouping things into a binary tree structure is a common problem in many fields. The general graph produced by this algorithm is called a *dendrogram.* During each iteration, the program finds the two nodes whose values are closest and clusters them. A new label is produced that represents the pairing of the two constituent labels, and also forms the basis for standard Newick format output. For example, the labels "mouse" and "squirrel" would be combined into the label "(mouse,squirrel)". This looping is complete when only one label remains, representing the Newick form of the tree structure. For this program, three different sets of input and output are presented, so that the reader can see how the structure of the output changes as single changes are made to the input.

Sample input:

```
a 12 b 14 c 14.5 d 17 e 18
a 0 b 14 c 14.5 d 17 e 18
a 0 b 14 c 16.5 d 17 e 18
```

Corresponding output:

```
Input sequence: a 12 b 14 c 14.5 d 17 e 18
Final structure: ((a,(b,c )),(d,e))
Input sequence: a 0 b 14 c 14.5 d 17 e 18
Final structure: (a,((b,c),(d,e)))
Input sequence: a 0 b 14 c 16.5 d 17 e 18
Final structure: (a,(b,((c,d),e)))
```

Code:

```
#!/usr/local/bin/perl

# Example program 4. Perform a UPGMA (unweighted-pair-group
# method with Arithmetic Mean) analysis of a set of data.
# The input consists of label/value pairs, all inputs
# separated by spaces. Therefore, a label cannot have
# blank spaces inside of it.

# Make sure there are an even number of inputs (matched
# value/label pairs).

if ( ( @ARGV % 2 ) != 0 )
{
  die( "$0: Place the label/value pairs on the commmand"
      . " line.\n\n" );
} # if

printf( "Input sequence: @ARGV\n\n" );

# Store the arguments in a hash, all at once. Recall that
# the first element of ARGV is a key, the second is a
# value, etc.

%labelhash = @ARGV;

@labelkeys = keys( %labelhash );

# Loop through all pairs of nodes, at each cycle finding
# and combining the two nodes which are the closest to each
# other. Halt when there is only one label remaining; that
# label will be the final tree structure.
```

```
while ( @labelkeys > 1 )
{
  $smallest = Infinity;

  for ( $i = 0; $i < @labelkeys; $i++ )
  {
    for ( $j = $i+1; $j < @labelkeys; $j++ )
    {
      $a = $labelkeys[$i];
      $b = $labelkeys[$j];
      $distance = abs( $labelhash{$a} - $labelhash{$b} );

      if ( $distance < $smallest )
      {
        $smallest = $distance;
        $smallesta = $a;
        $smallestb = $b;
      } # if
    } # for j
  } # for i

  # Merge $smallesta and $smallestb into one, deleting the
  # old labels, and placing the new label in the following
  # format: ($smallesta,$smallestb).

  # Note that when we compute the new value, we CANNOT
  # simply add the values at $smallesta and $smallestb and
  # divide by two: $smallesta might represent several
  # nodes that have already been merged; weight them by
  # their respective counts. The number of nodes
  # represented by $smallesta or $smallestb is equal to the
  # number of commas in the string, plus one.

  # The regular expression "$smallesta =~ tr/,//" counts
  # the number of commas in the string referred to by
  # $smallesta.

  $counta = ( $smallesta =~ tr/,// ) + 1;
  $countb = ( $smallestb =~ tr/,// ) + 1;
  $newval = (    $labelhash{$smallesta} * $counta
               + $labelhash{$smallestb} * $countb
             ) / ( $counta + $countb );

  delete $labelhash{$smallesta};
  delete $labelhash{$smallestb};
  $labelhash{"($smallesta,$smallestb)"} = $newval;

  # Recompute the list of the keys in the label hash.
  @labelkeys = keys( %labelhash );
} # while

printf( "Final structure: @labelkeys\n\n" );
```

Example 5: Common Ancestor

Given a tree structure in Newick format containing labeled leaf nodes, this program finds the most likely common ancestor. Strangely enough, although this is one of the easiest algorithms to do by hand, the code is the longest, partially due to the fact that Perl does not have an easy way to represent tree structures or perform intersection and union on strings. It is up to the user to ensure that all leaf node labels are of equal length; unpredictable results can occur if this rule is not adhered to. Note that since parentheses have special meaning to UNIX, we must put the input string in quotation marks.

Sample input:

```
"((AAA,TGA),GAG)"
```

Corresponding output:

```
Initial tree: ((AAA,TGA),GAG)

Common ancestor: (ATG)A(AG)
```

Code:

```
#!/usr/local/bin/perl

# Example program 5. Perform a common ancestor analysis of
# a tree, given a structure in the standard Newick format.
# Each leaf node specifies a nucleotide, or string of
# nucleotides. All leaves must have the same length.

# Unfortunately, Perl does not have a simple and efficient
# way to represent trees. In this example program, we
# store all of the elements in an array. The root is at
# index 0. For any node X in the tree, its left child
# resides at index 2X + 1, the right child at index 2X + 2.
# Define a way to mark internal nodes in the tree (as
# opposed to leaves) while the input is being read. Thus
# when the tree is evaluated, we know that these nodes must
# be computed. Define the root and current location.

$internal_node = "-";

$root = 0;
$pos = 0;

# Store and echo the input.

$input = @ARGV[0];

printf( "Initial tree: $input\n\n" );

# Parse the input string, one character at a time. The
# characters "(", ",", and ")" have special meaning to the
# parsing. "(" indicates we must move down a level in the
```

```
# tree, to the next left child. "," means to switch from
# the left child to the right child. ")" means to move
# upward a level in the tree. Any other character must be
# part of the label for the current node.

for ( $i = 0; $i < length( $input ); $i++ )
{
  $char = substr( $input, $i, 1 );

  if ( $char eq "(" )
  {
    $pos = $pos * 2 + 1;
  } # if
  elsif ( $char eq "," )
  {
    $pos++;
  } #elsif
  elsif ( $char eq ")" )
  {
    $pos = int( $pos / 2 ) - 1;
    $tree[$pos] = $internal_node;
  } # elsif
  else
  {
    $tree[$pos] = $tree[$pos]. $char;
    $leaf_len = length( $tree[$pos] );
  } # else
} # for i

# If the input has at least matching parentheses, we should
# end up back at the root node after parsing the input.

if ( $pos != $root )
{
  die( "$0: invalid input string.\n\n" );
} # if

# Compute the common ancestor, one character at a time.
# Thus we compute and print the common ancestor for the
# first character of all leaf nodes, the second character
# of all leaf nodes, etc. If the result is a single
# letter, simply print it. If it is a longer string, print
# them in parentheses.

printf( "Common ancestor: " );

for ( $i = 0; $i < $leaf_len; $i++ )
{
  $ancestor = evaluate( $root, $i );
```

```
  if ( length( $ancestor ) == 1 )
  {
    printf( "$ancestor" );
  } #if
  else
  {
    printf( "($ancestor )" );
  } # else
} # for i

printf( "\n\n" );

# A recursive function to evaluate a subtree and return
# the common ancestor string. The first input is a pointer
# to the current node being evaluated, the second is which
# input character is currently under consideration.

sub evaluate
{
  my( $ptr, $pos ) = @_;
  my( $left, $right, $eval_str );

  # If this is an internal node, compute it from its
  # children. As stated in Chapter 5, if the intersection
  # of the child nodes is nonempty, it is used. If that
  # intersection is empty, the union of the child nodes is
  # used.

  if ( $tree[$ptr] eq $internal_node )
  {
    $left = evaluate( left_child( $ptr ), $pos );
    $right = evaluate( right_child( $ptr ), $pos );

    $eval_str = intersection( $left, $right );

    if ( length( $eval_str ) == 0 )
    {
      $eval_str = union( $left, $right );
    } # if
  } # if
  else
  {
    # This is a leaf node: it simply evaluates to the
    # correct character position in this string.
    $eval_str = substr( $tree[$ptr], $pos, 1 );
  } # else

  return $eval_str;
} # evaluate
```

```
# Function to return the pointer to the left child of a
# given node.

sub left_child
{
  my( $ptr ) = @_;

  return $ptr * 2 + 1;
} # left_child

# Function to return the pointer to the right child of a
# given node.

sub right_child
{
  my( $ptr ) = @_;

  return $ptr * 2 + 2;
} # right_child

# Function to calculate the intersection of two strings.

sub intersection
{
  my( $a, $b ) = @_;
  my( $str, $i, $char );

  for ( $i = 0; $i < length( $a ); $i++ )
  {
    $char = substr( $a, $i, 1 );

    if ( index( $b, $char ) != -1 )
    {
      $str = $str . $char;
    } # if
  } # for i

  return $str;
} # intersection

# Function to calculate the union of two strings.

sub union
{
  my( $a, $b ) = @_;
  my( $str, $i, $char );

  $str = $a;

  for ( $i = 0; $i < length( $b ); $i++ )
  {
    $char = substr( $b, $i, 1 );
```

```
    if ( index( $a, $char ) == -1 )
    {
      $str = $str . $char;
    } # if
  } # for i

  return $str;
} # union
```

Example 6: Splice Junction Identification

This program automates splice junction identification. All locations where "GT" occurs in the string are noted and the probability of a splice junction occurring there is computed from the preceding and succeeding nucleotides. The potential splice junctions are printed in order from most to least likely. Note that although the indices in Perl have the first character being at index 0, the more common method is used for printing, with the first character being index 1. The program makes use of the Perl treatment of strings, in that stepping off the left or right end of a string is allowed; the result is simply the null string.

Sample input:

```
AAGTAACAAGGTAAACAGGTAAGT
```

Corresponding output:

```
Nucleotide sequence: AAGTAACAAGGTAAACAGGTAAGT

Potential splice junctions and associated probabilities:

Index     Probability
-----     ------------
19        0.1070931456
11        0.0119801856
3         0.0039933952
23        0.0011517952
```

Code:

```
#!/usr/local/bin/perl

# Example program 6. Find all possible splice junctions in
# a nucleotide sequence, determine the probability of each,
# and print them out in order from most to least likely.
# Define some constants that we'll need later.

$minlength = 2;
```

```
@offsets        = ( -2,    -1,     2,     3,     4,     5   );
@nucleotides    = ( "A",   "G",   "A",   "A",   "G",   "T"  );
@probabilities  = ( 0.64, 0.75, 0.62, 0.68, 0.84, 0.63 );

# Retrieve and check the command line parameter.

$input = @ARGV[0];

if ( length( $input ) < $minlength )
{
  die( "$0: Place the nucleotide string on the commmand"
      . " line.\n\n" );
} # if

printf( "Nucleotide sequence: $input\n\n" );
printf( "Potential splice junctions and associated"
       . " probabilities:\n\n" );
printf( "Index \t Probability\n" );
printf( "----- \t -----------\n" );

# Find all positions where "GT" occurs, and build a list
# of these indices.

$pos = index( $input, "GT" );

while ( $pos != -1 )
{
  push( @indices, $pos );

  # Note that we need to start searching at $pos+1 to make
  # sure that we skip the "GT" we just found; otherwise
  # we'd find the first occurrence over and over again.
  $pos = index( $input, "GT", $pos+1 );
} # while

# For each of the indices where "GT" occurs, compute
# the probability of a splice junction occurring at that
# point.

foreach $index ( @indices )
{
  $prob = 1.0;

  # Run through all of the offsets where important
  # nucleotides occur.

  for ( $i = 0; $i < @offsets; $i++ )
  {
    # Check to see if the nucleotide gives the maximum
    # probability, or the reciprocal thereof. Note: we are
    # counting on Perl's treatment of strings: if we step
```

```
    # off the left or right end of the string, the "eq"
    # test will fail, as the nucleotide at that position is
    # "", the empty string.

    if (  $nucleotides[$i]
          eq substr( $input, $index + $offsets[$i], 1 )
        )
    {
      $prob *= $probabilities[$i];
    } # if
    else
    {
      $prob *= 1.0 - $probabilities[$i];
    } # else

  } # for i

  # Add an entry to a hash table, where the probability is
  # the key and the index where it occurs is the associated
  # value. We do it in this order, since we will later
  # sort on the key values.

  @hash{$prob} = $index;
} # foreach

# Sort the list of probabilities in descending order, then
# print out that sorted list, along with the index into the
# string where the potential splice junction occurs.

@sortedprobs = sort descending_numerical keys( %hash );

foreach $prob ( @sortedprobs )
{
  $printindex = $hash{$prob} + 1;
  printf( "$printindex \t $prob\n" );
}

printf( "\n" );

# Subroutine used for sorting a list in descending
# numerical order.

sub descending_numerical
{
  if ( $a < $b )
  {
    return 1;
  } # if
  elsif ( $a == $b )
  {
    return 0;
```

```
    } # elsif
    else
    {
      return -1;
    } # else
  } # reverse_numerical
```

Example 7: Hydrophobicity Calculator—The 2D-HP Model

One major simplification that can be made to the process of protein folding is to consider only hydrophobic interactions. Taking things a step further, suppose that all amino acids must fall onto the intersections of equally spaced lines in a grid. For a further simplification, consider only a two-dimensional grid, because it is easy to draw examples for such a grid. The program here reads a sequence consisting of classifications of amino acids and directions to get to the location of the next amino acid. It maps these locations into a hash table; counts those amino acids, which **must** be adjacent since they were adjacent in the input sequence; counts those that are adjacent at all; and thus finds the number of adjacent pairs that are not adjacent in the initial ordering. The negative value of this count is a score, representing an energy minimization; more negative scores are thus better.

Sample input:

```
HRPUHUPLPDHLHUPLHUPLPDHDPRHDHLPDPRHRPUH
```

Corresponding output:

```
Input sequence: HRPUHUPLPDHLHUPLHUPLPDHDPRHDHLPDPRHRPUH

Score: -9
```

Code:

```
#!/usr/local/bin/perl

# Example program 7. Perform a 2DHP (2-dimensional
# hydrophobic/hydrophilic) analysis of a given input
# sequence. The input starts with either H (hydrophobic)
# or P (hydrophilic), and is then followed by pairs of
# either U, D, L, R (up, down, left, right), then H or P.
# For example, a simple input would be: HUPRPDH
# (H up P right P down H).

# Define some constants that we'll need later. Put the
# allowed input characters here, for easy modification.
```

```
$minlength     =  3;
$hydrophobic  = 'H';
$hydrophilic  = 'P';
$up           = 'U';
$down         = 'D';
$left         = 'L';
$right        = 'R';

# Start out at coordinates 0,0.

$xpos = 0;
$ypos = 0;

# Retrieve and check the command line parameter.

$input = @ARGV[0];

if (     ( length( $input ) < $minlength )
      || ( ( length( $input ) % 2 ) != 1 )
   )
{
  die( "$0: Place the input string on the commmand"
      . " line.\n\n" );
} # if

printf( "Input sequence: $input\n\n" );

# Run through each input character. If it is invalid,
# print an error and exit. Otherwise process the character
# according to position: even indices are H or P, odd are
# movement directions.

for ( $i = 0; $i < length( $input ); $i++ )
{
  $char = substr( $input, $i, 1 );

  if ( ( $i % 2 ) == 0 )
  {
    # H or P character.

    if (     ( $char ne $hydrophobic )
          && ( $char ne $hydrophilic )
       )
    {
      die( "$0: Input character $i ($char ) must be"
          . " $hydrophobic or $hydrophilic.\n\n" );
    } # if

    # Check to see if the new grid position already exists
    # in the hash. If it does, then the protein has folded
    # back on itself, and is invalid.
```

```
    if ( exists( $grid{"$xpos,$ypos"} ) )
    {
      die( "$0: Input sequence has a collision at"
          . " character $i.\n\n" );
    } # if

    # Save this grid position. If we've seen two
    # hydrophobics in a row in the input sequence, count
    # them for later use in the scoring.

    $grid{"$xpos,$ypos"} = $char;

    if (     ( $char eq $hydrophobic )
          && ( $previous eq $hydrophobic )
       )
    {
      $input_adjacencies++;
    } # if

    $previous = $char;

  } # if
  else
  {
  # Movement character. Modify the current position
  # accordingly, exit if the input character is
  # unrecognized.

  if ( $char eq $up )
  {
    $xpos++;
  } # if
  elsif ( $char eq $down )
  {
    $xpos--;
  } # elsif
  elsif ( $char eq $left )
  {
    $ypos--;
  } # elsif
  elsif ( $char = $right )
  {
    $ypos++;
  } # elsif
  else
  {
    die( "$0: Input character $i ($char) must be $up,"
        . " $down, $left, or $right.\n\n" );
  } # else
```

```
  } # else
} # for i

# Run through all PAIRS of keys in the grid (the
# locations), and check whether any locations that are
# adjacent in the grid are both hydrophobic. If they are,
# count them for scoring.

@keys = keys( %grid );

for ( $i = 0; $i < @keys; $i++ )
{
  for ( $j = $i+1; $j < @keys; $j++ )
  {
    $a = $keys[$i];
    $b = $keys[$j];
    if (     ( $grid{$a} eq $hydrophobic )
          && ( $grid{$b} eq $hydrophobic )
          && ( adjacent( $a, $b ) )
       )
    {
      $adjacencies++;
    } # if
  } # for j
} # for i

# Compute the score. Note that when we counted hydrophobic
# positions that were adjacent in the grid, we also counted
# those that MUST be adjacent because they are adjacent in
# the input sequence. Thus we subtract these from the
# score. Also, make the final score a negative value, to
# be consistent with the energy minimization concept.

$score = $adjacencies - $input_adjacencies;
$score = - $score;
printf( "Score: $score\n\n" );

# Function to determine if two locations are adjacent
# in a grid. Note that only adjacencies along the grid
# axes are considered. Returns 1 if they are adjacent,
# 0 otherwise.

sub adjacent
{
  my( $a, $b ) = @_;
  my( $ax, $ay, $bx, $by, $commapos );

  # Find the position of the comma; everything before that
  # is 'x', everything after it is 'y'. Then compute the
  # Euclidean distance. Don't bother with the square root.
```

```
    $commapos = index( $a, "," );
    $ax = substr( $a, 0, $commapos );
    $ay = substr( $a, $commapos+1 );
    $commapos = index( $b, "," );
    $bx = substr( $b, 0, $commapos );
    $by = substr( $b, $commapos+1 );

    $distance = ( $ay - $by ) ** 2 + ( $ax - $bx ) ** 2;

    # Return the result of the boolean comparison, which
    # results in 1 if the comparison is true, 0 if false.

    return ( $distance <= 1 );
  } # adjacent
```

Example 8: Atomic Density Calculation

This program is fairly straightforward, scanning a PDB file for ATOM lines, saving the x, y, and z coordinates of each one, counting the number of other atoms within a specific distance of each one, sorting this list, and printing it. The 3apr.pdb file was retrieved from the Protein Data Bank (http://www.rcsb.org/pdb) and used as input to this program. Although the program prints the counts for all atoms, only the top 10 atoms are shown here, rather than all 2,464.

Sample input:

```
3apr.pdb
```

Corresponding output:

```
Input file: 3apr.pdb

Atom #      Count
------      -----
 191          16
1008          16
1027          16
1035          16
1201          16
1727          16
2003          16
2319          16
 476          15
 758          15
  .            .
  .            .
  .            .
```

Code:

```
#!/usr/local/bin/perl

# Example program 8. Given the name of a PDB file, read
# the file and locate all ATOM locations. For each atom,
# count the number of neighboring atoms within a threshold
# distance. As PDB files can contain thousands of atoms,
# we want to do this as efficiently as possible.

# Define some constants that will be needed later. Define
# the threshold as the square of the value we want (in
# angstroms). This way, when we compute Euclidean
# distance, we don't have to use the square root function
# on EVERY pair of distances, cutting down on computation
# time.

$threshold       = 3.6 ** 2;
$atom_keyword    = "ATOM";
$atom_start      = 0;
$label_start     = 6;
$label_length    = 5;
$x_start         = 30;
$x_length        = 8;
$y_start         = 38;
$y_length        = 8;
$z_start         = 46;
$z_length        = 8;

# Retreive and check the command line parameter.

$inputfile = @ARGV[0];

if ( @ARGV == 0 )
{
  die( "$0: Place the PDB file name on the commmand"
     . " line.\n\n" );
} # if

printf( "Input file: $inputfile\n\n" );

unless( open( PDB, "$inputfile" ) )
{
  die( "$0: Cannot open PDB file $inputfile.\n\n" );
} # if

while ( <PDB> )
{
  $input = $_;
  chop( $input );
```

```
    if ( index( $input, $atom_keyword ) == $atom_start )
    {
      $label[$records] = substr( $input, $label_start,
                                 $label_length );
      $x[$records] = substr( $input, $x_start, $x_length );
      $y[$records] = substr( $input, $y_start, $y_length );
      $z[$records] = substr( $input, $z_start, $z_length );
      $records++;
    } # if
  } # while

  for ( $i = 0; $i < $records; $i++ )
  {
    for ( $j = $i+1; $j < $records; $j++ )
    {
      $distance = ( $x[$i] - $x[$j] ) ** 2
                + ( $y[$i] - $y[$j] ) ** 2
                + ( $z[$i] - $z[$j] ) ** 2;

      if ( $distance < $threshold )
      {
        $count[$i]++;
        $count[$j]++;
      } # if
    } # for j
  } # for i

  # Sort the labels and counts in order by count, from
  # highest to lowest. Note that we cannot simply use 'sort'
  # here, since the information resides in two different
  # arrays. The method presented here is a simple bubble
  # sort: check each pair of counts, swapping them (and the
  # labels) if the later one has a larger value.

  printf( "Atom #   Count\n" );
  printf( "------   -----\n" );

  for ( $i = 0; $i < $records-1; $i++ )
  {
    for ( $j = $i+1; $j < $records; $j++ )
    {
      if ( $count[$i] < $count[$j] )
      {
        ($count[$i], $count[$j])  = ($count[$j], $count[$i]);
        ($label[$i], $label[$j]) = ($label[$j], $label[$i]);
      } # if
    } # for j
```

```
    # Print out each record as it becomes sorted.
    printf( "%6s %5d\n", $label[$i], $count[$i] );
  } # for i

  # Print out the final record, which was skipped by the
  # for i loop going only up to $records-1.
  printf( "%6s %5d\n", $label[$i], $count[$i] );
```

Example 9: Enzyme Kinetics—Linear Regression

As a tool for producing Lineweaver–Burk plots, this program calculates a simple least-squares linear regression of the two-dimensional input data. The data are fitted onto the general formula $y_i = mx_i + b$, where x_i and y_i are paired input data. The values of m and b are computed and printed, as well as the x and y intercepts of the fitted line, along with r, the correlation coefficient. Note that the program handles cases where the fitted line is horizontal ($m = 0$), but not a vertical line, because then m would be infinity.

Sample input:

```
1 1.1 2 3.4 3 4.2 4 5.0 6 7.1 7 9.6
```

Corresponding output:

```
Input coordinates: 1 1.1 2 3.4 3 4.2 4 5.0 6 7.1 7 9.6

a = 1.2583850931677, b = 0.242857142857142

X-intercept = -0.192991115498519
Y-intercept = 0.242857142857142

Correlation coefficient: 0.983229656640151
```

Code:

```
#!/usr/local/bin/perl

# Example program 9. Perform a linear regression on the
# input data. The input consists of x/y pairs of numerical
# data, all inputs separated by spaces. The output values
# are m and b, such that y = mx + b is the form of the best
# fit, as well as the x- and y-intercepts of this line,
# along with the correlation coefficient.

# Define some constants that we'll need later.

$mininputs = 6; # With less than 3 input pairs, linear
                # regression is either unnecessary or
                # inappropriate.
```

```
# Make sure there are a sufficient number of evenly matched
# x/y inputs.

if (     ( @ARGV < $mininputs )
      || ( ( @ARGV % 2 ) != 0 )
   )
{
  die( "$0: Place the x/y pairs on the commmand"
      . " line.\n\n" );
} # if

printf( "Input coordinates: @ARGV\n\n" );

# Store the input values in two arrays.

$numinputs = @ARGV / 2;

for ( $i = 0; $i < $numinputs; $i++ )
{
  $x[$i] = $ARGV[$i*2];
  $y[$i] = $ARGV[$i*2 + 1];
} # for i

# Compute the values that are needed for the linear
# regression.

for ( $i = 0; $i < $numinputs; $i++ )
{
  $sumx   += $x[$i];
  $sumy   += $y[$i];
  $sumxsq += $x[$i] ** 2;
  $sumxy  += $x[$i] * $y[$i];
} # for i

# Compute and print m, b, x-intercept, and y-intercept,
# checking for possible division by 0 if the slope is 0
# (in which case there is no x-intercept).

$m = ( $numinputs * $sumxy - $sumx * $sumy )
   / ( $numinputs * $sumxsq - $sumx ** 2 );

$b = ( $sumxsq * $sumy - $sumxy * $sumx )
   / ( $numinputs * $sumxsq - $sumx ** 2 );

printf( "m = $m, b = $b\n\n" );

if ( $m == 0 )
{
  $xintcp = "DNE";
} # if
else
```

```
{
  $xintcp = - $b / $m;
} # else

$yintcp = $b;

printf( "X-intercept = $xintcp\n" );
printf( "Y-intercept = $yintcp\n\n" );

# Compute the values needed for the correlation
# coefficient.

$avgx = $sumx / $numinputs;
$avgy = $sumy / $numinputs;

for ( $i = 0; $i < $numinputs; $i++ )
{
  $xminusavgx = $x[$i] - $avgx;
  $yminusavgy = $y[$i] - $avgy;

  $sum1 += $xminusavgx * $yminusavgy;
  $sum2 += $xminusavgx ** 2;
  $sum3 += $yminusavgy ** 2;
} # for i

$sum2 = sqrt( $sum2 );
$sum3 = sqrt( $sum3 );

# Compute and print the correlation coefficient.

$r = $sum1 / $sum2 / $sum3;

printf( "Correlation coefficient: $r\n\n" );
```

Glossary

1*s* orbital Spherical orbital closest to the nucleus of an atom, where electrons with the lowest energy are found.

2*p* orbital A set of three dumbell-shaped orbitals in the second sublevel of electron orbitals.

2*s* orbital Lowest energy, spherically shaped orbital in the second sublevel of electron orbitals.

50% majority-rule consensus A consensus tree in which any internal node that is supported by at least half of the trees being summarized is portrayed as a simple bifurcation, and those nodes where less than half of the trees agree are shown as multifurcations.

A Adenine. One of two purines that are used as nitrogenous bases.

absolute direction representation A representation for protein configurations in which the position of each residue is specified according to the direction moved from the previous position. In absolute representation, the previous move's direction is not considered. For example, in a two-dimensional model, the possible values for each position are: up, down, left, and right.

activation energy (E_{act}) The amount of energy required to excite a molecule or molecules into a reactive state that allows new molecules to be made.

activator Any substance that increases the velocity of an enzyme-catalyzed reaction.

active site The region on the three-dimensional surface of a protein where catalysis occurs.

additive Term applied to a scaled tree if the physical length of the branches connecting any two nodes is an accurate representation of their accumulated differences.

alignment A pairing of two homologous nucleotide or protein sequences for the purpose of identifying the location of accumulated changes since they last shared a common ancestor. *See also* global and local alignment.

alleles Different versions of any given gene within a species of organism.

alpha carbon The central carbon atom in an amino acid to which side chains (R-groups) are bound.

alternative splicing The production of two or more mRNA molecules from a single hnRNA by using different splice junctions.

amino terminus (N-terminal) In a polypeptide, the end of the molecule that has an unbound amide group (NH_2) and corresponds to the 5' end of a gene.

anti-parallel Showing opposite orientation; in the case of double-stranded DNA, this means that if one strand is 5' to 3', its complementary strand will be in the opposite, 3' to 5' orientation.

arguments *See* parameters.

aromatic Compounds that have molecular structures based on the six-carbon ring of benzene.

array Variables that can store multiple values. Each value is retrieved using an integer index.

backbone (of an amino acid) Consists of an amide, an alpha carbon, and a carboxylic acid, or carboxylate group.

basal promoter A set of nucleotide sequences such as the "TATA-box" that serves as a minimal promoter within eukaryotes and to which basal transcription factors bind.

basal transcription factor Proteins such as the TATA-binding protein that are needed for RNA polymerase assembly around promoter regions of eukaryotic genes.

base pair (1) The interaction between purines and pyrimidines (specifically between A and T and between G and C) in double-stranded DNA.

(2) The smallest unit of measure of double-stranded DNA length.

beta turns U-turn-like structures within proteins formed when a beta strand reverses direction in an anti-parallel beta sheet.

bifurcating Refers to a point in a phylogenetic tree in which an ancestral taxon splits into two independent lineages.

binding operator In Perl, the =~ operator, which matches a string to a regular expression.

block A set of statements enclosed in curly braces.

blotting and hybridization The transfer of molecules (often nucleic acids) from a gel onto a membrane followed by washing with a labelled probe that binds specifically to a molecule of interest.

blunt end End of a double-stranded DNA molecule that has no single-stranded overhang.

bootstrap test Test that allows for a rough quantification of confidence levels.

branch and bound method A clever and efficient means of streamlining the search for trees of maximum parsimony. Consists of two steps: First, determine an upper bound to the length of the most parsimonious tree for the data set and, second, incrementally grow a tree by adding branches one at a time to a tree that describes the relationship of just a few of the sequences being considered.

branch swapping Pruning and regrafting of branches from phylogenetic trees.

branches In a phylogenetic tree, the graphical representation of an evolutionary relationship between two (bifurcating) or more (multifurcating) lineages arising from a single ancestral node.

bump A steric collision.

C Cytosine. One of two pyrimidines that are used as nitrogenous bases in DNA and RNA molecules.

CAAT box A short segment of many eukaryotic promoters that is typically located approximately 80 nucleotides upstream of the transcriptional start site. A variety of factors bind to this segment that contains the bases C-A-A-T.

capping Refers to a set of chemical alterations, including the addition of a terminal G and methylation, at the 5' end of all hnRNAs.

carboxy terminus In a polypeptide, the end of the molecule that has a carboxylic acid group (—COOH) and corresponds to the 3' end of a gene.

cDNA Complementary DNA. DNA synthesized from an RNA template by a reverse transcriptase enzyme.

cDNA library A collection of DNA sequences generated from mRNA sequences. This type of library contains only protein-coding DNA (genes).

central dogma Process by which information is extracted from the nucleotide sequence of a gene and then used to make a protein (DNA→RNA→protein).

chain-termination method The basis of most DNA sequencing strategies. Dideoxy-nucleotides (missing both the 2' and 3' OH) prevent the addition of any additional nucleotides.

character In a phylogenetic tree, a well-defined feature that can exist in a limited number of different states.

charged amino acid Amino acid that carries a positive or negative charge at biological pH.

chloroplast transit sequence A string of roughly 25 charged amino acids at the N-terminus of a protein destined for delivery to a chloroplast.

Chou-Fasman parameter A set of numeric parameters indicating the empirically observed tendency of an amino acid to be involved in an alpha helix, a beta strand, and in each position of a hairpin turn.

chromatin The roughly equal mixture by mass of DNA and closely associated histones within eukaryotic nuclei.

chromosome In prokaryotes, the DNA molecule containing a cell's genome. In eukaryotes, a linear DNA molecule complexed with proteins that contains a large amount of genetic information.

cladist An evolutionary biologist who is generally more interested in the evolutionary path taken by organisms since they last shared a common ancestor than their relative placement on phylogenetic trees.

cladogram A graphical representation that conveys the evolutionary relationship between organisms.

cloning Insertion of specific DNA fragments into chromosome-like carriers that allow their maintenance and replication within living cells.

codon Group of three nucleotides in an RNA copy of the coding portion of a gene, corresponding to a specific amino acid.

comment In computer programming, nonexecutable notes in a computer program that describe a function; comments make long, complex programs easier to understand.

compiler A computer program that translates a symbolic programming language into machine language so that the instructions can be executed by the computer.

complementary (1) The pairing of specific nucleotides (G and C; A and T; A and U) through hydrogen bonding. (2) The antiparallel pairing of strands of nucleotides.

conceptual translation Using the universal genetic code to convert the nucleotide sequence of the open reading frame of a gene into its corresponding amino acid sequence.

conditional code Code that will only be executed if a test performed on a conditional statement is true.

conditional execution Execution of a program block that may or may not occur, depending on the result of a particular test. For example, the "if" statement allows conditional execution in Perl.

conditional (if) statement Statements that are only executed if certain conditions are met.

conformational parameter Numeric values representing the empirically observed tendency of a particular amino acid to be found in a specific conformation (alpha helix, beta sheet, or turn).

consensus sequence A sequence that represents the most common nucleotide or amino acid at each position in two or more homologous sequences.

consensus tree A single tree that summarizes the graphical representations of a set of trees.

constitutive A gene or operon that is expressed continuously.

contig A set of clones whose sequences can be assembled into an array that is longer than what can be obtained from any single sequencing reaction.

convergent evolution The independent evolution of similar genetic or phenotypic traits. For example, eyes evolved independently in a variety of organisms such as mammals, molluscs, and insects and are not homologous structures.

cot equation An equation that relates the fraction of single-stranded DNA remaining as a function of time (with genomic complexity and concentration as the only variables) in genomic renaturation experiments. The time required for half of a genome to reassociate ($cot_{1/2}$) is a useful measure of genomic complexity.

covalent bonding The sharing of electrons in overlapping orbitals.

CpG island A stretch of 500 to 3,000 bp in which the dinucleotide CpG is found at higher than normal levels relative to the rest of a mammalian genome. Usually associated with the promoters of eukaryotic housekeeping genes.

crystal Solid structure formed by a regular array of molecules.

C value Measure of a cell's total DNA content.

C-value paradox The absence of a perfect correlation between organismal complexity and genome size.

declaring (a variable) Most computer programs require the programmer to list at the start of a program all variables that will be used within the program; this is the process of declaring a variable.

degeneracy The ability of some amino acids to be coded for by more than one triplet codon.

deleterious mutation A mutation that has an adverse effect on the fitness of an organism.

denatured protein A protein that has lost its normal tertiary and quaternary structure usually because of exposure to heat or chemicals such as detergents or urea.

deoxyribonucleic acid (DNA) A usually double-stranded biopolymer of linked nucleotides in which the sugar residue is deoxyribose. The molecular basis of heredity.

dipeptide Two amino acids joined by a peptide bond.

dissociation constant (K_I) A measure of an enzyme's affinity for an inhibitor.

distance In a phylogenetic tree, a measure of the overall, pairwise difference between two data sets.

disulfide bond Cross-linking residues that are far removed from one another in the primary structure of a protein.

DNA *See* deoxyribonucleic acid.

dot plot A graphical method of comparing two sequences. A series of diagonal lines within the graph correspond to regions of sequence similarity.

dynamic programming Program that allows computers to efficiently explore all possible solutions to certain types of complex problems; it breaks a problem apart into reasonably sized subproblems and uses these parts to compute the final answer.

electronegativity Measure of an atom's need to acquire or to donate atoms to fill or empty its outermost shell of orbitals.

element Something that cannot be further reduced by chemical reactions.

endergonic reaction Chemical reaction that requires a net input of energy to convert reactants into products; has a positive Gibbs free energy.

endoplasmic reticulum (ER) A web-like network of membranes that is intimately associated with the Golgi apparatus.

enhanceosomes An assembly of transcription factors bound to the promoter of a eukaryotic gene.

enhancer Any of a number of DNA sequences to which eukaryotic transcription factors can specifically bind. Enhancer sequences function in either orientation and act cumulatively to increase transcription levels.

enzyme A biological catalyst (usually a protein) that causes a specific chemical reaction to proceed more quickly by lowering its activation energy.

equilibrium constant (K_{eq}) The point in a reversible chemical reaction at which reactants are converted to products at the same rate that products are converted to reactants.

escape sequence Multi-character sequence that allows programmers to insert special characters (such as newlines and tabs) into strings. In Perl, escape sequences start with the backslash (\) character.

ESI Electrospray ionization.

EST Expressed sequence tags; short DNA sequences obtained from either the 5' or 3' ends of cDNAs.

euchromatin Open chromatin characterized by high levels of histone methylation and low levels of DNA methylation in eukaryotes.

evaluate In Perl, to replace a variable name with the contents of the variable.

exergonic reaction Chemical reaction that releases energy; has a negative Gibbs free energy.

exhaustive search An evaluation of all possible problem solutions.

exon Parts of an hnRNA molecule spliced together to form mRNA.

exon shuffling The creation of proteins with new functions by the process of recombining exons corresponding to functional domains of existing genes at the level of DNA. Strictly, exon duplication and insertion.

family Consists of proteins that are more than 50% identical in amino acid sequence across their entire length.

file pointer A special type of Perl variable that indicates a file to read from or write to.

fixation A condition in which an allele's frequency within a population reaches 100%.

fold Often used synonymously with the term "structural motif" but typically used to connote large regions of similar secondary structure found in two or more proteins.

fourfold degenerate site Codon position where changing a nucleotide to any of the three alternatives has no effect on the amino acid that ribosomes insert into protein.

four-point condition A situation in which two pairs of taxa are grouped together on a tree with four terminal branches in a way that the distances separating the paired taxa are shorter than both alternatives.

function A subroutine that returns a value.

functional constraint The tendency in particularly important genes to accumulate changes very slowly over the course of evolution.

G Guanine. One of two purines that are used as a nitrogenous base.

gap penalty A reduction in the score for an alignment that is invoked to minimize the introduction of gaps.

gaps A dash or series of dashes introduced to an alignment to reflect the occurrence of an indel in one of two aligned sequences since they last shared a common ancestor.

GC content The measure of the abundance of G and C nucleotides relative to A and T nucleotides within DNA sequences.

gel electrophoresis Process in which an electric field is used to pull charged molecules through a polyacrylamide, starch, or agarose gel to separate them by their size and or charge.

gene A specific sequence of nucleotides in DNA or RNA that is essential for a specific function; the functional unit of inheritance controlling the transmission and expression of one or more traits.

gene expression Process of using the information stored in DNA to make an RNA molecule and then a corresponding protein.

gene tree Phylogenetic tree based on the divergence observed within a single set of homologous genes.

genome The sum total of an organism's genetic material.

genomic equivalent The amount of DNA that corresponds to the size of an organism's complete set of genetic instructions.

genomic library A set of clones containing genomic DNA inserts.

genotype All or part of the genetic constitution of an individual or group.

global alignment A sequence alignment method that provides a score for aligning two sequences in their entirety.

global variable A variable that is active throughout an entire program; contrast with *local variable.*

GU-AG rule Associated with eukaryotic protein-coding genes, this rule states that the first two nucleotides at the 5' end of the RNA sequence of introns are invariably 5'-GU-3' and the last two at the 3' end of the intron are always 5'-AG-3'.

hairpin turn Place in an RNA chain where it reverses to allow intramolecular base pairing.

hash A Perl variable that can store multiple values. Unlike an array where values are retrieved using integer indices, a hash can use any type of value (including strings) as an index.

Henri–Michaelis–Menten equation A mathematical model that uses changes in the initial velocity of an enzyme catalyzed reaction when substrate concentration is varied to determine the enzyme's V_{max} and K_m.

heterochromatin Transcriptionally inactive, densely packed chromatin; associated with high levels of DNA methylation and low levels of histone methylation.

heuristic methods Trial-and-error, self-educating techniques for parsing a tree.

hnRNA Heterogeneous RNA; primary RNA polymerase II transcripts in eukaryotes, converted to mRNAs after capping, splicing, and polyadenylation.

homologs Sequences that share a common ancestor.

homoplasies Character states that have arisen in several taxa independently and not from a common ancestor.

horizontal gene transfer The process of passing genes from one species to another. The mechanism for this movement of genes is unknown, though pathogens and transposons are often suspected as the cause.

housekeeping gene Gene that is expressed at a high level in all tissues and at all times in development.

H-P (hydrophobic-polar) model Simple lattice model that represents each amino acid residue in a protein as a single atom of fixed radius.

hydrogen bonding Interaction between molecules resulting from the slight separation of charges that results from polar covalent bonds.

hydrophilic Easily dissolved in a watery solution; literally, "water friendly."

hydrophobic Having limited interaction with water molecules; literally, "afraid of water."

hydrophobic amino acid Amino acid having an R-group composed mostly or entirely of carbon and hydrogen; it is unlikely to form hydrogen bonds with water molecules.

hydrophobic collapse The process of folding a polypeptide into a compact conformation that isolates hydrophobic residues from solvent.

hydrophobic zipper A theoretical mechanism for the formation of secondary structure in proteins. According to the theory, alpha helix and beta sheet formation is largely driven by the mutual attraction of hydrophobic residues.

if statement *See* conditional (if) statement.

indel event An insertion/deletion event.

index The number of a particular array element.

induced fit docking Changing shape of a receptor's surface as it specifically interacts with a ligand.

inferred ancestor In a phylogenetic tree, an ancestor for which empirical data are no longer available.

inferred tree A depiction of the phylogenetic relationship of three or more homologous sequences that is a close approximation of their true relationship.

informative Diagnostic position for a parsimony analysis; contrast with *uninformative*.

ingroup A species or set of species that is not the most divergent of a set of species; contrast with *outgroup*.

inhibitor Any substance that decreases the velocity of an enzyme-catalyzed reaction.

initiation complex A set of transcription factors interacting with themselves and the promoter region of a gene that facilitates the initiation of transcription.

initiator (Inr) sequence The nucleotides associated with the transcriptional start site of eukaryotic genes that are necessary and essential; the consensus sequence within humans is: 5'-44CARR-3'.

insertion sequence A transposable element containing no information content beyond what is needed for its own transposition; when inserted into a gene, it disrupts the normal structure and function of that gene.

internal node In a phylogenetic tree, a node for which no actual data have been collected; graphical representation of a common ancestor that gave rise to two or more independent lineages at some point in the past.

intractable A problem for which all algorithms require an unacceptable amount of computation time as the problem size grows large.

intrinsic terminator A specific signal for the termination of transcription in prokaryotes; a string of nucleotides in a newly transcribed RNA capable of forming a secondary structure followed by a run of uracils.

intron Internal sequence excised in splicing; present in the primary transcripts (hnRNAs) of eukaryotic genes but not in their mRNAs.

invariant A position within a sequence alignment in which all sequences contain the same character.

isochores Long regions of homogeneous base composition within eukaryotic genomes.

isoelectric focusing The process by which differences in the pI values of proteins are exploited to allow their separation.

isoelectric point (pI) The pH at which a protein has no net charge.

junk DNA Disposable DNA sequences; sequences for which no function is currently known.

kinase Enzymes that catalyze phosphorylation reactions.

labeled feature vector A representation for the examples used to train a pattern recognition algorithm. A labeled feature vector consists of a list of feature values for the example, along with a label indicating the correct classification of the example.

lead compound A molecule that is a viable candidate for use as a drug.

length (of a tree) The total number of substitutions required at both informative and uninformative sites.

length penalty Used by sequence alignment algorithms to penalize the introduction of long gaps.

Levinthal paradox Observation that the number of potential three-dimensional conformations for even a small protein is so large that an exhaustive comparison of them all cannot be accomplished by nature during protein-folding.

ligase Enzyme that catalyzes the formation of a phosphodiester bond between two DNA molecules.

LINE Long interspersed nuclear element.

local alignment A sequence alignment method that searches for subsequences that align well.

local variable A variable that exists only inside a specific subroutine; that is, it is not active throughout the entire program; contrast with *global variable*.

lock and key approach A docking approach in which the conformations of the two docked molecules are rigid and fixed; contrast with *induced fit docking*.

log odds matrix Matrix in which the entries are based on the log of the substitution probability for each amino acid.

loop statement Statement that allows a computer program to repeat a block of code either a certain number of times or until a certain condition is met.

machine language The set of symbolic instruction codes, usually in binary form, that is used to represent operations and data in a computer.

MALDI Matrix assisted laser desorption ionization; a method of mass spectrometric analysis.

match score The amount of credit given by an algorithm to an alignment for each aligned pair of identical residues.

Maxam-Gilbert method An early DNA sequencing strategy that relies on chemical degradation to generate the DNA subfragments.

maximum likelihood approach Phylogenetic approach in which probabilities are considered for every individual nucleotide substitution in a set of sequence alignments; a purely statistically based method of phylogenetic reconstruction.

methylation The attachment of a methyl group ($—CH_3$) to either a nucleotide's nitrogenous base or to a protein.

microarray An ordered grid of DNA probes fixed at known positions on a solid substrate.

microsatellite A region in the genome where relatively short nucleotide sequences such as 5'-CA-3' are tandemly repeated; typically highly variable between individuals.

minisatellite A region in the genome where nucleotide sequences ranging in size from 5 base pairs to a few tens of base pairs long occur multiple times in a tandem array; likely to be highly variable between individuals.

mismatch score The penalty assigned by an algorithm when nonidentical residues are aligned in an alignment.

mitochondrial signal sequence A string of amino acids (specifically an amphipathic helix 12 to 30 amino acids long) that causes a eukaryotic protein to be delivered to a cell's mitochondria.

mixed inhibitor An inhibitor that displays aspects of both noncompetitive and competitive inhibition.

molar A measure of the concentration of solute dissolved in a solvent; i.e., a one molar solution contains one mole of solute in one liter of solvent.

molecular clock A controversial hypothesis that, for a given DNA sequence, mutations accumulate at a constant rate in all evolutionary lineages.

molecular clones Numerous identical copies of a DNA sequence, typically associated with a vector such as a plasmid or virus that allows their maintenance and propagation in bacterial cultures.

Monte Carlo algorithm A method that samples possible solutions to a complex problem such as energy minimization as a means of estimating a general solution.

multifurcating A graphical representation of an unknown branching order involving three or more species in a phylogenetic tree.

multiple sequence alignment An alignment of three or more homologous sequences.

mutation Change in a nucleotide sequence that occurs due to mistakes in DNA replication or repair processes. Strictly, changes prior to passage through the filter of selection.

native structure Unique structure into which a particular protein is usually folded within a living cell.

natural selection (selection) Differential success between individuals in passing on genes to subsequent generations due to differences in fitness; leads to changes in allele frequencies (evolution).

nearest neighbor classifier A statistical method that classifies objects or concepts according to similarity of their features.

nearest neighbor energy rules In computing the energy of an RNA structure, rules that only consider base pairs that can potentially interact. The use of nearest neighbor rules can significantly reduce the computation time required to determine the energy of a particular conformation of an RNA molecule.

negative regulation The binding of a regulatory protein that prevents transcription from occurring.

neighbor-joining method Phylogenetic approach that starts with a star-like tree in which all species come off of a single central node regardless of their number. Neighbors are then sequentially found that minimize the total length of the branches on the tree.

neighborliness approach Phylogenetic approach that considers all possible pairwise arrangements of four species and determines which arrangement satisfies the four-point condition.

neighbors In a phylogenetic tree, the pairs of species that are separated from each other by just one internal node; sister taxa.

nested block A block of code embedded within another block. For example, nested "if" blocks can be used to enforce multiple conditions.

neural network A computer program that learns by emulating the function of a small set of neurons;

can be used to predict specific properties of data sets based on statistical similarity.

neutral mutation A mutation that has no effect on the fitness of an organism.

Newick format In a computer program, the format in which basic information about the structure of a phylogenetic tree is conveyed in a series of nested parentheses. For example, (A, (B,C)) means that taxa B and C are more like each other than either are to taxa A.

N-linked glycosylation The addition of an oligosaccharide to asparagine residues during post-translational modification of proteins within the Golgi apparatus.

NMR Nuclear magnetic resonance; technique for resolving protein structures.

nodes In a phylogenetic tree, a distinct taxonomic unit.

nondegenerate site Codon position where mutations always result in substitutions within the amino acid sequence of a protein.

nonsynonymous substitution Any nucleotide substitution that alters a codon to one for a different amino acid.

nuclear localization sequence A string of amino acids (specifically 7 to 41 basic residues) that cause a eukaryotic protein to be delivered to a cell's nucleus.

oligosaccharide A short chain of sugars.

O-linked glycosylation A post-translational process in which the enzyme *N*-acetylglucosaminyl-transferase attaches an oligosaccharide to the oxygen atom of a serine or threonine residue of a protein.

open reading frame (ORF) Any nucleotide sequence that contains a string of codons that is uninterrupted by the presence of a stop codon in the same reading frame.

operator sequence Nucleotide sequence, associated with the promoter of a gene, to which prokaryotic regulatory proteins bind.

operon A group of closely linked genes that produces a single mRNA molecule in transcription and that consists of structural genes and regulating elements.

origination penalty Penalty assessed as a result of starting a new series of gaps; part of the gap penalty.

orthologs Sequences that share similarity because of a speciation event that allowed them to evolve independently from an ancestral sequence.

outgroup A species or set of species that is least related to a group of organisms.

PAM unit A unit of evolution; specifically, the amount of evolutionary time required for an average of one substitution per 100 residues to be observed.

paralogs Sequences that share similarity because they are descendants of a duplicated ancestral gene.

parameters In computer programming, the values on which a subroutine will operate; also referred to as arguments.

parsimony The process of attaching preference to one evolutionary pathway over another on the basis of which pathway requires the invocation of the smallest number of mutational events.

pathogen A disease-causing agent.

peptide A chain of several amino acids.

peptide bond The covalent chemical bond between carbon and nitrogen in a peptide linkage.

peptide mass fingerprint A representation of the sizes of peptide fragments obtained when a specific protein is digested to completion by a protease.

Perl interpreter The program that reads and executes Perl code.

Perl script A text file that contains a list of Perl commands.

peroxisomal targeting signal A string of amino acids (specifically Ser-Lys-Leu at the carboxy-terminus) that causes a eukaryotic protein to be delivered to a peroxisome.

pH Unit of measure used to indicate concentration of hydrogen ions in a solution; specifically, the negative log of the molar concentration of H^+.

pharmacogenomics Field that uses information about an individual's genetic makeup to maximize the efficacy of treatments, while at the same time minimizing the unwanted side effects.

pheneticist Someone who studies the relationships among a group of organisms or sequences on the basis of the degree of similarity between them.

phenotype The visible properties of an organism that are produced by the interaction of its genotype and environment.

phosphatase Enzyme responsible for removing phosphate groups from phosphorylated residues.

phosphodiester bond The covalent chemical bond that connects the phosphate group of one nucleotide to the deoxyribose sugar of another.

phylogenetic tree A graphical representation of the evolutionary relationship among three or more genes or organisms.

pKa Unit of measure of the relative ease with which an amino acid releases its dissociable protons.

point accepted mutation (PAM) A mutation that has been "accepted" by natural selection in the sense that organisms bearing the mutation have survived.

polar amino acid Amino acid that often contains oxygen and/or nitrogen in its side chain and readily forms hydrogen bonds with water.

polar bond Interaction between a molecule with a full positive charge and another with a full negative charge.

polyadenylation The process of replacing the 3' end of a eukaryotic hnRNA with a stretch of approximately 250 A's that are not spelled out in the nucleotide sequence of a gene.

polycistronic Containing the genetic information of a number of genes (cistrons).

polymerase chain reaction An *in vitro* technique for rapidly synthesizing large quantities of a given DNA segment that involves separating the DNA into its two complementary strands, using DNA polymerase to synthesize double-stranded DNA from each single strand, and repeating the process.

polynucleotide A polymeric chain of nucleotides; DNA or RNA molecules.

polypeptide A polymeric chain of amino acids; protein.

position-specific scoring matrix A matrix of values that represent the frequency with which a particular amino acid type occupies a certain position in a set of aligned homologous sequences.

positive regulation When the binding of a regulatory protein makes it easier for an RNA polymerase to initiate transcription.

primary structure Sequence in which the various amino acids are assembled into a protein.

probe A piece of labeled DNA or RNA or an antibody that can specifically interact with a molecule of interest.

program control The ability to tell a computer to execute one set of instructions if a certain condition is true and another set of instructions otherwise.

promoter sequence Sequences that are recognized by RNA polymerases as being associated with a gene.

proteasome A multiprotein structure involved in protein degradation and immune response.

protein backbone The non-side-chain atoms in a polypeptide chain.

protein electrophoresis Method of using an electric field to separate and compare related proteins on the basis of superficial features such as size and charge.

protein sequencing Determining the order in which amino acids are linked together to make a given protein; often obtained by the Edman procedure in which amino acids are removed one at a time from a polypeptide's carboxy-terminus.

protein threading A process by which the conformation of a polypeptide is assumed, and then the energy of the resulting structure is calculated. By calculating the energy for a variety of known structures, the conformation that best "fits" a particular protein sequence can be determined. Because the structure is assumed rather than calculated, threading is sometimes referred to as "reverse protein folding."

proteome The sum total of an organism's proteins.

pseudogene A gene that acquires mutations that make it nonfunctional and transcriptionally inactive.

pseudoknot Created when bases involved in a loop pair with other bases outside of the loop; the most difficult type of RNA structure to predict.

purine Nucleotides whose nitrogenous bases have a two-ring structure; usually guanine and adenine.

pyrimidine Nucleotides whose nitrogenous bases have a one-ring structure; usually cytosine, thymine, and uracil.

quantifier In regular expressions, a special character that specifies the count for a specific character or character class. For example, in the regular expression /A*/, which matches zero or more A characters, the * character is a quantifier for A.

quaternary structure The intermolecular interactions that occur when multiple polypeptides associate; overall structure formed by interacting proteins.

$R_0/t_{1/2}$ value A measure of the time required for half of an organism's RNA polymerase II transcripts to hybridize to complementary sequences; a measure of transcriptome complexity.

reading frame Linear sequence of codons in a protein-coding gene starting with the start codon and ending with a stop codon.

regulatory Allowing or preventing the expression of genes under particular circumstances; contrast with *constitutive.*

relative direction representation A representation for a protein conformation in which the position of each successive residue is encoded relative to the position and direction of the previous two residues. In a two-dimensional square lattice, the possible directions for such a representation are forward (F), left (L) and right (R).

relative mutability A measure of the number of times an amino acid was substituted by any other amino acid within an alignment of homologous protein sequences.

relative rate test A check for the constancy of the rate of nucleotide substitutions in different lineages; see also *molecular clock.*

residue The portion of an amino acid that remains as a part of a polypeptide chain. In the context of a peptide or protein, amino acids are generally referred to as residues.

restriction enzymes Proteins that introduce double-stranded breaks in DNA molecules whenever they encounter a specific string of nucleotides.

restriction mapping Using simultaneous digestions with two or more restriction enzymes to determine the relative positions of restriction enzyme recognition sequences within a DNA molecule.

restriction site String of nucleotides recognized by a restriction enzyme; restriction enzyme recognition site.

retroposon A transposable element that is propagated by an RNA-intermediate but not as part of a virus and does not have terminally redundant sequences.

retrotransposition Transposition that involves an RNA-intermediate.

reverse transcriptase A special enzyme used to convert RNA to DNA.

reversible reaction Chemical reaction in which the products can be converted back into the reactants.

ribosome Complex of proteins and rRNA that are responsible for catalyzing translation.

ribozyme RNA molecules that are capable of catalyzing specific chemical reactions such as self-cleavage.

RNA polymerase Enzyme responsible for transcription; converts the information in DNA molecules into RNA molecules.

rooted tree Phylogenetic tree in which a single node is designated as a common ancestor and a unique path leads from it through evolutionary time to any other node.

rotamer A commonly observed conformation of an amino acid side chain.

satellite DNA Eukaryotic DNA fragments with unusual densities and little information storage capacity relative to other genomic data.

saturation mutagenesis The process of making all possible changes to the nucleotide sequence of a gene to determine which alter the gene's function.

scalar variable A variable that can hold only one value.

scaled tree Phylogenetic tree in which branch lengths are proportional to the differences between pairs of neighboring nodes.

scoring matrix Matrix used to score each nongap position in the alignment.

secondary structure Structural features such as alpha helices and beta sheets of a protein that arise from primary structure.

selectively neutral See *neutral mutation.*

semiglobal alignment A sequence alignment in which gaps at the start or end of the sequences do not contribute to the alignment score.

sequence (1) The linear order of nucleotides in a DNA or RNA molecule or the order of amino acids in a protein. (2) The act of determining the linear order of nucleotides or amino acids in a molecule.

serial analysis of gene expression (SAGE) An experimental technique used to assess gene expression levels.

Shine–Delgarno sequence A ribosome loading site on prokaryotic mRNAs; specifically, a string of nucleotides whose consensus sequence is 5'-AGGAGGU-3' that is complementary to a short sequence at the 3' end of the 16S rRNA found in the 30S ribosomal subunit.

side chain A short chain or group of atoms attached to the central carbon of an amino acid that confers a distinctive chemistry.

signal cascade A signal amplification strategy used by many biological systems in which an event causes one protein to interact with and activate the next in the cascade and so on until a large number of the final proteins in the cascade have been activated.

signal peptidase An enzyme that specifically removes the signal polypeptide (sequence) of a eukaryotic protein.

signal sequence (polypeptide) A string of 15 to 30 amino acids at the amino-terminus of a eukaryotic protein that causes it to be translated by a ribosome associated with the endoplasmic reticulum.

SINE Short interspersed nuclear element.

species tree Phylogenetic tree based on the divergence observed in multiple genes.

spliceosomes Enzyme complexes responsible for splicing in eukaryotes.

splicing Process of excising internal sequences of eukaryotic hnRNAs, introns, and rejoining the exons that flank them.

start codon Triplet codon (specifically, AUG) at which both prokaryotic and eukaryotic ribosomes begin to translate an mRNA.

statement Basic unit of execution in Perl; represents one instruction and is generally terminated with a semicolon.

stem A region of intramolecular base pairing in an RNA molecule.

steric collision (bump) The physical impossibility of two or more atoms occupying the same space at the same time.

sticky ends Single-stranded DNA at the cleaved end of a double-stranded fragment.

stop codon One of three codons (specifically, UGA, UAG and UAA) that does not instruct ribosome to insert a specific amino acid and, thereby, causes translation of an mRNA to stop.

strict consensus tree A consensus tree in which all disagreements are treated equally even if only one alternative tree is not consistent with hundreds of others that are in agreement regarding a particular branching point.

structural protein A term used to describe proteins generally involved with maintaining a cell or tissue's shape such as those that provide rigidity and support in bones and connective tissues.

structured programming A method of computer programming in which consistent indentation, liberal comments, and the use of subroutines are employed to create readable code.

subroutine A re-usable portion of a computer program. Subroutines are usually supplied with one or more values ("arguments", or "parameters") from the calling routine. Subroutines that return a value to the calling routine are often called functions.

substitution Mutation that has passed through the filter of selection on at least some level.

superfamily Groups of protein families that are related by detectable levels of sequence similarity that are reflective of an ancient evolutionary relationship.

synapomorphies Informative sites that support the internal branches in an inferred tree; a derived state that is shared by several taxa.

synonymous substitution Change at the nucleotide level of coding sequences that does not change the amino acid sequence of the protein.

T Thymine. One of two pyrimidines that are used as a nitrogenous base in DNA molecules.

target identification The process of identifying biological molecules essential for the survival or proliferation of a particular pathogen.

taxonomist One who studies the general principles of scientific classification including the naming and placement of taxonomic groups.

terminal node In a phylogenetic tree, a node at the tip of a branch for which data have been collected.

tertiary structure The overall three-dimensional shape of a folded polypeptide chain.

thermodynamically favorable A chemical reaction that has a large negative Gibbs free energy value is said to be thermodynamically favorable and can therefore occur without an input of energy.

topology The topographical features of a molecule; its configuration.

transcription The first step in the process of gene expression; making an RNA copy of a gene.

transcriptome The complete set of an organism's RNA sequences.

transformed distance method A distance-based method of phylogenetic reconstruction that takes the different rates of evolution within different lineages into account.

transition Mutation in which a purine (A or G) is replaced with another purine or in which a pyrimidine (C or T) is replaced by another pyrimidine.

transition state theory Theory that states that products are formed only after reactants have (1) collided in an appropriate spacial orientation and (2) acquired enough activation energy to reach a transition state.

translation Process of converting the information from the nucleotide sequences in RNA to the amino acid sequences that make a protein.

transliteration Replacement of one or more of a set of characters with corresponding characters from another set as performed by the tr operator in Perl.

transversion Mutation in which a purine (G or A) is replaced with a pyrimidine (C or T) or vice versa.

triplet code A set of three nucleotides that can be used to specify a particular amino acid during translation by ribosomes.

turnover number The number of substrate molecules converted to product by a single enzyme in a unit of time (usually seconds or minutes).

twofold degenerate site Codon position where two different nucleotides result in the translation of the same amino acid, but the two other nucleotides code for a different amino acid.

uninformative In a parsimony analysis, a position within a sequence alignment that does not allow alternative trees to be differentiated on the basis of the number of mutations they invoke; contrast with *informative.*

unrooted tree Phylogenetic tree that specifies the relationship among nodes, but does not make any representation about the direction in which evolution occurred.

unscaled tree Phylogenetic tree that conveys information about the relative kinship of terminal nodes, but does not make any representation regarding the relative number of changes that separate them.

unweighted-pair-group method with arithmetic mean (UPGMA) The simplest method of tree reconstruction, it employs a sequential clustering algorithm to build trees in a stepwise manner.

UPGMA *See* unweighted-pair-group method with arithmetic mean.

upstream promoter element Nucleotide sequences associated with the promoters of eukaryotic genes to which proteins other than RNA polymerase bind.

valence The number of unpaired electrons in an atom's outermost orbital.

vector An agent, such as a virus or a plasmid, that carries a modified or foreign gene. When used in gene therapy, a vector delivers the desired gene to a target cell.

word In sequence searching, a subsequence of fixed length. Some database search algorithms divide a query sequence into fixed-sized words, and then search for instances of these words in a sequence database.

Solutions to Odd-Numbered Questions and Problems

Chapter 1

1.1 The structure of deoxyribose is shown in Figure 1.1 while the structure of ribose shown in Figure 1.2.

1.3 See Figure 1.4.

1.5 The hydrophilic (or polar) amino acids all have either oxygen, nitrogen, or sulfur in their R groups, while the hydrophobic (non-polar) amino acids generally do not (methionine with its internal sulfur and tryptophan with its internal nitrogen are slight exceptions).

1.7 The answer to this question is the smallest value for *n* that satisfies this equation: $3{,}000{,}000{,}000 < 4^n$. That value is 16.

1.9 Using the genetic code provided in Table 1.1, the nucleotide sequence would be translated into the following string of amino acids: Met-Gly-Cys-Arg-Arg-Asn. Changing the sequence to 5'-UGG GAU GUC GCC GAA ACA-3' would cause it to code for the following string of amino acids: Trp-Asp-Val-Ala-Glu-Thr.

1.11 cDNA libraries contain clones that each contain a DNA copy of an mRNA from a cell. Since intergenic sequences and most promoter sequences are usually not transcribed by RNA polymerase (and since introns are removed from eukaryotic RNA polymerase II transcripts) sequences that correspond to those regions are not usually found in cDNA libraries. Genomic libraries are a group of clones that contain portions of an organism's genomic DNA.

Chapter 2

2.1 A pairwise sequence alignment might be useful in determining whether genetic sequences from two species are evolutionarily related. Pairwise alignments are also a useful step in database searching, since the alignment scores provide a measure of similarity between sequences. Likewise, pairwise alignment scores can be used as a distance metric for constructing phylogenetic trees (see Chapters 4 and 5). Multiple sequence alignments are useful for identifying regions of conserved sequence between more than two nucleotide or amino acid sequences. They are useful in constructing phylogenies, protein modeling, understanding substitution rates and tendencies, etc. Sequence database searches are useful for finding sequences similar to a particular target sequence. Database searches can be invaluable in identifying the functional role of a new genetic sequence, or inferring the structure of a new protein sequence.

2.3 A region of strong identity is revealed starting at position 3 of each sequence, and continuing until nearly the end of the two sequences. The region of similarity is much more clearly identified than in the previous plot.

	G	C	T	A	G	T	C	A	G	A	T	C	T	G	A	C	G	C	T	A
G									•					•						
A																				
T			•																	
G				•					•											
G					•															
T						•							•							
C							•													
A								•												
C									•											
A										•										
T						•					•									
C												•								
T													•							
G														•						
C															•					
C																•				
G																				
C																				

2.5

		A	C	A	G	T	C	G	A	A	C	G
	0	−1	−2	−3	−4	−5	−6	−7	−8	−9	−10	−11
A	−1	1	0	−1	−2	−3	−4	−5	−6	−7	−8	−9
C	−2	0	2	1	0	−1	−2	−3	−4	−5	−6	−7
C	−3	−1	1	2	1	0	0	−1	−2	−3	−4	−5
G	−4	−2	0	1	3	2	1	1	0	−1	−2	−3
T	−5	−3	−1	0	2	4	3	2	1	0	−1	−2
C	−6	−4	−2	−1	1	3	5	4	3	2	1	0
C	−7	−5	−3	−2	0	2	4	5	4	3	3	2
G	−8	−6	−4	−3	−1	1	3	5	5	4	3	4

One of the optimal alignments is:

```
ACAGTCGAACG
ACCGTC---CG
```

As there are multiple paths from the lower right to the upper left corners of the partial scores table, other optimal alignments are also possible.

2.7 Assuming the same match, mismatch, and gap scores as in problem 2.5, the partial alignment scores table would be:

		A	C	G	T	A	T	C	G	C	G	T	A	T	A
	0	−1	−2	−3	−4	−5	−6	−7	−8	−9	−10	−11	−12	−13	−14
G	−1	0	0	0	0	0	0	0	0	0	0	0	0	0	0
A	−2	0	0	0	0	1	0	0	0	0	0	0	1	0	1
T	−3	0	0	0	1	0	2	1	0	0	0	1	0	2	1
G	−4	0	0	1	0	1	1	2	2	1	1	0	1	1	2
C	−5	0	1	0	1	0	1	2	2	3	2	1	0	1	1
T	−6	0	0	1	1	1	1	1	2	2	3	3	2	1	1
C	−7	0	1	0	1	1	1	2	1	3	2	3	3	2	1
T	−8	0	0	1	1	1	2	1	2	2	3	3	3	4	3
C	−9	0	1	0	1	1	1	3	2	3	2	3	3	3	4
G	−10	0	0	2	1	1	1	2	4	3	4	3	3	3	3
G	−11	0	0	1	2	1	1	1	3	4	4	4	3	3	3
A	−12	0	0	0	1	3	2	1	2	3	4	4	5	4	4
A	−13	0	0	0	0	2	3	2	1	2	3	4	5	5	5
A	−14	0	0	0	0	1	2	3	2	1	2	3	5	5	6

The highest value in this table is 6, in the lower right. Following this value back towards the upper left until we reach a 0 value, we can obtain the following local alignment:

```
TCGCGTATA
TCTCGGAAA
```

Chapter 3

3.1 The mean time for fixation for a new neutral mutation is equal to $4N$ generations. For this question, $N = 6{,}000{,}000{,}000$ and the generation time is 30 years, so the fixation time for a neutral mutation would be (4)(6,000,000,000)(30 years) = 72,000,000,000 years (more than 20 times longer than the current age of our solar system and over 2,000 times longer than the age of our species).

3.3 93.3% of all possible substitutions (126/135) at the first position will be nonsynonymous (so about 7% will be synonymous), 100% of all possible substitutions at the second position will be nonsynonymous, and 25.2% of all possible substitutions at the third position will be nonsynonymous (75% will be synonymous). Based on these calculations, the second position is likely to be the most highly conserved position, as any change at this position will result in a nonsynonymous substitution.

3.5 The substitution rate is calculated from the simple formula $r = K/2\mathrm{T}$ where K is the estimated frequency of substitutions and T is the amount of time the two sequences have been diverging independently (in this case 100 million years). Here, r = (0.13 substitutions/site)/[2(100 million years)] or 0.065 substitutions per site per 100 million years.

3.7 The rate at which a sequence accumulates mutations is always greater than the rate at which it acquires substitutions. Substitutions are a subset of mutations—those that have passed through the filter of selection.

Chapter 4

4.1 Molecular data is less likely to suffer from problems associated with convergent evolution. It is also generally easier to find molecular characters for the purpose of comparison in distantly related or morphologically simple organisms. Randomly chosen nucleotide sequences are less likely to be subject to natural selection than morphological characters.

4.3
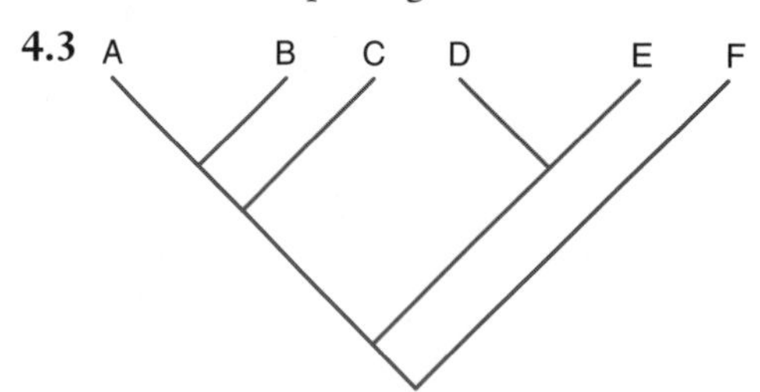

4.5 A total of 135,135 different rooted trees describe the possible relationships between eight organisms, so the chances of choosing the correct one randomly are 1 in 135,135. A total of 10,395 unrooted trees are possible for the same set of species so the chances of randomly choosing the correct unrooted tree are 13 times better than randomly choosing the correct rooted tree.

4.7 Three numbers are shown in each cell below. The first corresponds to the pairwise number of transitions, the second to the number of transversions, and the third to a weighted number of substitutions.

Species	A	B	C	D
B	9, 0, 9	—	—	—
C	6, 2, 10	8, 2, 12	—	—
D	12, 0, 12	15, 0, 15	9, 2, 13	—
E	14, 1, 16	17, 1, 19	10, 3, 16	4, 1, 6

Note that the distance between C and D is now further than the distance between A and C when this weighting scheme is applied.

4.9 A and B are closest so (A,B) and:

Species	A	B	C
B	9	—	—
C	8	11	—
DE	13.5	16.5	11.5

Now (A,B) is closest to C so ((A,B),C) and:

Species	B	AC
AC	10	—
DE	16.5	12.5

Here ((A,B),C) and D are closest so ((A,B),C),D) and then ((((A,B),C),D),E).

4.11 $x = (d_{AB} + d_{AC} - d_{BC})/2$
$y = (d_{AC} + d_{BC} - d_{AB})/2$
$z = (d_{AB} + d_{BC} - d_{AC})/2$

Chapter 5

5.1 Informative sites have at least two different nucleotides, and each of these nucleotides has to be present at least twice. They are underlined in the following alignment:

```
1   GAATGCTGAT ATTCCATAAG TCACGAGTCA AAAGTACTCG
2   GGATGGTGAT ACTTCGTAAG TCCCGAGTCG AAAGTACTCG
3   GGATGATGAT ACTTCATAAG TCTCAAATCA AAGGTACTTG
4   GGATGCTGAC ACTTCATAAG TCGCGAGTCA AAAGTACTTG
5   GGATGCTGAC ACTCCGTAAG TCCCGAGTCA AATGTACTCG
```

5.3 Only one tree invokes just one substitution. The remaining two trees both invoke a minimum of two substitutions.

5.5 Strict consensus tree for six taxa:

Chapter 6

6.1 The –35 and –10 sequences of the lactose operon's promoter are 5'-TTTACA-3' and 5'-TATGTT-3', respectively. The consensus sequence for promoters recognized by prokaryotic RNA polymerases containing the σ^{70} σ-factor at –35 is 5'-TTgACA-3' and at –10 is 5'-TATaaT-3'. The three differences are shown in lower case letters. Changing the nucleotides in the lactose operon so that they more closely match those consensus sequences would increase the expression levels of this operon.

6.3 Primers could be designed based on the sequences flanking the gap and then used with PCR to amplify and then clone the intervening sequence. Probing a library with sequences based on the ends of either of the two flanking clones could yield new clones that bridge the gap. A third, much less efficient alternative would simply involve the random sequencing of additional clones in the hopes of eventually finding one that yields the required sequence information.

6.5 The three substitutions that would stabilize the formation of a secondary structure would be to replace the mismatching nucleotides within the inverted repeats and to extend the region capable of base pairing.

6.7 The beta (β) subunit of prokaryotic RNA polymerase is responsible for binding to nucleotides and linking them together. As a result, the beta subunit would need to be mutated in order to be able to distinguish between nucleotides and their analogs (probably at the cost of making the enzyme significantly slower).

6.9 The C's of 5'-CG-3' dinucleotides (on both strands) are methylated in transcriptionally inactive regions of eukaryotic genomes but not in transcriptionally active regions. Therefore, 10 nucleotides would be methylated in a transcriptionally inactive region and none would be methylated in a transcriptionally active region.

6.11 Only 11,057 of roughly 2,400,000 nucleotides (<0.5%) in the primary transcript of the human dystrophin gene correspond to coding information. 85 to 88% of the nucleotides of prokaryotic genomes are associated with the coding regions of genes.

6.13 From the consensus sequences shown in Figure 6.6, it can be seen that most of the nucleotides scrutinized by spliceosomes during splicing are within the introns. The only nucleotides within exons that are scrutinized to some extent by spliceosomes during splicing are the last two nucleotides (AG) at the 3' end of an exon. The only triplet codons that contain the dinucleotide 5'-AG-3' are: the two codons for serine (AGU and AGC); two of the six codons for arginine (AGA and AGG); one of the three stop codons (UAG); one of the two codons for glutamine (CAG); one of the two codons for lysine (AAG); and one of the two codons for glutamic acid (GAG). Taken all together, the only amino acid that is not likely to be found at the end of a eukaryotic exon is serine—and, that is only if the 5' splice junction is a very good match to the consensus for that feature.

6.15 Open reading frames do not have to begin with a start codon. By definition, they are simply long runs of triplet codons that are not interrupted by a stop codon (UAG, UAA or UGA). The longest ORF in the sequence provided is 13 codons long and has two occurrences of the GUU codon within it. It is underlined in the sequence below:

```
5'-GAGCGGAAGUGUUCGAUGUACUGUUCCAGUCAUGUGUUCACC-3'
```

Chapter 7

7.1 At a low pH, such as 3.5, the side chains of the polypeptide are likely to be protonated, resulting in a net positive charge. At pH 8, more of the side chains will be deprotonated, resulting in a more negative overall charge. This change can result in a reduced binding affinity between the polypeptide and the column, allowing the protein to be eluted.

7.3 Since the calculations for this problem are extensive, it is easiest to write a short Perl script or use a spreadsheet to determine the values of P(t), etc. for each residue in the sequence. The following table shows the value of P(t) at each position, as well as the average P(turn), P(a), and P(b) values for the four consecutive residues starting at each position. The final column shows the predicted turn positions, where P(t) > 0.000075, and the average P(turn) value is greater than the average P(a) and average P(b) values:

Res	P(t)	Avg P(turn)	Avg P(a)	Avg P(b)	Turn
C	7.9E-05	103.75	107.5	82	
A	6.5E-05	99.25	118.5	70.75	
E	2.3E-05	97.5	113.25	82.5	
N	5.4E-05	115.5	100.75	86.75	
K	1.3E-05	100.25	109	86.25	
L	3.3E-05	87.5	107	110.25	
D	1.1E-05	89.25	112.25	98.5	
H	1.9E-05	89.25	112.25	98.5	
V	1.1E-04	95.25	104.75	106.5	
A	9.9E-05	112.5	95.75	93.75	T
D	4.8E-05	107.75	87.25	113	
C	6.8E-06	86	92.25	132	
C	1.2E-05	71.25	103	136.75	
I	3.8E-06	56.5	121.75	133.25	
L	2.8E-06	68.75	115.5	123	
F	5.3E-05	78	112.25	124.75	
M	5.9E-05	91.5	101.25	127	
T	1.2E-05	115.5	81.75	123	
W	7.7E-05	128	86.25	106.75	T
Y	1.9E-04	143	73.5	91.25	T
N	2.3E-04	152.5	70.5	68.25	T
D	5.4E-05	143.25	71.25	75.75	
G	2.0E-05	118.5	73	102.25	T
P	4.3E-06	94.5	87	118	
C	1.8E-05	68.25	99.75	144.25	
I	2.9E-06	67	99.5	151.25	
F	1.9E-05	91.75	97.75	124.75	
I	4.6E-05	115.75	86.25	112.5	
Y	2.6E-04	143	73.5	91.25	T
D	1.6E-04	152.5	70.5	68.25	T
N	#N/A	#N/A	#N/A	#N/A	#N/A
G	#N/A	#N/A	#N/A	#N/A	#N/A
P	#N/A	#N/A	#N/A	#N/A	#N/A

7.5

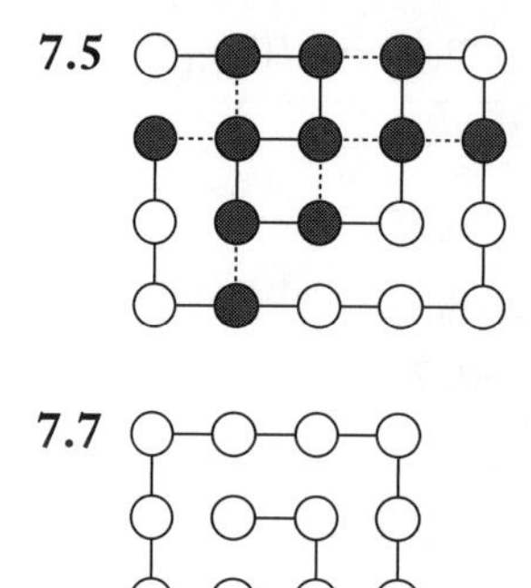

7.7

Chapter 8

8.1 cAMP activates protein kinase A. Protein kinase A (PKA) then phosphorylates and activates phosphorylase kinase, which, in turn, phosphorylates and activates phosphorylase—the enzyme responsible for glycogen breakdown. PKA simultaneously phosphorylates and *deactivates* glycogen synthase, the enzyme responsible for glycogen synthesis. Since each PKA molecule can activate multiple phosphorylase kinase molecules, the rate at which phosphorylase kinase is activated is faster than the rate at which PKA is activated. Likewise, each phosphorylase kinase can activate multiple phosphorylase molecules, so phosphorylase is activated at an even higher rate. This increase in activation velocity at each step in the cascade allows a rapid response to increased cAMP levels.

8.3 The following solution includes the entire program from Example 8, Appendix 3. Boldfaced regions have been modified from the example program:

```
#!/usr/local/bin/perl

# Example program 8. Given the name of a PDB file, read
# the file and locate all ATOM locations. For each atom,
# count the number of neighboring atoms within a threshold
# distance. As PDB files can contain thousands of atoms,
# we want to do this as efficiently as possible.

# Define some constants that will be needed later. Define
# the threshold as the square of the value we want (in
# angstroms). This way, when we compute Euclidean
# distance, we don't have to use the square root function
# on EVERY pair of distances, cutting down on computation
# time.

$threshold    = 3.6 ** 2;
$atom_keyword = "ATOM";
$atom_start   = 0;
$label_start  = 6;
$label_length = 5;
```

```
$x_start       = 30;
$x_length      = 8;
$y_start       = 38;
$y_length      = 8;
$z_start       = 46;
$z_length      = 8;
$atom_count    = 0;
$sum_x         = 0;
$sum_y         = 0;
$sum_z         = 0;

# Retrieve and check the command line parameter.
$inputfile = @ARGV[0];
if ( @ARGV == 0)
{
  die( "$0: Place the PDB file name on the commmand"
     . " line.\n\n");
} # if

printf( "Input file: $inputfile\n\n");

unless( open( PDB, "$inputfile"))
{
  die( "$0: Cannot open PDB file $inputfile.\n\n");
} # if

while ( <PDB>)
{
  $input = $_;
  chop( $input);
  if ( index( $input, $atom_keyword) == $atom_start)
  {
    $label[$records] = substr( $input, $label_start,
                               $label_length);
    $x[$records] = substr( $input, $x_start, $x_length);
    $y[$records] = substr( $input, $y_start, $y_length);
    $z[$records] = substr( $input, $z_start, $z_length);
    $records++;
    $atom_count++;
    $sum_x += $x[$records];
    $sum_y += $y[$records];
    $sum_z += $z[$records];
  } # if
} # while

for ( $i = 0; $i < $records; $i++)
{
  for ( $j = $i+1; $j < $records; $j++)
  {
```

```
      $distance = ( $x[$i]-$x[$j]) ** 2
                + ( $y[$i]-$y[$j]) ** 2
                + ( $z[$i]-$z[$j]) ** 2;
      if ( $distance < $threshold)
      {
        $count[$i]++;
        $count[$j]++;
      } # if
    } # for j
  } # for i

  # Sort the labels and counts in order by count, from
  # highest to lowest. Note that we cannot simply use 'sort'
  # here, since the information resides in two different
  # arrays. The method presented here is a simple bubble
  # sort: check each pair of counts, swapping them (and the
  # labels) if the later one has a larger value.

  printf( "Atom # Count\n");
  printf( "------ -----\n");
  for ( $i = 0; $i < $records-1; $i++)
  {
    for ( $j = $i+1; $j < $records; $j++)
    {
      if ( $count[$i] < $count[$j])
      {
        ($count[$i], $count[$j]) = ($count[$j], $count[$i]);
        ($label[$i], $label[$j]) = ($label[$j], $label[$i]);
      } # if
    } # for j

    # Print out each record as it becomes sorted.
    printf( "%6s %5d\n", $label[$i], $count[$i]);
  } # for i

  # Print out the final record, which was skipped by the
  # for i loop going only up to $records-1.
  printf( "%6s %5d\n", $label[$i], $count[$i]);
```

```
  # Compute the centroid position
  $centroid_x = $sum_x / $atom_count;
  $centroid_y = $sum_y / $atom_count;
  $centroid_z = $sum_z / $atom_count;

  # Keep track of the closest and furthest atoms
  # from the centroid
  $closest_dist = Infinity;
  $furthest_dist = -1;
```

```
# Loop through the list of atoms again, finding the atoms
# closest to and furthest from the centroid
for ( $i = 0; $i < $records; $i++)
{
    # Compute the distance from the current atom to the
    # centroid:
    $distance = ( $x[$i] - $centroid_x ) ** 2
         + ( $y[$i] - $centroid_y ) ** 2
         + ( $z[$i] - $centroid_z ) ** 2;

    # Keep track of the closest and furthest
    if ($distance < $closest_dist) {
        $closest_dist = $distance;
        $closest_atom = $i;
    }
    if ($distance > $furthest_dist) {
        $furthest_dist = $distance;
        $furthest_atom = $i;
    }
}

# Print the centroid position, and the closest and
furthest atoms from the centroid:

printf ("Centroid position: (%f, %f, %f)\n", $centroid_x,
     $centroid_y, $centroid_z);

printf("Closest atom to centroid is atom %d",
     $label[$closest_atom]);

printf("at position (%f, %f, %f)\n", $x[$closest_atom],
     $y[$closest_atom], $z[$closest_atom]);

printf("Furthest atom from centroid is atom %d",
     $label[$furthest_atom]);

printf("at position (%f, %f, %f)\n", $x[$furthest_atom],
     $y[$furthest_atom], $z[$furthest_atom]);
```

8.5 First, one would need a database of cysteine residues to serve as a training and testing example for the empirical algorithm. To obtain this database, one might select high-resolution structures from the PDB, and identify both a set of cysteine residues involved in disulfide bridges and another set of cysteine residues not involved in disulfide bridges. Next, one would identify and measure features of each cysteine residue in each set. Potential features might include sequence features (neighboring residues, overall hydrophobicity over a window, and so on), and structural features (number of neighboring atoms, number of neighboring water molecules, surface exposure, local hydrophobicity, local charge, and so on). Finally, one would select a classifier algorithm, and train it using a subset of the examples from each class (disulfide and non-disulfide cysteines). Possible classifiers include neural networks, nearest neighbor methods, Bayesian methods, decision trees, and many others. Once the classifier is trained, the accuracy can be measured using the cysteine residues from the database that were not used to train the classifier.

8.7 A signal peptide of length 5 would have $4^5 = 1024$ possible variations, insuring that there is a unique sequence for each possible destination within the cell.

Appendix 1

A1.1
```
# printnums – print the integers from 1 to 100
for ($i = 1; $i <= 100; $i++)
{
    print("$i\n");
}
```

A1.3
```
# average – print the average of the numbers in
#           values.txt to the screen
# Open the file:
open(VALUES, "<values.txt");

# Keep track of the number of values we have read
$count = 0;

# Read each value into an array
while(<VALUES>)

{
   # Remove the \n from the end of the line and save
   # it to the array @varray:
   chomp;
   $varray[$count] = $_;
   $count++;
}

# Now that all the numbers are stored in @varray,
# compute the average:
$sum = 0;
foreach $value (@varray)   # We could use an ordinary
{                          # for loop instead here,
   $sum += $value;         # but this is shorter.
}
$average = $sum/$count;
print ("The average value is: $average\n");
```

A1.5 The following solution allows you to place the filename containing the data on the command line. For example, if we name the program `findclose` and the coordinates above are the file `coords.txt`, then we can type `findclose coords.txt` to run the program.

```
# Findclose - count the number of atoms within 20
#             angstroms of the point (45.0, 45.0,
#             100.0).

# Retreive and check the command line parameter:
if ( @ARGV == 0)
{
  die( "$0: Place the coordinates filename on the"
      . " command line.\n\n");
}

$inputfile = @ARGV[0];
open (INFILE, "$inputfile")
  or die ("Cannot open $inputfile");

# Count the number of atoms within the distance cutoff
$count = 0;
while ( <INFILE>)
{
  $input = $_;
  chop( $input);
  ($x, $y, $z) = split('', $input);

  # $distance is actually the squared distance, no
  # need to take the square root
  $distance = ( $x - 45.0) ** 2
            + ( $y - 45.0) ** 2
            + ( $z - 100.0) ** 2;

  # again, we are checking the squared distance
  if ($distance < (20.0 ** 2))
  {
    $count++;
  }
}

# Print the results:
print("The number of atoms < 20 angstroms from\n");
print("the point (45.0, 45.0, 100.0) is: $count\n");
```

For the input given above, the program would print out the following:

```
The number of atoms < 20 angstroms from
the point (45.0, 45.0, 100.0) is: 4
```

Appendix 2

A2.1 $K_{eq} = k_1/k_{-1} = 60$

A2.3 Lineweaver-Burk plots have the straight-line form of an equation in the format of $y = mx + b$. The two variables (y and x; $1/v_0$ and $1/[S]$, respectively) are described in terms of each other, a slope ($m = K_m/V_{max}$) and the intercept of the y-axis ($b = 1/V_{max}$). With a direct plot, V_{max} must be estimated from an asymptote. Eadie-Hofstee plots also yield a straight line, but now both $m = -K_m$ and $b = V_{max}$.

A2.5 Noncompetitive activators have no effect upon the rate at which substrate binds the enzyme so they do not change the enzyme's K_m. However, since they do increase the rate at which the ES complex is converted to EP and, eventually to E + P, noncompetitive activators do increase an enzyme's V_{max} in a fashion that is dependent upon the amount of inhibitor present.

Index

Nonprofit Kit

5th Edition

by Stan Hutton and
Frances N. Phillips

Nonprofit Kit For Dummies®, 5th Edition

Published by: **John Wiley & Sons, Inc.,** 111 River Street, Hoboken, NJ 07030-5774, www.wiley.com

For general information on our other products and services, please contact our Customer Care Department within the U.S. at 877-762-2974, outside the U.S. at 317-572-3993, or fax 317-572-4002. For technical support, please visit https://hub.wiley.com/community/support/dummies.

Wiley publishes in a variety of print and electronic formats and by print-on-demand. Some material included with standard print versions of this book may not be included in e-books or in print-on-demand. If this book refers to media such as a CD or DVD that is not included in the version you purchased, you may download this material at http://booksupport.wiley.com. For more information about Wiley products, visit www.wiley.com.

Library of Congress Control Number: 2016958912

ISBN 978-1-119-28006-4 (pbk); ISBN 978-1-119-28008-8 (ebk); ISBN 978-1-119-28009-5 (ebk)

Manufactured in the United States of America

C10011893_070119

Contents at a Glance

Table of Contents

Introduction

It may sound corny, but we feel a certain sense of mission when it comes to nonprofits. We've started them, directed them, raised funds for them, consulted for them, volunteered for them, given money to them, and written about them. We've worked with nonprofits in one way or another for more years than we care to remember.

Why have we continued to work for nonprofit organizations? Yes, we care about others and want to see the world become a better place — our values are important to us. But, to be honest, that's not the only reason we've worked for nonprofit organizations for so many years. We believe the reason is that we can't think of anything more interesting or more challenging to do.

Starting a new program is exciting. Getting your first grant is thrilling. Working with the multifaceted personalities that come together on a board of directors is fascinating. Learning a new skill because no one else is there to do it is fun. Seeing the faces of satisfied clients, walking along a restored lakeshore, hearing the applause of audiences — all are gratifying.

That's why we do it.

About This Book

When we refer to nonprofit organizations, unless we say otherwise, we're talking about organizations that have been recognized as 501(c)(3) nonprofits and are considered public charities by the IRS.

We try to cover the gamut in this book — everything you need to know to start and manage a charitable organization, from applying for your tax exemption to raising money to pay for your programs. We include supplemental information at Dummies.com, including forms to help you create a budget, examples of grant proposals, and links to websites where you can find more help.

We also attempt to give you a bird's-eye view of the economy's nonprofit sector. When you look at financial resources, for example, nonprofits are much like the rest of the world: Most of the wealth is held by relatively few nonprofit organizations, a certain number of them are in the middle, and many, many more struggle to make ends meet.

We try to be honest about the difficulties you'll sometimes face. You probably won't be able to achieve everything you set out to accomplish, and you'll always wish you had more resources to do more things. Still, we can't imagine doing anything else. Maybe you'll feel the same way after you jump into the nonprofit world.

As you're reading, you may note that some web addresses break across two lines of text. If you're reading this book in print and want to visit one of these web pages, simply key in the web address exactly as it's noted in the text, pretending the line break doesn't exist. If you're reading this as an e-book, you've got it easy — just click the web address to be taken directly to the web page.

Foolish Assumptions

When writing this book, we made some assumptions about who may be interested in reading it. Here are some of the readers we imagined:

- You have an idea that will help solve a problem in your community, and you believe that starting a nonprofit organization is the best way to put your idea into action.
- You serve on a board of directors and wonder what you're supposed to be doing.
- You work for a nonprofit and need some ideas about fundraising, managing your organization, or working with your board of directors.
- You're simply curious about the nonprofit sector and want to find out more about it.

If you're one of these people, we're confident that this book will answer your questions and give you the information you're seeking.

Icons Used in This Book

We use the following icons throughout the book to flag particularly important or helpful information.

REMEMBER

The Remember icon emphasizes important information that you should be ready to put into practice.

TECHNICAL STUFF

You may not need this technical stuff today (and can skip over it), but — who knows? It may be invaluable tomorrow.

TIP

This icon is posted next to little hints and suggestions gleaned from our experience over the years.

WARNING

Warnings are just what you think they may be. We alert you to information that can help you avoid problematic situations.

Beyond the Book

In addition to the material in the print or e-book you're reading right now, this product also comes with some access-anywhere goodies on the web. Check out the free Cheat Sheet for a list of steps that are necessary for securing nonprofit status from the IRS, a rundown of the roles and responsibilities of people who sit on the board of directors for a nonprofit organization, and ideas for raising money for your nonprofit organization. To get this Cheat Sheet, simply go to `www.dummies.com` and type **Nonprofit Kit For Dummies Cheat Sheet** in the Search box.

You can also go to `www.dummies.com/go/nonprofitkitfd5e` for samples, forms, and lists of helpful websites. We mention many of these files within the chapters; we also include a file of web resources for most chapters. All digital files are labeled with the chapter number and the order in which the element appears in the chapter. For instance, the first digital file in Chapter 2 is labeled File 0201. For a complete list of digital files, turn to Appendix B at the back of the book.

Where to Go from Here

If you're new to the nonprofit world, we suggest beginning with Chapter 1, where you find fundamental information to get you moving in the right direction. If you're familiar with nonprofits already but want to better understand your responsibilities as a board member, you can find the answers you need in Chapter 6. If you're a new board member and want to understand the organization's finances when spreadsheets are passed out at board meetings, we provide guidance about both making a budget and understanding financial statements in Chapter 11. If you need help to publicize and market your programs, we offer some suggestions in Chapter 12.

If you're like many nonprofit workers or volunteers, you want to know how to get money for your organization. Chapters 13 through 18 cover this topic, so they're good places to begin.

Whether you are new to the nonprofit world or a seasoned professional, we think you'll find helpful and valuable information in this book to get you started or continue your good work.

1

Getting Started with Nonprofits

IN THIS PART . . .

Peek inside the nonprofit sector and get a glimpse of the role that these organizations play in our society and economy.

Get an overview of everything that goes into starting and running a nonprofit organization.

See what goes into a mission statement and get pointers on how to write one that will serve your organization well into the future.

Discover what you need to do to incorporate your new nonprofit. After that task is completed, apply for tax-exempt status from the IRS.

Make sure you maintain your nonprofit status by filing the required IRS reports.

IN THIS CHAPTER

» Getting started with a nonprofit

» Encouraging volunteerism

» Getting the resources your nonprofit needs

Chapter 1

Getting to Know the World of Nonprofit Organizations

It's a typical day in your hometown. Your alarm wakes you from a restful sleep, and you switch on your radio to hear the latest news from your local public radio station. You hear that a research institute's study reports that economic indicators are on the rise and that a health clinic across town is testing a new regimen for arthritis. Plato, your Golden Retriever/Labrador mix, adopted from the animal shelter when he was 5 months old, bounds onto your bed to let you know it's time for breakfast and a walk. Plato is followed by Cynthia, your 4-year-old daughter, who wants to help you walk Plato before she's dropped off at her preschool housed in the community center. You remember that you promised to bring canned goods to the food bank that's next door to Cynthia's school. You haven't even had your coffee yet, but already your morning is filled with news and services provided by nonprofit organizations.

You know that your public radio station is a nonprofit because you hear its pledge drives three or four times a year, and you volunteer a few hours each month for the food bank, so clearly it's a nonprofit. But you may not know that the research institute is probably a nonprofit organization, just like the health clinic where the arthritis research is being tested and the animal shelter where you found Plato. It's likely that Cynthia's preschool and the community center where the preschool rents its space are nonprofit organizations, as well. Whether we realize it or not, all of us — rich, poor, or somewhere in between — benefit from the work of nonprofit organizations every day.

Nonprofits get revenue from a variety of sources in order to provide services. Because most nonprofits serve a need in the community, tax-deductible donations are an important revenue source. Sometimes nonprofits charge a fee for what they do. Other nonprofits may sign contracts with your city or county to provide services to residents. Usually, nonprofit organizations get their income from a combination of all these revenue sources.

The nonprofit sector isn't a distinct place — it isn't some plaza or district that you come upon suddenly as you weave your way through your day. It's more like a thread of a common color that's laced throughout the economy and our lives. No matter where we live or what we do, it's not easy to get through the day without being affected by the work of a nonprofit organization. So, in this chapter, we help you understand exactly what a nonprofit organization is and how to start and manage one.

TIP

Check out File 1-1 at `www.dummies.com/go/nonprofitkitfd5e` for a list of web resources related to the topics we cover in this chapter.

What Is a Nonprofit?

People hear the term *nonprofit* and picture Mother Hubbard's cupboard, as in awfully bare with a zero bank balance. But in fact, some nonprofit organizations turn profits on their operations, and that's good, because surplus cash keeps an enterprise humming, whether it's a for-profit business or not.

Comparing for-profits to nonprofits

REMEMBER

The main difference between a for-profit and a nonprofit enterprise is what happens to the profit. In a for-profit company like Walmart, General Motors, FedEx, or your favorite fast-food chain, profits are distributed to the owners (or shareholders). But a nonprofit can't do that. Any profit remaining after the bills are paid has to be plowed back into the organization's service programs, spent to strengthen the nonprofit's infrastructure, or kept in reserve for a rainy day. Profit can't be distributed to individuals, such as the organization's board of directors.

What about shareholders — do nonprofits have any shareholders to pay off? Not in terms of a monetary payoff, like a stock dividend. But in a broad, service (not legal) sense, nonprofits do have "shareholders." They're the people who benefit from the nonprofit's activities, like the people who tune in to public radio or enroll their children in a nonprofit preschool. These people are often called stakeholders because they are committed to the success of the nonprofit.

Introducing the one and only 501(c)(3)

When we use the term *nonprofit organization* in this book, for the most part, we're talking about an organization that has been incorporated (or organized formally) under the laws of its state and that the Internal Revenue Service (IRS) has classified as a 501(c)(3) and determined to be a public charity. If the term 501(c)(3) is new to you, add it to your vocabulary with pride. In no time, "five-oh-one-see-three" will roll off your tongue as if you're a nonprofit expert.

A SECTOR BY ANY OTHER NAME

Not everyone thinks that *nonprofit sector* is the best name. That's because of the array of organizations with different types of nonprofit status. Some of these organizations are formed to benefit their members — such as fraternities and labor unions — and don't share a broad public-serving intent. Another reason *nonprofit sector* may not be the best choice of terms is its negative connotation. After all, what's worse than not making a profit? But, as we point out earlier, and we remind you again in later chapters, not making a profit isn't the determining factor. Alternative terms that you may hear include the following:

- **Voluntary sector:** This term emphasizes the presence of volunteer board members and the significance of voluntary contributions and services to the work of 501(c)(3) organizations. In this definition, the organizations alone don't represent the meaning of *nonprofit;* the definition includes the vast web of supporters who participate as volunteers and donors.
- **Independent sector:** This term emphasizes the public-serving mission of these organizations and their volunteers and their independence from government. (Independent Sector is also the name of a nonprofit organization that provides research, advocacy, and public programs for and about the nonprofit sector.)
- **Charitable sector:** This term emphasizes the charitable donations these organizations receive from individuals and institutions.
- **Third sector:** This term emphasizes the sector's important role alongside government and the for-profit business economy.

We use the term *nonprofit sector* throughout this book, but we want you to understand its limitations and be familiar with other commonly used terms.

Private foundations also have the 501(c)(3) classification, but they aren't *public charities*. They operate under different regulations, and we don't cover them in this book.

Other kinds of nonprofit organizations *do* exist; they're formed to benefit their members, to influence legislation, or to fulfill other purposes. They receive exemption from federal income taxes and sometimes relief from property taxes at the local level. (Chapter 2 discusses these organizations in greater detail.)

Nonprofit organizations classified as 501(c)(3) receive extra privileges under the law. They are, with minor exceptions, the only group of tax-exempt organizations that can receive contributions that are tax deductible for their donors.

The Internal Revenue Code describes the allowable purposes of 501(c)(3) nonprofit organizations, which include serving religious, educational, charitable, scientific, and literary ends.

Check out File 1-2 at `www.dummies.com/go/nonprofitkitfd5e` for a more detailed list of the activities that 501(c)(3) nonprofits take on.

Being a nonprofit organization doesn't mean that an entity is exempt from paying all taxes. Nonprofit organizations pay employment taxes just like for-profit businesses do. In some states, but not all, nonprofits are exempt from paying sales tax and property tax, so be sure that you're familiar with your local laws. Also, check with the appropriate office in your state to learn if you're required to apply for a state tax exemption.

Knowing Your Mission Before Entering the Nonprofit World

People form nonprofit organizations in order to work toward changing some condition in the world, either for a specific group of people or for society in general. The overall goal or purpose of a nonprofit is known as its *mission*. Taking the time needed to clearly outline a nonprofit's mission is time well spent because the mission guides the activities of the organization, helps the nonprofit's directors decide how to allocate resources wisely, and serves as a measure for evaluating the accomplishments of the group. We think developing a mission statement is so important that we devote an entire chapter (Chapter 3) to guiding you through this process.

It's also important to examine your personal mission before launching a nonprofit. You're creating a legal entity that has responsibilities for reporting to both the state and federal governments. If the organization grows to the point where you must hire employees, you're responsible for paying regular salaries and providing adequate benefits. And although you can be compensated for your work as a nonprofit staff member, you can't develop equity in the organization or take away any profits at the end of the year. Chapter 2 has more information to help you make this important decision.

Setting up a nonprofit

Nearly all nonprofit organizations are established as corporations under the laws of a particular state. If you're located in Iowa and you plan to do most of your work in that state, you follow the laws in Iowa to set up the basic legal structure of a nonprofit corporation. Although you'll find some differences from state to state, in general, the process requires writing and submitting articles of incorporation to the state and developing *bylaws,* the rules under which the corporation will operate.

After your nonprofit is established under your state laws, the next step is applying for 501(c)(3) status from the IRS. This step requires completing and submitting IRS Form 1023 or Form 1023-EZ. If you submit Form 1023, you will need to specify in some detail the proposed activities of the new organization, and you're asked for projected revenue and expenses for the year in which you apply and two years into the future. To be honest, you can't complete this form in one afternoon. It requires substantial time and thought to develop the necessary material and should be reviewed by an accountant and legal representative before filing. We discuss the incorporation and IRS application process in Chapter 4.

Making plans and being flexible

After you start managing a nonprofit organization, you'll discover that planning is your best friend. Every task from budgeting to grant writing requires that you make plans for the future. And you need to do a substantial amount of planning before you're ready to send in your IRS application for tax exemption.

Don't be frightened by this recommendation to plan. The act of planning fundamentally comes down to thinking through what you're going to do as well as how and when you're going to do it. Your plan becomes the map that guides you toward achieving your goals and your nonprofit mission. Planning is something that you should pay attention to every day.

REMEMBER

You should always begin with a plan, but that doesn't mean that plans shouldn't be altered when the situation calls for it. Circumstances change; flexibility and adaptability are good traits to nurture if you're running a nonprofit organization. Chapters 7 and 11 cover planning and budgeting. Chapters 12 and 13 discuss planning for marketing and fundraising. Chapter 8 addresses how to evaluate your work and know if your plans are achieving the results that you want to see.

BIGGER THAN A BREADBOX

The nonprofit sector is larger than many people realize. Here are some figures from the National Center for Charitable Statistics, based on IRS data, and the Independent Sector, regarding 501(c)(3) public charities in the United States:

- Nearly 1.2 million organizations were registered as public charities with the IRS in 2015.
- Assets held by these groups in 2013 totaled more than $3 trillion.
- The number of public charities increased by nearly 23 percent between 2008 and 2015.
- Nearly 30 percent of public charities that reported to the IRS in 2012 had annual expenses of less than $100,000.

Being Inspired and Inspiring Volunteers

The nonprofit sector is exciting. It encourages individuals with ideas about solving social problems or enhancing arts, culture, the environment, or education to act on those ideas. It creates a viable place within our society and economy for worthy activities that have little chance of commercial success. We think that it combines the best of the business world with the best of government social-service programs, bringing together the creativity, zeal, and problem solving from the business side with the call to public service from the government side.

We also find volunteerism inspiring. Everyone has heard stories of tightly knit communities where neighbors gather to rebuild a barn. That spirit of pitching in to help is the best part of living in a community in which people share values and ideas.

Many people these days live in diverse communities with neighbors who come from a wide variety of places and cultures. The nonprofit sector provides institutions and opportunities where everyone can come together to work toward the common good. Volunteerism gives everyone the chance to pitch in to rebuild "the barn."

Applying the term *voluntary sector* to nonprofit organizations came about for a good reason. The Urban Institute estimates that in the United States, 62.8 million people volunteered at least once in 2014.

When you're working in a nonprofit, you'll likely be supervising volunteers — and they'll likely supervise you. What we mean is that (with very few exceptions) nonprofit boards of directors serve as unpaid volunteers. And if you're the executive director, your supervisors are the trustees or board members of the organization. At the same time, you likely depend on volunteers to carry out some or all the activities of the organization. You may serve as a volunteer yourself.

The word "supervision" sounds harsh, and we don't mean to suggest that nonprofits are or should be run with an iron hand. The board of directors does have ultimate responsibility, however, for the finances and actions of a nonprofit organization, and, therefore, people serving in that capacity have a real duty to make sure that the organization has sufficient resources to carry out its activities and that it's doing what it's supposed to be doing.

We prefer to think of nonprofits as organized group activities. You need to depend on others to reach your goals, and they need to depend on you. We talk about boards of directors in Chapter 6 and working with volunteers in Chapter 9. If your nonprofit employs paid staff or hopes to someday, Chapter 10 provides some guidance in hiring and managing employees.

Finding the Resources to Do the Job

One distinctive feature of the nonprofit sector is its dependency on contributions. We devote many pages of this book — most of Part 3 — to advice about getting contributions through fundraising.

Gifts from individuals of money, goods, services, time, and property make up the largest portion of that voluntary support. This portion is also the oldest of the voluntary traditions in the United States and goes back to colonial times. Since the late 19th century, private philanthropic foundations have emerged as another source of support, and more recently — particularly after World War II — the federal government and corporations have become important income sources. Earned income through fees for service, ticket sales, and tuition charges also is an important revenue source for many nonprofits; in fact, in 2013 nearly three-quarters of the revenues for public charities was earned.

Seeing where the contributions come from

Among private, nongovernmental sources of support, gifts from living individuals — as opposed to bequests from people who have died — have always represented the largest portion of total giving, but philanthropic giving by foundations and corporations has been growing. Table 1-1 gives estimates of the sources of private contributions in 2014. The best fundraising strategy for most organizations is to take a balanced approach that includes individual giving as well as grants and corporate contributions.

TABLE 1-1 Sources of Private Contributions, 2014

Source of Income	Amount of Total Giving in Billions	Percentage of Total Giving
Individuals	$258.51	72%
Foundations	$53.97	15%
Bequests	$28.13	8%
Corporations	$17.77	5%
Total	$358.38	100%

Source: Giving USA: The Annual Report on Philanthropy for the Year 2014 (2015). Chicago: Giving USA Foundation.

Fundraising for fun and profit

Nearly every nonprofit organization depends on generous donors for the cash it needs to pay its bills and provide its services. Even if you have income from ticket sales, admission charges, or contracted services, you'll find that raising additional money is necessary to keep your organization alive and thriving.

You can see from Table 1-1 that individual giving is the largest single source of contributed income to nonprofit organizations. But you can't just sit waiting by the mailbox for the donations to begin arriving. Two basic rules of fundraising are that people need to be asked for donations and thanked after giving one. Chapter 14 focuses on raising money from individuals, Chapter 15 covers raising money with special events, and Chapter 18 discusses campaign fundraising, which is used when you need to raise extra money for your building or your endowment.

Grants from foundations and corporations make up a smaller percentage of giving to nonprofits, but their support can be invaluable for startup project costs, equipment, technical support, and sometimes general operating costs. Keep in mind that the figures given in Table 1-1 don't apply equally to every nonprofit. Some organizations get most of their income from foundation grants; others get very little. Chapter 16 introduces you to resources to help you find potential grant sources. Chapter 17 takes you through the process of crafting a grant proposal.

Fundraising works better if people know you exist. That knowledge also helps get people to your theater or to sign up for your programs. Here's where marketing and public relations enter the picture. Chapter 12 helps you figure out what your message should be and how to circulate it to the world.

REMEMBER

Make no mistake about it, fundraising is hard work, but if you approach the task with a positive attitude and make your case well, you can find the resources you need.

» Starting off your nonprofit on the right foot

» Understanding who owns a nonprofit

» Surveying the different types of nonprofit organizations

» Comparing and contrasting nonprofits and for-profits

» Enlisting a fiscal sponsor to help you provide a service or complete a project

Chapter 2

Deciding to Start a Nonprofit

Maybe you've been thinking about starting a nonprofit organization for years, or maybe an idea to solve a social problem or provide a needed service just popped into your head. It could be time to make your idea a reality. But before you file your incorporation papers, you need to understand the positive and not-so-positive factors that can make or break your new organization. Like opening any business, starting and managing a nonprofit organization isn't a simple matter.

Take a look at the economy around you. Are existing nonprofit organizations in your community thriving, or are they struggling to find financial and volunteer support? In addition, are you equipped to manage money and raise funds — which isn't an easy task even when business is booming — and can you inspire others to work with you whether they're board members, staff, or volunteers?

In this chapter, we pose some questions that you should think about (and answer) before you begin the process of incorporating and applying for tax exemption. If some of your answers point to the conclusion that your idea is worth pursuing but that you want to test the idea first, we suggest you consider using a fiscal sponsor. As we point out later in this chapter, the benefits of fiscal sponsorship are many.

When we use the term *nonprofit* in this book, we're referring to organizations that have been recognized by the IRS as exempt under section 501(c)(3) of the IRS Tax Code and described as public charities. Later in this chapter, we describe some of the other kinds of nonprofits to point out the distinct attributes of these 501(c)(3) public charities.

Check out File 2-1 at `www.dummies.com/go/nonprofitkitfd5e` for a list of web resources related to the topics we cover in this chapter.

Weighing the Pros and Cons of Starting a Nonprofit

Before you jump headfirst into making your nonprofit dream a reality, you need to understand some basic facts about nonprofit organizations. We begin with some of the pros:

- You'll receive exemption from taxes on most income to the nonprofit.
- For most nonprofits formed under section 501(c)(3) of the IRS Tax Code, you'll have the ability to receive contributions that are deductible for the donor.
- You'll have the opportunity to receive grants from foundations and corporations.
- You'll get the feeling that you're contributing to the solution of a problem or to the improvement of society.

Just about everyone would consider these facts to be positive, but they aren't the whole story. If you're thinking of starting a nonprofit to get rich or to avoid paying taxes, consider the following list of cons:

- Nonprofit employees' salaries are subject to income tax like all other types of compensation and the organization must pay employment taxes.
- You'll be required to file an annual report with the Internal Revenue Service (IRS). The complexity of the report increases as your nonprofit income

increases. (See Chapter 5 for more information about reporting requirements.)

- You can't start a nonprofit organization to benefit a particular individual or family member.
- Competition for grants from foundations, corporations, and government agencies is tough, and so is getting donations from individuals. You'll be up against more-established nonprofits with successful track records.
- If you decide to move on to other pursuits down the road, you can't take any assets accumulated by the organization you've built with you. Others will need to continue running the nonprofit or it will need to be dissolved.

REMEMBER

The bottom line is this: Think carefully about your motivation for launching a nonprofit organization. Remember that nonprofit organizations are given special privileges because they're formed to benefit the public, not specific individuals.

Doing Your Homework First

Beyond thinking about the challenges you'll face in starting and running a nonprofit organization, you also need to apply some common sense. Nonprofits don't operate in a vacuum, and neither should you. Personal commitment and inspiration can take your organization far, but you also need to find out how your community will receive your particular idea. So before going full steam ahead, investigate your competition, get community support, decide how to fund your organization, determine whether you're really ready to run a nonprofit, and develop a game plan. Read on to find out how to get your nonprofit off to a great start.

Assessing the competition

Just as if you were starting a business, you should examine your competition before starting your nonprofit organization. If you wanted to open a grocery store, you wouldn't choose a location next to a successful supermarket, because the market can bear only so much trade. This principle holds true for nonprofits, too. You may have the best idea in the world, but if someone else in your community is already doing it well, don't try to duplicate it.

On the other hand, if your area doesn't have a similar program, ask yourself why. Maybe your community doesn't have enough potential clients or audience members to support the project. Or maybe funders don't perceive the same needs in the community as you do.

TIP

Assessing the needs of your area is a good way to evaluate the potential market for your nonprofit's services. You may want to use some or all of the following methods to determine your community's needs:

- Online surveys or written questionnaires to a random sample of residents in your community
- Interviews with local foundation and civic officials
- Focus groups with people who are likely to benefit from the organization

For more details on assessing your community's needs, see Chapter 7.

Finding people to help you

Your chances of success increase if you begin with support from others, and the more help you have, the better. Sure, you can probably find an example of a single-minded visionary who battles alone through all sorts of adversity to establish a thriving nonprofit, but starting and running a nonprofit organization is essentially a group activity.

When starting a nonprofit, you need to find people who will serve on the board of directors and support your efforts with donations of money and volunteer time. The first people you usually identify as supporters are family and friends. How can they turn you down? In the long run, however, you need to expand your supporters to others who believe in the organization's mission (and not because you personally created it).

REMEMBER

Some people hesitate to share their idea with others because they believe someone may steal it. We think this fear is largely unfounded. No matter what your idea is, you'll be better off if you invite others to join you in making it a reality.

To find people to help you and support your organization, take every opportunity to speak about your idea before civic groups, religious groups, and service clubs. You can pass out fliers or set up a booth at a volunteer fair. Set up a Facebook page and invite your contacts to spread the word. Talk to your friends and coworkers. Put on your salesperson's cap and convince the community that it needs your program.

WARNING

If you're having a difficult time drumming up support, it may be a sign that you need to refine your idea or that others view it as impractical. You may need to go back to the drawing board.

Figuring out how you'll pay the bills

Funding your nonprofit organization is a big issue. Even if you begin as a volunteer-run organization and work from a home office, you still need funds for letterhead, a website, supplies, equipment, postage, and insurance. You also have to pay filing fees for your incorporation and tax-exemption applications. Many nonprofit startups are funded by the founder in the beginning. Are you able to pay all the startup expenses before revenues start flowing to your new nonprofit?

Putting together a budget can help you decide whether the startup expenses are manageable for you. You'll be creating lots of budgets sooner or later, so you may as well get an early start. Flip to Chapter 11 for details on budgeting and other financial issues.

If you can't fund the operation by yourself in the early months, you need to make a compelling case that your new organization will provide an important service to the community and then convince donors that you have the knowledge and experience to provide it. You can solicit contributions from individuals before the IRS grants the organization tax-exempt status, as long as you reveal that your exemption is pending (and will be deductible to the donor if the exemption is granted), and you have met state charitable solicitation registration requirements. These contributions become deductible to the donor if you file for your exemption within 27 months of the date you incorporated and receive tax-exempt status. If your exemption is denied, the contributions won't be deductible and you may be liable for income tax on the money you've received. Startup grants from foundations or corporations are rare and next to impossible to obtain before the IRS recognizes your organization's tax-exempt status, so don't plan to receive any grants from outside organizations.

TIP

New organizations can avoid the awkward period between starting up and receiving tax-exempt status by beginning as a sponsored program of an existing nonprofit organization. This arrangement is known as *fiscal sponsorship.* We discuss fiscal sponsorship in detail later in this chapter.

You also can try out your program on a small scale before filing for tax-exempt status by partnering with an existing nonprofit organization to test the success and need of your idea. For example, if you want to start a summer arts program for low-income children, talk to a local church or community center that serves that population and ask if you can teach an art class one day a week for a month. This enables you to try out your idea, demonstrate the need, and set up benchmarks for success.

Taking a long, hard look in the mirror

Ask yourself whether you're the right person to start a nonprofit organization, and try to answer honestly. If your organization offers a service, especially in the health and social-service fields, do you have the educational background, qualifications, or license necessary to provide those services? In addition to being professionally qualified, you need to consider whether you feel confident about your management, fundraising, and communication skills.

When starting and working in a new nonprofit organization, you need to be able to stretch yourself across many different skill areas. You may be dressed to the nines one day to pitch your project to the mayor or to a corporate executive, and the next day, you may be sweeping the floor of your office or unplugging a clogged toilet. In other words, you need to be versatile and willing to take on just about any task that needs to be done.

REMEMBER

When potential donors are evaluating grant proposals, they certainly look to see whether the organization's leadership has the background, experience, and knowledge necessary to carry out the proposed program. This doesn't necessarily mean that you need to be an experienced nonprofit manager, but try to assess your background to see how you can apply your experience to the nonprofit that you hope to start.

Planning — and then planning some more

If there were ever a time to plan, this is it. Planning is what turns your initial idea into a doable project. Planning is also a good way to find potential holes in your thinking. For example, you may believe that your community doesn't have adequate animal rescue services. You may be right, but when you begin to break down the idea of starting an animal shelter, you may find that the project costs more money or requires more staff or facilities than you first thought it would. When armed with that knowledge, you can adjust your plan as necessary or scrap the idea altogether.

To begin planning, write a one- or two-page synopsis of your nonprofit idea. In your synopsis, include why your organization should exist, what you're trying to do, and how you plan to do it. It's a good idea to outline both short-term and long-term goals and the resources needed to meet those goals. The list of resources should include money, volunteers, and an appropriate space to carry out your activities. After you've prepared your synopsis and list of resources, talk to as many people as you can about your idea, asking for help and honest feedback about your project. The purpose of this planning process is to think through your nonprofit idea step by step. (If you need help in the planning process, take a look at Chapter 7.)

Understanding Nonprofit Ownership

We once received a telephone call from a man who was shopping for a nonprofit. "Do you know if there are any nonprofits for sale in New Hampshire?" he asked. Although this question doesn't come to us often, it illustrates a misconception about the status of nonprofit organizations. No one person or group of people can *own* a nonprofit organization. You don't see nonprofit shares traded on stock exchanges, and any "equity" in a nonprofit organization belongs to the organization itself, not to the board of directors or the staff. Nonprofit assets can be sold, but the proceeds of the sale must benefit the organization, not private parties.

If you start a nonprofit and decide at some point in the future that you can't or don't want to manage it anymore, you have to walk away and leave the running of the organization to someone else. Or, if the time has come to close the doors for good, any assets the organization owns must be distributed to other nonprofits fulfilling a similar mission. You'll need to follow the laws of your state to close the nonprofit organization, including selection of an appropriate and, in some cases, an approved nonprofit that will receive the assets.

REMEMBER

When nonprofit managers and consultants talk about "ownership" of a nonprofit organization, they're using the word metaphorically to make the point that board members, staff, clients, and the community all have a stake in the organization's future success and its ability to provide needed programs.

Benefiting the public

People form nonprofit organizations to create a public benefit. In fact, nonprofit corporations are sometimes referred to as *public benefit corporations*. A nonprofit organization can't be created to help a particular individual or family, for example. If that were possible, we'd all have our separate nonprofit organizations. You can start a nonprofit to aid a specific group or class of individuals — everyone suffering from heart disease, for example, or people living below the poverty level — but you can't create a nonprofit for individual benefit or gain.

REMEMBER

Just because you're working for the public's benefit doesn't mean you can't receive a reasonable salary for your work. And despite the name *nonprofit*, such an organization can have surplus funds — essentially, a profit — at the end of the year. In a for-profit business, the surplus money can be distributed to employees, shareholders, and the board of directors; however, in a nonprofit organization, the surplus funds are used to strengthen the organization or are held in reserve by the organization to respond to emergency needs or invest in future programming.

Being accountable

Although nonprofit organizations aren't public entities like government agencies and departments, their tax-exempt status and the fact that contributions are tax deductible require them to be more accountable to the public than a privately owned business is.

It only takes a few media reports about excessive salaries or concerns about how a nonprofit has spent donated funds to prompt donors, legislators, and the general public to begin asking questions regarding the nonprofit's finances and management.

A few nonprofit organizations have taken on the task of collecting information about other nonprofits and sometimes rating them in various categories so prospective donors can use this information to help them decide which organizations to support. Charity Navigator (`www.charitynavigator.org`), CharityWatch (`www.charitywatch.org`), GuideStar (`www.guidestar.org`), and BBB Wise Giving Alliance (`www.give.org`) are four prominent organizations providing information about nonprofits. If you're just starting out and your nonprofit is small, it's unlikely that your organization will be evaluated by one of these organizations. But it will serve you well to remember that the degree of operational transparency, financial stability, and program results in your organization can be put under scrutiny.

TIP

We discuss nonprofit disclosure requirements in more detail in Chapter 5, but at a minimum, federal law requires that nonprofits file a report (Form 990) every year with the IRS. The amount of detail required in the report depends on the size of the nonprofit organization. States have their own reporting requirements, so contact your appropriate state office to learn what is required.

TECHNICAL STUFF

Most nonprofits with annual gross receipts equal to or less than $50,000 can file the 990-N; nonprofits with gross receipts less than $200,000 and assets less than $500,000 can file the 990-EZ. Nonprofits with gross receipts equal to or greater than $200,000 or assets equal to or greater than $500,000 must file the long-form 990.

Federal law also requires nonprofits to make their three most recent 990 reports, as well as their application for tax exemption and supporting documents, available for public inspection. State and local laws in your area may require additional disclosures. Posting your 990 reports and other required documents on the web is an acceptable way to meet disclosure requirements.

TIP

To get familiar with the 990 report, download a copy of it from the IRS website at `www.irs.gov`. You also can view completed 990 forms from other nonprofit organizations at GuideStar (`www.guidestar.org`).

A WORD ABOUT EXCESSIVE COMPENSATION

Although nonprofit employees have no dollar limit on the amount of compensation they can earn, the IRS does have the authority to penalize individuals (and organizations) who receive (or pay) excessive compensation. Whether the IRS considers benefits "excessive" depends on the situation. For instance, a staff member earning $100,000 annually from an organization with a budget of $125,000 may need to worry, but someone earning $100,000 from a nonprofit with a $5 million budget probably doesn't.

An employee who's found to be receiving excessive compensation may be required to return a portion of his compensation and to pay an excise tax, and, in dire cases, the nonprofit organization may lose its tax-exempt status. Chapter 5 offers more information on excessive compensation.

So when setting your nonprofit's executive director's salary, make sure the amount of compensation is justified by salary surveys of similar organizations. Also factor in the local cost of living, the size of the nonprofit's budget, and the type of services being provided. Community foundations and nonprofit consulting firms can offer salary guidelines to aid in the compensation decision. Also, many states have a statewide nonprofit organization that collects information about and provides information to other nonprofits.

Looking at the Many Varieties of Nonprofits

The words *nonprofit* and *charity* go together in most people's minds, but remember that not all nonprofits are charitable organizations. The most common examples are business and trade associations, social welfare organizations, labor organizations, political advocacy groups, fraternal societies, and social clubs. Although these nonprofits enjoy exemption from corporate income taxes, people who donate to them can't claim a tax deduction for their contributions.

Most nonprofits, charitable or not, are incorporated organizations that are formed under the laws of the state in which they're created. Some nonprofits have other legal structures, such as associations or trusts, but these are in the minority. The IRS grants tax-exempt status to a nonprofit after reviewing its stated purpose. (See Chapter 4 for information about incorporating and applying for a tax exemption.) Nonprofit types are identified by the section of the IRS code under which they qualify for tax-exempt status.

In this section, we provide an overview of the types of nonprofit organizations and some of the rules and regulations that you'll be subject to if you decide to incorporate and seek tax-exempt status from the IRS. You may discover, for example, that your idea will have a better chance for success if you create a social-welfare organization — a 501(c)(4) — or a for-profit business.

Identifying nonprofits by their numbers

Nonprofit organizations can be formed under the IRS code for many reasons other than charitable ones. To give you a taste of the variety of these organizations, the following list summarizes various classes of nonprofit organizations:

- **501(c)(3):** These organizations are formed for educational, scientific, literary, charitable, or religious pursuits and also to test for public safety and to prevent cruelty to children and animals. Nonprofits in this category may be classified as public charities or private foundations; the source of support for the organization usually determines the classification. Foundations are subject to additional rules and reporting requirements. This book focuses on 501(c)(3) organizations that are classified as public charities. Contributions to these organizations are tax deductible for their donors. Public charities may engage in limited lobbying. More information about lobbying can be found in the following section.
- **501(c)(4):** These organizations are known as *social-welfare organizations* because they're formed for the improvement of general welfare and the common good of the people. Advocacy groups tend to fall into this category because organizations with this classification are allowed more leeway to lobby legislators as a part of their mission to improve the general welfare. Contributions to these organizations aren't deductible for the donor.
- **501(c)(5):** Labor unions and agriculture and horticulture organizations formed to improve conditions for workers are in this category. These groups also may lobby for legislation.
- **501(c)(6):** Business and trade associations that provide services to their members and work toward the betterment of business conditions are placed in this classification. This category includes chambers of commerce and real estate boards, for example. Again, lobbying for legislation is allowed.
- **501(c)(7):** This section covers social clubs formed for recreation and pleasure. Country clubs and organizations formed around a hobby come under this classification. These organizations must be funded primarily by memberships and dues.

» **501(c)(9):** This type of nonprofit is an employees' beneficiary association that's created to pay insurance benefits to members and their dependents. It must be voluntary, members must have a common bond through employment or a labor union, and in most cases, it must meet nondiscrimination requirements.

Several other 501(c)-type organizations are so specialized in nature that we won't go into them here. One of our favorites is the 501(c)(13), which covers cemetery companies.

Political action committees (PACs) and parties have their own special classification, too. They're recognized under IRS code section 527 and are organized for the purpose of electing persons to office. They have special reporting requirements and aren't required to be incorporated. Donations to these organizations are not tax-deductible and the names of their donors must be disclosed.

TIP

To get the full flavor of the various categories of nonprofit organizations, check out the IRS website (`www.irs.gov`) or review IRS Publication 557.

Rules and regulations to add to your file

Entire volumes have been written about IRS regulations and laws pertaining to nonprofits. But don't worry — we just want to give you an overview of some facts that may help you decide whether starting a 501(c)(3) nonprofit organization is your best choice.

WARNING

IRS regulations can change from year to year, so be sure to look at the most recent version of IRS Publication 557 for the latest information.

Nonprofits and political activities

Nonprofits can't campaign to support or oppose the candidacy of anyone running for an elected office. However, the stipulations on lobbying for specific legislation are less clear. In the following list, we break down the rules so you know what you can and can't do, depending on what type of nonprofit you set up:

» **Social-welfare organizations and labor unions:** These organizations have more leeway when it comes to legislative lobbying than 501(c)(3) organizations do. They can engage in political intervention activity as long as it is not their primary activity. Groups that lobby must inform their members what percentage of dues they use for lobbying activities, and they can't work toward a candidate's election.

- **Charitable organizations, or nonprofits that the IRS considers public charities under section 501(c)(3):** These nonprofits may participate in some legislative lobbying if it isn't a "substantial" part of their activities. The IRS doesn't define the term *substantial,* so it determines this question on a case-by-case basis. These nonprofits can generally spend a higher portion of their budgets on lobbying activities if the organization chooses to elect the *h designation* (IRS Form 5768), which refers to section 501(h) of the IRS code. The IRS allows more expenditures for *direct lobbying* (when members of the nonprofit talk with a legislator about an issue at hand) than for *grassroots lobbying* (encouraging members of the general public to contact legislators to promote an opinion about a piece of legislation).
- **Private foundations:** Although they, too, are recognized under section 501(c)(3), these organizations may not participate in any legislative lobbying. The only exception to this rule is when pending legislation may have an impact on the foundation's existence, tax-exempt status, powers, duties, or the deductibility of its contributions.

REMEMBER

Penalties for engaging in too much political activity can include loss of your organization's tax exemption. However, going deeper into the details of these laws and reporting requirements is really beyond the scope of this book. So if you're contemplating involving your nonprofit in serious legislative activity, consult an attorney or tax specialist for advice. One place to find more information is National Council of Nonprofits. (www.councilofnonprofits.org).

The situation with churches

Churches are in a category all by themselves. The IRS doesn't require them to file for a tax exemption, nor does it require them to file annual reports to the IRS. Some churches, however, do apply for an exemption because their social-service programs often include anything from preschools to soup kitchens. These programs seek 501(c)(3) status so they can more easily apply for foundation funding and government grants or contracts to help pay the costs of providing the services.

REMEMBER

Churches that haven't been officially recognized as being tax-exempt are highly unlikely to receive foundation grants or government contracts.

Taxes, taxes, taxes

Nonprofit organizations may be subject to *unrelated business income tax,* also known as UBIT. When a nonprofit makes $1,000 a year or more in gross income from a trade or business that's regularly carried on and that's unrelated to its exempt purpose, this income is taxable. In addition, some corporate sponsorship funds may be subject to UBIT if they're perceived by the IRS as advertising dollars.

IRS Publication 598 tells you all you need to know about this subject; visit `www.irs.gov` to take a look at this publication.

Some states exempt *some* nonprofits from paying state sales and use taxes. Check the laws in your state to see whether your organization is exempt from paying these taxes. The same is true of property taxes — it depends on your local jurisdiction. Your nearest tax assessor can tell you whether you have to pay property taxes.

Nonprofit employees must, of course, pay income tax on their salaries and other taxable compensation.

Nonprofits owning for-profits

Nonprofits can own for-profit businesses. We don't recommend it, because you're going to have enough on your plate, especially when you're starting out, but it is possible. The business is subject to all regular taxes, just like all other for-profit businesses. Profits from the business may be distributed to the nonprofit, and the nonprofit must use them to further its goals and programs.

Very small organizations

If your nonprofit has less than $5,000 in annual revenues, it doesn't need to apply for a tax exemption. You can even go a bit over $5,000 in a year if your average annual income over a three-year period is less than $5,000. When your income averages more than that amount, however, you have 90 days following the close of your most recent tax year to file for a tax exemption.

Even if your organization has revenues under $5,000, you may still need to file the IRS 990-N form. This form can only be filed online. Check the IRS website at `www.irs.gov` for the latest information about this requirement.

Nonprofit compensation

Nonprofit organizations have one common feature, regardless of their type: No board member, staff member, or other interested party can benefit from the earnings of a nonprofit. Instead, assets are forever dedicated to the purpose of the organization. If the organization dissolves, the nonprofit must transfer the assets to another organization that performs a similar function.

Just because assets are dedicated to fulfilling an organizational mission doesn't mean that people are required to work for nonprofit organizations for free. Nonprofits can and should pay reasonable salaries to their staff members, if they have any. But keep in mind the difference between paying a salary and splitting the profits at the end of year.

Comparing Nonprofits and For-Profits

Believing that nonprofit organizations have special advantages isn't uncommon. And to some extent, it's true. How many for-profit businesses, for example, get help from volunteers and generous donors? On the other hand, the advantages to owning your business exist, too. In this section, we discuss the similarities and differences between nonprofits and for-profits to help you decide which direction you should go.

How they're alike

We start with the similarities between the nonprofits and for-profits because, believe it or not, there are several. For instance, consider the following:

- Sound business practices are important to both organizations.
- Strong financial oversight, including budgeting for revenue and expenses, is a key factor for both organizations.
- Good planning based on good information is a critical factor in the success of both nonprofits and for-profits.
- Management skills, the ability to communicate clearly, and attention to detail make a difference whether you're working in a nonprofit or somewhere else.
- A little bit of luck doesn't hurt either type of organization.

A final similarity involves the term *entrepreneur*, which usually describes someone who starts a new business. But any person or group that sets out to establish a nonprofit organization is entrepreneurial as well. After all, you're starting out on a path that may lead to great success, and you'll assume some risks along the way. We hope that you won't risk your house or your savings account to get a nonprofit going, but you may have uncertain income for a while.

How they differ

Although for-profits and nonprofits require similar professionalism and dedication from their leaders, they differ when it comes time to interpret their bottom lines and successes.

REMEMBER

The biggest difference between nonprofits and for-profits is the motivation for doing what you do — in other words, the *mission* of the organization. For-profit businesses exist to make money (you know, a *profit*). Nonprofits exist to provide a public benefit.

Evaluating the success of a for-profit endeavor is easy: Did you make money, and, if so, how much did you make? We're not saying this to cast stones at the capitalist system or to in any way disparage the millions of folks who work for profit-making endeavors. After all, the nonprofit sector depends on profits and wealth from the for-profit sector for its support. And of course, nonprofits have to balance the books, too. Even nonprofits prefer to end the year with more money than they had when they started. They just don't call it *profit;* they call it a *surplus.*

For a nonprofit to be successful, it needs to change some aspect of the human condition; it needs to solve a problem, provide education, or build a monument. Because the goals of nonprofits are so lofty and progress toward achieving them is often slow, evaluating nonprofit success is sometimes difficult. See Chapter 8 for information about evaluating the results of your work.

HYBRID CORPORATIONS: NEITHER FISH NOR FOWL?

Some states have passed legislation that allows for the formation of corporations that can earn and distribute profits to shareholders but also must pay attention to the social and public benefits of their business. Efforts to create these new corporate entities, which blend the best of the nonprofit and for-profit worlds, arise largely from the corporate social responsibility movement, which has grown substantially during the last 30 years.

You may have heard the term *double bottom line,* a concept that essentially means businesses should not only try to make profits but also work toward improving the social condition of their workers and customers and the environment in which they operate. To that end, many companies have corporate giving programs that support nonprofit organizations in their communities, as well as programs that encourage employees to volunteer for local charitable groups. In the case of the new hybrids, however, corporate responsibility isn't optional; companies are obliged to consider the social impact of their activities and work toward making the world a better place.

The following three corporate models fall into the hybrid category. As such, they all have one thing in common: In the eyes of the IRS, they're viewed as for-profit entities, and their income is taxable as in any for-profit business. That means if you wanted to make a donation to the work of a hybrid corporation, the IRS wouldn't allow you to claim your contribution as a charitable gift.

- **Low profit limited liability companies (L3Cs):** The L3C model was designed to better attract capital investments from investors seeking a return on their money, as well as program-related investments (PRIs) from foundations. (PRIs are typically loans

(continued)

(continued)

from foundations, not grants.) Foundations can count the funds used to provide PRIs toward their charitable-spending requirements as long as the charitable purpose of the project furthers the foundations' charitable purposes. The governing documents of the L3C are designed to make this approval easier to receive. So far, the IRS hasn't approved blanket acceptance of all PRIs awarded to L3C companies. Find more information about the L3C hybrid corporation on the Americans for Community Development website (`www.americansforcommunitydevelopment.org`).

- **B-Corps:** B-Corp organizations must adhere to certain accountability and transparency standards and have positive impacts on society and the environment. The nonprofit organization known as B-Labs (`www.bcorporation.net`) is leading the B-Corp accountability standards and advocacy efforts toward encouraging state lawmakers to pass legislation establishing this corporate model. B-Labs compares the B-Corp movement to Fair Trade Certification, which ensures that products are produced in an equitable, environmentally sound manner.
- **Flexible Purpose Corporations (FPCs):** Sometimes called social purpose corporations, these are the newest form of hybrid corporation and are similar to the B-Corp, except that they must have a more specific charitable purpose. Accountability and transparency standards for FPCs are developed internally by the corporation rather than by an outside certifying agency.

Although the notion of doing good in the world while still being allowed to make a profit is attractive, we recommend approaching these new corporate forms with caution. It's still very early in this movement, and it's difficult to predict how effective these new hybrid corporations will be. If you're thinking of establishing an L3C or other hybrid organization, consult an attorney who's knowledgeable in this field before moving forward.

Using a Fiscal Sponsor: An Alternative Approach

If you're simply interested in providing a service, maybe you don't want to waste your time with the bureaucratic and legal matters that can complicate a new nonprofit startup. Or maybe you have a project that will end after a year or two, or you simply want to test the viability of an idea. Why bother to establish a new organization if it's going to close when you finish your project?

You may not need to start a nonprofit to carry out the program you're thinking of starting. Instead, *fiscal sponsorship* may be the best route for you to take. In this approach, your new project becomes a sponsored program of an existing 501(c)(3) nonprofit organization. Contributions earmarked for your project are tax-deductible because they're made to the sponsoring agency.

FISCAL SPONSORSHIP AS A FIRST STEP

Using fiscal sponsorship as a temporary solution while establishing a new nonprofit corporation and acquiring a tax exemption can be an effective approach for the following reasons:

- You have an opportunity to test the viability of raising funds for your idea.
- You have time to establish an organizational infrastructure and to create a board of directors in a more leisurely manner.
- You can pay more attention to building your program services in the crucial beginning stages of your project.
- Your fiscal sponsor can provide bookkeeping, human resources, and other types of expertise, enabling you to focus primarily on developing your programs and activities.
- You have time to determine if your program is effectively meeting the needs you intend and can develop benchmarks to support the organization if and when you pursue your own 501(c)(3) entity.

A fiscal sponsor is sometimes called a *fiscal agent*, but this term doesn't accurately describe the relationship between a fiscal sponsor and the sponsored project. The term *agent* implies that the sponsoring organization is acting on behalf on the project, when really the project is acting on behalf of the organization. After all, the project is technically a program of the sponsoring nonprofit.

REMEMBER

This distinction may seem nitpicky, but it's an important one to keep in mind because you must satisfy the IRS requirements for this type of relationship. The 501(c)(3) sponsoring organization is responsible to both the funders and the IRS to see that the money is spent as intended and that charitable goals are met.

Examining common details of a fiscal sponsorship relationship

Here are some important points to keep in mind if you decide to go the fiscal sponsor route:

» The mission of the fiscal sponsor must be in alignment with the project. In other words, if you have a project to provide free food to the homeless, don't approach your local philharmonic orchestra as a potential sponsor. Find a nonprofit that has similar goals in its mission statement.

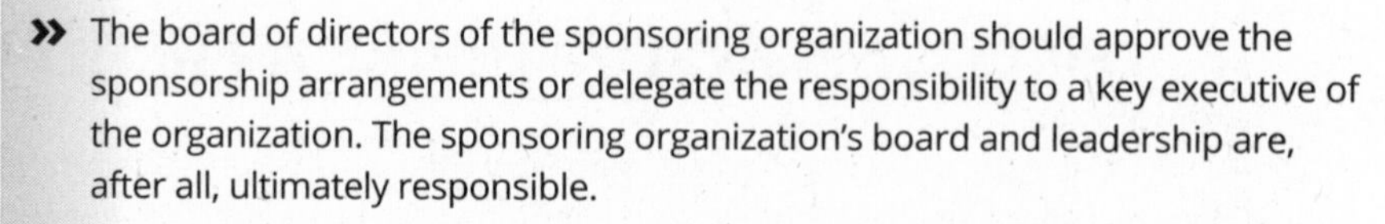

- The board of directors of the sponsoring organization should approve the sponsorship arrangements or delegate the responsibility to a key executive of the organization. The sponsoring organization's board and leadership are, after all, ultimately responsible.

TIP

- Both parties should agree to and sign a contract or memorandum of understanding, detailing the responsibilities of each. See File 2-2 at `www.dummies.com/go/nonprofitkitfd5e` for a sample fiscal sponsorship agreement.
- The fiscal sponsor customarily charges a fee for sponsoring a project — usually between 5 percent and 15 percent of the project's annual revenues, depending on the services it provides to the project.
- Some fiscal sponsors provide payroll services, bookkeeping, office space, group insurance coverage, and even management support, if needed. Be sure to ask if these additional services are included in the fiscal sponsor's fee.
- Contributions to the sponsored project should be written to the sponsor, with a note that instructs that they be used for the project.

REMEMBER

Some foundations are reluctant to award grants to fiscally sponsored projects, even announcing in their guidelines that they won't do it. One reason for this reluctance is their concern that the board of the sponsoring organization exercises less oversight toward fiscally sponsored projects than it does toward their agency's other programs. Those foundations also may be concerned that the sponsoring nonprofit is providing convenient access to 501(c)(3) status to entities engaged in activities that don't qualify for that tax status from the IRS. Not all foundations share these prohibitions, however. In fact, some are proponents of fiscal sponsorship as a way of supporting new ideas and timely programs. You can read much more about foundations and grant proposals in Chapters 16 and 17.

Finding a fiscal sponsor

You may be able to find a fiscal sponsor near you by using the Fiscal Sponsor Directory (`www.fiscalsponsordirectory.org`). Another place to search is at your local community foundation. Community foundations have wide connections in the areas they serve and likely are aware of qualified fiscal sponsors.

TIP

If your area doesn't have a community foundation nearby, find another nonprofit in your area that provides referrals and ask for help in finding the right agency to sponsor your project.

You don't want to go with just any fiscal sponsor. You have to do your homework to find one that fits your needs. First determine whether the sponsor's mission covers the type of program you'll be offering. Then do a little research to find out whether the sponsor is trustworthy and financially healthy. For example, you can perform an Internet search for the fiscal sponsor's name. Does its name appear in news stories detailing nonprofit misconduct or other skullduggery? Ask others in your community, including individuals who are knowledgeable about nonprofit activities in your town. While you're at it, read its 990 tax form posted on GuideStar (`www.guidestar.org`) to see if it's financially sound.

When you're vetting a fiscal sponsor, ask the sponsor these questions to determine whether it's a good fit for your project:

- Do your board of directors and accounting and legal advisors approve of each fiscal sponsorship?
- Do you charge for specific services, such as access to insurance programs, over and above your basic sponsorship fees? What additional services do you offer?
- Do you allow sponsored projects to hire salaried employees, and do you provide payroll services and access to health insurance?
- Do you provide coaching and mentoring in nonprofit management and fundraising?
- How frequently do you write checks to pay bills? What's the frequency and format of financial reporting for the sponsored program?
- Do you require projects to maintain a minimum annual income?
- Do you formally acknowledge gifts and donations?
- Do you help sponsored projects raise funds through your website?

TIP

The National Network of Fiscal Sponsors (`www.fiscalsponsors.org`) has developed guidelines for best practices in fiscal sponsorship. If you're considering using a fiscal sponsor, we suggest reviewing these guidelines to help you make a choice about which fiscal sponsor is best for your project.

» Creating an effective mission statement with the help of your board, staff, and volunteers

» Using your statement to make organizational decisions

Chapter 3
Creating Your Mission Statement

A good *mission statement* clearly states a nonprofit's purpose — including who benefits from its work — and how it works to fulfill that purpose. The process of developing your mission statement is important because it can help you refine your ideas, test them with other people, and inspire those involved in the mission-writing process.

Mission statements can be one-liners or long declarations that go on for two or more pages. We suggest aiming for something between these two extremes. A mission statement contained in one line resembles an advertising slogan, and a long, rambling statement is rarely read or remembered, even by the board of directors and staff members.

Take some time to think about what you want to include in your mission statement because it defines what your organization hopes to accomplish. After you've decided on your organization's mission statement, you can use it as your go-to reference when making decisions about your nonprofit's activities. You'll also include your mission statement in your 990 tax report to the IRS, in brochures, and in grant proposals. You may even print it on your business cards or coffee mugs.

In this chapter, we give you some guidance about how to create a simple yet compelling mission statement.

TIP

Check out File 3-1 at www.dummies.com/go/nonprofitkitfd5e for a list of web resources related to the topics we cover in this chapter.

Mission Statement Basics

The mission statement is an organization's center. We were tempted to use the word *heart* rather than *center*, but we think that's stretching the metaphor a little. We also could have said that mission statements are living, breathing organisms from which all organizational life flows, but that's really going too far. People are at the heart of and bring life to an organization. Mission statements just help give this human energy direction.

Can organizations operate without good mission statements? Yes, and some do. We're sure that some nonprofits out there haven't looked at their mission statements since the first Bush administration, and they're still doing good things. But the world has changed since the early 1990s, and organizations have likely adapted to those changes. An organization's chances of success in making that adaptation are better if the nonprofit and the people associated with it know exactly why it exists, what they're trying to do, and how they're going to do it.

REMEMBER

A mission statement should state what the organization's purpose is, how the purpose will be achieved, and who will benefit from the organization's activities. It may also include organizational values and vision. In addition, the mission should be

- **Memorable:** You want to carry it around with you at all times.
- **Focused:** You want it to be narrow enough to focus the activities of your organization but broad enough to allow for growth and expansion.
- **Compelling:** You want to communicate the need your organization addresses and the importance of doing something about it.
- **Easy to read:** Your statement should be written in plain language so folks don't need a set of footnotes to decipher it. Be sure to limit your use of adjectives and try to avoid jargon.

Homing in on your purpose

When thinking of your organization's purpose, think of your desired end result. What would you like to see happen? What would the world (or your community) be like if your organization were to succeed?

JARGON IS NOT YOUR FRIEND

Jargon is a term for words that have specialized meaning within an industry or profession. When you're communicating with colleagues, everyone knows the meaning of the specialized words you use. But when you're writing for the general public — the people who are the target of your mission statement — you should use words that have a common meaning for everyone.

For example, if you're an educator, you're probably familiar with the term *scaffolding*. In education, the word is used as shorthand to describe the process of using the skills a student already has to help him learn new skills. If you're a housepainter, it means something entirely different.

Think about the words you use to describe your mission and make sure everyone knows what they mean.

To say that you have to have a purpose seems almost too basic. Maybe you're thinking to yourself, "Of course I have a purpose. Why do you think I bought this book? I want to start a nonprofit to [*fill in the blank*]." But we bring up this point because clarifying the purpose is basic to a mission statement. Why should your nonprofit exist?

For example, you may know that you love cats and dogs and have always wanted to work with them, but that isn't the same thing as identifying a nonprofit organization's purpose. The mission statement for a fictitious humane society might be:

> Friends of Animals provides temporary shelter and medical care for homeless puppies, dogs, kittens, and cats until responsible, loving homes can be found.

This sentence doesn't describe the humane society's facilities or how it recruits and trains volunteers, but it does clearly state which animals it serves and that it doesn't intend to foster them as long as they live but rather to place them in good homes. And if someone came to Friends of Animals with a ferret, a pony, or a tarantula, its staff would know to refer that person to another shelter.

REMEMBER

Knowing and understanding your organization's purpose is essential to making important organizational decisions. It's also a fundamental tool to use when asking for money, recruiting board members, hiring and motivating staff, and publicizing your activities.

Specifying your beneficiaries

If you've determined your purpose, you probably know the primary beneficiaries of your activities. Their needs — whether they're kittens or refugees — make your mission compelling. Defining who will benefit from your nonprofit helps to focus your organizational activities and is an essential ingredient in your mission statement.

Some organizations have a more general audience than others. If your purpose is preserving historic buildings, the beneficiary of this activity may be current and future residents of a city, a county, or even a state. It may also be the workers you train in the crafts needed to complete the building restorations.

Explaining how you'll accomplish your goals

After you know your organization's purpose and its beneficiaries, the next step is deciding how you're going to make it happen. Mission statements usually include a phrase describing the methods your organization will use to accomplish its purpose. Think about the activities and programs you'll provide to achieve your goal. Take a look at these examples:

- To indicate how it will accomplish its goals, the Friends of Animals' mission may say, "This is accomplished by veterinary professionals and dedicated volunteers who provide temporary shelter for homeless animals."
- The mission of a national wilderness preservation organization may state that its mission is to conserve and rebuild natural ecosystems for the benefit of future generations and the planet's biological diversity. How will the organization accomplish its mission? The statement can explain that it maintains a national network of volunteer activists, provides educational materials to the public, and organizes advocacy campaigns on behalf of threatened wilderness areas.

REMEMBER

When describing how your organization addresses its purpose, you don't want to be so specific that you have to rewrite your mission statement every time you add a new program. At the same time, you want your mission statement to be concrete enough that people reading it (or hearing you recite it) can picture what your organization does.

Imagining your future with a vision statement

Simply put, a *vision statement* is your dream — your broadly described aspiration for what your organization can become. Vision statements can describe a future desired condition as a result of the organization's activities, but they're more typically applied to the organization itself. Usually, the statement includes phrases like "the best" or "recognized as a leader."

Here's an example of a vision statement from a fictitious agricultural policy think tank:

> The Agricultural Economics Institute will encourage excellence in its staff by providing opportunities for collaboration and professional development. The Institute's research projects and position statements on agricultural matters will be widely reported in the media and referred to in the setting of state and national agriculture policy.

Some nonprofits include their vision statements in their mission statements, and others don't. We believe that holding a dream is a good thing, and we see some truth to the statement, "If you can't imagine doing something great, you probably can't do it." Still, we encourage you not to spend an excessive amount of time shaping your vision statement. Focusing on a clear purpose and concrete means for addressing it is more important.

Keeping your focus narrow at first and broadening over time

REMEMBER

The narrower your mission statement, the easier it is to convey your organization's purpose and activities and to focus your attention and resources on achieving success. If you start out with a mission to benefit every man and woman on Earth, you're likely to frustrate yourself and your board. In this situation, you'd likely be trying to work at a scale that far exceeds your organization's resources.

Keep in mind, though, that you can always tweak your mission statement. As your organization achieves initial success or wants to modify its approach, you may want to review your mission as a part of your organizational planning process. However, be aware that your basic purpose should stay the same. For example, altering your mission from providing arts education to children to providing counseling for homeless families might create problems.

Keeping your statement short and sweet

In keeping with our notion that a good mission statement is easily expressed, we tend to favor short, pithy statements over long, rambling ones.

We understand, however, that some organizations want to describe their aims and activities more fully than what can be captured in one or two short sentences. For those organizations, we suggest that you begin your mission statement with a short summary that states simply what your nonprofit hopes to accomplish, how it plans to do it, and for whom. You can print this succinct version on brochures or use it as an introductory statement on your website. Then you can prepare your longer statement, which includes all your organization's programs.

Writing the Mission Statement

You probably already have a good idea of your purpose (if you don't, check out the earlier section "Homing in on your purpose"). Now you should refine that idea, state it simply in a few words, and get it down on paper. It's probably easier to just do it yourself, but we recommend that you invite your board of directors, staff, and volunteers to participate in identifying the statement's key content — your organization's goals, purpose, programs, and vision. The act of creating or refining a mission statement can motivate and even inspire them.

While inspiration may be your goal, as you gather your group's ideas and begin to write your statement, you need to make sure your written statement is clear. In the following sections, we recommend a process for inviting participation *and* achieving clarity.

Gathering input from your group

Whether you're working with a newly formed nonprofit or a long-running institution, everyone on your board or planning committee needs to agree on the mission statement. We recommend holding a meeting to solicit their input. The biggest advantage to this kind of group activity is achieving full buy-in from everyone involved. After all, you want people to believe in and accept the organization's mission statement. If they don't, they likely won't stick around to help uphold that mission (or won't do a good job of upholding it while they're there).

TIP

For groups that are working to establish a mission statement for a new nonprofit organization, we recommend that you find an outside facilitator to guide the group through the inevitable discussions about priorities and the direction of the new organization. Finding a neutral person who can bring an outsider's perspective to the group's deliberations is extremely helpful. A facilitator also takes responsibility for managing the group so you and your colleagues can be full participants in the meeting. If you aren't near a *nonprofit support organization* (a nonprofit that helps other nonprofits with technical assistance), ask other organizations near you for suggestions.

Bring a few prewritten suggestions to the group meeting. Present them as drafts and ask for feedback. After an initial discussion among the group, give each member index cards or sticky pads and have all of them write down three ideas they feel strongly about keeping as a part of the mission.

After you collect everyone's written ideas, read them all aloud and — as a group — organize them so that similar ideas are grouped together. These notes should identify the key ideas that belong in your mission statement. Also be sure to ask whether anyone thinks an important idea is missing. If not, you're ready to assign someone to draft the mission statement.

Drafting the statement

A group process is essential to identifying the core ideas that belong in the mission statement, but when it comes to putting words on paper, we recommend that you choose your best writer and turn her loose.

We aren't fans of committee-written prose. We've sat in meetings where committees discussed word choices and the placement of commas without apparent end. The result of such efforts is usually murky writing that requires several readings to interpret the meaning. After a draft is on paper, feel free to bring it back to your group for their final thoughts and approval on content, grammar, and word choice. Until then, however, assign only one person to work on the draft.

TIP

When finalizing the wording of your statement, the best advice we can offer is to stay away from jargon and flowery rhetoric. Avoid the buzzwords that are currently popular in your field. In fact, this is good advice for any kind of writing — grant proposals, memos, letters, and so on. You don't want your audience scratching their heads and wondering, "What does that mean?"

Here's an example of a vague, unclear mission statement:

> The Good Food Society works to maximize impact of the utilization of nutritious food groups to beneficially help all persons in their existence and health by proclaiming the good benefits of balanced nutrition.

You probably get the idea that this organization wants people to have better eating habits so they can enjoy better health. But can it realistically help all people everywhere? Also, try reciting this statement to someone you're trying to convince to contribute to your organization. Can you say tongue-tied?

Long, multisyllabic words don't make a mission statement more impressive. If anything, they have the opposite effect. Instead of using the preceding statement, try something like this:

> Believing in the value of good nutrition, the Good Food Society aims to improve public health by providing information about the benefits of a balanced diet to parents of school-age children through public education programs.

This mission statement may not be perfect, but it states the organization's values, its long-term goal, its targeted beneficiaries, and a general method for accomplishing the goal — all the ingredients of an effective mission statement (see the earlier section for details).

TIP

Imagine that you're riding in an elevator with someone who knows nothing about your nonprofit. You have 60 seconds to describe your organization's purpose and activities. Doing so is easy if you have a clear, short mission statement. Even if you have a longer mission statement, develop a 50- to 75-word spiel that you can recite from memory.

Living by Your Mission

You've brainstormed, drafted, and refined a short statement that clearly identifies your organization's mission. Congratulations! Now that you've put considerable thought and time into this exercise, what are you going to do with it? You'll use your mission statement in practical ways, of course. For instance, you'll likely

- Incorporate its description of the organization's purpose in your articles of incorporation (see Chapter 4).
- Include it in brochures, on your website, and in other marketing materials (see Chapter 12).
- Use it to help complete IRS Form 990 — the tax statement you'll file in one form or another each year (see Chapter 5).

Your mission statement also resembles the pin that holds the needle of a compass: You'll use it to chart your organization's direction. Here are some examples:

- When you have to make a decision about creating new programs or setting priorities, your mission statement should guide you as to whether they're appropriate for your organization.
- When your board and staff sit down to create a new three-year plan, they first need to revisit and commit to the mission statement. All other discussion about setting goals and refining programs should be tested for appropriateness against the statement.
- When you need to cut your budget and eliminate programs or activities, your mission statement should guide you to protect those programs that are core to your organization's purpose and vision.

IN THIS CHAPTER

» Writing your organization's articles of incorporation

» Applying for an Employer Identification Number

» Creating bylaws — the rules of your nonprofit

» Scheduling board meetings

» Applying for IRS tax exemption

Chapter 4
Incorporating and Applying for Tax Exemption

If you want to provide programs and services as a nonprofit organization, you must set up the legal structure for your organization and apply for its tax exemption. This process usually consists of forming a nonprofit corporation under the laws of your state and then submitting an application to the Internal Revenue Service (IRS), requesting that your organization be recognized as tax-exempt. You must take care of these tasks, which require attention to detail and ample planning, before you can begin to fulfill your mission.

REMEMBER

In this chapter, we provide you with a guide to the incorporation and exemption process for a 501(c)(3) public charity. We also give you suggestions about where you can go for help. Keep in mind that we aren't attorneys and the information in this chapter isn't meant to be legal advice. Although many nonprofits are formed without the aid of legal counsel, we think that consulting an attorney is a good idea, even if it's only to review your work. After all, you're taking on legal responsibilities. Why not be certain that you've done everything right?

TIP

Check out File 4-1 for a list of web resources related to the topics we cover in this chapter. Also, for quick reference, we've put together a checklist for forming a nonprofit organization; check out File 4-2. Both of these files are available at www.dummies.com/go/nonprofitkitfd5e.

Creating a New Entity: The Corporation

In almost all cases, the first legal step in creating a nonprofit organization is forming a corporation. A *corporation* is an entity that has legal standing. It's established by a group of individuals — the incorporators — under the laws of the state in which it's formed.

TECHNICAL STUFF

We say the first step is forming a corporation "in almost all cases" because exceptions do exist. In the United States, for example, associations, trusts, and sometimes limited liability companies can operate as tax-exempt nonprofit organizations. And charitable groups with less than $5,000 in annual revenues, as well as churches, aren't required to apply for tax exemption. In this chapter, we focus on the most common legal structure for nonprofit organizations: the corporation.

One advantage of creating a corporation is that the individuals who govern and work for it are separate from the abstract entity they create. Although board members can be liable for the corporation's actions if they don't exercise their duties and responsibilities carefully, in most cases, corporations protect individuals from personal liability. So you can think of a corporation as separate from the people who start it, because, well, it is.

REMEMBER

When you establish a corporation, you're creating something that's expected to continue in perpetuity. In other words, the corporation you create goes on living after you decide to do something else or after your death. Corporations can be closed or dissolved, but you must take legal steps to do so. You can't just take down your shingle and walk away.

Following your state law

In the United States, corporations are created and regulated under the laws of the state in which they're formed. Although the way a corporation is formed from state to state has more similarities than differences, you do need to create your nonprofit so that it conforms to the peculiarities of your state. Some states, for example, require a minimum of three members on a board of directors; others require only one. (Flip to Chapter 6 for more information about boards of directors.)

The best way to find information about incorporating in your state is to do a web search using the phrase "incorporating nonprofit corporation in (your state)." Usually, the secretary of state's office handles incorporations. Some states have an incorporation package that includes samples of the formation documents you need to file. You also can find state contact information on the IRS website at www.irs.gov.

Understanding your corporation's governing documents

Think of a corporation as a tiny government with a constitution and laws. To set down the rules under which the organization will operate, you need to prepare the following important documents:

- **Articles of incorporation:** The *articles of incorporation* (in some states called the *certificate of incorporation*) make up the document that creates the organization. It names the organization and describes its reason for existence. In the case of a nonprofit corporation, it specifies that the corporation won't be used to create profit for its directors. The articles are signed by the corporation's incorporators, usually three people, or its initial directors.
- **Bylaws:** A corporation's *bylaws* typically specify how directors are elected and the length of their terms, the officers and their duties, the number of meetings to be held, whether there are voting or nonvoting members, how many members or directors are required to be in attendance for a *quorum* (the minimum number of people required to be at a meeting so official business can be conducted, often a simple majority of members or directors) to be present, the rules for director attendance at board meetings, and the process through which the bylaws may be amended. They also may list the standing committees of the board and grant or limit particular powers of the directors.

REMEMBER

As you draft these documents (see the later sections "Writing the Articles of Incorporation" and "Developing Your Organization's Bylaws" for more specific information about writing them), remember that you're creating the legal rules under which your nonprofit will operate. You can change your articles of incorporation and bylaws by following the laws of your state (in the case of articles of incorporation) and, if applicable, the rules described in the bylaws.

Deciding whether to have members in your corporation

Corporations may have members. In fact, they may even have different classes of members — voting and nonvoting, for example. Generally speaking, though, having voting members in your corporation adds additional responsibilities to the governance of the organization. If you have voting members, for instance, you need to have membership meetings, probably at least one per year, and the voting members will be involved in choosing directors for the organization.

Depending on state law, you usually have the option to create membership conditions in either the articles or the bylaws. If you have a choice, we recommend adding these conditions to the bylaws, which are easier to amend than the articles of incorporation.

We don't advise having voting members because it adds work and responsibilities to satisfy the obligations to these individuals. However, you may want to get as many people involved in your organization as possible, and having members is one way to achieve this goal. If your nonprofit is a neighborhood-improvement group, for example, including as many people as possible in the governance of the organization may be important.

Many nonprofit organizations have "members," who get membership cards and special rates on admissions to performances or exhibits. Don't confuse this kind of membership, which is a marketing and fundraising strategy, with statutory membership in a corporation. You're free to start a membership program of this type without amending your bylaws or your articles of incorporation. See Chapter 13 for more information.

Finding the best name

Choosing a name for your new corporation may be one of the most important things you do as you set it up. A nonprofit's name is a little like a mission statement (see Chapter 3 for more on mission statements). Like your mission statement, your organization's name needs to suggest the types of programs and services you offer and the people you serve.

So if your programs provide home health services to people over 65, don't name your nonprofit something generic like Services for the People. Also stay away from names that are so abstract that they have no meaning at all. A name like the Reenergizing Society prompts more questions than it answers. What or who is being reenergized? How are clients being reenergized and why? Instead, use concrete, descriptive terms.

To avoid possible embarrassment, check the acronym that results from your organization's name. The Associated Workers for Union Labor, for example, isn't a title you want to abbreviate on your letterhead. After all, we don't know many people who'd want to associate with the AWFUL organization.

Also be careful that you don't select a name that's easily confused with that of another organization. Before you decide on a name, do a web search to see whether any other companies or organizations already have that name. The state agency that accepts your application for incorporation has procedures for ensuring that two corporations in your state don't end up with exactly the same name. However, these procedures can't help you uncover organizations with the same name in other states or with names that are similar and could be confused with your name. To help ensure that you have a distinctive name, you may want to include the name of the city or region where your organization is located — Tap Dancers of Happy Valley, for example.

You also can search the Trademark Electronic Search System (TESS) found at `www.uspto.gov` to determine whether the name you have chosen is trademarked by another organization. If you want, you can register your own trademark with the federal government and with the state in which you incorporate.

If you incorporate your organization under one name and then decide that you aren't happy with it, you can amend the incorporation papers you filed with the state. Often, however, organizations find it easier to register another name as a DBA, which stands for "doing business as." The original corporate name continues to be your organization's legal name, but you can use the DBA name on your letterhead, annual reports, and press releases — everywhere except legal documents. A county office usually handles this type of transaction. Check your local laws for more information.

Writing the Articles of Incorporation

For this section, we assume you have the articles of incorporation papers you need from the appropriate state office. Your state office may even have provided sample articles of incorporation and instructions about how to prepare your own. Pay close attention to the instructions and follow them step by step. The whole process may be as simple as filling in the blanks. Although you can amend articles of incorporation, it requires filing additional forms and paying more fees, so you may as well spend time getting them right the first time.

In this section, we give you some general guidance on drafting your articles. However, note that the sample articles we include here may not coincide exactly with your state's requirements.

TIP

IRS Publication 557 contains information about the language needed in the articles of incorporation, along with some sample articles. You can download this publication from the IRS website (`www.irs.gov`).

Crafting a heading

You must put a heading on your articles so people can identify them. The heading should be something like this:

> *Articles of Incorporation of the XYZ Theater Company, Inc.*

Sometimes you're required to add a short paragraph after the heading, stating that the incorporators adopt the following articles under the [cite the state code number under which you're filing] of [give the state name].

Article I

You insert the name you worked so hard to choose here in Article I. Simply write a sentence like this one:

> *The name of the corporation is the XYZ Theater Company, Inc.*

Could it be any easier than that?

Article II

Some states require that you affirm that your corporation is *perpetual* (meaning it's intended to exist forever). If your state requires that affirmation, put it in Article II. You can include something like this:

> *This corporation shall exist in perpetuity unless dissolved.*

However, chances are that the state will give you the language to use if it's needed.

Article III

Article III is a good place to state the organization's purpose. This article is probably the most important because state authorities and the IRS review it to determine whether your organization qualifies as a charitable entity.

Remember that 501(c)(3) organizations must be organized for a charitable, religious, educational, literary, or scientific purpose or other qualifying exempt purpose. (See Chapter 2 for more on various classes of nonprofits.) You've already created your mission statement, right? If so, stating your purpose shouldn't be too difficult. (If you haven't created your statement yet, check out Chapter 3.) Using the XYZ Theater Company as an example, your purpose may look like this:

> *This corporation is established to present theatrical productions of new and classic plays. It also will work to strengthen the theater arts, support emerging playwrights, and encourage persons to enter the acting profession by providing scholarships and grants to theater arts students and by promoting the benefits of dramatic entertainment to the general public.*

This article also must include a statement of exempt purpose under the IRS code, as in this example:

> *This corporation is organized and operated exclusively for charitable purposes, within the meaning of Section 501(c)(3) of the Internal Revenue Code or any corresponding section of any future federal tax code.*

You must state that no proceeds of the corporation will enrich any individual, except that reasonable compensation may be paid in exchange for services to the corporation. Finally, in this article, you must note that if the corporation is dissolved, any assets remaining will be distributed to another corporation that serves a similar purpose and qualifies as a tax-exempt, charitable organization under the provisions of 501(c)(3) of the Internal Revenue Code. You don't need to identify a particular nonprofit corporation; you just need to affirm that assets will be distributed to one serving a purpose similar to yours.

WARNING

Article III may be the most critical for getting your nonprofit corporation established and, ultimately, approved for tax exemption by the IRS. If your state doesn't provide good examples of the language required in this article, ask a lawyer about the requirements in your state.

Article IV

All articles of incorporation identify the name and address of an *agent of the corporation*, someone to whom mail can be addressed. This address is considered the address of the corporation until changed. Include the person's name and street address. Post office boxes aren't allowed to be used as addresses.

REMEMBER

The agent of the corporation doesn't need to be a director or incorporator of the corporation. This person can even be your attorney.

Article V

If you have initial directors, put their names and addresses in Article V. Most nonprofits start with three initial directors. If you're incorporating in a state that requires only one director, we still recommend having three. Because nonprofit organizations are formed to provide public benefit, demonstrating that several people are involved as volunteers may strengthen your application to the IRS for tax-exempt status.

Article VI

In Article VI, you list the incorporators' names and addresses. *Incorporator* simply refers to the person or people who are creating the corporation. Often, the incorporators and the initial directors are one and the same. Again, whether you need one or more depends on your state requirements.

Article VII

If you want your corporation to have members, you define the qualifications for membership in Article VII. You can define classes of membership — voting and nonvoting, for instance. If you don't want members, all you have to say is, "This corporation has no members." Better yet, refer the question to your bylaws, which are easier to amend if you change your mind. If that's what you decide to do, you can use this language:

Membership provisions of this corporation are defined in the bylaws.

REMEMBER

Corporate members aren't the same as the subscribers to a PBS station or the members of a museum or zoo, for example. Members of a corporation have the right to participate in governing the organization.

Article VIII

You may not need an Article VIII in your articles of incorporation. Some forms have a blank space here to add additional provisions. We don't recommend adding any unless you're sure you know what you're doing. Maybe your group is adamant that all future directors must be elected by 85 percent of the membership. Such a provision probably would ensure that you'd never elect new directors, but who knows? Use this blank space cautiously.

Signed, sealed, and delivered

After you finish writing your articles of incorporation, you must have the incorporator(s) sign them. In some states, you need three incorporators; in others, you need only one. If the articles need to be notarized, the signatures must be added in the presence of a notary public.

Most states charge a fee for filing for incorporation. If your state requires a fee, include a check or money order with the articles and any other required forms, and then mail everything to the appropriate state office. Usually you mail only the original articles to the state office, but sometimes a state office requires one or more additional copies.

Your next step is to wait. It's hard to say how long the response will take — it depends on the efficiency of the state office and the volume of incorporation papers it receives. For a surcharge, some states offer an opportunity to expedite processing.

WARNING

Simply creating a nonprofit corporation doesn't make your organization tax-exempt. You also need recognition from the IRS (see the section "Applying for Tax Exemption" later in this chapter) before your nonprofit is a "real" nonprofit. You may need to complete a tax-exempt application in your state, too.

If your articles are in order and your corporate name passes muster (meaning that no other organization in the state has the same name), you receive a certified copy of the articles, stamped with an official seal. Guard this piece of paper as if it were gold. Make copies and put the original away for safekeeping in a fireproof box. You've taken the first step toward starting your nonprofit. Congratulations!

Getting Your Employer Identification Number (EIN)

The first thing to do after you complete your incorporation is to apply to the IRS for an *Employer Identification Number,* or EIN. Even if you don't plan to hire employees anytime soon, you need this number for your application for tax exemption and for all your state and federal reports. The EIN is like a Social Security number for organizations; it'll be attached to your nonprofit forever.

Getting an EIN is easy and free. All you have to do is submit IRS Form SS-4. You can either download and print the form from `www.irs.gov` or complete the online application. As IRS forms go, this one is simple and straightforward and only one page long. If you apply online, we recommend that you download Form SS-4 beforehand to get an idea of the questions you'll have to answer.

Choose only one method to apply for your EIN. Don't mail IRS Form SS-4 and apply online. You may end up with two EINs, a confusing situation for everyone.

The name of the applicant isn't your name; it's the name of your new organization. As with the incorporation papers, you need to identify an individual as the principal officer and include that person's Social Security number on the form. Here are some of the other items you need to fill out, line by line:

- **Line 9:** If your organization is a church or church-controlled organization, check that box in section 9a. If it's not, check the "Other Nonprofit Organization" box. Specify what sort of nonprofit organization you are. In most cases, "charitable" is sufficient. In section 9b, fill in the state where your organization is incorporated.
- **Line 10:** Most likely, your reason for applying for an EIN in line 10 is "started a new business."
- **Line 11:** In line 11, the date on your incorporation papers is the date the business was started.
- **Line 12:** This line asks for the closing month of your organization's accounting (fiscal) year. Many organizations choose June 30 as the end to their fiscal year. December 31 (the same as the calendar year) is also popular. It's up to you. (See the later section "Dealing with financial information" for more on the fiscal year.)
- **Lines 13 and 14:** These lines relate to the number of employees you intend to hire and your expected payroll tax liabilities over the coming 12 months. You can enter "0" and "No" if your organization doesn't plan to hire staff in the year ahead. If you do intend to hire people for whom you'll have to pay payroll taxes, you need to check IRS withholding tables (`www.irs.gov`) and estimate the amount of payroll taxes that will be due to the IRS.
- **Line 16:** This line asks you to check the box that describes your organization's principal activity. You can check "Other" here, but try to be a little more specific when describing your activities in the blank space. You may say "Charitable — Arts," for example.

The IRS estimates that you'll receive your EIN in four to five weeks if you apply by mail. You'll receive your EIN immediately if you submit the online application.

Writing Your Organization's Bylaws

Bylaws are the rules by which your organization operates. As with articles of incorporation, different states have different requirements about what must be included in the bylaws, so make sure you contact the appropriate agency in your state to get general guidance about the information you need.

In general, bylaws guide the activities of your organization and the procedures of your board of directors — how many directors, how long they serve, how they're elected, what constitutes a quorum, and so on. Like the articles of incorporation, bylaws are divided into articles. However, because bylaws require more detail, the articles themselves are divided into sections (and subsections, if needed) to address various aspects of the articles.

REMEMBER

You can always change bylaws by following the rules you have set up for your organization in the bylaws.

If you were to review the bylaws of ten different organizations, you'd find variation in the order in which articles are presented. For example, you may find the board of directors specified in Article III or Article V. Bylaws also vary in how specifically they spell out what's required. Some bylaws specify the number and type of standing committees; others give the board president the responsibility of making those specifications. If you don't address a particular question in your bylaws — setting a quorum, for example — most states have a default position in their code that applies to the governing of nonprofit corporations.

TIP

Check out File 4-3 at `www.dummies.com/go/nonprofitkitfd5e` for a general guide to creating bylaws for your new nonprofit.

Holding Your First Board Meeting

Your organization's first board meeting is more or less a formality, but documenting it is important because it officially kicks off your new nonprofit corporation. If you've named directors in your articles of incorporation, each one should be present at the meeting. You should adopt the bylaws and then elect officers. You may also want to ratify the actions of the incorporator (including adopting the articles and appointing the board), adopt a conflict of interest policy, adopt the accounting year, and approve reimbursement of expenses of anybody who provided the initial monies to incorporate. It's also important to pass a resolution authorizing the board or its designate to open the necessary bank accounts. You need a copy of this resolution to open an account.

Prepare minutes of the meeting and keep them with your articles of incorporation. We say more about keeping records in Chapter 5, but now is a good time to start a *board book* — a binder containing a copy of your articles of incorporation, your bylaws, and the minutes of your first board meeting and every board meeting to follow. Also be sure to include a copy of the letter of determination that acknowledges your organization as tax-exempt when you receive it from the IRS. See the next section for how to make your application for tax exemption.

Applying for Tax Exemption

The final step in becoming a tax-exempt charitable organization is to apply for tax exemption from the IRS. Keep in mind that *tax-exempt* doesn't mean you're exempt from *all* taxes. You don't have to pay taxes on the organization's income from its charitable activities, and donors who contribute to your 501(c)(3) organization can claim a tax deduction. But if your nonprofit employs staff, it does have to pay payroll taxes like any other employer. And your liability for sales and property taxes depends on your state and local laws. If your nonprofit has income that's unrelated to its charitable purpose, you're required to pay taxes on that revenue as well.

To apply for tax exemption, you need to request what's known as a *determination letter* or *ruling* — a letter from the IRS stating that it has determined that your organization qualifies as a tax-exempt organization under the applicable sections of the IRS code. You'll send copies of this letter to foundations, government agencies, and state tax authorities — in short, to anyone to whom you need to prove that your nonprofit organization is indeed tax-exempt.

REMEMBER

Be sure to check with your state about what you need to do to register your nonprofit as a tax-exempt organization. Registration and reporting requirements vary from state to state. Chances are you received this information when you contacted your state office to begin your incorporation process; if you didn't, contact the appropriate state government office for more details. You can find contact information on the IRS website at `www.irs.gov` or by searching the web for the appropriate office in your state.

You request tax-exempt status by submitting IRS Form 1023 (Application for Recognition of Exemption) or IRS Form 1023-EZ to the IRS. You can download the form at `www.irs.gov`. Be sure to download the form's instructions at the same time and read the next section, "Tackling Form 1023."

WHO DOESN'T NEED TO APPLY FOR TAX-EXEMPT STATUS

If your organization is a church, a church auxiliary, or an association of churches, you don't need to apply for tax-exempt status. Also, if your nonprofit had gross receipts of less than $5,000 in any previous year in which you operated, and if you don't expect your revenue to grow beyond this limit, you aren't required to submit an application.

Keep in mind that you still may apply for tax exemption even though you aren't required to do so. Having a determination letter from the IRS acknowledging your tax-exempt status has some advantages. For example, you need to show proof of your exemption to get a bulk mail permit from the U.S. Postal Service. The determination letter also serves as a public acknowledgment that contributions to your organization are deductible to the donor. If you hope to receive foundation grants, your organization should be recognized as tax-exempt by the IRS as well.

Tackling Form 1023

Yes, Form 1023 is long, and yes, it's a little scary. If you can get help from your accountant or attorney (or both), we recommend that you do so. Don't discount getting help from your friends and associates, either. Two heads are usually better than one.

Read the instructions carefully. The application package includes line-by-line explanations as well as the various schedules that you may need to submit if you're applying as a church or school, for example. If a schedule isn't required for your type of nonprofit, don't submit a blank one. Toss it into the round file (also known as the trash can).

If you use the interactive 1023 application form, definitions and explanations for many of the questions can be accessed by clicking the question mark icon.

Your organization may be eligible to submit an electronic 1023-EZ application. The good news is that the application is only three pages long. The not so good news is that you must answer "no" to 26 questions on a seven-page eligibility quiz. But if you don't expect to have more than $50,000 in gross revenue in each of the next three years and if you can answer no to the remaining 25 questions, you may want to submit the 1023-EZ.

Understanding the difference between public charities and private foundations

Public charities and private foundations are both 501(c)(3) organizations (see Chapter 2 for more on these types of organizations), but you need to know the important differences between them before you begin completing the application for tax exemption.

REMEMBER

Private foundations have different reporting requirements from public charities and may be required to pay excise taxes on their investment income. Contributions to private foundations are deductible but are subject to lower deduction limits.

One important factor the IRS determines based on your application is whether your organization is classified as a public charity or a private foundation. Part X of Form 1023 addresses this question. We assume that you're striving for public charity status, so check the "No" box in line 1a and skip to line 5.

Establishing public charity status

The IRS applies several measures to determine whether an organization is a public charity. Generally, for a nonprofit to be considered a public charity, it must receive one-third of its revenues from public sources. It's complicated, and we can't cover all the nuances and technicalities here. But, fundamentally, it comes down to how much of your organizational income you get from the public.

A 501(c)(3) nonprofit organization wanting to be considered a public charity will undergo what's referred to as the *public support test.* To pass this test, the nonprofit needs to demonstrate, over a five-year average, that one-third of its revenue comes from contributions from the general public, support from government agencies, or grants from organizations that get their support from the public, such as United Way. So if your organization's average revenue is $60,000 per year and at least $20,000 a year comes from donations, state grants, and support from United Way, you're home free. Your nonprofit will be considered a public charity under code sections 509(a)(1) and 170(b)(1)(A)(vi).

But what if your organization doesn't get a third of its revenue? Fortunately, it may be able to qualify under another test that gives more leeway on the percentage of public support. Under this test, only 10 percent of total revenue needs to come from those public categories. But your organization also must demonstrate that it has an ongoing fundraising program that's reaching out to the public for more donations. Other factors also are considered. If the nonprofit's contributions come in many small gifts rather than a few large ones, it's more likely to be given public charity status. It's even better if your organization's board of directors is broadly representative of the community. And if the organization makes its facility available to the general public, that's another feather in its cap.

A 501(c)(3) organization that expects to get a substantial portion of its revenues from fees related to its exempt purpose (but less than one-third of revenues from investment income) may qualify as a public charity under code section 501(a)(2).

When you get to Part X of Form 1023, you need to check a box indicating which public charity classification you seek. Take schools as an example. To apply as a school, you have to submit Schedule B. Your school must provide regular instruction and have a student body and a faculty. In other words, you can't just start an organization and call it a school.

If you aren't sure — and, frankly, who would be? — you can let the IRS decide which category you fit by checking the box in Part X, 5i.

Every IRS rule has an exception or two, and qualifying as a public charity is, well, no exception. Churches, schools, hospitals, public safety testing organizations, and a few others are automatically considered public charities. If you have an intense interest in this rule, read Chapter 3 of IRS Publication 557, which is available at `www.irs.gov`. If you think your organization falls into one of these categories, check the appropriate box in Part X, line 5 (a through f) and answer the questions on the appropriate schedule included with Form 1023. But we don't recommend doing so unless you're absolutely certain that you know what you're doing. If you need further advice regarding these categories, we suggest you consult an attorney.

When you download Form 1023 and its instructions, be sure to include any change notices that describe new regulations and new instructions. Sometimes revisions to IRS forms lag behind changes in the regulations. Always be certain you're working with the most recent information.

Describing your activities

In Part IV of IRS Form 1023, you need to attach a narrative description of your charitable activities and how you plan to carry them out. You must provide more detail here than you have in your articles of incorporation or in your mission statement. The IRS wants you to list what you're going to do in order of importance, approximately what percentage of time you'll devote to each activity, and how each activity fulfills your charitable purpose.

When you fill out this form, chances are you haven't started operating yet, so you're simply listing *proposed* activities. If you've created an organizational plan (which we recommend), refer to the plan to make sure you cover all your activities. (Refer to Chapter 7 for help with planning.) If you've been operating your organization, explain what you've been doing and how it relates to your charitable purpose.

A nonprofit that provides family counseling and public education about strengthening families may say something like this in Part IV:

> *The charitable purpose of the Claremont Family Counseling Center is to promote the benefits of positive family interaction in the following ways:*
>
> *The primary activity will be to provide professional family counseling services, on a sliding scale fee basis, to families in Claremont County and surrounding counties. No family will be refused service due to the inability to pay for services, and services will be available to all members of the general public. (75 percent)*
>
> *The center will maintain a website that provides referral information, suggestions for strengthening family interactions, reference materials concerning families, and email and bulletin board posting capabilities. (10 percent)*
>
> *Staff members will present information about families to local schools, service clubs, and churches through multimedia presentations. (10 percent)*
>
> *Brochures will be prepared and distributed on request. (5 percent)*

You're asked more questions about your specific activities in Part VIII of Form 1023, including, for example, how and where you intend to raise funds, whether you operate a bingo game or other gaming activities, and the extent of your involvement with foreign countries and organizations. Answering yes to these and other questions in Part VIII means that you must attach explanations.

In Part VIII, you're also asked whether your organization will attempt to influence legislation. If you answer yes, you're given an opportunity to elect section 501(h) of the IRS code by filing IRS Form 5768. If you do so, your organization's expenses for allowable lobbying activities are measured by a percentage of your revenue. Chapter 5 has more information about this subject.

Reporting salaries and conflicts of interest

The IRS revises its forms from time to time. In the past few years, Form 1023 has been revised to include more questions about staff salary levels, board member compensation, and business and family relationships of board members and staff. If you pay or plan to pay any staff member or consultant more than $50,000 per year, you need to report this fact. If you compensate or plan to compensate members of your board, you also have to share that information on Form 1023.

If you plan to compensate employees, directors, or consultants with *non-fixed payments*, that is, bonuses or revenue-based compensation, be prepared to describe these arrangements in detail.

You're also asked whether the organization has a conflict of interest policy and whether compensation has or will be set by comparing salaries of your staff to salaries of similar organizations. Although the IRS says these practices aren't required to get a tax exemption, they are recommended. The instructions for Form 1023 contain a sample conflict of interest policy statement. We suggest having your board adopt a conflict of interest policy based on the IRS model that also is consistent with your state's laws.

Part V of the application stretches over two pages and is clearly a reaction to increased public and legislative concern about financial abuse and self-dealing in both the nonprofit and business sectors. We say more about the increased scrutiny aimed at the nonprofit sector in Chapter 5. Keep in mind that nonprofits are, in a sense, "quasi-public" organizations because the government is granting them special status as tax-exempt organizations.

Dealing with financial information

The IRS wants to see financial information. (Surprise, surprise.) New organizations have to estimate their income and expenses for three years — the current year and two years following. Making financial projections sends shivers down the backs of many folks, but it's really not that difficult. Ideally, you've made plans for your nonprofit already, and you can take the figures from your organizational plan. If you haven't written a plan, now's a good time to do so. See Chapter 7 for planning information and Chapter 11 for help in creating a budget.

When you estimate your income, keep in mind the requirements for qualifying as a public charity (check out the earlier section "Establishing public charity status" for details). Diverse sources of income are important, both for qualifying as a public charity and for the stability of your nonprofit.

You also need to choose your annual accounting period. This period is usually referred to as your *fiscal year*. It can be any 12-month period you desire. Most organizations choose as their fiscal year either the calendar year (January 1 through December 31) or the period from July 1 to June 30. Most government agencies operate on a July 1–through–June 30 fiscal year, and nonprofits that get support from government grants and contracts often prefer to operate on the same schedule. Some organizations offering services to schools set the fiscal year to correspond with the academic year. But the choice is yours. You can set your fiscal year from November 1 through October 31, if you want.

Your first accounting period doesn't need to be a full 12 months; in fact, it probably won't be. If you form your organization in September and select a calendar year as your fiscal year, your first accounting period covers only four months, September through December. If you do decide to begin with a short first year, don't forget to file a 1023 form at the appropriate time to report your activities during that period.

TIP

If you have an accountant, seek advice about the best accounting period for your organization. Remember that your fiscal year determines when future reports are due to the IRS and also when you prepare year-end financial reports for your board of directors. The annual 990 report (see Chapter 5) is due, for example, four and a half months after the close of your fiscal year. So if your fiscal year is the same as the calendar year, the report is due on May 15. If you always spend early May traveling to the Caribbean, you may want to pick another accounting period.

Collecting the other materials

In addition to the completed tax-exemption application, you need to submit as attachments *conformed* (exact and certified) copies of your articles of incorporation and the certificate of incorporation, if your state provides one. You also need to submit the articles of incorporation and bylaws of your corporation, but remember that bylaws alone don't qualify as an *organizing document*, as do the articles of incorporation.

TECHNICAL STUFF

Every nonprofit applying for tax exemption must have an organizing document. Usually, this document is the articles of incorporation, because most nonprofit organizations are incorporated. However, in the United States, associations, trusts, and, in some cases, limited liability companies, also may apply for tax-exempt status. An organizing document for an association may be the articles of association or a constitution; a trust is usually organized by a trust indenture or deed.

Put your organization's name, address, and EIN on each attachment and specify the section of the application that each attachment refers to. (The earlier section "Getting Your Employer Identification Number (EIN)" provides more details on obtaining your organization's number.) The application package you receive from the IRS has a checklist of the materials that must be included. Be sure to include this checklist with your completed application package.

Paying the fee

The fee for filing IRS Form 1023 is $850 for organizations that have had or anticipate having revenue of more than $10,000 per year. If your organization has been operating without a tax exemption and has income of $10,000 or less per year, or if you anticipate having revenues of $10,000 or less per year in the future, the fee is set at $400.

We don't recommend trying to save $450 on the application fee unless you're certain that revenues to your organization will remain under $10,000. We understand that the $450 difference in the fees can loom very large for a startup nonprofit, but why play with fire? Although the penalties for misstating your income aren't clear, we wouldn't want you to find out the hard way.

Send the fee when you mail the form with all its appropriate attachments and schedules. Keep a copy of everything for your records. Then take a break! You deserve it.

Getting along until the exemption comes

After you file your tax-exemption application, your organization is in never-never land for a while. You don't have your exemption yet. Although making an accurate prediction is difficult, the IRS can take between two and six months to act on your application. The process may take longer if the application is returned for corrections or additional development.

During this period, your organization can operate and even solicit contributions, assuming you've registered with the appropriate state agency. However, you must tell donors that you've applied for a tax exemption and are waiting to hear from the IRS. Assuming that your application is approved, if you submitted Form 1023 and the other required materials to the IRS within 27 months from the time you established the organization, your tax-exempt status is retroactive to the date of incorporation. Another option is to use a fiscal sponsor during this period. (See Chapter 2 for information about using and finding fiscal sponsors.)

If your organization had been operating for a longer period before you submitted Form 1023, the exemption may be retroactive only to the date on which you submitted the application. In this case, you also need to include Schedule E with your application.

Check with your state officials to find out what steps you need to take to have your tax exemption recognized by your state's government. If you didn't receive this information when you incorporated your organization, you can find contact information for the appropriate state office by performing a web search.

» Avoiding trouble over pay and politics

» Sending Form 990 to the IRS

» Filing state and local reports

Chapter 5
Protecting Your Nonprofit Status

Nonprofits are private organizations, but because they're awarded special tax status and are acting on behalf of the public, they're required to disclose more information than privately held for-profit companies. Nonprofit status is a privilege.

One way to think about it is to compare nonprofits to companies that sell shares of stock to the public. These companies must follow the rules and regulations set forth by the Securities and Exchange Commission about disclosing financial information. Nonprofits in the United States must follow the rules of the Internal Revenue Service (IRS) to make financial information available for public scrutiny. In some cases, state and local governments also have disclosure rules.

Your nonprofit organization can get into trouble with authorities in a few ways. However, keeping your nonprofit status isn't difficult if you follow the rules. This chapter lays out the reporting requirements that you need to follow and some pitfalls to avoid in order to maintain your 501(c)(3) public charity status. (See Chapter 2 for more on 501(c)(3) organizations.)

REMEMBER

We don't aim to scare you! Just keep good financial records and stick to your mission, and you'll be fine.

Check out File 5-1 at `www.dummies.com/go/nonprofitkitfd5e` for a list of web resources related to the topics we cover in this chapter.

Disclosing What You Need to Disclose

Disclose is a funny word, isn't it? It seems to imply that you're hiding something that must be pried from your clutches. Don't think of it that way. The IRS regulations that lay out the rules for disclosure refer to "the public inspection of information." That's much more genteel. In this section, we briefly cover what you are and aren't expected to disclose to the public.

What you do need to show

What information must be disclosed to the public? (Sorry, we can't help ourselves. We like the word.) It's simple, really. Your three most recent filings of IRS Form 990 and your IRS Form 1023 application for tax exemption and supporting documents — if you filed for your exemption after July 15, 1987 — must all be available to the public.

If you have a question that this chapter doesn't cover, refer to IRS Publication 557, which you can find at `www.irs.gov`.

Form 990 is the annual report that you must file with the IRS. We talk more about filing the 990 later in this chapter, but in a nutshell, it's a report of your annual finances and activities as a nonprofit organization. The IRS has three versions of the 990 form, and the version you file depends on your nonprofit's gross receipts and total assets held. Don't panic! We get into that later.

Current IRS regulations state that the three most recent 990 reports and any schedules and attachments to the reports must be available for public inspection at the organization's primary place of business during regular business hours. A staff member may be present during the inspection, and copies may be made for the person requesting them. If you make copies on your own copier, you may charge what the IRS charges: $0.10 per page for black and white or $0.20 per page for color.

The information provided in this chapter is about 501(c)(3) organizations that are considered public charities. If your organization is a private foundation, special disclosure rules apply. Check with the Council on Foundations (`www.cof.org`) about disclosure requirements.

If you work out of your home or really have no primary place of business, you can arrange to meet at a convenient place or mail copies to the person who requested them. You have two weeks to do so. The information also can be requested in writing, and if it is, you have 30 days to respond.

If you filed and received your exemption after July 15, 1987, you must include your application for exemption and all the supporting materials submitted with it in the information you make available for inspection. Although the IRS doesn't say so, we assume that you include your articles of incorporation and your bylaws, if you attached them to your application.

TIP

You need to know the rules and regulations about public inspection of nonprofit materials, but frankly, in the many years we've been working for nonprofits, no one has ever asked to see our 990 forms. This doesn't mean that you won't be asked, however. Follow the Boy Scout motto and be prepared.

On your 990, you need to disclose any compensation paid to board members. This information is available for public inspection, as are employee salaries and contractor payments of more than $100,000 per year. Also, if you paid any former board member more than $10,000 or any former employee more than $100,000, this information must be reported.

REMEMBER

Check your state and local government requirements to see whether your nonprofit must make other information available under local laws.

What you don't need to show

You don't need to reveal your donor list. Donors who contribute more than $5,000 in your tax year must be listed on Schedule B of your 990-EZ or 990 report, but 501(c)(3) public charities aren't required to make the information public.

Other items that may remain private include the following:

- Trade secrets and patents
- National defense material
- Communications from the IRS about previous unfavorable rulings or technical advice related to rulings
- Board minutes and contracts

IRS regulations go into more detail, but we think the preceding four items sum up the major points. If you have questions about a particular item, consult your attorney or tax advisor.

USING THE WEB TO SATISFY DISCLOSURE REQUIREMENTS

You can avoid the hassle of photocopying your Form 990 by placing the required information on the web. If you do so, be sure that you post an exact copy of the materials, not a summary or a retyped copy. You can scan documents and put them on your website as graphics, but the best way to do it is to put the documents in PDF format using Adobe Acrobat (www.adobe.com).

Both GuideStar (www.guidestar.org) and the Foundation Center (www.foundationcenter.org) maintain a database of PDF files of 990 returns. You can add information about your nonprofit to the GuideStar database. Some states also maintain web databases of 990 forms of nonprofits incorporated in their jurisdictions. Remember that you need to make available your three most recent 990 returns and your application for exemption.

If someone arrives at your office and asks to inspect your 990 forms, you still need to allow that person the privilege. Web access to the materials only relieves you of the responsibility of mailing out copies.

The public disclosure rules offer a nice tool for anyone who gets a bee in his bonnet and wants to harass your organization. So if you start getting request after request after request, check with the nearest IRS office. IRS officials have the authority to relieve you of the disclosure responsibilities if they agree that harassment is occurring.

Avoiding Excessive Payments and Politicking

Paying excessive compensation and engaging in campaign politics can get a charitable nonprofit organization in trouble. It doesn't happen often, but you need to be aware of the rules.

Determining reasonable pay and benefits

In past years, if the IRS discovered wrongdoing in a nonprofit, it had little recourse but to take away the organization's tax-exempt status. However, when revised tax laws were passed in 1996, the IRS gained the authority to apply "intermediate sanctions" when nonprofits provide excess benefits to certain staff members or other disqualified persons. Board members and their family members — really anyone who can influence the organization's activities — are disqualified persons.

Excess benefits can include excessive salaries for staff members or a business deal arranged to benefit a disqualified person in which the nonprofit overpays for a service. So, if you sit on a board of directors that decides to rent office space from your uncle at two or three times the going rate, you and your uncle may be in trouble. If you do decide to rent from your uncle, be sure you can document that you're paying a market-rate rent (if not less). Also, you shouldn't participate in the board vote about renting the space.

An excessive benefit for a nonprofit staff member is any sum that's above a "reasonable amount." You're probably thinking, "What's a *reasonable* amount?" and you're right to ask the question. In the case of executive compensation, it's up to the board of directors to find out what a fair salary is for nonprofit managers in your area and for an organization of your size and scope. (See Chapter 10 for tips on determining salaries.)

WARNING

If the IRS finds that someone in your nonprofit has received an excessive benefit, the financial penalties are severe. The IRS levies a 25 percent tax on the excess amount, and the employee must pay the full excess amount (including interest and the 25 percent tax) back to the organization. If payment isn't received by that time, the tax can go up to a whopping 200 percent. Board members also may be liable for penalties for approving excessive compensation.

These three tips can help you avoid excess benefits problems:

- Don't allow board members to participate in decisions that may benefit themselves or their family members. And, certainly, don't let the executive director participate in setting her own compensation.
- Rely on credible independent information about reasonable costs in business deals and compensation matters.
- Document the reasons that you make the decisions you make.

REMEMBER

We don't want to frighten you with visions of huge tax bills. We're not saying that you can't pay nonprofit staff well or that you have to undertake a scientific study to determine fair compensation. Use your head, be reasonable, and exercise caution.

Document board decisions by keeping minutes of board meetings. You don't need to keep a verbatim record of board deliberations, but you want your minutes to reflect the discussions you have and the decisions you make. Maintain what's known as a *board book* — a binder that contains copies of your articles of incorporation, IRS letter of determination for tax exemption, bylaws, amendments to bylaws, notices sent to announce board and membership meetings, and a chronological record of your board-meeting minutes.

Using caution when getting involved in politics

We aren't saying that you shouldn't get involved in politics, because it's your right to do so. If you want to give your personal support and endorsement to a candidate, by all means do it. But be sure to separate yourself from your nonprofit organization when you do. Nonprofits in the 501(c)(3) category can't support or endorse candidates for political office.

If you want to talk to your legislator about the passage of a bill that benefits your clients, go ahead and make an appointment. But if you find yourself traveling to your state capital or to Washington, D.C., on a regular basis, step back and consider how much organizational time and money you're spending on the activity. Charitable nonprofits can spend an "insubstantial" amount on direct lobbying activities. If it's 5 percent or less of your organizational budget, you're probably within the limits allowed, according to the people who pay attention to these things.

REMEMBER

Be more cautious with what's known as *grassroots lobbying*, or attempting to influence the general public to vote in a particular way. If lobbying is important to your overall mission as a 501(c)(3) public benefit nonprofit, you can elect the "h" designation, which requires more financial reporting to the IRS but allows you to spend more money on these activities. To do so, file IRS Form 5768 after you have a look at the regulations in IRS Publication 577.

Why all the fuss? Understanding the increased scrutiny

Nonprofit organizations face increasing regulation and public scrutiny in the years to come. Why? Three reasons come to mind:

- Concern about how nonprofit organizations use contributions that are made for disaster relief efforts
- A series of widely reported cases in which some nonprofits, both public charities and private foundations, have paid very high salaries to executives and trustees and otherwise pushed the limits of ethical behavior

» The number and degree of corporate-accounting and insider-trading scandals that have been revealed during the past few years in the for-profit sector

These factors have focused more public and legislative attention on how nonprofit organizations (and for-profit corporations) operate and how their affairs are regulated.

Greater accountability is being asked of nonprofit leaders, both managers and board members, in how money is raised, how conflicts of interest are avoided, and how funds are spent, especially for salaries and expenses. The National Council of Nonprofits has compiled good principles of nonprofit management from several states that provide overall guidance in this area. You can find them at www.councilofnonprofits.org/resources/principles-and-practices.

If you're involved with a startup nonprofit that's operating on a shoestring, you may be asking yourself why you need to worry about getting paid too much. And you're right; you probably don't need to worry. But you should stay informed of changes in reporting requirements. You still need to have good, ethical operating policies in place; keep good financial records; and document your organizational decisions with board and committee meeting minutes.

If you're already involved with a medium or large nonprofit, be aware of the need for independent financial audits and board policies that address excessive compensation, self-dealing, whistle-blowers, and conflicts of interest. The IRS provides a sample conflict of interest policy in the instructions for filing Form 1023. You can also find examples in File 5-1 at www.dummies.com/go/nonprofitkitfd5e.

TIP

Because the regulatory situation is always changing, and because it varies from state to state, we suggest visiting the websites of Independent Sector (www.independentsector.org) and BoardSource (www.boardsource.org) to keep abreast of the latest developments. If you have an accounting firm performing a review or audit, it should be able to keep you informed about current legislation and verify that you're in compliance.

Reporting to the IRS

The formal name of the report that nonprofit organizations must file annually is *Return of Organization Exempt from Income Tax*, and the IRS calls it the "annual information return." Everyone else refers to it as "the 990." It *is* IRS Form 990, after all. Depending on a nonprofit's gross receipts and total assets held, a version of this report must go to the IRS each year.

The IRS uses gross receipts to determine which 990 Form your organization is required to file. *Gross receipts* refers to all funds that come into the nonprofit during the tax year before any expenses are subtracted. For example, if you have a special event that brings in $20,000, but you have expenses of $8,000 for a net income of $12,000, your gross receipts for that event are $20,000. This figure is added to grants, fees for services, contributions, and any other income to determine your gross receipts for the year.

No matter what report you file, it must be submitted no later than the 15th day of the fifth month after the end of your annual accounting period: in other words, after the close of your fiscal year. When you filed your application for exemption, you selected your accounting period. If you use the calendar year as your fiscal year, your 990 has to be out the door on May 15. You can apply for an automatic three-month extension. If you need even more time, you can apply for an additional three-month extension — if you have a good reason.

The IRS has three versions of the 990: The 990-N, the 990-EZ, and the long Form 990. The report you file depends on the total gross receipts (and total assets) of your nonprofit organization. We focus on the 990-N and 990-EZ in the next couple of sections, but we do provide some advice for the long Form 990 as well. Most important, if your nonprofit organization is big enough to require filing the long Form 990, we advise that you seek professional help.

Filing the 990-N

The 990-N is required for all 501(c)(3) nonprofit organizations (with the exception of churches and church-related organizations) that have annual gross receipts of normally $50,000 or less. This simple form must be filed electronically; in fact, the IRS refers to this form as an e-postcard, so you need a computer, Internet access, and an email address.

The due date for filing this form is the same as for the 990-EZ or the long Form 990: the 15th day of the fifth month after the close of the tax year. However, at present there's no penalty for late filing. You can't file before the end of the tax year.

If your organization fails to file a 990-N for three consecutive years, its tax-exempt status will be revoked automatically.

To file the 990-N, go to the IRS website at `www.irs.gov/filing/charities-non-profits`. Follow the link to the e-Postcard site. If your organization is new or if you haven't filed a 990-N before, you'll need to register to obtain a login ID before you file the form.

The 990-N asks for the following information:

- The legal name of the organization and any other names under which it's doing business.
- A mailing address and a web address, if you have one.
- The Employer Identification Number, or EIN, of the organization. (See Chapter 4 for more on EINs.)
- The name and address of the principal officer, which is likely either the executive director or board president. Use the organization's address as the address for this individual.

You'll also be asked for the organization's tax year (the same as the fiscal year), whether the nonprofit normally has $50,000 or less in gross receipts, and whether the organization is still in business.

Taking it easy with the 990-EZ return

If your organization has gross receipts of less than $200,000 and assets of less than $500,000, you can file the 990-EZ form, which is more complex than the 990-N. The 990-EZ includes a lot of fine print, and the instructions are a good test of the importance of close reading. That said, the instructions are complete, and most terms are well defined. You can make your task easier by getting help from your accountant or attorney.

TIP

Keep in mind when reading the instructions that the 990-EZ report is used for categories of tax-exempt organizations other than 501(c)(3) public charities. This can be confusing unless you attend closely to the instructions. Because the 990-EZ is used for other types of nonprofits, be sure to check the 501(c)(3) box.

Fill in the identifying information at the top of the form. Be sure to add your employer identification number (EIN). Also check the correct box to indicate whether you're using a cash or accrual accounting method. If you need help with that question, see Chapter 11.

REMEMBER

Be sure to include a phone number where the IRS and members of the public can contact your organization. Don't forget that the 990 is a public document. So, if you've had an address or name change since you filed your last report, check that box in the "B" section. If these changes required amendments to your articles of incorporation or if this is the first time your organization has filed the 990-EZ, check the corresponding boxes. If your tax-exempt application is still pending, indicate that as well.

If in the past year a single individual has given your 501(c)(3) organization a contribution of cash, a grant, or property that was valued at $5,000 or more, *or* if a gift totaled more than 2 percent of your total grants and contributions, you must file Schedule B with your 990-EZ. If a single person gives you two gifts of $2,500 within the year, you must file the schedule. If you don't have grants or gifts in this range, check the box in the "H" section to let the IRS know you aren't required to file Schedule B.

The IRS doesn't require Schedule B to be disclosed to the public. If your organization must file its 990 report with its state to satisfy state reporting requirements, check about its disclosure requirements.

Parts I and II

In Part I of 990-EZ, you report your financial activities over the past year. Chances are that the two lines that have the highest amount of revenue are line 1 (contributions, gifts, and grants) and line 2 (program service revenue, including government fees and contracts). You may have some income from membership fees for line 3 and income from investments or interest on a savings account for line 4.

Lines 5 through 8 deal with income from sales of materials or assets and income from special events. You're asked to report the costs and expenses associated with the materials sold and any direct expenses associated with special events. In other words, the IRS wants to see a net amount for sales and special events. The total you end up with on line 9 is your total revenue for the year.

If you earn more than $15,000 in gross receipts from special events (reported in line 6A), you need to complete Part II of Schedule G. If any part of these gross receipts came from gaming, such as bingo and pull tabs, you also need to complete Part III of this schedule.

Don't confuse your total revenue with your gross receipts. Gross receipts include all the income you receive from sales and special events *before* you subtract expenses. Annual gross receipts (and asset value) determine which 990 Form you submit to the IRS.

Report your expenses on lines 10 through 16 in their appropriate categories and total the amount on line 17. Subtract line 17 from line 9, and you have your surplus (or deficit) for the year. This amount goes on line 18. We come back to line 19 in just a minute.

Part II includes lines 22 through 24, which is where you report your assets — cash, savings, or investments, as well as property and equipment. (Equipment goes on line 24, other assets. Attach Schedule O with the equipment list.) Total these figures for line 25.

Enter your total liabilities on line 26. They may include accounts payable, outstanding loans, and vacation time owed to employees, for example. Describe these liabilities on Schedule O. Subtract line 26 from line 25, and you have your net assets or fund balance. Put this amount on line 27. If you pay off all your bills and sell all your assets, this amount is what's left over — in theory.

Notice that you need to report assets and liabilities for the beginning of the year and the end of the year. (If this 990 report is your first, you provide only end-of-year totals.) Refer to the form you submitted last year for the beginning-of-year figures.

Now go back to line 19 in Part I. Enter your net assets from last year's form on this line. If an adjustment to your net assets was made during the year (usually done by an accountant), enter the figure for this adjustment on line 20 and explain on Schedule O. On line 21, enter the amount of your net assets at the end of the current year.

That's it for reporting your financial activities on the 990-EZ form. It *was* easy, wasn't it?

Parts III and IV

Part III of 990-EZ asks you to state your primary exempt purpose. This statement doesn't need to be long like the one you wrote for your Form 1023 application; a few words will do. You also need to describe your three largest programs, their objectives, program accomplishments, how many people you served, and a total cost for each program. The form has spaces for three separate programs, but if you have only one program, that's fine.

TIP

Part III includes a space to report grants in each program area. This space is for grants made *by* your organization, not grants *to* your organization. You probably didn't make any grants, but if you did, report them here.

Part IV requires a list of your directors, with the board officers identified, and key employees and their addresses. Usually, for smaller nonprofits, the only key employee is the executive director. You also need to state how much time is devoted to each position each week and record any compensation received, including salary, retirement benefits, and expense allowances. If you need more space than what's provided on the form, attach an additional sheet. You don't need to list addresses for any individuals in this section.

Parts V and VI

You're likely able to check no to the questions in Part V of 990-EZ or leave the items blank. But several things may need your attention.

- » If you've amended your bylaws or articles of incorporation since you filed your last report, check that item and include a copy of the amended document when you send in the form.
- » If you've engaged in any significant activities not reported previously on the 1023 or an earlier 990, you should indicate that on Line 33 and describe those activities on Schedule O. The same is true if you've discontinued any significant activities.
- » Line 35 refers to income that may have come to your organization through activities not related to your charitable purpose. This is called *unrelated business income,* and if it totals $1,000 or more, you must report it on Form 990-T. You may be liable for taxes on this income. If you think your nonprofit has income in this category, you're wise to consult an accountant. (For more on unrelated income, check out the nearby sidebar "How the IRS decides whether income is unrelated.")
- » If your organization loaned or borrowed money during the year from a director, trustee, or key employee — or if it has outstanding loans from the previous year — check yes on line 38a, fill in the amount on line 38b, and file Schedule L, Part II.
- » In Part VI, lines 46 and 47 ask whether your organization engaged in political campaigns on behalf of a candidate or engaged in lobbying activities. If you can legitimately answer no to these questions, do so and move on. If you have questions about your nonprofit's engagement in these areas, seek advice from an attorney.

WARNING

Campaigning for a candidate under the banner of your nonprofit organization is a serious matter that is grounds for the loss of your tax exemption. An "insubstantial" amount of lobbying is allowed for 501(c)(3) public charities, but this is an area filled with shades of gray and shouldn't be taken lightly. Nonprofit organizations can engage in nonpartisan voter registration and public education activities, but to be on the safe side, you should seek professional counsel.

- » List all employees and contractors who were paid more than $100,000 during the last year. Chances are, especially if yours is a startup organization, and because you're filing the 990-EZ, you can put "NA" in these sections.

REMEMBER

Be sure to have an officer of the organization sign and date the form. Often this officer is the board chair or board treasurer. If you used a paid tax preparer, she must sign, as well.

HOW THE IRS DECIDES WHETHER INCOME IS UNRELATED

The IRS definition of *unrelated business income* is based on three questions: Does the income come from a trade or business? Is it regularly carried on? Is it not substantially related to the organization's exempt purpose?

To better understand what these questions mean, consider an extreme example. The Juniper Avenue Young People's Club, a 501(c)(3) nonprofit with the mission of providing recreational opportunities, mentoring, and job-skills training to adolescents, decides to open a shoe-repair business. It rents a storefront space, purchases equipment, hires employees, and begins soliciting business. The shoe-repair shop does very well. In fact, in the first year of operation, income exceeds expenses by $10,000. The club uses the money to pay some costs incurred in working with young people in the neighborhood. The problem is that the shoe-repair shop has no relationship to the club's charitable purpose. In this case, the club is required to file Form 990-T and pay appropriate taxes on the $10,000. That's okay, however, because the club still realizes a profit that can be applied to some of its program costs.

However, what if the club opens the shoe-repair shop with the purpose of providing training in the shoe-repair trade to adolescents in the neighborhood? In that case, chances are good that any income from the business will be considered "substantially related to the organization's exempt purpose," and the unrelated business income tax (known as UBIT) won't be due on any proceeds.

Schedule A

All 501(c)(3) public charities that submit the 990-EZ must complete and submit Schedule A so the IRS can apply the public support test to determine whether your organization qualifies as a public charity. (Refer to Chapter 4 for a discussion of the public support test.) Although it's a little more complicated and there are other ways to pass the test, in a nutshell, public charities must receive one-third of their support from the public.

REMEMBER

During the first five years of a nonprofit's existence, data is collected but the test isn't applied. So even if you're sending in your first 990-EZ, you need to provide the information requested on Schedule A.

In Part I of Schedule A, you're asked to check one box from lines 1–11. Most 501(c)(3) organizations that are the subject of this book check either line 7 or line 9. Line 7 applies to organizations that qualify as public charities under section 170(b)(1)(A)(vi); in this case, you also need to complete Part II of Schedule A. Line 9 applies to organizations that qualify as public charities under section 509(a)(2). Organizations checking this box must complete Part III of Schedule A.

TIP

The IRS refers to the appropriate code section for your nonprofit in the letter of determination (tax-exemption letter) it sends you after you file your IRS Form 1023 application for exemption.

Parts II and III both ask for information about sources of financial support for the organization for the current tax year and the four previous years. If your organization has been in existence for less than this five-year period, you only have to provide information for the year or years in which your nonprofit has been active. Also, you don't need to compute the percentage of public support in Section C in either part until you've completed your sixth year of operations, although it's a good idea to complete a personal copy of the calculation for the first five years so you can keep track of how you're doing.

Tackling the long Form 990

If your organization has gross receipts of $200,000 or more or assets of $500,000 or greater, you're required to tackle the full 990 experience. The Form 990 emphasizes accountability and transparency in nonprofit organizations, which reflects growing public scrutiny of nonprofits. The first page of the form is a summary of the more-detailed information that follows. You have to describe your mission or "most significant activities" in the first question of Part I. In addition to questions about the number of board members, employees, and volunteers, you must present a summary of the detailed financial information requested in the final pages of the report.

Make no mistake, completing this report requires more time and effort than the 990-EZ. The long Form 990 spans 11 pages and is accompanied by 16 separate schedules. Although you won't be required to complete all the schedules, you may need to submit some of them, depending on the circumstances of your organization. One section of the 990 is a checklist with questions that must be answered "yes" or "no." Usually a "yes" answer triggers a requirement to complete one of the 16 schedules.

REMEMBER

The IRS estimates that you need 16 hours to learn about Form 990 and nearly 24 hours to complete the form. We think this may be an underestimate. The instruction booklet contains 71 pages of dense reading. Answering the questions and providing the information required on this form should be a team effort. And, if at all possible, we recommend that a member of that team be a qualified accountant. It wouldn't hurt to have an attorney on board, either. If you decide to undertake the task of completing the long form without the assistance of professional help, start early and plan to spend lots of time with the instruction manual.

Providing financial information

When you use the long Form 990, you'll be asked for more-detailed financial information than is needed for the 990-EZ. The IRS wants to know more about where you get your money and where you spend it. Also, all your expenses must be allocated to one of three categories: program, management and general, and fundraising costs. You'll also need to provide a detailed balance sheet. (Chapter 11 includes more information about financial recordkeeping.)

The IRS wants to know what method of accounting you're using and whether you're following certain accounting standards. You'll also be asked whether your financial statements have been audited or reviewed by an independent accountant. If so, the IRS will ask whether your organization has a committee that provides oversight to the preparation of financial statements and audits.

If it looks like your organization will be required to file the full 990 report in the next year or two, start putting in place the accounting system you'll need to collect the appropriate information.

Describing your programs

In the long Form 990, you're asked to describe the program achievements of your three most expensive programs and whether any programs were added or eliminated in the year for which you're making the report. If you did make significant changes in your program offerings, you need to describe them on Schedule O.

The 990 report is a public document and is available for public review, so be sure to take the time to describe your activities and program achievements completely and clearly and without the use of excessive jargon.

Looking at management and governance

The sections relating to members of the board (the IRS refers to the board as the "governing body"), board policies, and highly compensated and key employees are detailed and require special attention.

For example, you're asked how many board members are independent. An *independent* board member can't be compensated as an officer or employee of the organization or receive more than $10,000 as an independent contractor of the organization. He or a member of his family also may not engage in a business transaction with the organization or a key employee that would trigger the requirement to submit Schedule L.

Also, although not required by IRS regulations, the 990 asks several questions about board governance. For example, you'll be asked whether the organization has written policies for the retention and destruction of documents, whether you have a conflict of interest policy, and whether you have a whistle-blower policy.

Also, the organization must note if a process exists for setting the compensation of the executive director and other key employees and if the organization provided board members with a copy of the 990 before its filing. Although not explicitly required, it's good practice to establish these policies and procedures. They'll strengthen and protect your organization.

Getting your 990 to the IRS

Your 990 report (whether it's the 990-EZ, 990-N, or long Form 990) needs to be submitted on the 15th day of the fifth month after the close of your financial year — unless, of course, you request an extension. If you chose to mail the form, check the most recent version of the instructions to find the address to which the report should be sent. You can file the 990 reports electronically if you want, and some larger nonprofits are required to file electronically. Go to `www.irs.gov/efile` for instructions on electronic filing. The 990-N can *only* be filed electronically.

REMEMBER

Large organizations that have assets worth $10 million or more and that submit more than 250 IRS reports in a year must file their 990 returns electronically.

Reporting to Your State and Local Governments

You'll almost certainly have reporting requirements for your state and possibly your local government, especially if you provide services under contract. Sometimes reporting to the state is as simple as completing a one-page form, attaching a copy of your 990, and paying a small fee.

WARNING

More and more states (and some local governments) require registration for fund-raising activities. Be sure to check local laws.

Some states require a separate financial form, although we think that most follow the federal form closely. You'll probably get all the needed reporting information when you ask for the incorporation packet, but if you didn't or you aren't sure, check with your state office that regulates nonprofit corporations. Find the appropriate office in your state by searching the web.

2 Managing a Nonprofit Organization

IN THIS PART . . .

Get the inside scoop on appointing people to the board and holding effective meetings.

See how plans come in handy for everything from laying the foundation for your organization to setting up programs. The best way to know whether your plans are on track is to evaluate the work your organization is doing. Find out how to conduct evaluations, interpret the results, and apply the findings.

People, whether volunteers or paid employees, are an essential part of every nonprofit organization. Discover how to recruit and manage volunteers and employees.

Nail down your nonprofit's numbers with a budget and other financial reports. Good management requires balancing your books while addressing your nonprofit's mission.

Let the world know about your nonprofit and its good works with a strong marketing strategy. Find tips on everything from contacting reporters to posting updates on social media.

» Recruiting the right board members

» Understanding why having staff members on your board can be tricky

» Getting the most out of your board

Chapter 6
Building Your Board of Directors

Some boards of directors do miraculous things. Others just muddle through. No nonprofit organization has an ideal board of directors all the time. You may think that your board is so disorganized that you may as well close up shop and go fishing. Don't despair. Boards can always get better, and many non-profits do excellent work with boards of directors that would send shivers down the backs of most nonprofit consultants. The purpose of this chapter is to help you understand how nonprofit boards work in a perfect world. (We don't live in a perfect world. So please keep that in mind while reading.)

REMEMBER

We're the first to agree that working with a board sometimes can drive you a little batty. Doing the work yourself may seem easier, but the strength and success of a nonprofit organization grows from its board of directors. It's surprising what a group of people working together can accomplish. So take the time to build a good board and work with its members to achieve your mission — trust us, you'll be glad you did.

TIP

Check out File 6-1 at `www.dummies.com/go/nonprofitkitfd5e` for a list of web resources related to the topics we cover in this chapter.

Understanding the Duties of a Nonprofit Board of Directors

A *board of directors* (which we refer to simply as *a board* from here on) is a group of people who agree to accept responsibility for a nonprofit organization. A board is responsible for ensuring that the nonprofit organization is fulfilling its mission. It makes decisions about the organization, sets policy for the staff or volunteers to implement, and oversees the nonprofit's activities. Raising money for the nonprofit is another important responsibility that many, but not all, boards assume. Board members almost always serve without compensation; they're volunteers who have no financial interest in the nonprofit's business. However, they do have responsibility for financial oversight and are held accountable for the accounting and financial reporting of the organization.

REMEMBER

Paid staff members generally may serve as board members and often do in startup and small nonprofit organizations. Check the laws in your state to determine any restrictions on paid staff members. However, if they're being compensated, they are being paid as employees, not as board members. Skip to the section "Putting Staff Members on Your Board" for details on why we don't recommend having staff serve on your board.

A nonprofit organization doesn't have owners like a for-profit business, but a board of directors guides and oversees the organization like an owner might. No one owns city, state, and federal governments either, so citizens hand over the responsibility of running the government to elected officials. In turn, they expect those officials to govern the affairs of their city, state, and nation. The job of a nonprofit board is similar; in fact, it's referred to as *nonprofit governance.* If you want more information on this type of governance than you find here, grab a copy of *Nonprofit Law and Governance For Dummies*, by Jill Gilbert Welytok and Daniel S. Welytok (Wiley).

Primary role: Preserving public trust

A board's primary governance responsibility is *fiduciary* — to uphold the public trust. Laws in the United States give special rights and privileges to nonprofits recognized by the Internal Revenue Service (IRS) as public charities. Primarily, these nonprofits earn the right to exemption from corporate income tax and the right to receive contributions that are tax-deductible for the donor. The government gives nonprofits this special status because they provide a public benefit. A board's leadership and oversight keep the organization's focus on that public benefit and make sure it doesn't abuse these rights and privileges.

Suppose that a nonprofit begins with a mission to rescue feral kittens. People who support this idea make contributions to the nonprofit with the belief that their money is being spent on programs to help the plight of feral kittens. But unknown to the donors, the nonprofit begins to spend its money on programs to support preschool education. Supporting preschool education is a worthy goal, but it's a long way from the original purpose of helping feral kittens. So in this example, even though the nonprofit is using its contributions for a good cause, it isn't using them as the donors intended or to fulfill the organization's original purpose.

A nonprofit that collected funds to help feral kittens but instead used the funds for the personal benefit of the board members and staff would be even worse. This dishonest act is serious and possibly a crime. Aside from potential felony fraud charges, such an activity violates IRS rules and can result in the revocation of the nonprofit's tax-exempt status.

In a for-profit business, the managers can share the net earnings (in other words, profit) at the end of the year among the company's owners or stockholders. If nonprofit board members decide to divide surplus funds among themselves, it's called *inurement*, a big word that basically means personal enrichment, which isn't allowed in the nonprofit world. Board members and staff can't personally benefit from nonprofit funds except for compensation for services provided or for reimbursement of expenses. A board must not only ensure that the nonprofit is doing what it set out to do but also make sure that it spends its funds properly. If a board does nothing else, it must make sure the organization adheres to these standards.

Keep in mind that a board's responsibilities are legal responsibilities. The three main duties that every board must uphold are care, loyalty, and obedience. Although they sound like vows one may take when entering a monastery, they actually describe established legal principles. Here's what these duties entail:

- The duty of care refers to the responsibility to act as a prudent board member. In other words, board members must pay attention to what's going on and make decisions based on information available with reasonable investigation.
- The duty of loyalty means that a board member must put the organization's welfare above other interests when making decisions.
- The duty of obedience requires that board members act in accordance with the nonprofit's mission, goals, and bylaws.

Basic information about board governance is available from BoardSource (`www.boardsource.org`) without charge. You can access more in-depth material by paying a membership fee.

EX OFFICIO BOARD MEMBERS

Ex officio is a Latin phrase that means " by virtue of an office." For example, if you had the necessary clout, you might have the mayor of your city sitting on your board as an ex officio member. When the current mayor leaves office, she would be replaced by her successor. Sometimes, in practice, the term is confused with *honorary* or *advisory* by nonprofit organizations that want to honor someone or add a prominent name to their letterhead.

Secondary role: Dealing with planning, hiring, and other board tasks

In addition to the legal and fiduciary responsibilities, a nonprofit board performs other roles, too, including those described in the following sections.

Providing a guiding strategy

Every nonprofit should have an organizational plan, and every board should play a part in creating and maintaining that plan. Thus, an important role of any board is to guide the overall planning and strategy of its organization. At the most basic level, this job means reviewing the organization's mission statement and goals on a regular basis. Turn to Chapter 7 for more information on planning.

Hiring and working with the executive director

A nonprofit board is expected to hire the organization's executive director. Of course, many nonprofit organizations operate without paid staff, but if your nonprofit does have employees, finding the right executive director is one of the board's most important tasks. See Chapter 10 for information about hiring paid employees.

A board works with its executive director to set goals and objectives for the year. Board members shouldn't look over the director's shoulder every day, but they should have a good idea of the director's work plan and should ensure that her efforts are in line with the agency's purpose.

Overseeing the organization's finances

A board must make sure the organization has the resources to carry out its goals. As part of this duty, many boards are active in fundraising. But a board also is responsible for reviewing the organization's budget and staying informed of the organization's financial situation. Nothing is more dismal than finding out, for example, that an organization hasn't been paying its payroll taxes, and could create personal liability for the directors. A word to the wise — insist on good financial reporting. At a minimum, have the board treasurer (or perhaps the executive director or bookkeeper, if you have one) prepare quarterly financial reports and distribute them to your board for review.

WARNING

Many people join boards because they care about and understand the nature of the service that the organization provides, but they may not be trained in bookkeeping and accounting. These board members must try not to let their eyes glaze over when the financial report is reviewed at the board meeting, because part of a board member's job is to understand finances. If members don't understand the financial fine points, they need to ask questions of staff and other board members until they do understand them. Of course, they also can study Chapter 11 on nonprofit budgets and financial statements.

REMEMBER

Sometimes the problem is more than a lack of comprehension. For instance, the financial information may need to be presented more clearly. If one board member doesn't understand the financial reports, chances are good that other board members don't understand them, either. If your organization uses an outside accountant or bookkeeper to keep track of finances, ask for a brief meeting with him to explain how he has presented the information. Many nonprofit service organizations offer affordable workshops on nonprofit finances and recordkeeping to assist board and staff members; others have programs that place volunteers from businesses into nonprofits.

ESTABLISHING AN ADVISORY BOARD

Some organizations form *advisory boards,* which don't have governance responsibilities. Advisory boards are optional, and how they operate and relate to the governing board of directors varies widely among different nonprofit organizations. Generally, their role falls into one of the following two categories:

- Advisory boards that have members who provide advice and guidance because of their professional expertise
- Advisory boards that are formed so that prominent names can be listed on the organization's letterhead

We favor advisory boards that actually give advice (which generally means you have to ask for it) — even if they do so only once a year. Some organizations also use advisory board appointments as a way of getting to know potential board members.

Sharing responsibilities among the board, staff, and volunteers

Defining roles causes problems for a number of nonprofit organizations. Should the board of directors be involved in day-to-day management decisions? No, probably not. An executive director and her staff don't need board members to approve every management decision that comes along. The board must trust the staff to run the organization. To put it simply, the board sets the overall goals and policies, and the staff implements them. But (and this is a big but) many nonprofits have limited staff or even no staff at all. What happens when the organization's work is done by the board and other volunteers?

In the case of volunteer-run organizations, board members must wear two hats. Most importantly, they must exercise their fiduciary responsibilities. When they meet as a board of directors, they must see the larger picture and make group decisions that benefit the organization and its programs and clients. But, at the same time, when they have to do the hands-on work needed to provide the services and perform the day-to-day tasks of running a nonprofit, board members must act as if they were employees or volunteer staff members. They may even hold regular, unpaid volunteer jobs with job descriptions and scheduled hours. Confusing, isn't it? Still, anyone who serves as both a board member and a volunteer or staff member needs to keep this distinction in mind.

TIP

In any organization, building practices that create financial checks and balances is a good idea. These practices are particularly important when board members also act as staff. For example, an organization has better financial oversight and control when one person approves bills for payment, a different person signs the checks, and yet another person reviews the canceled checks and monthly bank statements. You don't *have* to use this system — it's just one way to make sure that funds are used and accounted for properly. (For more on having staff members serve as board members, check out the later section "Putting Staff Members on Your Board.")

Recruiting the Right People for Your Board

You don't want just anybody to serve on your board. You want to choose the members of your community who believe in what you're doing, will come to all your meetings, will be advocates for your programs, will provide honest and ethical oversight to the organization, will make regular and generous donations, and will sweep the floor on weekends.

Perhaps, not surprisingly, you won't find many board members who fit this description. Even so, the following three traits are critical to the success of the organization:

- Believing in your mission
- Being a strong advocate on behalf of your programs
- Serving the organization as a careful and honest board member

Sure, having wealthy members who do the dirty work when needed is nice, but, most importantly, you must find board members who understand and believe deeply in your work. Showing up for board meetings is a nice habit, too.

You also need to think seriously about the skills that board members bring to your organization. Do you need an accountant to set up financial systems? A public relations specialist to help with media campaigns? An attorney to help with legal matters? Yes, you probably do. But don't expect the accountant to do your audit or the attorney to represent you in court. You need a disinterested professional to do that work.

REMEMBER

Your board should reflect your organization's character and mission. A community-organizing group dedicated to collective decision making may want board members who work well together. A neighborhood development organization clearly wants board members from its neighborhood. A youth leadership organization may want to invest in future leadership by creating positions for youth members on its board. (*Note:* Before adding young people to your board, check whether your state laws allow minors to serve on nonprofit boards. If your state prohibits minors from serving, consider inviting them to serve on advisory committees instead.)

WARNING

Although having a friend or two on the board is fine, be careful about overloading the board with golfing buddies and carpool partners. Boards need diverse opinions and honest feedback from members.

Keeping it fresh

Building a board should be a continuing process. Therefore, we highly recommend that your organizational bylaws specify terms of service. Two three-year terms or three two-year terms are the most common terms for board service. In most cases, bylaws allow reelection to the board after one year's absence. Limiting terms of service helps you maintain a fresh supply of new ideas, which are more likely to come from people who are new to your organization. Plus, limiting terms of service can help you recruit new board members because your potential recruits know their time commitment is of limited duration.

PERSUADING SOMEONE TO JOIN YOUR BOARD

Many people think a board member's primary role is to raise money. In fact, a popular slogan addressed to board members who aren't raising funds is "Give, get, or get off." Harsh, isn't it? Nonprofits can't fulfill their purposes effectively without money, and board members who take an active interest in the organization's financial vitality are important. But a board member's role is broader than fundraising. Other roles include staying well informed about the organization's work, selecting leadership, setting policies, planning, overseeing, and serving as an ambassador for the organization. Many highly skilled volunteers think they shouldn't serve on boards because they aren't wealthy, and this misconception represents a real loss to nonprofit organizations.

Some people are reluctant to serve on boards for other reasons, including personal and financial liability. However, unpaid volunteers who aren't grossly negligent and act in good faith have some protection under the Volunteer Protection Act of 1997 and may have additional protections under your state volunteer protection laws. If you need more information about the possible liabilities of board members and other volunteers, we suggest that you get in touch with BoardSource (`www.boardsource.org`) or your state's association of nonprofit organizations through the National Council of Nonprofits (`www.councilofnonprofits.org`).

Organizations can protect their boards by purchasing directors' and officers' insurance (see Chapter 21). Generally, the organization's creditors can't come after its board members' personal wealth for payment. An exception is that the IRS can hold board members financially accountable when organizations fail to pay payroll taxes. But even the IRS will work with an organization to develop a payment plan and schedule to catch up on taxes. People are right to take the responsibility of board service seriously, but nonprofit board work also can be fun and satisfying.

TIP

To avoid having all your board members leave in the same year, stagger the years when terms expire. You can allow someone to serve an extra year or ask others to serve shorter terms if necessary and if your bylaws allow it.

TIP

If you're thinking about recruiting new board members, create a spreadsheet to help you visualize the skills you need to seek out when you're looking for new members. Along the top of the grid, list the skills you think you need on your board. Along the side, list your current board members and place check marks under the skills they bring to the board.

TIP

Check out File 6-2 at `www.dummies.com/go/nonprofitkitfd5e` for a sample grid for planning board recruitment.

So where do you find new board members? Start with your organization's address book. Whom do you know who may make a good member and be willing to serve? Who benefits from your agency's work? Who are your agency's neighbors? Who is actively involved as a volunteer for your agency? Some cities have nonprofit support organizations that can help in this regard. Consider asking your funders for suggestions and look at former board members of other high-functioning nonprofit organizations with similar missions.

REMEMBER

Even if you don't specify board terms, continuously recruiting new board members is important. As time passes, board members' lives may change in ways that draw them away from your organization. Many nonprofit organizations lose vitality when their boards don't refresh themselves with new members.

Big boards or little boards

Opinions differ about the ideal number of board members. One school of thought holds that big boards are better because the members can divide work among more people, consider more diverse viewpoints, and reach into the community more extensively. Other people say that smaller boards are better because maintaining a working relationship with a smaller group is easier and decision making is better because the board has to consider fewer opinions. Those who support small boards also say that board members can easily become invisible in a large group, meaning that no one will notice if they don't do their share of the work.

In our opinion, there really isn't one right answer to the board-size question. According to a 2010 BoardSource survey of its members, the average size was 16 members and the median size was 15 members. Nonprofits vary widely in size, function, and type, so what works for one nonprofit may not work for another. Nonprofit board size is truly a case where the phrase "one size fits all" doesn't apply.

Following are some points to consider when setting your board's size:

- **Startup nonprofits tend to have smaller boards than more mature organizations.** Startup budgets tend to be smaller, and building a board of directors takes time. State laws generally specify the minimum size required (often three members).

- **Boards that are actively engaged in fundraising for major gifts and special events tend to be larger because both fundraising techniques are fueled by personal contacts and friendships.** The more board members you have, the more personalized invitations you can send. Some large cultural institutions have 50 board members or more.
- **Boards that govern nonprofits funded mostly by grants and contracts tend to have fewer members, perhaps an average of 10 to 16 members.** Board members in these nonprofits usually have fewer fundraising responsibilities for the organization and frequently are representatives of the communities or clients served. They also may have professional experience in the types of service provided by the nonprofit.

Choosing officers and committees

Typical officer positions of most nonprofit boards include president, vice-president, secretary, and treasurer. Sometimes the positions of secretary and treasurer are combined into one office. Your state laws may specify which officers are required. Seniority on the board, professional expertise, and skills at negotiating with and listening to others are common traits sought in a board's leaders. Ultimately, however, the board of directors chooses the officers. They're usually elected to two-year terms. The following list outlines the common responsibilities of board officers.

- **President (or chairperson):** Presides at board meetings, appoints committee chairpersons, works closely with the executive director to guide the organization, and acts as a public spokesperson for the organization (but also may assign this responsibility to the executive director)
- **Vice-president (or vice-chairperson):** Presides at board meetings in the president's absence and serves as a committee chairperson as appointed by the president
- **Secretary:** Maintains the organization's records, takes minutes at board meetings, and distributes minutes and announcements of upcoming meetings to board members
- **Treasurer:** Oversees the organization's financial aspects, makes regular financial reports to the board, and sometimes serves as chairperson of the board finance committee

TIP

Check out File 6-3 at www.dummies.com/go/nonprofitkitfd5e for more detailed job descriptions you can use.

If the board has *standing,* or permanent, committees, the board president may appoint committee chairpersons or they may be appointed by the board. Typical standing committees include finance, development or fundraising, program, and nominating committees. Other possible committees that may be either standing or *ad hoc* — a temporary committee organized to deal with time-limited projects — include planning, executive search, investment, special events, and facilities. The following list outlines the responsibilities of common standing committees.

- **Development:** Sets fundraising goals and plans fundraising activities for the organization in consultation with the board president and executive director
- **Finance:** Assists the treasurer in overseeing financial reports and official tax filings, making budgets, and maintains relationship with professional accounting firm, if applicable
- **Nominating:** Recruits new board members and nominates board officers for election to their positions
- **Program:** Oversees and advises on the program activities of the organization

Board committees make regular reports to the full board about the organization's activities in their particular areas. Board officer terms and the number and type of standing committees may be written into the organization's bylaws.

If your organization is large enough to conduct an annual financial audit, your board may be legally required to appoint an audit committee. Be sure to check the laws in your state. You may not be permitted to include the board chair or treasurer on the audit committee.

Executive committees are standard groups on some larger boards of directors. Usually the members of the executive committee are the officers of the board, but the committee sometimes also includes the chairs of the standing committees. The executive committee may hold regular meetings to set the agenda for the meetings of the full board and to advise the board president, or it may come together on an as-needed basis. Sometimes an organization's bylaws empower the executive committee to make decisions on behalf of the full board in an emergency or in other special circumstances.

Introducing new and prospective members to the board and the organization

Boards of directors exhibit all the characteristics of small groups, maybe even families: Friendships develop, alliances form, and disagreements occur. Over time, the group develops routines and habits that help make members feel comfortable with one another and help guide the board's work. When a new member joins the group or when a prospective new member visits, the existing members need to make that person feel comfortable and share with her the collective wisdom they have accumulated.

Invite a prospective board member to observe at least one board meeting before electing her to membership. That way, the new member gets a chance to see how the board operates, and the current members have an opportunity to size up the new person. Encourage the prospective member to ask questions. Also, if your nonprofit provides programs, such as childcare, a health clinic, or a music school, be sure to give your prospective board members a tour of the facilities before they join.

When asking someone to serve on your board, don't shy away from sharing a clear picture of the work to be done. You may be afraid that your prospect will say "no" if it seems like too much work. However, keep in mind that being asked is an honor, and contributing good work to a good cause is satisfying. Besides, if the person does decide you're asking too much of her, isn't it better to know now rather than later?

A packet of background materials about the organization and board procedures can help new members get up to speed quickly. The following information is useful for orienting new members:

- Board job descriptions
- Board minutes for the last two or three meetings
- Articles and bylaws
- Conflict of interest policy (if you have one)
- Calendar of the organization's events and scheduled board meetings
- Description of board member responsibilities and expectations
- Description of programs
- Financial audit or financial statement
- 990 Forms for the past three years
- Mission statement
- Names, addresses, and phone numbers of other board members
- News clippings about the organization
- Organizational plan (if one is available)

This information may seem like a lot of reading — and it is. But even if a new board member doesn't read everything from cover to cover, she at least has the reference material when she needs it.

We also suggest that the board president or the executive director meet with a new board member soon after she begins serving on the board, both to welcome her and to answer any questions.

Putting Staff Members on Your Board

As a general rule, we think paid staff shouldn't be board members. The situation can get too complicated. For example, *conflict of interest* is always a potential problem, especially when board and staff have different priorities, such as when employees want raises but the board says no.

Some exceptions to the rule do exist, though. In fact, many nonprofits have at least one staff member on their boards. In startup nonprofits, for example, founders frequently serve as both board members and staff members. This situation isn't surprising. Who's better suited to bring the vision and passion needed to create a new organization than the person who formed it in the first place? In many new nonprofits, of course, paying the staff isn't even possible; resources are so limited that all work is done on a volunteer basis.

WARNING

If a founder or other staff member serves on the board, we recommend that he not be elected board president because doing so tends to put all responsibility for vision and leadership in a single person's hands. Sharing that leadership can be an important first step toward broadening an organization's base of support.

Laws vary by state, but in many cases, having a paid staff member on your board is permitted. For example, under California nonprofit corporation law, up to 49 percent of the members of a nonprofit board are allowed to receive compensation from the nonprofit. But the standards of governance set forth by the BBB/Wise Giving Alliance say that a board should include no more than one compensated member (or 10 percent of members for larger boards). Be sure to check out and follow the nonprofit corporation law in your state.

Using Your Board to Full Capacity

If you spend any time around nonprofit staff, you'll probably hear a few complaints about the board of directors. They may say, "I can't get my board to do anything" or "I can't get them to face hard decisions or raise money." Worst of all, you may hear, "I can't get them to show up to meetings."

Motivating the board is an important part of any nonprofit leader's job. Keeping members well informed so they can make thoughtful, appropriate recommendations is essential. Sometimes your most important task is gently steering the board's attention back to the organization's mission and immediate needs. Whether or not your nonprofit has paid staff, you can take steps to help the board do its work well.

The working relationship between the executive director (if you have one) and the board president is a key factor to having an effective board and organization. Ideally, the relationship between these two leaders is one of respect and trust.

Encouraging commitment from board members

Getting members of a board to pull their weight sometimes seems like a problem that you can't solve. Not every board member will contribute equally to the work involved in governing a nonprofit organization. If everyone on your board shows up at every meeting, reads all the materials, studies the financial statements, and contributes to fundraising activities, consider yourself fortunate.

Here are some techniques you can use to encourage full board member participation:

- **Board contracts:** Some nonprofits ask new board members to sign an agreement that outlines expectations for board service. The contract may include a commitment to contribute financially to the organization, attend all board meetings, and serve on one or more committees. Although board contracts aren't legally binding, they provide a clear understanding of responsibilities for each board member. With a board contract, no one can say he didn't understand what he was getting into when he joined the board.

  Check out File 6-4 at `www.dummies.com/go/nonprofitkitfd5e` for a sample board contract.

- **Bylaws:** Organizational bylaws can state the requirements for board participation. For example, a board member may face dismissal from the board if he misses three consecutive board meetings.
- **Job descriptions:** Just like employees, board members often do better when they know exactly what they're supposed to do. Creating job descriptions for officers, committee chairpersons, and individual board members may clarify responsibilities and make them easier to fulfill. If a member isn't pulling his weight, a committee chairperson should speak with him and remind him of his responsibilities.

- **Reliance on board members:** Solicit opinions from members between board meetings. Use their expertise and recognize their contributions.
- **Self-evaluation:** Sometimes encouraging a board to look at itself motivates board members or encourages those who aren't pulling their weight to resign.

Board members fail to contribute equally to the work of the board due to time constraints, business travel, and just plain laziness. Give each member some slack. But if a board member's lack of participation impacts the full board, the decision is up to the board president (often in partnership with the executive director) to ask the member to reconsider his commitment to the organization.

REMEMBER

To a great degree, each board member's work reflects his commitment to the organization's mission. Board members who truly believe in what you're doing will do everything they can to help you succeed.

Holding effective board meetings

Most board work is done in meetings, either with the full board or in committees. The board president is responsible for ensuring that meetings are well organized and begin and end at a scheduled hour. We can't think of anything that damages board effectiveness more than poorly organized meetings that don't stay on topic and that continue late into the night. Nonprofit board members are volunteers; they aren't being paid by the hour.

If the organization has an executive director, the president may delegate some responsibilities for setting up meetings. Ultimately, however, part of the president's job is to see that board members have the information they need to make good decisions and that they do so in a reasonable amount of time.

Getting the members to show up

Saying how often a board of directors should meet is impossible. The only real answer is as often as it needs to. A meeting schedule depends on the organization's needs and the amount of business conducted at board meetings. The frequency of meetings should be specified in the bylaws. These tend to be small but stable organizations with one or two employees. Most nonprofit boards meet more frequently than once a year; some meet quarterly, some meet every other month, and others schedule monthly meetings. Of course, the board president may call a board meeting at any time if the board needs to handle special business.

The advantage of having more frequent board meetings is that board members are more engaged in the governance of the organization. The disadvantage — especially if the agenda doesn't include much business — is that board members may be more tempted to skip meetings.

Some boards schedule meetings at the beginning of the year for the entire year. By entering these dates in their appointment calendars months in advance, board members are less likely to schedule other events on the same days and are more likely to attend the meetings. For example, if you meet monthly, you may schedule your meetings for the second Tuesday of each month. If you aren't this organized, always schedule the next meeting before the end of the present meeting. Doing so is much easier than trying to schedule a meeting by telephone or email.

TIP

Some boards use online collaboration tools to communicate between board meetings and to compile documents in an easily accessible place.

Conducting efficient meetings

If you're looking for some tips to ensure effective board meetings, check out the following ideas:

- **Schedule a meeting between the executive director and board president prior to the meeting.** This can be done in person or via phone, but it's an important step in determining the agenda and the focus for the upcoming meeting. The meeting also allows the executive director to update the president on staff issues, funding opportunities, and any areas where the executive director needs particular guidance and support from the board.
- **Ten days to two weeks before a board meeting, send an announcement of the meeting to all board members.** Include the minutes from the last meeting and an agenda for the upcoming meeting. Also include any committee reports, financial statements, or background research that the board will discuss at the meeting.

  TIP

  If the meeting minutes include a list of tasks for board members to complete before the next meeting, try to send members a rough draft of the minutes as soon after the meeting as possible so they can get started. (File 6-5 at `www.dummies.com/go/nonprofitkitfd5e` contains an outline of meeting minutes that the board secretary can fill in during the meeting.)
- **Limit the length of meetings to two hours or less, if possible.** After two hours, especially if you're holding the meeting in the evening, attention begins to wane. If you must go longer than two hours, take a break. Offering refreshments is always a good idea.
- **Try to find a conference room for the meeting.** Holding a discussion around a conference table is much easier than sitting in someone's living room. The table offers a place to set papers, and people won't argue over who gets the recliner. It also sets the stage and implies that work is to be done.

WARNING

Avoid holding meetings in restaurants and cafes, if possible. The noise levels are too high to make good discussion possible, and all the activity is a constant distraction. You also have no privacy. Believe us — we've tried it!

TIP

- **Stick to the agenda.** Don't allow people to wander off topic. Some agendas set the time allowed for discussion after each item. You don't have to do this, but if your meetings have been veering off course, setting time limits may help control them. Files 6-6 and 6-7 at `www.dummies.com/go/nonprofitkitfd5e` are samples of common agenda types you can use as models.
- **Follow an orderly procedure.** You don't need to be overly formal in your meetings (many board meetings are very casual), but having a basic knowledge of when to make a motion and when to call the question is helpful.
- **Thank your board.** Board members are volunteers who give time and money to your organization. Take every opportunity during meetings to make sure they're appreciated. Mention their names when appropriate in newsletters and media releases. Small gifts are sometimes useful, but don't be extravagant. You don't want to be accused of wasting the organization's money.

» Jumping into the planning process

» Converting plans into action

» Making program plans

» Planning for facilities

Chapter 7

Planning: Why and How Nonprofits Make Plans

The word *planning* can be intimidating. It brings up images of daylong meetings in stuffy conference rooms with consultants wielding marker pens. Of course, not all planning takes that sort of effort. People make plans all the time, from organizing a vacation to deciding how to get all the errands done on a Saturday afternoon.

Organization planning is a group project that calls for research, brainstorming, discussion, and, in the end, agreement on a goal and the strategies and tactics needed to reach that goal. Simply put, organizational planning is deciding where to go and how to get there. The planning process helps ensure that everyone is headed in the same direction.

In this chapter, we cover planning for nonprofits in all its forms, from organizational planning to work planning to program planning to facility planning. Plan to join us!

TIP

Check out File 7-1 at www.dummies.com/go/nonprofitkitfd5e for a list of web resources related to the topics we cover in this chapter.

Understanding the Importance of Planning

No organization has unlimited funds. Even the largest, wealthiest nonprofit needs to decide how to allocate its resources effectively. Planning helps you make decisions about how to align your organization's mission with its resources by answering questions such as "Is now the best time to invest in a new program?" or "Is our tutoring program still filling a community need?" or "Is our service center convenient to the people we're trying to serve?"

A nonprofit organization undertakes planning for the following reasons:

- To build a structure that guides its activities in pursuit of its mission
- To allocate organizational resources in the most effective ways
- To create a framework against which the organization's performance can be evaluated
- To adapt to changes in the external environment
- To reach agreement among board, staff, and supporters on desirable goals for the organization and well-considered ways to meet those goals

Think of a plan as a blueprint or scheme describing what needs to be done to accomplish an end. In an ideal world, if your organization completes every step of its plan, you achieve your goals. However, not all plans are written as recipes that — if followed closely — will serve up a perfect meal. Scenario plans, for example, help an organization think through alternatives it may choose among; other planning approaches invite organizations to assess and revisit their strategies frequently.

REMEMBER

The *act* of planning has value apart from the written document that's ultimately created. That's because, in theory, the planning group comes to a shared understanding of the organization's mission, and the decision-making process ensures that everyone understands what needs to be done and agrees that it's worth doing.

Making Your Organizational Plan

Organizational planning is what people usually think about when they consider planning. An *organizational plan*, usually covering a three- to five-year period, sets goals for the organization and describes the objectives that must be accomplished to achieve those goals.

The steps for successful organizational planning are as follows:

1. **Decide whether it's the right time to plan.**
2. **Look at your mission.**
3. **Assess the external and internal situation.**
4. **Hear from stakeholders.**
5. **Make decisions about goals and strategies to meet them.**
6. **Write the plan.**
7. **Act on the plan.**

REMEMBER

Planning is an ongoing activity. A formal effort to develop an organizational plan may occur only once every three to five years, but planning for the purposes of refining objectives, creating a budget, and developing fundraising programs goes on all the time.

Getting ready

Make no mistake: An all-out organizational planning effort requires considerable time, energy, and commitment from everyone involved. Some nonprofits spend a year or more developing an organizational plan.

Don't jump into the planning process without understanding that it will add to your workload and complicate your life for a period of time. Also, keep in mind that you can't plan by yourself. If you're the executive director of a nonprofit organization who thinks that a full-bore planning effort is needed, but the board of directors doesn't agree, don't try to start the process on your own. Take a couple of steps backward and begin the work of persuading board members that planning is worth the effort.

WARNING

Don't start the planning process if your nonprofit is in crisis mode. It's tempting, for example, to launch an organizational planning effort if you just lost a major source of funding. But you have more immediate concerns to deal with in that situation. Delay organizational planning until you see a period of smooth sailing.

If you have the support, and foreclosure isn't hanging over your agency's head, the best way to get started is to form a *planning committee*. This small group of board members, staff, and one or two outside people can take on the role of guiding the planning process and, in the beginning, pull together the questions, facts, and observations that you need to make your planning decisions.

IS IT LONG-RANGE PLANNING OR STRATEGIC PLANNING?

People often use the terms *long-range planning* and *strategic planning* as if they describe the same thing, but many planning experts make a distinction between the two. *Long-range planning* is setting a path to accomplish goals several years in the future. It assumes that current knowledge about future conditions is reliable and can reliably guide future operations. *Strategic planning,* on the other hand, assumes dynamic changes in your nonprofit and the outside environment. It stresses successful adaptation to change.

We think both planning approaches are necessary. It should set goals based on what it has learned about current conditions alongside strategies for adapting when change occurs.

Starting from your organization's mission

Reviewing the organization's mission statement is one of the first tasks facing an organization that's developing a plan. (Chapter 3 covers mission statements.) Ask yourself these questions:

- Is the problem we set out to solve still a problem?
- Do the organization's current programs and activities address the mission in a meaningful way?
- Is the mission statement clear yet flexible enough to allow the organization to grow and adapt?

REMEMBER

Keep your mission statement in mind throughout the planning process. At every turn, ask yourself: "If we do this, will we be true to our mission?"

Surveying the external situation

Early in the planning process, you need to collect information about external factors that influence your nonprofit's operation. Someone, or a subcommittee of the planning committee, should find the answers to the following questions and distribute them to everyone on the committee before the formal planning meetings begin:

- What are the demographic and other trends in our area and in our organization? Will these trends have an impact on the number of people who may need or use our services in the future?

- What are the trends in the professional arena in which our nonprofit operates? Are new methods being developed? Are we complying with current laws regulating our field? Does the future show a shortage of professionally trained staff? Is technology changing the way people acquire knowledge or services?
- Are other nonprofits providing similar services in the community? If so, how are our services different? Should we be working together? How are we distinctive?
- How stable are the funding sources on which we depend? What about changes in government funding? In our earned revenue? Can we find new potential sources of funding?

The decisions you make are only as good as the information on which you base them. Therefore, it's important to find the best and most up-to-date data available that may have an impact on your organization and its programs, and it's important not to overlook unfavorable trends.

You may want to collect information from the general public or the constituents served by your nonprofit. We also recommend conferring with major donors, foundation representatives, and other nonprofits or government agencies that work closely with you. Surveys, interviews, and focus groups are ways to gather input from the public. Before you undertake any of these techniques, however, spend some time thinking about what you want to learn. Write a list of questions for which you want answers.

If possible, consult with someone experienced in preparing surveys or interview protocols, because the phrasing of questions and the way you distribute your surveys affect the answers you get. Look for expertise in this area from consulting firms that work with nonprofits, at local colleges and universities, or from marketing or market-research firms.

Bad information leads to bad decisions. Or, put another way, garbage in, garbage out. So when gathering background information to guide planning decisions, take the time to get the most accurate, up-to-date facts available.

Looking at the internal situation

In addition to surveying the environment in which the organization operates, you need to expend some effort assessing the organization itself. Consider the following factors when doing an internal analysis:

- What are the organization's major accomplishments? What are its milestones?
- Whom does it serve and what does it offer them?

PLANNING FOR UNCERTAINTY

No matter how well you plan, circumstances can, and probably will, change. So, of course, creating an organizational plan requires that you make assumptions about the future; sometimes these assumptions turn out to be wrong. Maybe, for example, a dependable funding source drops out of the picture; maybe a major industry in your city closes its doors; or maybe new government regulations affect your programs. Unexpected events within your nonprofit or in your community can dump cold water on your carefully designed plans.

But this doesn't mean that your nonprofit should abandon planning. Fundamentally, planning is about making decisions that are based on the best available information and the most careful thought you can muster. If conditions change or a new opportunity arises, the analytical muscles that your organization has developed through that thoughtful decision making should prepare it to respond quickly and appropriately to apply new strategies. Remember that all plans are living, breathing documents and can be adjusted as conditions change.

» Is the board of directors fully engaged with the organization? Should any weaknesses on the board be filled? Is the board engaged in fundraising?

» What are the staff's capabilities? Does the organization have enough staff — and the right staff that is sufficiently trained and licensed?

» Do our governing documents — such as our articles of incorporation — allow for our current and planned activities?

» Is the organization's office and program space in the right location and of the right size?

» Does the organization have adequate technology? Are its technology systems integrated with one another? Do employees need additional training in using technological tools? Are equipment needs anticipated?

» Has the organization operated within its budget? Is financial reporting adequate? Are appropriate financial controls in place?

» What does the organization's program cost and how does it pay for it?

» Does the organization have a variety of funding sources? Are those funding sources stable?

SCENARIOS FOR THE FUTURE

Scenario planning is an approach that involves imagining several futures in which your organization may find itself and then using those imagined futures to determine the strategies that your nonprofit needs to thrive in the years to come. The practitioners of scenario planning don't pretend that they're prophets or seers, but this planning method provides a framework for thoughtful conversations among the leaders of the nonprofits, corporations, and government agencies that engage in it.

Scenario planning pushes you to define the most pressing issue facing your organization. For example, if your organization works to provide equal access to information technology for everyone, you may want to ask what limits some people's access now — be it equipment, service costs, training, or available Internet access — and what conditions are changing that might solve or exacerbate the problem. After you've figured out the question you want to address — often the most difficult thing to do — you need to collect as much information as possible about trends that may impact access to technology. That may include future cost projections for computers; projected family income figures; trends affecting libraries, adult education, and after-school programs that have provided free access to technology; business trends in the technology industry; and efforts by local governments to provide wireless access to their communities, to name only a few.

Then you can begin thinking about how the future may play out by creating several descriptions — some would say stories — of possible future conditions. After you arrive at this point, you begin to talk about the roles your organization could play in each of those stories that would enable it to fulfill its mission.

Hearing from all your stakeholders

Unless you have a very small organization, you probably can't include every single person in the planning process. You do need to include all stakeholder groups, however. A *stakeholder* is someone who has a reason for wanting the organization to succeed. Paid employees and members of the board certainly qualify. But these two groups, who are the most closely connected to the nonprofit, are by no means the only people who have a stake in the organization's success.

In our view, you also should include a representative from the following groups in planning:

- Users of services
- Community leaders

DESIGN THINKING

Design thinking is an approach to solving creative problems from the design field. Its application to business settings is widely associated with the d.school: Institute of Design at Stanford. The key to design thinking is clearly defining a problem, considering present and future conditions, and identifying a number of solutions — always with the end user of a product or service in mind. In design thinking, you experiment with those solutions, intending to "fail fast" until the best result emerges.

Some nonprofit organizations and foundations are applying principles of design thinking to their planning work. It resembles traditional strategic planning in that it begins with an analysis of present conditions, but it's different in that it assumes that organizations can't think through the best solutions through discussion: testing prototypes and learning from failure are key to charting a future course. The d.school offers a virtual 90-minute crash course in design thinking on its website, and we've included a link in File 7-1 at www.dummies.com/go/nonprofitkitfd5e.

- Donors
- Volunteers

Check out File 7-2 at www.dummies.com/go/nonprofitkitfd5e for a sample planning-retreat agenda. It gives some guidance in organizing your own planning retreat.

One purpose of organizational planning is to bring stakeholders together in pursuit of a common goal. People work harder to achieve the goals when they're asked to help set the goals.

Guard against bias. Sometimes people get so close to the situation they're evaluating that they can't see the true picture with an objective eye. Include outside people who gain no personal benefit from the outcome.

Also, don't assume that all your stakeholders know what they're talking about in every instance. For example, if someone says that getting a grant to pay the costs of a program startup is a piece of cake, check with potential funders before you agree that it's an easy task.

Yes, honesty can create conflict. Be prepared for it. Set some rules when going into planning meetings. Make sure that all the participants have a chance to state their cases. Arguments can be productive if they exchange ideas about what is best for the organization and don't disintegrate into shouting matches.

Calling in the SWOT team

One common way to analyze the information you've collected is to perform a SWOT analysis. *SWOT* is an acronym for strengths, weaknesses, opportunities, and threats. SWOT analysis is usually done in a facilitated meeting in which the participants have agreed-upon ground rules. If you prefer, stakeholders can complete their individual versions of the analysis and then come together to discuss the results as a group.

TIP

We recommend using a professional meeting facilitator. A facilitator brings experience and a neutral viewpoint to the proceedings and can be effective in helping a group arrive at a consensus about organizational goals — even goals that may be unpopular with some staff or board members.

Be honest when looking at your organization's strengths and weaknesses. It's tempting, for example, to put the best face on the activities of the board of directors. They're volunteers, after all. How much time do they really have to give to the organization? But if you avoid identifying important deficiencies because you don't want to hurt someone's feelings, your planning efforts will be handicapped by bad information.

For example, the results of a SWOT analysis for an organization providing counseling services to unemployed adults may look something like the following:

Strengths

- » The program staff is highly qualified and committed.
- » Clients give program services a high-quality rating.
- » The organization has an accumulated budget surplus equal to approximately five months of operating expenses.
- » Program costs are funded largely by local government grants.

Weaknesses

- » The cost per client is higher than in similar programs.
- » Contributed income from individuals is low.
- » The programs aren't well known to the general public.

Opportunities

- » Closure and shrinking of nearby industries indicate that more members of the client base will be working as part-time independent contractors without employee benefits.

» The organization's marketing campaign has attracted additional visitors to its website.

» The organization has modest cash reserves to invest in growth.

Threats

» Local government (the primary funding source) has announced plans to open its own job training and placement programs in three years.

» Over the past five years, program costs have increased at a rate of 3 percent per year.

As this example shows, one or more items often can be listed as both strengths and weaknesses. Here, the fact that the organization is largely funded by the government is a strength, but it's also a weakness because it creates a situation in which the organization relies on a single source of funding.

A review of this SWOT analysis reveals an organization that has been successful in providing quality services and getting those services funded by government contracts. It has been financially prudent and developed modest cash reserves. However, the working conditions are changing for its clients and the program's future may be in jeopardy if it continues to rely heavily on government resources. A major new competitor is entering its field.

This organization has limited time and resources to diversify revenues. If its government funding is cut completely, it may have to close, spending its cash reserve on connecting its clients to other services, meeting with its donors to thank them, telling stories of its clients' successes on its website, providing modest severance packages to loyal employees, and taking necessary accounting and legal steps to close.

In the meantime, as its clients need new entrepreneurial and planning skills to manage their changed employment status, it has the opportunity to develop distinctive programs responding to current needs. It also has opportunities to develop an individual donor campaign led by the board of directors; meet with other nonprofits, local colleges, and business leaders about sharing counseling and training services; and seek coverage of its programs in the local media.

REMEMBER

A SWOT analysis can be a powerful guide to developing a plan because it looks at both your organization's inner workings and external environment. As your committee moves on to the next planning step, where you decide together on goals and strategies for coming years, it can refer to the SWOT analysis as a reminder to celebrate and further develop strengths, tackle weaknesses, and prepare for change.

MAPPING THE MATRIX

One approach to planning that may work well as a tool to guide your organization is the creation of a *matrix map*. The matrix map encourages nonprofits to take a hard look at their "dual bottom line" — both how well the goods and services they provide address their mission and each of those services' costs and earning potential. To create a matrix map, you and your planning committee follow four steps:

- Identify each of your "lines of business" or activities.
- Assess how well each of those activities addresses your mission.
- Analyze how well each of those activities covers its costs either through fees and admissions or by attracting grants and contributions.
- Create a chart or matrix that illustrates how programs compare to one another according to value in addressing your mission and contribution to the organization's financial stability.

The result of this process — easily understood illustrations comparing an organization's programs — can help a planning committee choose among options. Check out File 7-4 at `www.dummies.com/go/nonprofitkitfd5e` for a sample matrix map.

TIP

Check out File 7-3 at `www.dummies.com/go/nonprofitkitfd5e` for an example of how you can map your organization's strengths, weaknesses, opportunities, and threats to determine your overall SWOT.

Putting the plan in writing

After the research, analysis, and discussion, it's time to determine future directions and put the results into a final plan. Ideally, the plans of the organization become apparent after sifting through and discussing the material assembled. If the planning group can't reach a consensus, it may want to test its ideas on the board of directors or a few trusted peers to help it come to an agreement.

TIP

It can be easy to come up with a long "laundry list" of goals and then skim lightly over the strategies you'll use to achieve them. We recommend limiting your goals to a half dozen or fewer and paying full attention to the related strategies.

Ensuring your goals make sense

Organizational goals need to be specific, measurable, and attainable within a set period of time. Don't set a goal like "In five years, the XYZ Tutoring Project will be the best tutoring program west of the Mississippi." You have no way to evaluate whether the organization is the best tutoring program west of the Mississippi, and even if you can determine the "best" program, the goal is so general and vague that you can't set objectives to achieve it.

Instead, an organizational goal for this program might be, "In five years, the XYZ Tutoring Project will provide tutoring services that are recognized by classroom teachers as effective in improving student performance." This goal can be measured. It's also attainable through implementing a series of strategies that may include better training for tutors, improved communication with classroom teachers to determine student needs, acquisition of computers to aid in tutoring sessions, and so on.

Itemizing the parts that form your plan

When you reach a consensus about your organization's goals, assemble a written plan. The components of the plan should include the following items, in the order shown:

- An executive summary
- A statement of the organizational mission
- A description of the planning process
- Organizational goals
- Strategies to achieve those goals
- Appendixes that contain summaries of the background material used to determine the plan, including a list of the people who participated in creating the plan

Assign the task of drafting the document to your best writer. When he finishes the draft, the planning committee should review it to ensure that the plan is stated clearly. Submit the final draft to the full board for discussion and adoption.

WARNING

A common failing of planning documents is the use of jargon and vague language. An organizational plan that uses such wording and lacks clarity is a poor guide and does nothing to increase your credibility with your constituents or your funders. You don't want readers to finish reading it and ask, "What is it that they're going to do?"

TIP

Check out File 7-5 at `www.dummies.com/go/nonprofitkitfd5e` to see an example of how an organizational plan may be presented.

Adjusting your plan when necessary

Although having a plan is important, flexibility in its implementation is just as important. Things change. Reviewing your plan is an ongoing activity; at least once or twice a year, your board should revisit it to check on the progress you've made and identify whether assumptions and predictions were correct. You may find that you need to employ new strategies to achieve your carefully developed goals.

Putting Plans into Action

Unfortunately, too many well-crafted plans end up in a drawer or on a bookshelf. Participants may have expended great effort to create the document, and board and staff may have formed close bonds during the planning process. But if the decisions made during that board/staff retreat at the charming lakeside inn aren't translated into tasks, what's the point besides camaraderie?

REMEMBER

No matter what sort of plan you create — organizational, fundraising, marketing, program, or facility-related — you need to break the larger, all-encompassing strategies into a sequence of steps that enables you to chart your progress over time. The following sections help you do that.

Defining and setting goals, strategies, objectives, and outcomes

Getting lost in all the terminology of planning is easy. Here are brief definitions of four common terms — goals, strategies, objectives, and outcomes — using a simple example of traveling from Chicago to New York to attend a professional conference:

- *Goals* are your organization's aspirations. Goals can be set at the organizational level, the program or department level, or the individual employee level. Using a road trip as an analogy, a goal is to arrive in New York, having left from Chicago.
- *Strategies* are approaches or ways to achieve goals. Usually, more than one option exists. You can travel to New York by several methods: plane, train, automobile, bicycle, or on foot. After considering the costs, your schedule, and your hiking ability, you decide to travel by car.

- *Objectives* are smaller steps that one must accomplish to reach a goal, and they're always stated in a way that can be measured. So on a trip from Chicago to New York, an objective may be to drive 325 miles on the first day. When you pull into the motel parking lot, you can check your odometer to see whether you've achieved your objective.
- *Outcomes* describe the results of reaching a goal. In this example, you reach the goal — New York — and learn useful information at your conference.

To see how all four terms come into play, look at the example of a development plan in Table 7-1. Reading from top to bottom, you have the whole plan, from organizational goal to outcome.

TABLE 7-1 **Organizational Goal to Outcome**

What We Call It	Example
Organizational goal	Diversify income.
Strategy	Increase individual contributed income.
Strategic goal	Develop a reliable annual campaign.
Objective 1	Retain 80% or more of current annual fund donors.
Objective 2	Identify a minimum of 500 new prospects through board and advisory committee contacts and another 2,500 prospects through list exchanges.
Objective 3	5% of new prospects respond to annual fund letter.
Outcome	Organization is less dependent on two major donors and a handful of annual contributors.

REMEMBER

Don't get bogged down in terms. How you label the different steps in your plan is less important than clarifying what you need to accomplish and the steps you'll take to succeed.

Creating a work plan

You can see in Table 7-1 that the objectives are measurable results. Achieving each objective in the table requires several steps. Here's where *work plans* come into play. Work plans break tasks into small steps so they can be easily managed.

Work plans are the nuts and bolts of planning. They're also called *action plans.* They contain strategies for achieving specific objectives, identify deadlines for completion, and note who's responsible for completing the task. A work plan answers the following questions for each objective:

- What is the end result?
- How long will it take to do the job?
- Who will be responsible for doing the job?
- What resources are needed?

A typical work plan may look like Table 7-2.

TABLE 7-2 Sample Work Plan: Create an Appeal Letter

Objective	By When	By Whom	Resources Needed	Date Completed
Attend workshop on fundraising letters	February 28	Allen	Find workshop through the Foundation Center	February 26
Draft letter and seek feedback	March 15	Allen and board committee	Committee meeting for feedback	March 15
Revise and copyedit letter	March 20	Ashley and Gina	Experienced editor	March 25

Work plans require that a job be broken down into smaller tasks. In Table 7-2, for example, the three objectives can be split into even smaller tasks.

Be aware that you can take the creation of a work plan to the point of absurdity. Don't make work plans so detailed and specific that writing the plan takes more time than doing the work that the plan specifies.

Planning for Programs

Program planning is such an important part of nonprofit work that we think it needs its own section. Nearly all nonprofit organizations provide a service of one sort or another. The organization provides services through programs. A small nonprofit may have only one. Larger nonprofits may have dozens. No matter how many programs you have, you may be thinking of adding a new one or changing the ones you have. We walk you through the process in the following sections.

Assessing needs

A *needs assessment* is an important part of program planning. If you're thinking of starting a new program, for example, a needs assessment to determine whether the program is necessary should be the first step you take.

A needs assessment is more or less a research project. You don't necessarily need to hold to the strict requirements of scientific inquiry, but just as you do when collecting information to help guide organizational planning, you should do everything possible to ensure that the information is accurate and free of bias.

Determining the questions to be answered

Your needs assessment should evaluate the answers to the following questions:

- **Are other organizations providing the same service?** Obviously, you don't want to duplicate services if another organization is already doing the job. If you believe that your competition isn't doing a good job, that's another question. Jump to the second point.
- **How many people might use the service?** Making a good estimate of the number of people the new program will serve is important. Doing so helps you justify establishing the program and helps you plan for staff needs.
- **Can and will people pay for the service?** If so, how much? How many of the people you hope to serve will need discounted tuition or scholarships, for example?
- **Do we need to meet any special requirements for providing the program?** Does the program need to be near good public transportation? Is parking important? Will people come to the neighborhood where you're providing the services? Will the program require you to register your services, obtain licenses, or obtain operating permits?
- **What are the trends?** Will the number of people using the service increase or decrease in the future? Is the population in your community increasing, decreasing, or staying the same?

Finding the best data

Just as when you collect information for organizational planning, your goal in program planning is to get the most accurate, unbiased data available. Don't depend on only one source. Here are some ways to get the information you need:

- **Talk to colleagues in your community.** Ideally, you have relationships with the people and organizations that provide similar services. Ask their opinions about your ideas for new programming.
- **Look at census data.** The best sources for population information are the numbers collected through a census (www.census.gov). Some municipal and regional planning groups also publish population-growth projections. These projections are estimates, but census data doesn't represent an absolutely accurate count of the people in your community, either. Gather all the numbers you can get.
- **Visit a Foundation Center library or website.** By looking at funding trends, you can piece together a picture of what services are available and who is supporting them. You can find a Foundation Center library near you at www.foundationcenter.org or subscribe to its services online. In addition to guiding you to potential funding sources, this website's *Philanthropy News Digest* (PND) shares research and articles produced by foundations.
- **Get information from current or potential users of the service by distributing questionnaires and holding focus groups.** The questions you ask in large part determine the answers you get. If possible, find someone who has experience in preparing survey questions to guide you.
- **Look at similar programs in other communities.** Although you can't always depend on the experience of others fitting exactly with your particular situation, examining what others have done is always wise.

TIP

Check out File 7-6 at www.dummies.com/go/nonprofitkitfd5e for a sample needs assessment questionnaire.

REMEMBER

Some people say they don't want to share an idea with others because they're afraid someone may steal it. Although you can't rule out the chances of this happening, we believe that it's a rare occurrence. In almost all cases, being open about your plans is a good idea.

Going beyond the needs assessment

Just because a new program is needed doesn't necessarily mean that your organization should be the one to start it. You need to take other factors into account. Consider the factors in the following sections when assessing your ability to start a new program.

Paying attention to the budget

A new program is almost always going to add expenses to your organizational budget. If we thought about it long enough, we could think of an example in which increased program costs aren't a factor, but we don't have all year.

To be sure that you don't get yourself into a financial hole, carefully project the extra costs you'll have from additional staff, increased space, equipment, and insurance. After you have solid expense projections, you have to project where you're going to find the added revenue needed to pay these costs.

Evaluating organizational and staff capability

Does your organization have the knowledge and expertise to provide the program services? This question may not be a concern if the proposed program is merely an extension of what you've been doing. If your organization is branching out into new areas, however, be sure that you or someone else in the organization has the credentials to provide a quality program.

Also pay attention to hidden staff costs. For example, consider whether your current program director will have sufficient time to provide adequate supervision for the new program. Will you need to hire a new staff person or consultant to implement the program? Is this a one-time cost or recurring cost to continue the program? Will this person require a computer, cellphone, insurance, or office space?

Remembering special requirements

Check whether you need additional licensing, accreditation, or permits to provide the program. This is especially important for human-services programs. For example, if you've been working with teenagers and want to expand to elementary and preschool children, find out whether your program space must meet additional building-code requirements in order to serve a younger client group.

Fitting it into the mission

From time to time, we're all tempted by the idea of doing something new. Gee, wouldn't it be nice if we could sell goldfish in the front lobby? But you have to ask yourself what selling goldfish has to do with your organization's mission. If you go too far afield, you can detract from the hard work of addressing your organizational purpose. Also, if you go beyond what is permitted by your purpose statement in your articles and bylaws, you may be acting unlawfully beyond your organization's authority and perhaps outside of its tax-exempt status.

Thinking long term

An idea that looks good today may not look good next year or the year after. Don't forget that costs rise year after year. Your staff appreciates occasional raises. Try to imagine where the program will be five and ten years into the future.

REMEMBER

Funding sources may be available to launch a new program or expand services. But it's wise to have a plan for sustaining the program after it is launched by identifying funding sources for its ongoing support after the initial funding source is gone.

Working as a team

As with organizational planning, program planning should be done as a group exercise. It doesn't have to be as extensive as the organizational planning process, but you gain more acceptance of the new program and guard against omitting important details when you work with others to develop new programs.

TO EXPLAIN YOUR PROGRAM, TAKE A TIP FROM THE BUSINESS SECTOR

If you were starting a new business and looking for investors, one of the first things you'd do is create a business plan. Nonprofits are wise to follow this model when developing a new program. A business model is useful in explaining the program and can form the basics of grant proposals when you seek funding.

Business plans should include the following information:

- An executive summary that covers the main points of the plan.
- An explanation of the need for the program and who will use it. (This information is comparable to the market analysis in a for-profit business plan.)
- A description of the program and your strategies for implementing it.
- Why your organization is best poised to implement the program.
- Résumés and background information about the people who will provide and manage the program services.
- Three-year projections of income and expenses for the project, including an organizational budget for the current year.

Facility Planning: Finding a Place to Do Your Work

If your organization is grappling with a move to a new (or its first) office building, you're facing important decisions. And guess what? You need to write a plan!

This effort is likely to have three phases:

- Planning for your needs
- Identifying possible locations
- Analyzing the feasibility of the locations you find

How much space and of what kind?

Before you go out to seek a location, make a list of your organization's specific needs. If you've ever shopped for an apartment or house, you know that some features are critically important and some are desired but not essential. Breaking down your space needs by function and then including a list of general requirements helps. Consider current programs as well as programs you're planning to introduce in the near future.

TIP

For help in anticipating and specifying all your organization's facility needs, check out File 7-7 and File 7-8 at `www.dummies.com/go/nonprofitkitfd5e`.

Location, location, location

Many nonprofit organizations have learned the hard way that having a beautiful new facility doesn't necessarily mean that their students, patients, or audiences will go there. We suggest that an organization conduct a simple marketing test of a location it's considering. This "test" may take the form of a written survey, interviews, or an open house/walk-through at the proposed site followed by a discussion with current constituents. Also, talk to nearby residents, merchants, and the local police, and spend time observing the site at different times of day. Finally, check the zoning for the desired location and make certain that the use you propose for it is permitted.

Organizations often move to larger facilities when they want their programs to grow, and they discover the hard way that offering more seats, classes, or therapy sessions doesn't necessarily mean that they'll be used. Do you have clear evidence of growing demand for your services and that your organization's current physical space is inhibiting its growth? When you conduct your marketing test, you need to discern whether more people will go to your new location: Reach out to both potential clients and your current followers.

Your ideal location may change over time as neighborhoods change. Even if your organization has been based in one place for a long time, before signing a new lease, explore whether its location is still meeting its needs.

Owning, leasing, or taking a free ride

Stability, convenience, and cost — in addition to location — are key factors to consider when selecting your organization's home. In this section, we discuss the implications of ownership and leasing, or of accepting donated space.

REMEMBER

In considering your real-estate choices, also consider your context. Is the real estate market changing? Are interest rates rising or falling? How well are tenants' rights protected by law? Is the building that you want to lease likely to be sold?

The pros and cons of owning

Because of the tax benefits of private homeownership, many people assume that owning a building is best for a nonprofit organization. Although a nonprofit's building can be a valuable asset, remember that a nonprofit is already exempt from paying most business taxes, so any interest it may pay on a mortgage or building loan isn't a deductible expense — it's just an expense.

Two possible advantages of building ownership include:

- **Ownership stabilizes costs.** If your organization is based in a real estate market where prices are rising, purchasing a building may help to prevent steep rent increases or an untimely eviction.
- **Ownership improves the public image of your organization.** Organizations owning their own buildings appear in the public eye to be stable institutions. This perception may help them raise money.

Major disadvantages of building ownership for nonprofits can be that it increases the staff's workload and requires a continuing investment. If the organization buys a building that's larger than its needs, it may become a landlord to others and must be prepared to advertise the property, negotiate leases, and manage maintenance and repairs. Whether or not it has tenants, it becomes fully responsible for the building's care. Rental income it collects can be an unrelated business taxable income. Check with a tax attorney for advice before entering into this type of arrangement.

WARNING

If your nonprofit organization buys its own building, be sure to set aside a cash reserve for building maintenance. Otherwise, if a boiler explodes or the roof leaks, you may have to suspend operations for an extended period of time to fix the problem.

Considering renting

When you rent a home for your nonprofit organization, you're taking on costs that you need to cover month after month. Often these costs increase from year to year. Rent may not be the only such expense. Here are a few things you want to understand fully before signing a lease:

- What costs are covered? Is your nonprofit responsible for all or some of the utilities?
- How long is the lease and does it include options for you to renew it at a similar rate?
- If property or other taxes increase while you're a tenant, do you pay for the increase or does the landlord?
- Which repairs are the landlord's responsibilities and which ones are yours?
- Who's responsible for routine building maintenance?
- What will the landlord permit you to change about the building?

Deciding whether to take a freebie if it's offered

Taking a free ride through the donation of space sounds wonderful, doesn't it? Indeed, it lowers your operating costs and enables you to use more of your resources for programs. But you must be willing to look a gift horse in the mouth.

A free building is worthwhile only if it's in the right location, is the right size, and offers the right amenities. Doing effective work is very difficult in an inappropriate space. Ask yourself this: If the building weren't free, would you have chosen it for your nonprofit?

Making a move

Organizations with what seem to be straightforward plans for moving into new facilities often overlook the true costs of making such a move. Some spaces may need to be altered to suit your organization's needs. Even when you fit right into your new offices, you encounter one-time charges such as signs, cleaning deposits, phone and Internet hookups, and fees or deposits for starting up your utilities. Don't forget marketing costs to let the public know you have moved.

REMEMBER

The most important *things* for you to move are your constituents. You want to make a thoughtful, sustained effort to invite them into your new facility.

Taking on a capital project

What if no existing building suits your organization's needs? You may be in for a major effort to substantially renovate a space or construct a new building. If you're one of these brave and hardy types, you want to read this part of the chapter along with Chapter 18, which addresses planning and raising money for building projects in greater detail.

Even a small organization with the right board and campaign leadership can manage a successful capital campaign if its expectations are reasonable. So can organizations whose projects are happening at the right place and at the right time — such as those organizations based in community redevelopment areas or low-interest bank loans for community development.

To determine whether your organization can manage a capital campaign, you need to plan (no surprise, right?). Your facility plan should ask hard questions, including the following:

- What will the project cost?
- Are your board members in a position to contribute to a capital campaign above and beyond their usual annual gifts to your organization?
- Do public or foundation resources in your region support capital projects? Are they likely contributors?
- Do you have staff knowledge and time to contribute to this effort?

Having examined these preliminary questions, organizations that are considering capital campaigns often go through a planning step called a *feasibility study* — research most often led by a consultant who interviews people who support the organization and other generous donors in their communities whose grants and gifts are essential to its success. As with other types of planning, you're gathering information from key stakeholders, but you're focusing your attention on those who may become contributors. Through these interviews, the consultant estimates how much the organization is likely to raise with a capital campaign. Find out more about this practice in Chapter 18.

» **Knowing what the evaluation process involves**

» **Looking at the results and implementing them**

Chapter 8

Evaluating Your Work: Are You Meeting Your Goals?

How do you know if your nonprofit organization is meeting its goals? How do you know if your approach is valid, your staff is qualified, and your clients benefit from what you do? When your nonprofit's resources are scarce, you may want to put all of them into providing direct benefits to the people you serve. But, wait. When resources are scarce, it's especially important that they be used in the best possible way. That's why you want to set aside some of the organization's time, money, and attention for evaluation.

Most organizations gather information during the normal course of doing their work. A good evaluation can be one in which you collect information in a consistent, systematic way and ask the right questions so you learn from its patterns and particulars. This chapter outlines what goes into a meaningful evaluation and explains how to interpret the results and implement any necessary changes.

TIP

Check out File 8-1 at `www.dummies.com/go/nonprofitkitfd5e` for a list of web resources related to the topics we cover in this chapter.

We use the word "evaluation" as a general term in this chapter to mean any act of reflecting upon the quality and success of your nonprofit's work. Some professionals differentiate between "assessment" and "evaluation." An "Assessment" is the ongoing use of measures or tools (such as exams or surveys) to indicate progress being made, and an "evaluation" is a systematic, rigorous comparison of one's results to defined expectations, standards, or another program. Check out File 8-2 at www.dummies.com/go/nonprofitkitfd5e for more definitions of evaluation terms.

Knowing the Importance of Evaluation

Unlike for-profit businesses, nonprofits can't evaluate their performance solely by showing a profit at the end of the year or increasing the value of their stock. Indeed, they may generate a budget surplus to invest in their future work, but that isn't their primary purpose. Some people say that nonprofits have to achieve a "double bottom line," one in which their finances are healthy and their activities are meeting their goals — what a friend of ours calls "an appropriate balance of mission and resources." That's where evaluation comes in: It can measure how successful you've been at focusing on your purpose and achieving your challenging goals and objectives.

Evaluating your work goes hand in hand with every aspect of leading and managing your organization. An evaluation can:

- Tell you if your planning strategies are working and whether you need to adjust them
- Guide you in hiring the right staff for the work to be done and in making good use of volunteers' time and energies
- Help you draft your annual budget by pointing out where resources are most needed
- Convey the information you need to market your work to the constituents you wish to serve
- Strengthen your fundraising by arming you with information for reports on your work to foundations and donors

In short, to know where you're going, you need to know where you've been. Evaluation can tell you whether you're on the right road or need to recalibrate your route.

Working through the Evaluation Process

You've probably been part of an evaluation before, whether a personal performance evaluation or a program evaluation, but maybe you've never been in charge of the entire process. Not to fear. This section provides an outline of the process, gives you some pointers about what to evaluate, and helps you select who will perform the evaluation.

Evaluations have three main components:

- **Setting up an evaluation:** This step involves determining what type of evaluation you need to perform, asking the right questions, assigning someone the responsibility to gather and analyze the information you need to answer those questions, and deciding where and how you'll find that information.
- **Conducting the evaluation:** During this step, you gather the information you need in a consistent manner and measure what you're learning in an honest (perhaps self-critical) way. You may find that you need to devise tools to gather that information — such as surveys of constituents, focus groups, formal observations, or one-on-one interviews.
- **Interpreting the data:** In almost all cases, your understanding will be deeper (and your findings more valid) if you compare the information you compile to something else. That "something else" may be results for the people you serve before they participated in your program. It may be results for people who didn't take part in any comparable program (a placebo). Or it may be results for people who participated in a different kind of program. We address this stage in the "Analyzing Results and Putting Them to Work" section at the end of the chapter.

Selecting the right kind of evaluation

Evaluations come in many varieties. The type you choose will depend on whether a program is brand new or has been refined, how much is known about what works in the field in which you're operating, whether you want to prove that your approach is a model that can influence others, and how much money you have to invest in evaluation.

Your organization may consider these two basic types of evaluation:

- **Formative evaluation,** in which you analyze the progress of your nonprofit's work while your work is in process.
- **Summative evaluation,** in which you come to the end of a phase or a project and look back on its accomplishments.

Now that you can drop these terms easily into conversations, let us introduce three other kinds of evaluation you want to know:

- **Process evaluation:** Did the project do what it was supposed to do and on schedule? Did you acquire an office, hire three key staff, promote programs, or sign up participants? Very good. In your process evaluation, tell your readers about the steps you took — the story of your program activities.
- **Goal-based evaluation:** In a goal-based evaluation, you measure what took place and compare it to the original intentions. Did the project reach its goals? For example, your goal may be to "establish a tuberculosis awareness program in the southeastern quadrant of the city that reaches 500 individuals during its first year." Determining whether a program was established is simple; figuring out how many people the program actually reached is a little more difficult. This answer depends, of course, on what method the project is using to reach people. In other words, you must define what you mean by "reach" before you start measuring the program's results.
- **Impact or outcomes evaluation:** In an impact evaluation, you ask whether you met your objectives and outcomes. Did the project achieve the desired results and make a meaningful difference? Although it requires time and attention, this type of evaluation is relatively straightforward. For example, if you oversee a tuberculosis awareness program, a desired outcome may be increased numbers of people being tested for TB and, if they have contracted it, taking their medication consistently to improve their health and reduce the likelihood they will infect others. Evaluating such an outcome requires an in-depth study of the population in that section of the city to determine whether they acted on the knowledge they gained and — if they were infected — continued to follow a strict regimen for taking their medication. Public health department data about rates of tuberculosis infections can be tracked yearly for the neighborhoods served.

  One of the hardest and most important things to ascertain in an outcomes evaluation is whether your organization has set the most meaningful objectives and outcomes.

Planning for evaluation

Before you jump into launching a new organization or program, we strongly recommend that you decide how you're going to evaluate it. That way, from the beginning you can set up methods for collecting information, identify who will be responsible, and budget for the costs. By establishing an evaluation plan at a program's inception, you can collect initial information about the people or places you plan to serve so you have baseline data to which you can compare your results.

SHAPING AN OUTCOMES EVALUATION

As an example of shaping an outcomes evaluation, let's say your organization provides shelter to people who don't have homes. You may have identified the following aspirations:

> **Objective 1:** Provide temporary housing to an average of 40 people per night over the course of a year.
>
> **Objective 2:** Twenty people per month complete counseling to identify barriers to their finding and maintaining secure housing.
>
> **Objective 3:** Fifteen people per month enroll with local and federal subsidized housing programs.
>
> **Outcome 1:** Over the course of one year, 25 percent of the people who took advantage of nightly shelter beds, counseling, and connections to subsidy programs secure long-term, affordable housing.

If this were your organization and your evaluation to devise, you could measure whether you met the first three objectives by keeping intake records for people staying at your organization and taking advantage of its counseling and referral services. It would be harder to measure the outcome, but possible if you had good working relationships with the local and federal housing program staffs or if part of your organization's work involved sending out case workers to check on former clients to see how they're doing.

If you want to consider whether you've established the right objectives and outcomes or whether there may be a better way to serve your constituents, you can do the following:

- Survey your constituents to find out how they lost their housing and ask what they think they most need to maintain a stable place to live.
- Talk to colleagues (perhaps partners at the local and federal programs with which you're working) about gaps in services available to people who lack stable housing.
- Invest in research into your region's job and housing markets, and investigate the types of leases or mortgages that were common to people who lost their apartments or homes. You may find that your best goal is to address some of the root causes of your constituents' situations.

A Logic Model provides a visual outline of your program or project's resources, goals, and intended outcomes. Your logic model can guide you in understanding what you need to know about your nonprofit's work and its results. We provide a sample logic model in File 8-3 at `www.dummies.com/go/nonprofitkitfd5e`.

REMEMBER

You want to gather baseline information about the people you serve. Trying to backtrack later to re-create such information is profoundly difficult.

As you set up your evaluation, first identify the purpose and audience. Are you trying a risky new approach to offering a program? Do you need to fulfill requirements of a government contract? Are you expected by a foundation to share your results with peer organizations doing similar work? Are there standard practices or protocols in your field that your evaluation should follow? Are you hoping to publish your results to be read by a general audience? Or, do you just want to share what you learn with your board and staff to make program adjustments and strengthen your work?

Before you design elaborate surveys and interview plans, ask yourself what information your nonprofit already has at hand. It may include attendance records, referrals from other nonprofits, unique visitors to your website, instructors' progress reports, numbers of trees planted, and many other data points that can illustrate your program's progress and reputation.

REMEMBER

As you're evaluating your work, be sure to share and discuss what you discover with your board and staff. Strong nonprofit organizations remain self-critical, learn from their work, and press themselves to do better.

Does your evaluation need to be completed on a particular timeline? Whether that timeline is short or long can affect whether you undertake a simpler or more elaborate evaluation plan.

Also critical are the resources you have at hand to invest in your evaluation. "Resources" may include money to pay for consultants, invest in focus groups, or purchase data sets. The term also includes staff and volunteer time. We've heard estimates that an organization's taking on a rigorous evaluation will require 15 percent of the executive director's time. We can't swear that is an accurate estimate, but it's a valuable reminder that staff oversight and leadership are needed even if you hire experts to manage the work.

The W.K. Kellogg Foundation offers a primer on evaluation for its grantees. We provide links to it in File 8-1 at `www.dummies.com/go/nonprofitkitfd5e`.

Crafting valuable questions

The most important (and often the hardest) step to take in evaluating your nonprofit organization's work is to ask the right questions. What do you need to know to understand whether you've been successful?

If your organization has written a long-range plan, an obvious place to begin is by looking at the measurable objectives defined in that plan and asking how you will know whether you've met them. You may want to push yourself harder and ask whether you met your outcomes, which describe longer-term, significant change you intend to achieve.

Some other topics nonprofits often ask about when evaluating their work are:

- **Cost-benefit:** What does it cost per person to provide the service? Is that more or less than an industry standard? If you have multiple programs, does the one that is being studied cost more or less per person than others?
- **Sustainability:** Have you been able to raise the funds needed to continue the program? Have you been able to retain qualified staff?
- **Staff and board reflection:** What are you learning from your work? What have you done well, and what has surprised you?
- **Participation:** Are you attracting the people you intended to serve?
- **Client satisfaction:** What do your constituents say about the program? Do they refer others to it? Do they believe their suggestions and criticisms are taken seriously?
- **Volunteer engagement:** Are you attracting and retaining volunteers? How do they rate their satisfaction with their participation? Do they refer others to your volunteer opportunities?
- **Best practices:** If you work in a field in which studies have identified the traits of effective programs, does your program have those traits?
- **Context:** Has the context in which you offer your services changed? How has that affected the people you serve? Have their needs changed?
- **Critical feedback:** What do experts (theater critics, researchers, and so on) have to say about your work and how it compares to the work of others?
- **Model program:** Are you recognized by others for the quality of the program? Are you called upon to share your approach and advise others?

TIP

Don't be afraid to ask yourself, "What needs to happen for our nonprofit to no longer be needed?" Maybe it's that streams and creeks are running free, a healthy marsh system is in place, and your town's low-lying areas are no longer flooding. This answer may seem far away from what you can achieve in three or five years, but it's valuable to ask the question and remind yourself that your nonprofit wasn't created to operate forever, but rather to work toward achieving a meaningful outcome.

Choosing evaluators: Inside or outside?

One key decision you must make about your program evaluation is who will be responsible for conducting it. The work may involve gathering information from a number of staff members and clients and, after that information is compiled, analyzing what it means. You need one or more people to:

- Design the evaluation plan and clarify the questions to be asked
- Create the evaluation tools (surveys, interview protocols, and so on) if needed
- Collect the information to be analyzed (or train others to collect it)
- Organize and analyze the findings and write a report

If you're undertaking a simple process evaluation, a well-organized and fair-minded staff member can be put in charge. Who else can better understand the purpose of the organization and the nuances of what it's trying to do than someone who is directly involved in the work? To institute some checks and balances in the process, one person may be responsible for compiling the information, and then a team of staff or board members may discuss its interpretation before a report is written.

"Inside" evaluation offers several advantages. It's usually less costly than hiring an outsider, and the insider doesn't need to be briefed on the work of the nonprofit. On the other hand, the selected person may not be trained in evaluation, and some staff members may resent having their work judged by a fellow staff member. It may also be difficult for those who are steeped in the day-to-day work of a nonprofit to step back and cast a critical eye over it. Their closeness to the work may bias the findings.

REMEMBER

A person who works inside of your organization may do a good job of reporting on whether programs are meeting their goals. It can be harder for that person to recommend other approaches that may be more effective.

If you're undertaking a more complex evaluation, you may want to hire a professional evaluator who knows your field and who writes well. In hiring such a person, you gain the advantages of his or her expertise in research methods, knowledge of the work of comparable organizations, and lack of bias in interpreting the information collected. An assessment made by an independent evaluator may be taken more seriously by your funders and peers. Of course, we can't claim that anyone who is paid a fee by your nonprofit to work for you is 100 percent unbiased: He or she may hope to be hired again in the future.

What if you have a modest budget and still want good consulting practices to be worked into your evaluation? Three approaches we've used in the past have been to:

- Contact local colleges and universities to see if a graduate student or faculty member is interested in evaluating your work for academic credit, a dissertation, or a publication.
- Hire an evaluation firm to design your tools and protocols, train your staff to implement them, and then write an independent analysis of your findings after your staff has compiled them. Your staff will do the more time-consuming part of the project, but the professional adds design expertise and an outside perspective.
- Compile information to be evaluated internally, share it with an evaluator, and hire him or her to analyze and write about the findings.

Conducting Your Evaluation

After you have thought through the planning steps, refined the questions to be asked, chosen your tools for measuring results, and put the right person in charge of your evaluation, doing the work to study your program or nonprofit is relatively easy. You'll want to decide on a time frame (either one discrete period or a defined number of days or hours per month) for conducting your study and stick to it. If someone on your staff is responsible for the evaluation, be careful that the individual isn't pulled away by other tasks.

If you are using a survey, exam, or checklist for making observations, you'll want to make sure it's designed by someone who knows how to word and structure such a tool. (We write more about surveys in Chapter 12.) How that tool is used is just as important as its design: If several staff members or volunteers are helping to collect data, be sure to brief them as a group before they begin so everyone is recording the information in the same manner. Of course, you'll want to check for and insist on accuracy and thoroughness. Data is meaningless if it's gathered inconsistently.

Analyzing Results and Putting Them to Work

Imagine that you've collected and organized the information that will tell you what you want to know about your organization. It's at hand in lists and charts and graphs. Now you know whether your programs are outstanding and whether your clients' lives are enhanced by your work.

Wait! Not so fast. Just as you must decide on the right questions to investigate as you set out on your evaluation, you need to ask questions of the data you've compiled. Having data and descriptive information isn't the same thing as having knowledge and understanding. Now you must interpret the information and decide how to apply it.

TIP

Evaluation software can help you organize and review your findings. The website Idealware reviews nonprofit program evaluation software. You can find a link in File 8-1 at `www.dummies.com/go/nonprofitkitfd5e`.

Interpreting results

Although it may seem unscientific, we suggest that before you look closely at the information compiled, you ask yourself what your hunches are. Be honest with yourself: What do you think the findings are going to show, and what do you wish they would show?

Often a moment of insight comes when the pattern revealed in the data isn't what you expected to find. As you identify any such surprises, you can probe more deeply to ask why and how those results didn't meet expectations.

After you've read the data against your assumptions, it's time to look at it anew as if you had no expectations. What general picture does it present? What stands out? Is the information consistent over the course of the year, or does it vary — perhaps according to the season, age of participants, or changes in your staff?

We recommend that you include others in this exercise — ask a few key staff and board members and maybe even a few of your constituents. What do they see? What is surprising to them? Of course, if you hired an evaluation consultant, let his or her experience guide you at this phase.

REMEMBER

Although you don't want to confuse people with ambivalent or "anything goes" interpretations, seeking more than one point of view can be valuable. Bring a balanced perspective to the findings: Don't embrace all the praise and discount all the criticism (or vice versa). You can learn from both.

Using your evaluation to strengthen your work

You've looked at your evaluation findings with trusted colleagues; you've turned them inside out and upside down to see them from every angle. Now it's time to interpret them, to tell their story. Likely what you've discovered is not all positive and not all negative. Perhaps you've noticed that some services are rated more highly than others and some short-term objectives have been met, but others remain elusive. It's time to step back from your organization and programs, and acknowledge any disappointing results and set short-term and long-term goals to improve upon them.

A program may be flawed in its design, in how it's executed, or in who is leading it; and its effectiveness can be smothered by external circumstances that are beyond its control — a shift in the economy or demographics, a change in public policy, or even a natural disaster can derail it.

ASKING MORE OF THE DATA

Any finding, be it positive or negative, is worth probing more deeply. Let's take the case of an education policy and advocacy group as an example. The group invested in a communications manager and an improved website and data systems to expand its reach. When it wanted parents and other community members to know about legislation under review in the state senate or assembly, it sent alerts by email and through social media posts.

Because the group was using the web and social media as its programmatic tools, it made sense for the group to look at its online analytics, many of which looked good:

- Its email list was growing steadily, and the click-through rate to its website was higher than the industry average.
- The number of unique visitors to its website grew from month to month, and, on average, visitors read three pages when they went there.
- The number of people who "liked" the group on Facebook grew steadily, and many of them shared its posts with others.
- A small but gradually expanding number of people followed its tweets.

These are good signs of a growing online presence, but it's still worth asking whether the people who like the group on Facebook also take action and contact their representatives or talk to their school boards about the subjects raised. Social media friends and followers are a good thing, but do they include people with influence in local or state government or education policy?

As is true in many cases, the depth and quality of involvement of people in this organization were more important to its success than a total number of passive fans.

Some changes will be obvious and easy to do. Maybe a different program schedule would be more convenient for working parents, or better training for volunteers would make them feel more involved and successful. When your findings don't suggest easy responses, turn to ideas from others — best practices in your field, model programs, or scholarly research. Don't forget to ask the clients you serve.

Evaluation findings also should be shared with your board, which is ultimately responsible for your nonprofit's fulfilling its mission and purpose. It may decide that some programs aren't a good investment of the nonprofit's hard-won resources and recommend cutting them.

Telling the truth

If your evaluation results are disappointing, face them honestly and share what you've discovered. You may share them discreetly with your board or with the funders who know you well, but doctoring your data or analyzing the findings in a disingenuous way can hurt the reputation of your organization and of your evaluation consultant, if you hired one. And all that time and effort you put into your evaluation is lost if you allow it to sit on a shelf collecting dust.

TIP

These days the public deeply values transparency. Unless your report contains confidential information, we recommend that you create an executive summary of your findings and publish them as a brochure or PDF available through your website or shared with donors.

You may find that a poor evaluation lowers staff morale, but if those staff members are invited to participate in creative problem solving, improving customer service, and other goal-setting activities, working with the evaluation can contribute to building their teamwork and resolve.

Ultimately, it's important that you yourself embrace and attempt to understand the evaluation findings. Many of us start nonprofits based on our ideals, and many of us work long hours with limited resources to achieve very ambitious results. Achieving those results depends on collecting good information and interpreting that information with integrity and resolve.

At the end of the day, the purpose of your evaluation is to benefit your nonprofit organization and the constituents it serves.

» Setting up your volunteer program

» Finding volunteers

» Making sure a volunteer is right for your nonprofit

» Helping volunteers do their work

» Expressing your thanks

Chapter 9

Working with Volunteers

Volunteers are the lifeblood of the nonprofit sector. The Bureau of Labor Statistics reports that nearly 25 percent of the U.S. population over age 16 performed some kind of volunteer work during the 12-month period ending in September 2015. That's a lot of people — 62.6 million!

Just about every nonprofit charitable organization uses volunteers in some capacity. For example, in most cases, board members serve without compensation. And for many nonprofit organizations in the United States, volunteers do all the work, from planting the trees to paying the bills. Even if your organization employs paid staff, volunteers still provide valuable service. Organizations depend on volunteers to staff telephone hotlines, lead scout troops, provide tutoring, coach youth sports teams, serve hot meals, organize fundraising events, and stuff envelopes. So if you're going to manage a nonprofit organization, you need to know how to work with volunteers.

Of course, you may be sitting in your office asking, "Where are these millions of volunteers?" This chapter offers suggestions to help you decide what kind of volunteers you need, figure out how to find them, and keep them happy after they arrive.

TIP

Check out File 9-1 at `www.dummies.com/go/nonprofitkitfd5e` for a list of web resources related to the topics we cover in this chapter.

Knowing Why People Volunteer

The classic stereotype of a volunteer is someone who has lots of time to spare and is looking for something to do. Although this perception may have been true in the past, when many women stayed out of the workforce and gave their free time to charity, that stereotype no longer fits. It's true that more women volunteer than men, but people between the ages of 35 and 54 are the likeliest to volunteer. This is the age range when both men and women are probably balancing careers with raising families, not to mention taking care of aging parents, going to the gym, and keeping up with email and social media.

Why is it that people, even very busy ones, volunteer their time? We think it's because they've recognized the benefits of volunteering time to a favorite organization and because nonprofit organizations have become smarter about asking them.

REMEMBER

Understanding why people volunteer makes it easier for you to find volunteers, organize their work, and recognize their contributions. However, not everyone is motivated by the same factors. People volunteer for a variety of reasons, including their desire to

- **Help the community and others.** Helping others usually comes to mind first when people think of the reasons that people volunteer. But as you see when you read deeper into this list, volunteers' motives aren't always this simple.
- **Express their values.** Believing in an organization's mission is an important motivator for volunteers. If a person believes it's important for all children to have access to a quality education, she will be more likely to volunteer at her local school or tutoring program.
- **Increase self-esteem.** Volunteering makes people feel better about themselves. Giving a few hours a week, or even a month, to an organization creates good feelings.
- **Help out their friends.** Friends are often the first people we turn to when we need help. Volunteering is also a great way to get together with friends on a regular basis.
- **Make new friends.** Volunteering is usually a social activity. People often use this opportunity to meet interesting people who share their interests and values.
- **Try out a job.** People considering a job in the nonprofit sector often discover that volunteering is a good way to get a peek at what happens on the inside.
- **Polish their résumés.** Adding volunteer experience to a résumé shows a commitment to helping others or to working in a particular field.

- **Develop new skills.** A volunteer job often gives people an opportunity to learn how to do something they didn't already know how to do.
- **Enjoy something they love.** Many volunteer jobs come with intrinsic benefits for their participants. Ushers at the symphony get to hear the music. Gardeners removing invasive plants from a native plant preserve get to spend a day in a beautiful natural setting.

Keep this list in mind, and you'll realize that you don't have to focus your recruitment efforts exclusively on retired people or others who have a lot of leisure time. If you provide an environment in which volunteers can bring their friends, meet others who share their interests, and learn new skills, you can attract even the busiest people to volunteer work. Remember that you have no reason to be apologetic about asking for help: Volunteering benefits the people who step forward to assist you.

Designing a Volunteer Program

Most startup organizations depend on volunteers because money to pay staff is unavailable. But lack of resources isn't the only thing that drives a nonprofit to operate with an all-volunteer staff. Some nonprofits make a deliberate decision to operate solely with volunteers to contain their costs and to achieve results with a collective effort among people who care deeply enough to contribute their time and energy.

Considering a volunteer coordinator

Although volunteers don't expect to be paid every two weeks, that doesn't mean they come without costs. Recruiting, training, managing, retaining, and thanking volunteers require effort from someone in the organization. We recommend assigning someone the job of *volunteer coordinator*, a person responsible for overseeing or performing the following duties:

- **Recruiting:** Volunteers don't grow on trees. Depending on how many volunteers you need and the turnover rate of current volunteers, recruiting may be a continuous process.
- **Training:** Volunteers don't come to work knowing everything they need to know. They can do any job for which they're qualified, but don't expect them to know the ropes until they're told what to do and how to do it.
- **Scheduling:** Volunteers need a schedule. Scheduling is even more important if your organization uses volunteers to staff an office or manage other tasks that require regular hours.

» **Appreciating:** Volunteers need to know that their work is valuable to the organization. This item is the last on our list, but it may be the most important. You don't have to pass out plaques, but we do recommend heartfelt acknowledgment. Saying thank you and acknowledging the impact made by volunteers on a regular basis is essential to retaining your volunteers.

If your organization depends heavily on or is staffed exclusively by volunteers, consider recruiting your volunteer coordinator from among board members or more-experienced volunteers. You can create committees to take responsibility for many jobs, but some detail-oriented tasks — such as scheduling or bill paying — are better managed by a single responsible person.

Determining your need for volunteers

Look around your nonprofit organization and decide how many volunteers you need and what functions they can perform. We recommend creating (or helping your volunteer coordinator create) a schedule of tasks to be completed — planning what needs to be done and how many people it will take to do the work. Table 9-1 lists the kinds of volunteer assignments you may jot down. By having such a list and prioritizing the tasks, you know what to do when an unexpected volunteer walks in the door.

TABLE 9-1 Sample Volunteer Task List

Task	Number of People	Time
Data entry — donor list	1 person	3 hours per week
Social media posting and monitoring	2 people	5 hours per month
Answering the telephone	8 people	36 hours per week
Childcare	2 people	3 hours on Saturdays
Filing	1 person	2 hours per week

It's possible to have too many volunteers. Almost nothing is worse than asking people to help and then finding you have nothing for them to do. You may want to have both your chart of immediate tasks (such as Table 9-1) and a few back-burner projects — such as sorting team uniforms by size or taking inventory in the supply cabinet — in case you end up with more people than you need on a given day.

In the beginning, you may have to experiment before you know exactly how many volunteers you need for a particular job. For example, you may eventually discover that a 2,000-piece mailing takes about five hours for four people to complete. You also may find that preparing the soil and planting 200 seedlings takes two volunteers a full day.

Writing volunteer job descriptions

Volunteers perform better if they know what they're supposed to do. Preparing job descriptions for volunteer positions also helps you supervise better and know what skills you're looking for in volunteers. (Take a look at Chapter 10 for detailed information on writing a job description.)

Volunteer job descriptions should be even more complete than paid-employee job descriptions. If you can break jobs into small tasks, all the better, because volunteers often share the same job. For example, a different person may answer the office telephone each day of the week. In that case, to bring consistency to the job, you should keep by the telephone a job description that includes a list of telephone procedures, staff extensions, frequently used telephone numbers, and other important information.

TIP

Check out Files 9-2 and 9-3 at `www.dummies.com/go/nonprofitkitfd5e` for sample volunteer job descriptions.

Organizing volunteers

Many nonprofits invite their volunteers to join a committee. Committees enable volunteers to step forward, offer their best skills, and learn how to do new things. An advantage of forming committees is that it reinforces the social benefits of volunteering. As committee members get to know one another and figure out how to manage their tasks successfully, you or your volunteer coordinator can step back and let them take full responsibility.

Here's a fictitious example that gives you an idea of how to organize committees: The Healthy Diet Project provides telephone referral and information sources for people seeking help with weight loss. It was started by three people who had lost weight and decided to help others do the same. A ten-member board of directors provides governance for the Healthy Diet Project and assumes key volunteer roles in the organization. The nonprofit has five committees, which are each chaired by a board member but made up of individuals who provide volunteer services:

- **Telephone committee:** The Healthy Diet Project provides most of its services via telephone. The office receives about 100 calls each day from people seeking information about weight loss and referrals to health clinics and counselors. The telephones are answered 12 hours a day, from 9 a.m. to 9 p.m., Monday through Friday. Two volunteers share responsibility for the phones in three-hour shifts. The nonprofit needs 40 volunteers each week to answer phones and provide information. The Healthy Diet Project also needs backup volunteers in case someone is ill or can't make his shift for some other reason.

- **Program committee:** This committee researches programs to which callers can be referred, maintains the database containing referral information, and provides training to telephone volunteers. Committee members include one physician and two registered nurses, who provide professional oversight.
- **Publicity committee:** The Healthy Diet Project uses several methods to tell the public that its services are available. The publicity committee prepares and sends news releases and creates and distributes public service announcements to radio stations. In addition, the committee operates a speakers' bureau of people who have benefited from the Healthy Diet Project's services. The committee has also developed a website, a Twitter feed connected to any news that is posted on the website, and a Facebook page; each of these outlets offers basic information about weight loss and invites readers to sign up for a monthly email newsletter. The website maintains links to recommended programs in cities across the United States and Canada.
- **Fundraising committee:** The Healthy Diet Project raises funds in several different ways, including annual family-oriented walk-a-thons, email appeal letters sent to people who have joined its contact list through its website, and gifts from businesses promoting a healthy lifestyle. Committee members plan and coordinate the fundraising events, write and transmit the appeal letters, and make personal calls on the business sponsors. They also call upon all volunteers to make personal gifts, identify possible donors, and provide lists of contacts.
- **Administration committee:** The Healthy Diet Project receives individual donations from people who use its services, grants from foundations, and limited support from the health department in the city in which it's based. The administration committee is responsible for maintaining the organization's financial books, writing thank-you letters to donors, and maintaining a database of past donors.

REMEMBER

You may discover other tasks that can be assigned to additional volunteer committees. The kinds of jobs that need to be done vary, depending on the type of service your organization provides. The point to remember is that volunteer work needs to be organized (and supervised) in much the same way as paid work.

In an all-volunteer organization, the responsibility for ensuring that the work is done in a timely and effective manner resides with the board of directors. The board must be committed to finding new volunteers and supervising their work. And board members must be ready to step in to do a job if no volunteers can be found.

REMEMBER

Board members who also serve as program volunteers must remember to keep their roles as board members (governance and fiduciary) separate from their roles as program volunteers. In the latter case, the volunteers are operating like staff, not board members. Yes, there is a difference. See Chapter 6 for information about understanding and defining board members' roles.

Hunting for Volunteers

Most organizations are always on the lookout for volunteers. After all, volunteers move away, get tired, lose interest, or take new jobs with new hours. If your organization depends on volunteers, you probably need to maintain an ongoing recruitment process. This section shows you how.

Using the tried-and-true methods

To cast a wide net, you want to use more than one method to find volunteers, and you don't want to spend much money on those methods. After all, you're looking for free help. Persistence matters: Good volunteer recruitment is like a healthy habit that you want to repeat. Here are a few of the most common and most successful methods for recruiting volunteers:

- **Placing announcements in the media:** Newspapers and radio and television stations sometimes publish or air short public service announcements for nonprofit organizations. (Head to Chapter 12 for more information about writing and distributing news releases.)
- **Posting fliers:** Grocery stores, churches, coffee shops, college campuses, laundromats, schools, and civic buildings often have bulletin boards where you can post announcements. For best results, place them thoughtfully. For instance, put your call for foster homes for kittens at the pet food store and your community garden poster at the plant nursery.
- **Taking advantage of word of mouth:** Encourage your current volunteers to recruit others. Have a "bring a friend" day with time for socializing. Ask your volunteers to post your posters in their places of business, and don't forget to invite your own friends and associates.

- **Contacting schools and churches:** Both of these institutions look for ways for students and members to get involved in community service. *Service learning* — through which students learn about a topic by volunteering in their communities — is a growing practice. In fact, many high schools and colleges maintain centers for community relations and student volunteering.

  REMEMBER

  Don't forget to reach out to the young people in your area. By doing so, you benefit from their skills and ideas, and you also contribute to training the next generation of volunteers!

- **Relying on clubs and fraternal groups:** Many professional and social clubs include serving the community in their missions. From Kiwanis International and local Elks lodges to the Junior League, chamber of commerce, and campus-based sororities and fraternities, clubs and membership groups can be excellent volunteer resources.
- **Approaching corporations and businesses:** In some communities, businesses look for community involvement opportunities for their employees. If a company has a community relations, community affairs, or corporate giving department, it's likely to be a good place to begin asking about employee volunteers.

Going online

Finding volunteers may be as easy as booting up your computer. Besides posting volunteer listings on your organization's website, you also can post on other sites. Several organizations maintain databases where your organization can list its volunteer needs. Prospective volunteers can search the databases by ZIP code and the type of volunteer work available. Check out the following sites:

- **VolunteerMatch** (`www.volunteermatch.org`) invites nonprofits to set up accounts identifying the kinds of volunteers they need.
- **Points of Light** (`www.pointsoflight.org`) manages several projects linking volunteer centers to one another. Its HandsOn Network links volunteers at 250 centers in 30 countries to meaningful projects. It also operates networks for youth volunteers, alums of AmeriCorps who want to continue volunteering, and corporate volunteers.
- **Idealist** (`www.idealist.org`) and **Craigslist** (`www.craigslist.org`) are other widely used tools for finding volunteers.
- **Service Leader** (`www.serviceleader.org/virtual`) maintains the Virtual Volunteering Project, which can connect your organization with volunteers

who aren't physically located near your office. Maybe you can use online volunteers to tutor students, answer questions via email, help write grants or news releases, or consult on designing your website.

- **United Way** (www.unitedway.org) may have a volunteer tutoring or mentoring program in your area.
- **Youth Service America** (www.ysa.org) helps you involve young volunteers in your work. It organizes an annual Global Youth Service Day and other opportunities for volunteers between the ages of 5 and 25.

Nonprofit organizations also recruit volunteers through their pages on social networking sites, such as Facebook, Twitter, and LinkedIn. To get your name out there, create a page for your organization. These platforms work particularly well when you describe specific activities that your contacts can do to assist your nonprofit, and when people who know your organization well post photographs and news for their friends to read. Their networks extend your nonprofit's reach. Remember, opportunities to do things with friends motivate many people to volunteer. Chapter 12 provides more information about using social media to make friends and influence people.

USING THE WEB TO RECRUIT VOLUNTEERS

Take tips from other nonprofits about ways to reach out to and invite volunteers through your website. Remember, you can do much more than describe your nonprofit on your website or social media pages: You can even put people to work for you who don't step inside your nonprofit's doors. Among many fine examples, four sites that inspire involvement are:

- **DoSomething.org** (www.dosomething.org): This social networking site invites people to identify causes they care about and then volunteer. Its use of strong visual imagery, surveys, and clearly described activities is a good model for a social media site that succeeds at engaging volunteers.
- **The Humane Society of the United States** (www.humanesociety.org): The home page offers clear, prominent instructions on actions you can take to advance the well-being of animals.
- **The Surfrider Foundation** (www.surfrider.org): This site has a "Take Action" menu, which invites visitors to advocate for legislation and activities that protect shores and the ocean.
- **Water.org** (www.water.org): This site makes excellent use of photographs, graphics that illustrate the effectiveness of its work, and a "Get Involved" menu.

Looking for volunteers at other organizations

No, we don't suggest that you steal volunteers from other nonprofits, but some organizations do exist to provide volunteer help. Many communities have volunteer centers that participate in a national network of organizations that recruit and place volunteers in nonprofits.

Similarly, the Corporation for National and Community Service (`www.nationalservice.gov`) was established by Congress in 1993 to operate AmeriCorps, Senior Corps, the Social Innovation Fund, and the Volunteer Generation Fund. This agency is charged with encouraging national service through volunteering and helping nonprofits and public agencies make the best use of this resource. The corporation awards grants that help organizations strengthen their community through the use of volunteers.

REMEMBER

If your organization wants to take advantage of a program offered by the Corporation for National and Community Service, you may need to apply for a federal or state grant, which usually requires sound accounting procedures and extensive reporting of program activities. Turn to Chapters 16 and 17 for information about government grants.

Finding volunteers with special skills

If you're looking for volunteers with special training or experience, spend some time thinking about where you can find them. Limit your recruitment efforts to places where you're most likely to identify the people with the talents you need.

Suppose that your organization is seeking someone with accounting experience to help maintain your books. Local accounting firms, corporate offices, and professional accounting societies may be good recruiting grounds for someone who can assist with bookkeeping. If your organization needs help with a legal matter, some bar associations link nonprofits to attorneys who are willing to volunteer.

TIP

Sometimes your nonprofit competition won't mind sharing contacts who are loyal to a particular kind of volunteer activity. In our city, for example, several organizations put on film festivals once each year, and volunteers who love to see the films volunteer to sell tickets, greet the media, and manage other tasks, migrating from one festival to the next. Rather than detracting from their contributions to one organization, their "migrant volunteering" makes these volunteers more knowledgeable and strengthens their skills. When recruiting for short-term volunteer assignments, don't be afraid to ask your "competition" if you can invite some of their great bird-watchers, marathon runners, brownie bakers, or scarecrow makers.

Hiring interns

Interns are specialized volunteers who come to you as part of an education or training program. In most cases, a student intern's goal is to develop practical, hands-on work experience.

Sometimes internship programs require your organization to pay a fee or provide the intern with a modest stipend. If you do pay a stipend, be aware that you may be creating an employer-employee relationship that is subject to federal, state, and local laws, including minimum wage requirements, employment taxes, and other obligations. If you're working with an established internship program in your community, it's likely that these potential liabilities will be covered. If you're recruiting interns on you own, refer to Fact Sheet #71: Internship Programs Under the Fair Labor Standards Act, available on the Department of Labor website (`www.dol.gov/whd/regs/compliance/whdfs71.htm`). You may also want to consider including interns on your liability insurance policy. (Refer to the later section "Insuring your volunteers" for more details.)

As with employees and volunteers, you should provide the intern with clear expectations about duties, attendance, and so on. If you decide to go this route, be ready to spend time supervising and evaluating the intern's job performance. Don't forget that the intern's experience is part of his grade.

Interviewing and Screening Volunteers

Require potential volunteers to fill out job applications just as if they were applying for paid work. Ask for references and check them. Review résumés and conduct formal interviews. (See Chapter 10 for information about job interviews.) Avoid paranoia, but don't discount your gut feelings, either.

If you're using volunteers in professional roles, such as accounting, check their qualifications just as you would check the qualifications of an applicant for a paid position. It's possible that this process may offend potential volunteers, but it's far better to make sure that the person can do the job, even if she's doing it for free.

If you're placing volunteers in sensitive jobs, such as working with children or providing peer counseling, screen your applicants carefully. Criminal background screening, including a fingerprint check, is sometimes required by law, by licensing requirements, or by your insurance carrier. Some states and counties also require a test for tuberculosis. Check with local authorities about the requirements in your area.

A CHEAT SHEET FOR THE FAQs

Be prepared to answer questions when people call to volunteer. If you're already using volunteers to answer the telephone, prepare a list of common questions and answers and place that list near the telephone. Here are some sample FAQs that the Healthy Diet Project, a fictional nonprofit discussed in the section "Organizing volunteers," may need:

What are the hours I would be needed? We answer the phones five days a week from 9 in the morning until 9 at night. We ask people to work a three-hour shift once a week.

How will I know what to say? All volunteers receive one day of training. Training is offered once a month, almost always on Saturdays.

What kind of advice can I give? Our volunteers can't give medical advice or advice on specific diets. Volunteers refer callers to existing services and professionals. We ask volunteers to be positive and to offer general support to all callers.

How do I know where to refer people? We have an extensive database of weight loss-related counseling services. It's a simple matter of looking through a loose-leaf notebook or our computer database to find the appropriate phone numbers.

What if I get sick and can't cover my shift? We have volunteers on standby to cover unexpected absences. If you aren't able to volunteer on a regular, weekly basis, you may consider being a backup volunteer.

Are we asked to do any other work? Sometimes we ask volunteers to help with mailings between phone calls.

Will you pay my auto (or public transportation) expenses? We're sorry, but our budget doesn't cover reimbursing volunteers for expenses. Some expenses may be deductible on your income tax, however. You should check with your tax specialist.

Can I deduct the value of my time from my income taxes? No, the IRS doesn't allow tax deductions for volunteer time.

These questions and answers also can be printed in a brochure and mailed to potential volunteers who request more information. Be sure to include background information about your organization in the mailing.

Check out File 9-4 at `www.dummies.com/go/nonprofitkitfd5e` for a volunteer intake form.

We realize that screening can be a delicate issue. You're walking a tightrope between the right to privacy and the right of the organization to be sure that no harm befalls its clients. The failure to undertake a background check potentially could result in liability problems for the individual who "hires" the volunteer.

Some potential volunteers may be offended by background checks. Explain that the procedures aren't directed at them personally but are in place to ensure that clients are protected. Also, treat all volunteer applicants the same. In other words, don't pass up screening someone just because she's a personal friend.

Managing Your Volunteers

Just like managing paid employees, working with volunteers requires attention to management tasks. Volunteers need training and orientation as well as clear, written lists of responsibilities and expectations. Basic expectations for volunteers are easily outlined in a *volunteer agreement form.* You also want to maintain records of the time and tasks volunteers contribute to your organization and consider whether to include volunteers in your insurance coverage.

Check out File 9-5 at www.dummies.com/go/nonprofitkitfd5e for a sample volunteer agreement form.

Providing adequate training

The degree and extent of volunteer training depends on the type of job you're asking them to do. Volunteers who answer telephones, for example, may need more training than those who stuff envelopes for the publicity committee. To successfully answer phones, these volunteers need to know background information about the program or service, information about the types of services available, proper telephone etiquette, and emergency procedures, among other details.

If you need to provide a full day's training or training over a longer period, we suggest consulting with a professional trainer to either provide the training or help you design the curriculum. Although you may be concerned about investing too much of volunteers' valuable time in training, remember that key motivations for volunteering include meeting people and enjoying time with friends. Trainings can be great opportunities to introduce volunteers to one another and build camaraderie among them. It's a good idea to schedule refresher training sessions for ongoing volunteers, too.

In addition to offering on-site training, give volunteers written materials that restate the information covered in the training. Include with these materials attendance requirements, details about whom to contact in case of illness, and other necessary information that volunteers may need to know when carrying out their tasks.

Larger organizations that use many volunteers sometimes publish a *volunteer handbook.* Such a handbook doesn't need to be an elaborately printed document; it can be several typed pages stapled together, a simple loose-leaf notebook, or a PDF posted on the organization's website. The more information you provide, the better your volunteers can perform.

Keeping good records

Keep records of your volunteers and how much time they spend doing work for your organization. You may be asked to provide a reference for a volunteer who's working to develop job skills or providing a service through an organized volunteer program. You also may need to dismiss a volunteer who's unreliable, and having clear, written records of hours and tasks can justify that difficult act.

If you use professional volunteers to perform tasks that you'd otherwise have to pay for, you can include the value of the volunteer time as an in-kind contribution on your financial statement. Chapter 11 explains more about financial statements.

Insuring your volunteers

Typically, nonprofit organizations carry liability and property insurance. Almost all states require that workers' compensation insurance be in place to cover on-the-job injuries to employees (but not necessarily to volunteers). Beyond this basic insurance, coverage depends on the type of services provided and the degree of risk involved.

Keep in mind that volunteers usually aren't liable for their actions as long as they work within the scope of the volunteer activity to which they've been assigned, perform as any reasonable person would perform, and avoid engaging in criminal activity. Unfortunately, people these days have become more eager to file lawsuits. If someone sues you or one of your volunteers, you have to legally defend the case even if it's without merit. One advantage of having liability insurance is that your insurance carrier takes on the responsibility of defending the suit.

Workers' compensation may or may not be available to volunteers in your state. If you can include volunteers under your state law, consider doing so, because a workers' comp claim usually precludes the volunteer from filing a suit for damages against your organization. Plus, you want your volunteer to be covered if he suffers an injury.

Insuring volunteers is a subject of debate in the nonprofit sector. Some people take the position that insurance agents and brokers try to persuade you to insure anything and everything. Others believe that liability insurance and, in some cases, workers' compensation insurance should be provided. As is the case with all insurance questions, evaluate your risks and decide whether the cost of insuring against risks is a good investment. For example, if you fail to provide protection to a volunteer who is seriously injured, the reputation and future success of your organization could be harmed. This process is called *risk management*. To find information about risk management for nonprofit organizations, contact the Nonprofit Risk Management Center (`www.nonprofitrisk.org`).

Saying farewell to bad volunteers

If you work with lots of volunteers, especially volunteers who perform complex and sensitive jobs, you may discover one or more volunteers who don't have the skills or personalities to perform at an acceptable level. We hope you never face this situation, but if you do — for example, maybe someone is giving out inaccurate information or acting rudely — you shouldn't ignore the situation.

Discussing the problem behavior with the volunteer is the first step. Treat this meeting as if you were counseling a paid employee whose job performance was below par. Written job descriptions, written standards for performance, and records of volunteer time and contributed tasks are important when discussing problem behavior.

WARNING

Exercise caution when meeting with a volunteer about her unacceptable behavior, especially if you don't have clearly written performance guidelines. Volunteers who are released have been known to sue nonprofit agencies. If you have concerns about this possibility, consult an attorney before you do anything.

REMEMBER

Talking to someone, volunteer or not, about poor work is never pleasant. However, if someone working for your organization is being disruptive, giving out inaccurate information, or otherwise causing potential harm to your program or the people you serve, you have a responsibility to correct the problem.

Saying Thank You to Volunteers

Volunteers give their time and, in many cases, expertise to help your organization succeed. It's only right that you thank them and thank them often. Thank them in the hallway after they've completed their work for the day, and also formally recognize their contributions. Here are some standard ways of recognizing volunteers:

TIP

- **Annual recognition event:** This kind of event is the most formal (and probably the most expensive) way of thanking volunteers. Some organizations have a sit-down dinner or wine-and-cheese reception once a year to say thanks and give awards to volunteers who've made extraordinary contributions.

- **Gifts:** Although tokens of appreciation may be much deserved, we recommend caution when giving gifts to volunteers. Don't spend lots of money buying presents, because you can bet that some volunteers will ask why you're spending scarce nonprofit money on something that isn't necessary. Getting a local business to donate gift certificates or other items is a better way to go.
- **Admission to performances or events:** If your organization presents plays, musical performances, lectures, or readings, consider offering free admission to some events.
- **Public acknowledgment:** You can identify your volunteers in your newsletter or on your website. An alternative is an annual newspaper ad that lists the names of your volunteers.
- **Thank-you letters:** Don't underestimate the power of a simple thank-you note. Unless you have hundreds of volunteers, make sure you write your notes by hand. Most people appreciate a handwritten note more than a form letter or email. Try to let volunteers know how their work has made a difference.

In addition to thank-you letters and recognition events, you can increase volunteer satisfaction (and retention) by treating volunteers well on a day-to-day basis. Here are some easy tips to keep volunteer satisfaction high:

- **Don't make volunteers work in isolation if you can avoid it.** Many volunteers give their time because they enjoy socializing with others.
- **Vary the job to avoid boredom.** You may need help cleaning the storeroom or hand-addressing 1,000 envelopes, but try to assign jobs that offer more mental stimulation, as well.
- **Pay attention to the work done by your volunteers.** Your interest in what they're doing adds value to their work and recognizes that many of them are volunteering to develop new skills.
- **Help volunteers understand your nonprofit's work.** If they've been answering the telephones in the front office, give them a behind-the-scenes tour or a chance to observe or participate in other activities of the organization.
- **Bring in pizza or doughnuts once in a while as an impromptu thank you.** Food can provide a great break from a monotonous job or a celebration of a major task's completion.
- **Talk to your volunteers.** Get to know them as friends of your organization who are committed to its work.

IN THIS CHAPTER

» Preparing your organization for paid employees

» Taking care of the groundwork

» Interviewing and hiring

» Getting a new hire started

» Handling the day-to-day management

» Working with consultants

Chapter 10
Working with Paid Staff

Some nonprofits have paid staff from the very beginning. For example, nonprofits that start out with grant funding to operate a program may have paid staff. Other nonprofits may start more slowly, with the board of directors and other volunteers initially doing all the work and hiring paid employees later. And many nonprofits never have any paid staff. These organizations may use consultants or rely on volunteers.

No rules exist about when a nonprofit organization should start employing paid staff. The organization must determine whether it has enough work to justify employees and whether it has the resources to pay salaries and associated expenses. Hiring your first employee should be cause for celebration. It means that your nonprofit has reached a milestone in its development. But it also means that the organization (and the board of directors) will have more responsibilities to raise funds and ensure that proper personnel policies are put in place and followed. This chapter covers the details.

TIP

Check out File 10-1 at `www.dummies.com/go/nonprofitkitfd5e` for a list of web resources related to the topics we cover in this chapter.

Deciding That You Need Help

Knowing when to take the leap from being an all-volunteer group to being a boss or paid employee isn't easy, and it's not a leap to take without looking at where you're about to land. Hiring employees creates responsibilities for the board, not the least of which is making the payroll every two weeks or once a month. You also need to pay payroll taxes and provide a workplace, equipment, and — don't forget — guidance and supervision. Expect to have more bookkeeping duties and more complex financial reports because you need to keep track of payroll records, vacation time, and sick days and decide which holidays your organization will observe. (The last one should be a snap, right?)

To ease into the transition, a nonprofit may begin by hiring an independent contractor to handle a specific task, such as bookkeeping or grant writing, and then go from there. (We discuss working with independent contractors and consultants later in this chapter.)

REMEMBER

A variety of situations, such as the following, may signal that it's time to hire your first employee:

- Volunteers are growing tired, and the work isn't getting done as well or as quickly as it should.
- The demand for your organization's services has increased to the point that someone needs to focus consistently on administrative details.
- Resources have increased to the point that you can now pay a regular salary.
- The organization is starting a new activity that requires someone with a specific professional license or degree, and no volunteer is appropriately qualified.
- The nonprofit receives a major grant that both provides more resources and requires significant recordkeeping and program management.

Hiring salaried employees should be a long-term commitment. For this reason, you need to have sufficient cash flow to ensure regular payment of salaries, benefits, and payroll taxes.

TIP

A crisis can sometimes take place when a volunteer-run organization hires its first staff member. Knowing that they're now paying someone to be responsible, board members may decide to sit on their hands and let others do the work. Bringing on a first staff member is a good time to honor board and volunteer contributions to the organization (to keep motivation high) and to invest in a board retreat or training that reminds everyone of the work ahead and the board's important role in it.

WARNING

When you add paid employees to your organization, you assume legal responsibilities that begin with the recruitment process. We recommend that you consult *Human Resources Kit For Dummies*, 3rd Edition, by Max Messmer (Wiley), to be sure you cover all the bases.

Getting Your Nonprofit Ready for Paid Employees

Before you write a job description or place an ad online, you need to invest time in some upfront prep work, readying your organization to take on paid employees. This preparation, which we cover in the following sections, includes writing personnel policies, setting up payroll systems, and deciding on benefits.

Developing your personnel policies

Personnel policies and procedures outline how an organization relates to its employees. They're essential for both supervisors and employees because they provide guidelines about what's expected in the workplace and on the job. They lay out expectations for employees, ensure that all employees receive equal treatment, and provide the steps necessary for disciplinary action if it's needed.

Many startup and small organizations that have only one or two full-time employees give personnel policies a low priority. We suggest, however, that you begin early to formalize your rules with an employee handbook. Doing so really doesn't take much time, and it can save you headaches down the road.

WARNING

You must follow federal and state, and sometimes local, labor laws when establishing personnel policies. The U.S. Department of Labor website (`www.dol.gov`) contains the latest information on federal laws. If you're uncertain about whether you can require certain behavior or work hours from your employees, consult an attorney.

When forming policies, begin with the easy stuff: Decide on your organization's office hours, holidays, vacation policy, sick pay, and other basic necessities.

Determining work and off time

Most organizations follow the lead of others when setting holiday and vacation policies: Although you find a lot of variation, many nonprofits in the United States grant two weeks' vacation per year to new employees. Employees typically receive more vacation time after a longer period of service, such as three weeks after three years, and four weeks after five to eight years. The most common sick-leave policy is ten days per year. While you're at it, you want to give some thought to bereavement and maternity/paternity-leave policies.

Paid vacation time is a benefit to both employees and employers. Employees return from vacation rested and ready to give their best efforts to the organization. For this reason, you should encourage people to take vacations during the year in which they earn the vacation. Most organizations and businesses don't allow employees to accrue vacation time beyond a certain amount. This policy ensures that employees use vacations for the purpose for which they're intended. Such a policy also limits the need to make large cash payments for unused vacation time when employees resign or are terminated.

TIP

Consider reviewing the personnel policies of other nonprofit organizations in your area. A few telephone calls to other executive directors may help answer questions that arise as you refine your policies. Community Resource Exchange has sample personnel policies in its Tools for Nonprofits section at www.crenyc.org/resources_tools.

Covering other important items

In addition to vacations and holidays, which are discussed in the preceding section, you should cover these other basics in your personnel policies:

- A statement that the board of directors may change the policies.
- A statement of nondiscrimination in employment, usually presented as a policy established by the board of directors — especially important if your organization is seeking government grants or contracts — and other policies established by the board.
- A statement of procedures to encourage and protect employees and volunteers if they report wrongdoing — a whistle-blower policy. The National Council of Nonprofits provides guidance in this area at www.councilofnonprofits.org.
- A statement about parental leave and long-term disability policies.
- A statement of hiring procedures and the probationary period (see "Evaluating your new hire's progress" later in this chapter).
- A statement of the employment termination policy, including a grievance procedure.

Many organizations also include a statement of the organization's mission and values and a brief outline of its history.

Setting up a payroll system

When an organization begins employing workers, it first must establish a payroll system. The organization needs to decide how often to distribute paychecks, for example. Some states specify how frequently employees must be paid. Check with your state's department of labor about possible rules governing payment frequency. If your state doesn't specify payment periods, you can disburse paychecks on any schedule you choose — weekly, biweekly, semimonthly, or monthly. (*Biweekly* means every two weeks and results in 26 paychecks per year; *semimonthly* means twice a month and results in 24 paychecks per year.) In our experience, semimonthly is the most common schedule for payment.

TIP

Although in-house staff can handle payroll, contracting with a payroll service is a better option. Most banks either provide payroll services or can recommend a service. Payroll services are inexpensive — they're almost always cheaper than assigning a staff person the job of handling payroll. They make the proper withholding deductions based on income level, number of dependents, and state laws; make tax deposits; and maintain a record of vacation and sick leave. Usually when you contract with such a service, it requires that you keep enough funds on deposit with the company to cover two or three months of salaries and benefits.

WARNING

If you decide to handle your own payroll, be sure to make tax deposits on time. Failure to pay federal and state payroll taxes can get your organization in serious trouble and can even result in personal liability for board members. Check the Internal Revenue Service (IRS) website (`www.irs.gov`) for the latest information.

Typical deductions include the following, although state tax laws vary:

- Federal income tax withholding
- Social Security and Medicare (must be matched by the organization)
- State income tax withholding
- State unemployment and disability insurance

Providing benefits and perquisites

Health insurance is a benefit that your organization may or may not be able to provide to your paid employees. The cost of health insurance decreases based on the size of the group to which you belong. If your organization has only one employee, purchasing health insurance may be costly unless you can join a larger group. Check with your state association for nonprofit organizations to see whether they have a health insurance program that will cover your employees. Be sure to stay up to date on provisions of the Affordable Care Act. The U.S. Small Business Administration website (www.sba.gov) has the latest information.

Once your nonprofit organization is thriving, you may want to consider offering more benefits, such as retirement plans and long-term disability insurance.

Preparing to Hire

After you write your personnel policies, put your payroll system in place, and decide on your benefits, you're finished with the groundwork needed to hire one or one hundred employees. You're probably just hiring one person for now. To set off on that path, you want to clearly describe what the job entails, set salary levels, and announce the position. This section walks you through those important steps.

First things first: Writing a job description

One of the first things you should do when looking to hire a paid employee is to write a job description for the position you want to fill. Going through this exercise helps you clarify the skills needed for the job and guides you in selecting among applicants. The final description serves as a job blueprint for the new employee.

A job description usually includes the following information:

- A short paragraph describing the job and the work environment
- A list of duties and responsibilities
- A list of skills and abilities needed for the job
- Experience and education required
- Special qualifications required or desired

TIP

We include several standard job descriptions at www.dummies.com/go/nonprofitkitfd5e. See File 10-2 for an executive director job description, File 10-3 for a development director job description, and File 10-4 for an office administrator job description.

When writing your job description, keep in mind that work in nonprofit organizations can be split into three broad areas:

» **Services:** Services are the reason the organization exists in the first place. They may include developing protected open space on a coastal bluff, providing home visits and hot meals to seniors, or organizing after-school activities for children. This list is almost unlimited.

» **Administrative functions:** These functions include bookkeeping and accounting, office management, property or building management, marketing, website design, clerical services, benefits administration, and contract management. You can add to this list as needed.

» **Fundraising:** This area falls under various names, including *resource development* and *advancement*. Depending on the size of your organization, one person may be in charge of all aspects of raising money, or different people may specialize in writing grants to foundations and government agencies, creating sponsorships with corporations, or raising money from individual donors.

The larger these areas (or departments) are, the greater the specialization within them. But when you're looking to hire your first staff member, that one person may have job responsibilities in more than one area, perhaps even in all three areas. It may seem like too much to describe in a single document, but that complexity and the high level of responsibility this person will have make it particularly important to create a job description that's crystal clear.

Considering necessary qualifications

Some nonprofit jobs require various levels of formal education and special training. If you're hiring someone to provide counseling services, for example, that employee probably needs to meet certain education and licensing requirements to provide the services legally. If you're hiring someone to work with children, the applicant may need to pass various background checks, depending on the laws of your state.

TIP

Professional and business associations can provide helpful information about job qualifications. In addition to any degrees and certifications, you may want to specify that applicants have a certain amount of experience doing the work you're going to ask them to do. However, if you do so, be prepared to pay a higher salary to fill the position.

WARNING

You can't, of course, require that an applicant be of a particular ethnic background, race, age, creed, gender, or sexual orientation. You can't deny employment to a woman because she's expecting a child. You also can't refuse employment to a person with a disability as long as he can perform the job with reasonable accommodations.

Establishing salary levels

Deciding on a fair salary for your employees isn't easy. Compensation levels in nonprofits range from hardly anything to six-figure salaries at large-budget organizations. (Keep in mind, however, that large nonprofits represent only a small percentage of active nonprofits.) We guide you through the process in the following sections.

Considering factors that affect salary

Although exceptions always exist, the following factors may determine salary levels:

- **Geographic location:** Salary levels differ from place to place because of the cost of living. If your nonprofit is in a major metropolitan area, expect to pay more to attract qualified staff than if you're located in a rural area.
- **Experience and education:** Someone with ten years of experience can command higher compensation than someone just beginning his career. Education levels also affect salary, as does having specialized knowledge or skills.
- **Job duties and responsibilities:** Employees who direct programs and supervise others typically earn more than employees who have fewer responsibilities.
- **Nonprofit type:** Compensation levels vary from one nonprofit to the next. Organizations providing health services, for example, typically have higher salary levels than arts organizations.
- **Union membership:** Labor unions set standards for salaries and benefits in many fields, from musicians to nurses and educators.
- **Organizational culture:** This category, which is more difficult to define, is connected to the organization's traditions and values. For instance, nonprofits with boards of directors filled with business and corporate members often offer higher salary levels than organizations with boards that don't have the corporate perspective.

Finding out salaries of comparable positions

A salary survey, which you can conduct by phone or mail, is a good way to assess the current salary levels in your area and for your nonprofit type. Telephone surveys probably should be done from board president to board president because most people are reluctant to reveal their own salaries. Mail surveys can be constructed so respondents remain anonymous.

The simplest method of finding salaries of comparable positions is to look at other job listings. Not all ads give a salary level, but those ads that do can give you a general idea of what others are paying for similar work.

Some nonprofit management organizations conduct annual or biannual surveys of salary levels in the areas in which they work. More often than not, access to the surveys requires payment, but it's probably a good investment because these surveys tend to be the most complete and up to date. A good place to inquire about salary surveys is through your state association of nonprofits (see the National Council of Nonprofits website at `www.councilofnonprofits.org`). Idealist Careers also has links to nonprofit salary survey information at `www.idealistcareers.org/salary-surveys`. Local community foundations also may be a good resource for salary data in your geographic area and for the budget size of your organization.

Using a search firm can be helpful in hiring, especially if the position requires a national search. Be prepared to pay a hefty fee, however. Fees are often based on a percentage of the first year's salary.

Announcing the position

After you decide on the qualifications and skills needed for the job you want to fill, advertise its availability. Here's a list of places to publicize your job opening:

- **Professional journals:** If you're hiring for a professional position, professional journals are the place to advertise the job. Search the web to track down the addresses of appropriate trade-specific journals.
- **Websites:** The web has become a valuable marketplace for job seekers. Many websites charge a fee for posting a job opening. Work for Good (`www.workforgood.org`) and Idealist Careers (`www.idealistcareers.org`) are popular sites. Also consider craigslist (`www.craigslist.org/about/sites#US`), Monster (`www.monster.com`), CareerBuilder (`www.careerbuilder.com`), and LinkedIn (`www.linkedin.com`).
- **Word of mouth:** Spread the word to other nonprofits — especially those in your field — places of worship, and anywhere else people congregate. Place a job description on your website and announce it on your Facebook page and Twitter feed. Don't forget to email the announcement to colleagues.

Making the Hire

When it comes time to make the big decision, sifting through résumés, conducting interviews, and deciding on an employee can be a daunting task. Sometimes the right choice jumps out at you; other times, you have to choose between 2 or 3 (or 20!) candidates who have equal qualifications. That's when making up your mind becomes difficult.

Looking at résumés

Résumés and cover letters give you the first opportunity to evaluate candidates for a position. Respond quickly with a postcard or email to tell applicants that you have received their materials. If possible, give a date by which they can expect to hear from you again. Doing so reduces the number of phone calls you get asking whether you've received the résumé and when you plan to make a decision. It's also the polite thing to do.

REMEMBER

Résumés come in various formats, and we don't really have an opinion as to which one is best. Regardless of how the résumé is organized and whether it's on paper or online, here are the questions we ask ourselves when reviewing a résumé:

- **Is it free of typographical errors and misspellings?** A typo may be excused if everything else appears to be in order, but more than one or two errors implies that the candidate may be careless in her job.
- **Is the information laid out in a logical, easy-to-follow manner?** The relative clarity of the résumé can give you insight into the applicant's communication skills.
- **Does the applicant have the proper job experience, education, and licenses, if needed?** We like to give a little slack on experience because sometimes highly motivated and effective employees are people who have to "grow into" the job. Also, nonprofit organizations often receive résumés from people who are changing careers. They may not have the exact experience you're looking for, but maybe what they learned in their previous jobs easily transfers to the position you're trying to fill.
- **How often has the applicant changed jobs?** You can never be guaranteed that an employee will stay as long as you want him to, but you have to ask yourself, "If I hire this person, will he pack up and move on even before he finishes job training?" At the same time, don't automatically let higher-than-average job switching turn you off to an excellent candidate. Maybe he has an explanation. Ask.

Cover letters can also be good clues to an individual's future job performance. For one thing, you get an idea of the applicant's writing abilities, and you may even get some insight into her personality. Some job announcements state that a cover letter should accompany the résumé, and some even go so far as to ask the applicant to respond to questions such as "Why are you well suited to this job?" or "What do you think are the major issues facing so-and-so?" It's up to you to decide whether asking a list of questions enables you to more easily compare applicants to one another.

TIP

If good writing skills are required for a job you're posting, ask applicants to include writing samples with their cover letters and résumés.

You'll probably reject at least half the résumés out of hand. We never cease to be amazed by how many people apply for jobs for which they don't have even the minimal qualifications. We understand that searching for a job is difficult and frustrating, but we also wonder whether applicants read our "position available" ads as closely as they should.

Separate your résumés into two groups — one for rejected applications and one for applications that need closer scrutiny. Send the rejected candidates a letter or email thanking them for their interest. From the other pile, decide how many candidates you want to interview. Reviewing the résumés with one or two board members to bring several perspectives to the choice is often helpful. One technique is to select the top three applicants for interviews. Reserve the other applicants for backup interviews if the first three aren't suitable or if they've already accepted other jobs.

Interviewing candidates

After you've chosen the top three to eight résumés, invite the applicants in for an interview. Interviewing job candidates is a formidable task. Big companies have human resources departments with trained interviewers who spend their days asking questions of prospective employees. We're neither human-resources specialists nor trained interviewers, but here are some tricks we've learned over the years:

- **Prepared lists of three or four standard questions that you ask all applicants enable you to compare answers across applicants.** The interview shouldn't be so formal that it makes both the candidate and you uncomfortable, but standardizing it to some degree is beneficial. Here's a short list of typical questions:
  - Why are you interested in this position?
  - What do you see as your strengths? As your weaknesses?

- How would you use your previous work experience in this job?
- What are your long-term goals?

» **Group interviews with three or more people can give interviewers good insight into how the applicant will perform in board and community meetings.** Also, different people notice different things about each applicant. Avoid making the candidate face a large group, which can make her unnecessarily nervous.

» **If an employee isn't your first hire and the job to be filled is for a director or supervisor position, have each finalist meet at least some of the staff he'll be supervising.** Giving staff members a chance to meet their potential new boss is courteous, and their impressions are helpful in making the final selection.

You can't ask an applicant personal questions about his age, religious practice, medical history, marital status, sexual orientation, or racial background. You can't ask whether he has been arrested or convicted of a felony without proof of necessity for asking. You also can't ask whether he has children. Nor can you ask about any physical or mental conditions that are unrelated to performing the job. Workforce has a useful article about interview questions at `www.workforce.com/articles/interview-questions-legal-or-illegal`.

Taking notes during the interview is acceptable, and you may also find that preparing a checklist on which you can rate the applicant in different areas is a helpful exercise. If you do, try to rate applicants discreetly. Job interviews are stressful enough without letting the applicant know that you rated him a three on a scale of one to ten.

Digging deeper with references

Letters of recommendation can be helpful starting points in evaluating candidates, but we assume that no one would include a negative recommendation in her application packet. Therefore, we do think it's necessary to check references by telephone or even in a personal meeting, if possible.

But sometimes even talking to references doesn't provide much useful information. A job applicant wants to put her best face forward, so naturally she chooses people with favorable opinions as references. Also, employment laws are such that speaking to a former employer often yields little more than a confirmation that your applicant was employed between certain dates.

If you do have the opportunity to have a good conversation with the candidate's former employer, pay close attention to what she's not saying as well as attending to her description of the candidate's abilities. For example, if you need someone who is attentive to financial details, but the former employer only talks about the applicant's friendliness and phone manner, you may not have found the right person for your position. Some typical questions ask what duties the applicant handled in her previous jobs, what her strengths and weaknesses were on the job, whether the candidate had a positive attitude toward work, and whether the reference would rehire the candidate.

If the applicant is working for another organization, don't contact the current employer. You don't want to breach her confidentiality.

Check out File 10-5 at www.dummies.com/go/nonprofitkitfd5e for a sample reference-checking form.

You can conduct formal background checks of education credentials, criminal history, and other information as long as you get the applicant's permission. If you feel that this type of research is necessary, we suggest hiring a reputable company that specializes in this sort of work. However, you can conduct informal Internet searches about the applicant.

The lengths to which you go to collect information about an applicant depend on the magnitude of the position. If you're hiring someone to lead a large and complex organization, you likely want to dig deeper into a candidate's background than if you're hiring a data entry clerk. But for any position, you want to know as much as possible about an individual's previous job performance before hiring her.

Making your decision

We can't tell you how to make the final decision about whom to hire. You have to weigh qualifications, experience, poise, and desire. These decisions can be difficult, and, frankly, you may not be certain that you've made the right choice until the new employee performs on the job. If possible, get more than one opinion.

If qualifications and experience are equal, intangible factors come more into play. Will the candidate fit well into the organizational culture? Will the candidate's style of work fit with the organization's management style? Do the applicant's professional goals fit with the organization's goals? If this employee is going to be your first, is the candidate a self-starter type or does he need active supervision to do a good job?

WARNING

Make sure you document your reasons for making a personnel decision: They should be based primarily on the candidate's performance in previous jobs and experience that's relevant to your position. Be cautious when making a personnel decision based on intangible factors. A candidate who's charming may be a treat to have around the office, but that doesn't always mean that he'll do the job well.

Bringing a New Hire Onboard

Much hard work is behind you after you've made a hiring decision. But keep in mind that any employee's first days and months are challenging, and you want to give careful attention to helping your new staff member make a good start.

Confirming employment terms in writing

After a new employee accepts a position orally, we recommend that you send a letter to put the details in writing. Enclose a copy of the personnel policies (see the section "Developing your personnel policies" earlier in this chapter) and place a signature line near the lower-right corner of the letter so the new hire can acknowledge receipt of the letter and the personnel policies. Ask the employee to return a copy of the signed letter to you, and keep the letter in the employee's personnel file.

The letter should include the employee's starting date, job title, and salary as well as other information that you agreed to in the pre-hire discussions held between the organization and the employee. For example, you may include a brief statement about the employee's responsibilities and agreed-upon work schedule.

TIP

Check out File 10-6 at `www.dummies.com/go/nonprofitkitfd5e` for a sample hire letter.

Getting your new hire started on the job

REMEMBER

In the United States, one of the first things a new employee must do is complete a W-4 form (for income tax withholding) and an I-9 form (to show proof of the employee's legal right to work in the country). These forms are required by law and are available on the IRS website (`www.irs.gov`).

After all the necessary paperwork is out of the way, you need to spend some time getting your new hire acclimated to the working environment. New employees don't begin producing at top form on the first day of work. Absorbing the details of the organization and discovering the ins and outs of new job duties take time. This fact is particularly true when the person hired is the organization's first employee and has no model to follow.

Whether you're a board member for an organization that's hired its first employee, or the director of an organization bringing someone new onto the staff, here are some ways to ease an employee's transition to a new job:

- **Provide good working conditions.** You may think that a reminder to purchase the basic furniture and tools someone needs to perform his work is too basic, but we've heard about new employees who didn't even have a desk on their first day.
- **Show the new person around.** Provide a tour of the office and programs and introduce him to volunteers and board members. Review office emergency procedures on day one. Oh yeah, and don't forget to show the new employee where the restroom is!
- **Give the employee information about the organization.** Make available the organization's files and records, including any policies and guidelines that affect the employee's duties and the performance of his work. Reading board minutes, newsletters, solicitation letters, donor records, and grant proposals will steep him in the organization's work.
- **Answer questions.** Encourage new employees to ask questions, and provide the answers as soon as possible. Particularly if he's the organization's one and only employee, board members should check in regularly, making themselves available as resources. Being a one-person staff can be lonely and overwhelming.
- **Offer special training.** A new employee may need special training — for example, about a software program or laws and regulations specific to your nonprofit — to perform his job. Sometimes you may have to send the employee to a workshop; other times, he can be trained by a board member, a volunteer, or another staff member.

Evaluating your new hire's progress

REMEMBER

Conduct a performance evaluation within the first six months of employment. The evaluation should be written (and added to the employee's personnel file) and discussed in a meeting with the employee. Rating employees on a number scale on various aspects of the job was once the common format. Today, a narrative evaluation that addresses performance in achieving previously agreed-upon goals and objectives is much more helpful.

TIP

Many employees are happier in their jobs if they have ongoing opportunities to learn new things and develop new skills. An annual employee review is an excellent opportunity to review professional development goals with the employee. It's also a good time to review ways — through workshops, training, or trying on new roles — that she may continue to grow in the job.

Managing Employees

Much of this chapter focuses on small organizations that are hiring their first staff members. In these organizations, the board oversees the staff member, and someone taking on the coordinator role — either board or staff — oversees the volunteers.

What if your organization has grown more complex? Everyone needs a boss. In nonprofit organizations, the board assumes that role for the executive director, who in turn provides supervision to other employees, either directly or through a management team. The common way to visualize these relationships is through an *organizational chart,* a schematic drawing showing the hierarchical management relationships in an organization.

A chart such as the one in Figure 10-1 may be overkill for your nonprofit, especially if you're the only employee, but for larger organizations, charts help delineate management responsibilities and clarify who reports to whom.

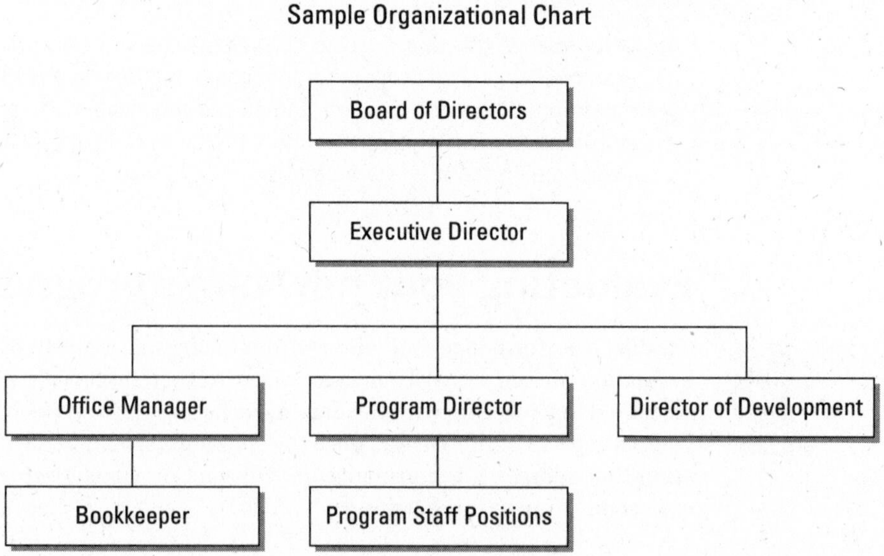

Sample Organizational Chart

Board of Directors
Executive Director
Office Manager
Program Director
Director of Development
Bookkeeper
Program Staff Positions

FIGURE 10-1: Organizational chart of a typical larger nonprofit agency.

Understanding what a manager does

The responsibility of managing employees includes the following aspects:

- **Planning:** Planning occurs at all levels, beginning with the board of directors, which carries out organization planning, and ending with the custodian, who plans how best to complete cleaning and maintenance tasks. It's best if managers work closely with the employees they're supervising to develop department and individual goals.
- **Leading and motivating:** You may want to add "inspiring" to this category. Good management grows out of respect and cooperation. Be sure that staff members are familiar with the organization's goals and that they know how their work helps foster those goals.
- **Gathering tools and resources:** Don't ask someone to dig a hole without giving him a spade. In other words, you can't expect employees to do a good job if they don't have the means to do it. Time, equipment, proper training, and access to information are necessary.
- **Problem solving:** This is one of the most important aspects of good management. You can bet that problems will arise as you manage your organization and employees. Be understanding and creative in solving these problems. Ask for help in the form of ideas and suggestions from the employees you supervise.
- **Evaluating:** Employees should receive formal evaluations once each year. Effective evaluation begins with goal setting, and you want to help subordinates set clear goals and objectives.

Communicating with your staff

We can't say enough about the importance of good communication. If your nonprofit has only 1 or 2 staff members, your job is easier than if you need to communicate with 50 or more employees. Either way, good communication is essential.

Of course, confidentiality is important in some matters. For example, if the organization is contemplating a major change, such as a merger with another organization, some information needs to be withheld from the staff. On the other hand, letting rumors circulate about changes that may affect staff can create worse problems than being forthcoming about the details of a potential change. Give people as much information as you can and be sure they have a chance to tell you how they feel.

Communication is a two-way street. A good manager keeps an open ear and devises ways to ensure that his employees have a way to voice complaints, offer suggestions, and participate in setting goals and objectives.

Holding regular staff meetings

Too many meetings can be a waste of time, but having regularly scheduled staff meetings is a good way to transmit information to employees, give them an opportunity to offer input and feedback, and keep everyone working toward the same goal. Keep these points in mind when arranging staff meetings:

- **Try to schedule the meetings at a specific time on a regular basis.** Hold meetings no more often than once a week and no less often than once a month, depending on the needs of the organization.
- **All meetings should have an agenda.** Nothing is worse than going to a meeting that has no point and no direction.
- **Keep to a schedule.** Unless you have big issues to talk about, one hour is usually long enough to cover everything you need to discuss.
- **Provide time on the agenda for feedback.** Be sure that everyone has a chance to speak.

Writing memos to staff

Use memos to introduce new policies and other important information so you don't have any misunderstandings. By putting the information in writing, you can clearly explain the situation. If the policy is controversial, distribute the memo shortly before a scheduled staff meeting so employees have an opportunity to respond.

Some larger organizations create staff newsletters that cover organizational programs and achievements. People work better if they receive recognition for their work. Stories about client successes, gains made by the organization, and announcements about staff comings and goings help instill feelings of accomplishment and organizational loyalty.

For very large organizations, a dedicated website or intranet that's accessible only to employees can transmit information and offer staff the opportunity to provide feedback and communicate with one another.

Talking around the water cooler

Although formal written communication is important, nonprofit leaders also need to communicate informally by being accessible to staff in the hallways and around the water cooler. Some people communicate concerns better in an informal setting than in a staff meeting. Managers who sit behind closed doors all the time often have a difficult time relating with their staff. However, don't force the camaraderie. Let it develop naturally.

Letting a staff member go

Some people enjoy barking, "You're fired!" We don't. We've spent sleepless nights before and after terminating an employee who wasn't right for a job. Although you may not lose sleep over letting an employee go, it's obviously not something to take lightly.

When do you need to consider cutting ties with an employee? Sometimes the employee isn't producing the quality of work required by the job or isn't producing work in a timely manner. Other times the employee may be fired for inappropriate behavior, such as not following your organization's guidelines for appropriate interaction with clients.

If you're a boss and facing the need to terminate an employee, we recommend that you

- **Review the written personnel policies that you provide to all employees.** Document in writing the employee's poor performance or inappropriate behavior and make clear the connection between your concerns and those policies.

  REMEMBER

  Many people deserve an opportunity to improve. Sit down with the employee and speak candidly and firmly about the level of improvement or change in behavior that you expect. Set written goals for an improved performance and a date for a follow-up consultation.

- **Respect the employee's privacy.** Don't share your complaints with others except, if necessary, your board chair.
- **Encourage the employee to leave as soon as she's terminated, watch as she collects her personal items, and walk her to the door.** Disgruntled former employees can damage your organization's records and documents in a few days, hours, or even minutes.

Wrongful discharge lawsuits are common and can be very expensive. If you want to be cautious, you may want to investigate the possibility of employment liability coverage. If you need to terminate an employee and aren't sure how to proceed, consult with a human-resources expert, your insurance carrier, or an attorney. The upfront investment may save you much time, expense, and trouble later. If your personnel policies specify the steps for ending employment (and they should), be sure to follow them to the letter.

Sometimes organizations have to let employees go because they no longer have sufficient money to pay the employees' salaries. If you're a boss in this situation, try to do the following:

- **Give the employee as much warning as possible about the date of her termination.** Don't keep it quiet while pulling out all the stops to raise money to save her position. Surprising employees with the bad news is inconsiderate.
- **Try to plan ahead and provide some severance pay to help her while she seeks a new job.** Although you may want to keep her on the job until the last possible minute, good employees deserve good treatment. If she leaves your agency with good feelings, you're in a better position to hire her back in the future.
- **Offer to serve as a reference or write letters of recommendation.** If the money isn't coming in, and you can't find a way to keep the employee on, this gesture is one of the best things you can do.

Using Independent Contractors

Maybe your organization isn't quite ready to take the leap into the employment waters. However, if you have work that needs to be done that volunteers can't do, working with an *independent contractor* may be a way to accomplish some organizational goals.

According to employment and tax laws in the United States, you can hire people in two ways: as salaried employees or as independent contractors. Rarely do you say to yourself, "I need an independent contractor to design my web page" or "I need an independent contractor to write more grants for us." More commonly used terms are *consultant* or *freelancer.* For example, small organizations may contract with an accountant or bookkeeper to maintain financial records and prepare financial reports.

Independent contractors aren't just a resource for small organizations: Large-scale, well-heeled nonprofits also use them. A good organizational consultant, hired as an independent contractor, can bring a fresh perspective to assessing an organization's work, a depth and breadth of experience the organization hasn't yet developed, and focused attention to a project — say the writing of a new strategic plan — that staff can't give while managing day-to-day operations.

Differentiating an independent contractor from an employee

Technical differences between employees and independent contractors are reflected in how you hire, pay, and manage them. Independent contractors are almost always paid a flat fee or hourly rate for their work, ideally on a schedule that's set out in a contract that specifies the work to be done and the fee. Although you don't have to withhold federal and state payroll taxes, you need to file an IRS Form 1099 that records the amount paid over a full year (usually a calendar year) and the contractor's Social Security number or federal tax ID number (EIN).

REMEMBER

Proceed with caution, because there's often a thin line between an independent contractor and an employee. Just because someone works a limited number of hours each week doesn't mean that he should be considered an independent contractor. The IRS doesn't look kindly on trying to pass off employees as independent contractors. Here's a short list of factors that differentiate an independent contractor from an employee:

- **Independent contractors are just that — independent.** Although setting time parameters for the job is fine, contractors typically are free to set their own schedules and work with little direction from the organization. They typically provide their own offices and equipment.
- **Duties should be written into a contract and the contractor should provide invoices for services provided.** The contract should be limited to a specific period of time and vacation and sick leave should not be paid.
- **If you have to provide extensive training for the contractor to do the job, chances increase that the contractor may be considered an employee.** In hiring an independent contractor, you're supposed to be engaging someone with specific expertise.
- **If a contractor is working only for your organization and is putting in many hours each week, he may be considered an employee.** If you hire a contractor with the idea that the relationship will continue indefinitely, rather than for a specific project or period, or if that contractor provides services that are a key part of your regular business activity, such as executive director, the IRS believes that you likely have the right to direct and control that worker's activities. To the IRS, this association looks like an employer-employee relationship.

WARNING

If you think you may be pushing the envelope on this question, consult an attorney or tax specialist who can give you proper advice. Authorities are giving increasing scrutiny to the distinction between regular employees and independent contractors. The IRS may require employers who pay individuals as contractors, when they're really employees, to pay back payroll taxes and penalties. Board members and responsible managers could be held personally liable for these taxes.

The IRS has defined *Common-Law Rules* for determining whether someone is an employee or an independent contractor. Check IRS Publication 15A, *Employer's Supplemental Tax Guide*, for a definition of these rules. The publication is available on the IRS website at `www.irs.gov`.

TIP

If you want to hire someone for a short-term assignment or a limited number of hours, but if the work is taking place in your offices, using your equipment, and requiring your regular supervision, consider using a temporary employment agency. The agency handles the employee's benefits, including his employment taxes, and reassigns the employee if he's a poor fit for the job. Although using a temp agency may cost more than hiring the person directly, the agency can help you find someone who's qualified.

Seeing what an independent contractor can do for you

Independent contractors can help you with just about any aspect of your organization, from managing personnel matters to hooking up a computer network. But nonprofits most commonly bring them in to help with

- **Fundraising:** Fundraising consultants help with grant writing, planning for fundraising, special events, direct mail, and major gift and capital campaigns, among other methods of raising funds.
- **Organizational development:** Organizational development consultants may guide your board and staff through a planning process or help you develop tools needed to evaluate your organization's work, to name just two examples.
- **Marketing and public relations:** Consultants can help you spread the word about your organization's good work by handling a public relations campaign directed at the media, developing your website, creating compelling brochures and newsletters, creating short promotional films, and many other strategies.
- **Evaluation and assessment:** Evaluators can take an in-depth look at an organization's programs, bringing specialized skills to that assessment and the value of an unbiased point of view.

Have a clear idea of what you want to accomplish before you seek help. Sometimes you don't know exactly what you're aiming to accomplish, but you should still try to articulate as clearly as possible what your aim is in hiring a consultant before you go looking for one.

Finding a consultant: Ask around

We don't have a single best way to find consultants. You can certainly search the web for nonprofit consultants who work in your area, but we think one of the better ways is to make a few calls to other nearby nonprofits or funding agencies. You may also be able to find a consultant through a nonprofit support organization.

We prefer consultants who have more than one way of doing things. Not all nonprofits are alike, and what works for one may not work for another. Also, as your project evolves, you may find that you need a different kind of help than you originally imagined. Believe us, changes happen all the time. Ask the consultant whether she's willing to consider revisiting the project goals along the way and adjusting her approach if the need arises.

Some consultants work as sole practitioners; others work in consulting companies. Working with a company or consulting group may give you access to more varied expertise. On the other hand, consultants working alone tend to have lower overhead expenses, so their fees may (and we stress *may*) be lower.

Interviewing consultants

Interviewing a consultant is similar to interviewing a regular job applicant. You want to review the résumé carefully to see whether the candidate's experience and expertise match your needs. Don't be intimidated just because the person sitting across the table from you has more experience than you do. He'll be working for you. Ask any questions that you feel are necessary so you can be sure you've chosen the right person.

Interview more than one person before you make your decision. As with interviewing for staff positions, having more than one person present at the interview is important because it lets you see how the prospective consultant interacts with a small group and gives you multiple perspectives on the consultant.

You may have the opportunity to use a volunteer consultant supplied by a program that provides free or low-cost assistance to nonprofits. Interview these individuals just as you would paid consultants. Remember that you'll be investing time in working with this person, so you need to be certain that she's right for your organization.

Ask interviewees to describe their method or approach to the problem or project that you're seeking to solve or complete. You shouldn't expect them to give you a full-blown plan in the interview, but you should expect a good picture of the initial steps they'll take. Also, if your project culminates with a written report, ask the applicant for writing samples.

Signing the contract

You need to have a signed contract with every consultant with whom you work. These points should be clearly stated in the contract:

- **The scope of work and expected results:** You can include more or less detail here, depending on the type of project. If you're hiring the consultant to facilitate a one-day retreat for your board of directors, be sure that the contract includes the preparation time needed. Also, will the consultant be writing a report after the retreat? Try to touch on as many details as possible. The more specific, the better. Sometimes if the scope of work is very detailed, that detail is included as an attachment to the main part of the contract.
- **Fees, of course:** Some consultants charge an hourly rate plus expenses; others charge a flat rate that may or may not include expenses per project. The contract should state that you must approve expenses over a certain amount. Or you can write into the agreement that expenses are limited to a certain sum each month. Consultant fees vary, just like salaries, from one geographic area to another and depend on the type of work to be done and the consultant's experience.

  We don't recommend paying fundraising consultants a percentage of money raised. Although some consultants, grant writers, and, in particular, telemarketers work under this arrangement, percentage payment isn't considered good practice by most fundraisers and nonprofit managers. See the Association of Fundraising Professionals' code of ethics (`www.afpnet.org`) for more information about this issue.
- **The schedule of when you'll pay the fees:** Some consultants who work on an hourly basis may send you an invoice at the end of the month. Consultants working on a flat-fee basis may require advance payment on a portion of the fee. This is fine. It's protection for the consultant, who probably has as many cash-flow issues as you do. The final payment shouldn't come before the project is completed. Be sure to ask the consultant for an estimate of her out-of-pocket expenses and how she plans to bill you for such costs.
- **Special contingencies:** What if the consultant gets sick? What if your organization faces an unforeseen crisis and you don't have time to work with the consultant? What if you've chosen the wrong consultant? You should include a

mutually agreed-upon way to end the contract before the project is completed. A 30-day cancellation notice is common.

- **A timetable for completing the work:** You and the consultant need to negotiate the timetable. Have you ever had remodeling work done on your house? You know that contractors can get distracted by other work, right? That's why having a schedule is so important.
- **The organization's role in the project:** If the consultant needs access to background materials, records, volunteers, board members, or staff, you must provide this access in a timely manner so the consultant can complete the job on schedule.
- **Required insurance:** What if your consultant injures someone in the course of working on your project? If she's not insured, the liability could become your organization's responsibility. Just to be on the safe side, many consultant contracts include standard language about the consultant's responsibility to carry liability insurance and pay workers' compensation insurance and other legally required benefits if the consulting company has employees who will be working on your project.
- **Ownership of the finished product:** This point doesn't apply to every consulting project, but if yours focuses on developing a study or report, specify in the contract whether your organization, the consultant, or both have the rights to distribute, quote, and otherwise make use of the results. Often the organization has the right to publish and produce the report if it gives proper credit to the author, and the consultant owns the copyright and has permission to make use of lessons learned in articles or future studies. Sometimes you agree to seek approval from one another before using the work.

TIP

We think that having the consultant prepare the contract is a good idea. It's the final chance to make sure he understands what you want done. You may suggest changes when you see the first draft.

» Constructing additional budgets for specific activities

» Using your budget

» Anticipating revenue and expenses

» Understanding financial statements

» Adopting good financial practices

Chapter 11

Showing the Money: Budgets and Financial Reports

Nonprofit organizations are expected to spend wisely and honor the trust placed in them by their donors. As a result, they need to be especially good at budgeting and living within their means.

Creating budgets is a critical part of program planning, grant writing, and evaluation. Maintaining a financially stable organization is one of management's most important tasks. To achieve that stability, a nonprofit needs to keep clear records and base its decisions on accurate financial information. Asking the right questions about financial reports is one of the board's key responsibilities.

The best things in life may be free, and volunteers can accomplish a lot, but frankly, no money equals no program.

Check out File 11-1 at `www.dummies.com/go/nonprofitkitfd5e` for a list of web resources related to the topics we cover in this chapter.

Making a Budget = Having a Plan

Because a budget is composed of numbers and because we're taught in grade school that arithmetic problems have right and wrong answers, you may panic when faced with making a budget. What if you don't get it absolutely right? It's true that you need to add and subtract your numbers correctly, but you should relax. It really isn't so hard.

Making a budget is yet another form of organizational planning (the topic of Chapter 7). Often it's done hand in hand with other types of planning. It estimates how you intend to gather and disburse money on behalf of your organization's mission. In the course of a year, the cost of utilities or postage may rise, and the cost of gasoline and computer equipment may fall. You're not expected to employ extrasensory perception in making a budget — just to be reasonable and thoughtful.

If your organization doesn't spend more than it takes in, and if it holds onto a little emergency money at the bank, why does it need a budget? Writing and agreeing upon a budget is an important process for your staff and board because it sets priorities. It's a discipline that keeps your organization on solid ground.

Beginning with zero

In many businesses, the annual budget is made by looking at what happened in the preceding year and adjusting numbers up or down based on the work that lies ahead. When you're starting something new, however, you have no preceding year's results to consider. In that case, you start with zero and carefully consider each number you use to build your plan. Zero is a hard place to start, but it's where every new organization begins.

A budget has two key sections — income and expenses. (See Figure 11-1 for typical line items.) Because dreaming up expenses that exceed your organization's means is easy, we suggest that you begin with income, making conservative estimates for what you may earn and what you may attract in contributions. As you work on the contributed income estimates, you also may want to look at Chapter 13, which discusses developing a fundraising plan.

Anatomy of a basic budget

INCOME

Earned Income

- Government contracts
- Product sales
- Memberships
- Interest from investments
  - Subtotal

Contributed Income

- Government grants
- Foundation grants
- Corporate contributions
- Individual gifts
- Special events (net income)
  - Subtotal
  - TOTAL Income

EXPENSES

Personnel Expenses

- Salaries
- Benefits @ _% of salaries
- Independent contractors
  - Subtotal

Nonpersonnel Expenses

- Rent
- Utilities & Telephone
- Insurance
- Office supplies
- Program materials
- Local travel
- Printing
  - Subtotal
  - TOTAL Expenses
  - Balance (the difference between Total Income and Total Expenses)

FIGURE 11-1: A sample budget showing how line items usually are named.

Examining income

It's common to separate your income statement into two general categories, "earned" and "contributed," also called "revenue" (contract, sales, or fee income) and "support" (grants and contributions), respectively.

Ask yourself these common questions as you begin developing the income section of your budget:

- Will you offer services or products for which you'll charge money? How many services and how often? How many people are likely to use them? What can you reasonably charge? How soon will you be ready to offer them?
- Can you sell memberships to people and give them premiums or discounts in exchange for paying those fees?
- Are the founders and board of your organization able to contribute some startup funds? Are they willing to ask their friends and associates to contribute?
- Is your organization well positioned to receive a grant or grants?
- Are you capable of sponsoring a fundraising event?
- Can you provide visibility to a business sponsor in exchange for a contribution?

Evaluating expenses

In anticipating expenses, start with anything concerning the payment of people. That list may include the following:

- **Salaries for employees, both full time and part time:** On your budget, list each position by title and identify the full-time salary and the percent of full time that person is working for you.
- **Benefits for salaried employees:** At a minimum, you need to pay approximately 12.5 percent of salaries to cover federally required benefits, such as Social Security, Medicare, and unemployment tax contributions. Some states also require disability and worker compensation insurance, and local jurisdictions may require paid sick leave. Many organizations provide paid leave to employees for vacations, illnesses, and jury duty. Your organization also may provide health and dental insurance or a retirement plan for its employees. If so, you want to compute those costs as percentages of your total salaries and include them in your budget as benefits. Benefits are listed immediately after your salary expenses.

  TIP

  When setting up your employee benefits, check on both federal and state requirements. The U.S. Department of Labor website, `www.dol.gov`, is a good guide to federally required employee benefits.

WARNING

Employee benefits may cost more than you think. According to the Bureau of Labor Statistics in March 2016, benefits — across all employment sectors — represented 31.3 percent of total compensation to employees.

- **Fees for services to consultants or service agencies:** You may hire a publicist, grant writer, evaluator, or other consultant to handle important tasks. Such consultants are responsible for paying their own tax and insurance costs. Show fees paid to consultants after salaries and benefits in your budget. Check out Chapter 10 for information about working with consultants.

At this point, compute a subtotal for all your personnel costs.

Next, identify all non-personnel expenses, beginning with ongoing operating costs that allow your organization to do its work. These expenses may include the following:

- Rent
- Utilities
- Telephone and Internet expenses
- Office supplies
- Printing
- Insurance

New organizations often have special startup costs for the first year, including nonprofit registration and filing fees. For example, you may need to purchase desks and chairs, computers, signs, shelving, office cubicles, and you may need to pay for training for new staff members, first and last months' rent, the telephone and Internet hookup costs, and a photocopier or postage machine lease. Finally, you may have costs associated with the specific nature of the work of your organization. These expenses range widely but can include diagnostic tests, carpentry tools, classroom supplies, and printed materials.

TIP

As much as possible, keep notes in your budget files about the estimates you made while drafting your budget. Keeping a worksheet helps you remember your assumptions and follow your budget.

TIP

We present three different organization budgets at www.dummies.com/go/nonprofitkitfd5e — Files 11-2, 11-3, and 11-4. Two of the examples show how organizations compare budgets to actual costs.

TIP

In 2013, the Wallace Foundation created a section of its website at www.strongnonprofits.org, featuring resources to help nonprofits improve their financial management. We've provided a link in File 11-1 at www.dummies.com/go/nonprofitkitfd5e.

Defining a good budget

In a good budget, the income and expenses are equal to one another, which is what the term "balanced budget" means. Although balance is a good state to achieve, we recommend that you work to produce modest surpluses each year so you can cover any unexpected costs. Many organizations slowly develop a *cash reserve,* money put aside in case unexpected expenses arise. If they need to use their cash reserves, they take care to replace them as soon as possible. Often their boards set policies for how large a reserve they aspire to set aside and how quickly expended reserve funds must be replenished. A common rule of thumb is that organizations should create cash reserves that are equal to at least three months of their operating costs.

TIP

In File 11-9 at www.dummies.com/go/nonprofitkitfd5e, you can find "Monthly Information Every Nonprofit Needs to Know."

As a general rule, your budget looks healthy when you show multiple sources of income. Why? Wouldn't it be easier to keep track of one or two major grants and contracts? Or one annual special fundraising event?

This rule is based on the adage "Don't put all your eggs in one basket." What if a power outage strikes on the night of your gala? What if one of your major grant or contract sources changes its guidelines? Your programs could be jeopardized.

Finally, a good budget is realistic and is based on an honest assessment of the resources a nonprofit can earn and raise. In leading a nonprofit, you want to be forward-thinking and optimistic about producing high-quality, meaningful work but you also want to bring a sober brand of that optimism to creating financial plans.

Budgeting based on your history

As your first year passes and your organization develops programs, it also develops a financial track record. After you have a financial history, anticipating the future can be easier. You need to analyze every assumption you make, but at least you have a baseline of revenues and expenses from the first year.

Here are some questions to consider when drafting future years' budgets:

- What are your earned income trends? Are they likely to continue or change? Do you face any new competition or opportunities?
- How many of your current year's grants or contracts may be renewed and at what levels?
- How healthy is the current financial environment? Will it affect your previous donors' abilities to give?
- How likely are you to increase individual giving or special-events revenues in the coming year?
- What's the duration of all employees' employment periods? Anticipate the timing of possible annual raises and the need for additional staffing.
- Do you offer benefits that kick in after an employee has worked with you for three months or six months or five years? If so, don't forget to include these increased costs.
- When does your lease obligate you to pay for rent increases or taxes?
- Have rate increases been scheduled for utilities, postage, or other services?
- Does your organization need new technology — perhaps additional or upgraded hardware, website development, or software?

Understanding and isolating general administrative and fundraising costs

Some of your organization's expenses are defined as general administrative or fundraising costs and aren't considered program expenses. The expense of having an audit, for example, is allocated to general administrative costs, and mailing costs for your end-of-the-year fundraising letter are identified as fundraising. IRS Form 990, which nonprofits file annually, asks organizations to present their expenses in a format that shows which annual costs are for general administration and which are for fundraising. (See Chapter 5 for more about preparing a 990.)

Accounting for in-kind contributions

Some nonprofit organizations benefit from donated goods and services rather than, or in addition to, contributed and earned cash. Suppose, for example, that a local business provides office space for your organization so you don't have to pay rent, 50 volunteers contribute labor to your organization each week, and a major advertising firm sponsors a free marketing campaign to promote your work. These gifts of goods and services are called *in-kind contributions.*

How can you show these valuable resources in your organization's budget? First, we encourage you to make the effort. If you don't show these contributions, your budget doesn't truly represent the scope of your organization, and you're underplaying how much the community values its work. On the other hand, mixing the in-kind materials and services with the cash can make following and managing your budget confusing.

So what's the solution? We prefer creating in-kind subheadings within the budget or summarizing all the in-kind contributions at the end of the cash budget in a separate section.

REMEMBER

If you choose to include in-kind items with the cash in your budget, don't forget that when you receive an in-kind good or service, it's a source of income, and when you make use of it, it's an expense. Often nonprofits fail to show the expenditure of an in-kind contribution, and it can make their budget appear to have more available cash than it really has.

You'll want to keep records of volunteer time if you're going to include it in your budgets or financial reports. *Generally accepted accounting principles* (*GAAP*, a framework outlining standards and rules for accounting) say you should show the value of volunteers' time in your financial statements if they improved a financial asset of the organization (for example, if volunteers rebuilt a greenhouse owned by a nonprofit botanical society) or if the organization needed a task to be performed that required a special skill, the volunteer had that skill, and the nonprofit would have paid for that help if it had not been volunteered (for example, if an attorney reviewed the nonprofit's lease pro bono).

TIP

Annually, Independent Sector computes a standard average value for volunteers' labor, both nationally and by state. (In 2015, the national average was $23.56 per hour.) Check out File 11-1 at `www.dummies.com/go/nonprofitkitfd5e`, where you can find a link to these tables on the Independent Sector website.

Creating Budgets for Programs or Departments

If your organization focuses on a single service, you can skip ahead to read about projecting cash flow. If you manage several programs or departments, each of these programs has a budget that must fit within the overall organizational budget.

Suppose that when your nonprofit organization begins, you offer only one program — an after-school center that provides tutoring and homework assistance for low-income children. The program really clicks, and more and more children begin showing up after school — some because they need tutors, but others because their parents are at work and the kids need a safe place to go. To serve these new participants, you add art classes and a sports program.

So now, instead of one program, you have three. Each has its own budgetary needs. A program coordinator trains and recruits the tutors and purchases books, notebooks, and school supplies. The art program requires an artist's time, supplies, space for art making, and access to a kiln. To offer a sports program, your agency rents the nearby gym, hires four coaches, employs a volunteer coordinator to recruit parents and other assistant coaches, and purchases equipment. The specific costs of a program — books, art supplies, coaches — are called its *direct* costs.

But each of the three programs also depends on materials and services provided by the people who work on behalf of the entire organization, such as the full-time executive director of your agency, the part-time development director, and the bookkeeper.

Each of the three after-school programs also uses the organization's offices, utilities, telephones, and printed materials. The costs that the various programs share — such as the executive director's salary, bookkeeping services, rent, and telephone bills — are called *indirect* costs. You can think of these shared costs as the glue that holds the nonprofit together.

REMEMBER

When you prepare a program or department budget, you want to include both direct costs and indirect costs.

Direct costs are pretty straightforward. For example, you know what you have to pay your tutors per hour, and you know how many hours they work. But indirect costs can be sticky, and we don't say that because they're the organizational glue. They're sticky to deal with because determining how to divide them accurately among your organization's activities is difficult. Many grant makers are reluctant to pay true indirect costs associated with programs, and, if your organization depends heavily on grants, that reluctance can put you in the awkward financial position of being able to pay for a famous basketball coach but not to turn on the lights in the gym.

TIP

In recent years, foundations and government organizations seem to be recognizing the short-sightedness of under-supporting nonprofits' overhead (or indirect) costs. For example, a group of foundations commissioned researchers at Rand Corporation to write the report, "Indirect Costs: A Guide for Foundations and

Nonprofit Organizations," which recognizes their importance. You can find a link to it in File 11-1 at www.dummies.com/go/nonprofitkitfd5e. In December 2013, the federal Office of Management and Budget published new guidance that clarified that a nonprofit's indirect costs were legitimate expenses and should be reimbursed. Further, the Real Cost Project and Nonprofit Overhead Cost Project (see links in File 11-1 at www.dummies.com/go/nonprofitkitfd5e) illustrates the conversation among nonprofits and foundations to recognize that supporting indirect costs is critical to nonprofits' heath and abilities to deliver effective programs.

If your organization is relatively small, with just a few programs, you may be able to keep accurate records for how staff members spend their time, how many square feet in a building a program uses, and even how many office supplies each staff member checks out of the supply cabinet. Based on those records, you can fairly estimate your indirect costs.

As your organization grows in complexity and number of programs, you may compute indirect costs by using a formula you derive based on your direct costs. We outline sample formulas and steps to derive them in Table 11-1.

TABLE 11-1 Computing Direct Cost Percentages for an After-School Center

First, add all your direct costs:	
Program	Direct Costs of Each Program
Tutoring and homework assistance	$46,000
Arts and crafts	$60,000
Athletics	$164,000
Total direct costs:	$270,000
Then compute the percentage of total direct costs that each program represents:	
Tutoring and homework assistance	$46,000 ÷ $270,000 = 17%
Arts and crafts	$60,000 ÷ $270,000 = 22%
Athletics	$164,000 ÷ $270,000 = 61%

In many cases, you can reasonably assume that the indirect costs of a program correlate to its direct costs, or that a program that's more expensive requires more oversight of staff and more use of office space. In the case of the after-school center, all the indirect costs — which include a salary for the part-time director, a contracted bookkeeper, rent, utilities, telephones, and printing — total $130,500.

Using this figure, we can put together the direct and indirect costs into a true financial picture of each program by multiplying the $130,500 in indirect costs according to the percentages we computed for direct costs in Table 11-1. We outline the proposed allocation in Table 11-2.

TABLE 11-2 Allocating Indirect Costs for an After-School Center

Program	Formula	Indirect Cost Amount
Tutoring and homework assistance	$130,500 × 17%	$22,185
Arts and crafts	$130,500 × 22%	$28,710
Athletics	$130,500 × 61%	$79,605

Sometimes organizations write a project budget by identifying all the direct costs in detail and then showing the indirect costs as one lump sum at the bottom of the list of expenses. But some funding sources balk at paying for indirect costs when they're presented this way. The funders want to know what actual expenses contribute to those indirect costs. We recommend that you show the specific indirect costs item by item rather than lumping them all together (unless some of those items are quite small). For the tutoring program, for example, you would show 17 percent of the executive director's salary, 17 percent of the bookkeeper's fees, and so on.

TIP

File 11-3 at `www.dummies.com/go/nonprofitkitfd5e` shows the tutoring program budget with the indirect costs specified alongside the direct costs.

Working with Your Budget

Your budget isn't capable of getting up and walking out of the room, but it needs to be an "active document." If you simply create it once a year to keep in a file folder and submit with grant proposals, it doesn't do you much good. A good budget is reviewed often. A good budget guides and predicts.

REMEMBER

Numbers in your budgets are meant to be compared. One critically important comparison looks at actual income and expenses alongside the original projections you made in your budget. Most organizations create a spreadsheet that includes columns for:

- Annual budget
- Year-to-date income and expenses
- Current month's budget (1⁄12 of the annual budget)
- Current month's income and expenses

Pay close attention to where you're exceeding your budget and where you're falling short. Use your progress as a guide to adjust your fundraising or your expenditures.

TIP

At `www.dummies.com/go/nonprofitkitfd5e`, in Files 11-6 and 11-7, you can see two examples of tracking actual income and expenses in comparison to budgeted income and expenses. To take a critical, long view of your financial progress, use the Five-Year Financial Trend Line in File 11-8.

Other useful exercises to make sure your budget is "living and breathing" include the following:

- Involve your staff (paid and volunteer) in the early stages of drafting the coming year's budget. If you can't afford all their dreams, involve them in setting priorities.
- Three months before the beginning of your new fiscal year, meet with a small committee of board members to review and refine a budget draft. Present the draft with options and recommendations to the entire board for discussion and formal approval.
- If your organization's situation changes significantly, prepare and adopt a formal budget revision. Budget revisions are time consuming, but don't try to proceed with a budget that doesn't reflect the size and scope of your organization. You don't want to try to find your way through Maine with a map of Utah.
- Provide budget copies monthly or quarterly to your entire board or board finance committee. Encourage your board treasurer to summarize the organization's financial situation and invite questions and discussion at each meeting. One important board action is reviewing the financial report at each meeting.
- Keep notes in your budget file about changes you recommend for years ahead.

Because revenues aren't earned, grants aren't received, and expenses aren't incurred in equal amounts every month throughout the year, we recommend that you also project your cash flow (see the following section).

Projecting Cash Flow

A *cash-flow projection* is a subdocument of your budget that estimates not only how much money you'll receive and spend over the course of a year but also *when* you'll receive and spend it. It breaks down your budget into increments of time. Some organizations create quarterly (three-month) projections, some monthly, some weekly. Our personal preference is to do cash-flow planning monthly.

Although grant-making organizations and major donors are likely to want to see your budget (and may want to see the version that compares projections to actual amounts), your cash-flow projection generally is a document for you, your board, and possibly a loan source.

If you think of a budget as the spine of an organization — supporting all its limbs — the cash-flow statement is its heart and lungs. A good cash-flow statement is in constant motion, anticipating and following your every move. Based on the careful way you developed your budget, you may know that you're likely to have enough revenue to cover your organization's expenses in the coming year. The cash flow helps you figure out whether you'll have that money at hand when you need it.

Constructing your cash-flow projection

To set up a cash-flow projection, begin with a copy of your budget and add details to the names of all the various categories. For instance, under "Foundation Grants," write down the names of every foundation from which you now receive money and of any from which you anticipate receiving a grant. Add the same kinds of details for your expense categories. For example, under "Utilities," add separate lines for each bill you receive, such as water, electricity, gas, and sewer service.

TIP

We recommend that you project forward by going backward. Sounds contradictory, doesn't it? Create columns for the most recently completed three months of your year. Go through your records. Put any income received into the appropriate periods, and write down all your expenses in the right categories and time periods. If you've forgotten any categories, you have a chance to add them. Having these actual figures for the recent past helps you see patterns of income and expenses.

Now, begin projecting: Go back to each line item and write down the estimated amount that's due during each monthly period. Begin with the "easy" items — like the rent that's a constant amount due on a certain day or the employer's share of federal payroll taxes. Then look at the consistent bills that vary over time. If your utility bills are high during winter months because your agency is using the furnace more, don't forget to project that increase. If you manage a community vegetable garden during the summer that increases water usage, project for that increase as well.

As you get into the flow of making predictions, it's easy to become too optimistic about your anticipated income. If you've applied for a grant that you're just not sure about, don't put it into your cash-flow projection. If your annual fundraising event raises between $25,000 and $32,000 each year, project $25,000 in income. The reason? You won't have any problem knowing what to do when you have more money than anticipated, but you may have a problem making up a shortfall! Your cash-flow projection is supposed to warn you if you're about to fall short.

You're almost finished at this point. As a next step, look back at your financial records to see how much money you had at the beginning of the first month of your cash-flow projection. Place this figure in a "Balance Forward" row as the first income item for your first month period. Add it to all the income for that month and subtract that month's expenses. The difference gives you the next "Balance Forward" amount that belongs at the top row for the next column of income for the *next* monthly period. And keep going. The balance for each month is steadily carried over to the top.

Table 11-3 shows a highly simplified sample of a cash-flow projection.

See File 11-10 at `www.dummies.com/go/nonprofitkitfd5e` for a detailed sample cash-flow projection, and check out File 11-11 for a blank form you can use to begin your own cash-flow statement.

Deciding what to do if you don't have enough

You probably won't have more income than expenses in every single month throughout the year. During some periods, you get ahead, and at other times, you fall behind. Your goal is to sustain a generally positive balance over the course of time and be able to cover your most critical bills — payroll, taxes, rent, insurance, and utilities — in a timely manner.

TABLE 11-3 **Simple Cash-Flow Projection**

Income	1/1–1/31	2/1–2/28	3/1–3/31
Balance forward	$21,603	$9,845	$43,445
Government contract		$65,000	
Williams grant	$15,000		
Power company contribution			$2,500
Board giving	$4,000	$750	
Total income and balance forward	$40,603	$75,595	$45,945
Expenses	**1/1–1/31**	**2/1–2/28**	**3/1–3/31**
Payroll	$22,197	$22,197	$22,197
Health benefits	$2,111	$2,111	$2,111
Payroll taxes	$2,775	$2,775	$2,775
Rent	$2,500	$2,500	$2,500
Electric company	$675	$675	$580
Telephone	$435	$450	$475
Office supplies	$65	$120	$120
Travel and transportation		$1,322	$8,700
Total expenses	$30,758	$32,150	$39,458
Balance:	$9,845	$43,445	$6,487

REMEMBER

Some government agencies and foundations pay grants as reimbursements for expenses after you've incurred them. You may need to prepare your cash-flow plan accordingly while waiting to be reimbursed.

We want to promise that you'll always have enough resources to sustain your organization's good work. But some times will be lean. Then what? First, your cash-flow projection should help you anticipate when you may fall short. It enables you to plan ahead and solicit board members who haven't yet made gifts in the current year, send letters to past donors, cut costs, or delay purchases.

We also recommend being proactive about contacting your creditors. If you think you can't pay a bill on time, call and ask for an extension or explain that you're forced to make a partial payment now with the balance coming in a few weeks.

Your ability to do business depends on your earning and sustaining other people's trust. If you can't have perfect credit, being honest and forthright is the next best thing. It's hard to do, but important.

WARNING

Don't "hide" behind your bills, thinking that if you don't say anything, nobody will notice. If you're facing a period of debt, tell your board. Also, call anyone to whom you owe money and explain the situation and your timeline for paying your bills. Among other things, your honesty helps them with *their* cash flow. And when your expenses are exceeding your income, always try to cover your federal, state, and local tax obligations, because fees and interest on unpaid taxes add up quickly. If covering those obligations isn't possible, don't forget to call those agencies. Setting up a payment plan, both over the phone and in writing, can prevent your assets (or a board member's) from being frozen.

Borrowing to make ends meet

Another route to managing cash shortfalls is to borrow the money you need to cover your bills. Your cash-flow statement can help you plan the size and duration of the loan you need. Here are some of your options if you need to borrow money:

- **Ask a board member for a loan.** If the board member can help, you can probably secure the loan quickly, but borrowing from a board member is only appropriate to do if you act in accordance with your organization's conflict-of-interest policies and applicable law, and if the board member can provide the loan at market rates or below market rates. Make sure to sign a promissory note with the lender and record the board's formal approval of accepting the loan in its meeting minutes.
- **Ask local foundations whether they know of a loan fund for nonprofit organizations.** Some associations of grant makers and government agencies offer loans at low interest to their grantees facing cash-flow problems. Such a program is likely to be more sympathetic to your needs than a commercial lending institution may be.
- **Apply for a small-business loan at your bank.** If your organization doesn't have a credit history, getting a bank loan can be challenging. However, foundations sometimes make this process easier by guaranteeing these bank loans for nonprofits.
- **Check to see whether you qualify for a line of credit from your bank.** A *line of credit* allows you to borrow up to a certain sum for a specified period of time. When the organization repays the borrowed amount, often it can't borrow from its line of credit again for a designated period of time. Your organization may want to apply for a line of credit even if it doesn't expect immediate cash-flow problems. Doing so can provide a safety net for emergencies.

» **As a last resort, borrow the money from an organizational credit card.** Do this only if you're positive that you can repay the loan quickly and cover the interest.

Borrowing money requires time and preparation and usually costs your nonprofit in the form of interest payments. However, borrowing is better than damaging your nonprofit's reputation or incurring severe penalties and interest charges — particularly on tax obligations.

Putting money away to make a nonprofit strong

Over the last five years, nonprofits, foundations, and consultants have talked and published articles about the importance of nonprofit "capitalization." The idea is simple: If a nonprofit has savings, it can respond to urgent needs and take advantage of opportunities. Just as a family is better positioned to manage a flooded basement or invest in a small business if it has set some money aside, nonprofits should set goals for saving working capital.

Any business — including a nonprofit — needs money to operate and it also needs money to grow, to test ideas, and to change its programs or structure. Working capital is that "grow and test" money that makes a nonprofit adaptable and also that allows it to invest in long-term needs (such as gradually replacing aging equipment).

Many grants and government contracts limit how you can spend project money, so if you want to develop working capital, we recommend turning to individual donors or setting aside a small portion of your earned revenue each month. We also recommend setting tangible goals for this effort — perhaps the cost of replacing essential equipment or operating for four months without any new income while you set up programs and write grants.

Keeping Your Books Organized and Up to Date

After you've created a budget, you need to set up a system for keeping consistent, organized records of your financial information. You may choose to do your bookkeeping by hand or with accounting software. If you're interested in bookkeeping by hand, *Nonprofit Bookkeeping & Accounting For Dummies*, by Sharon Farris (Wiley), can guide you. If you prefer to have a computer do the math for you, we name some software options later in this section.

Accounting, like the world of nonprofits, has some specific ways of doing things. In the next section, we explain the two fundamental accounting approaches.

Understanding the different accounting systems

You'll encounter two standard systems for compiling and presenting information about your organization's financial history: cash and accrual. Here are some ways they differ:

- **A cash system closely resembles the way most people keep their checkbooks.** You enter income into the books when you receive and deposit a check or cash. You enter expenses when you pay a bill. Many find a cash system to be a comfortable approach because it's straightforward. When your books show a positive balance, you have money in hand or in the bank.
- **An accrual system recognizes income when it's promised and expenses when they're obligated.** Suppose you receive a grant award letter from a foundation promising your organization $60,000 over a two-year period. This grant is going to be paid in four checks of $15,000 each. In an accrual system, you enter the entire $60,000 as income in your books when that money is promised — when you receive the grant award letter. On the expense side, you enter your bills into your books when you incur them — which may be before you pay them.

WARNING

Each of these approaches has possible disadvantages:

- In cash books, your organization may owe money in unpaid bills, but those debts aren't apparent because they aren't on your books until they're paid.
- In cash books, your organization may have been awarded a large grant but may look poor because you haven't received the check.
- In cash books, it's more difficult to tell whether you owe payroll taxes.
- In accrual books, you may have been promised a large contribution but not yet received it. You may have no cash in hand, but your books look as if you have surplus income.

Keeping books by an accrual system is standard accounting practice for nonprofits and is recommended by the Financial Accounting Standards Board, so this section of the book assumes you use an accrual system. To be honest, however, small organizations can often get by with a cash system on a daily basis and then transfer them to an accrual presentation once each quarter or at the end of their fiscal year.

Considering accounting software

If you decide to use software to keep your books, you can purchase types that are designed specifically for nonprofits, which have some different needs and use a few different terms than for-profit businesses. We recommend that you choose one of these products meant for nonprofits, though accountants differ on whether it's necessary. Some accounting software packages that many small nonprofits use include QuickBooks or QuickBooks Online by Intuit, Sage 50, and Fund E-Z.

If more than four or five people at your organization will be using the accounting software, if your budget exceeds $1 million, or if you need to track many programs, you'll likely want to invest in more-complex accounting software. Two options are Abila MIP Accounting and The Financial Edge by Blackbaud.

Also important, if your nonprofit has multiple full-time staff members, it will need systems to manage payroll and track time and attendance. Intuit's Online Payroll and Paychoice, a cloud-based service that integrates well with QuickBooks, are popular among small organizations. Larger nonprofits may prefer ADP (Automatic Data Processing), Sage 50 Payroll, or Paychex.

The online journal Idealware (`idealware.org`) is an excellent source of articles, blogs, webinars, and forums about technology for nonprofits. We include a link to it at `www.dummies.com/go/nonprofitkitfd5e` in File 11-1.

When creating your startup budget, don't forget to include accounting software — the cost of purchasing the package or online service, of setting it up, and of being trained to use it. As your organization grows and you need more sophisticated software, the setup time will be greater.

Reviewing the Numbers: Financial Statements and Audits

Budgets and cash-flow statements predict your organization's financial future. But you also need to review each year's annual income and expenses in a financial statement. When your nonprofit organization is small and lean, your staff or bookkeeper can prepare its financial statements. As it grows and becomes more complex, you'll likely hire an outside accountant to prepare and audit its financial statements.

Preparing financial statements

If a budget is a document about the future, a financial statement tells the story of your organization's past. Nonprofit organizations keep and review their financial records throughout the year, and they prepare a financial statement at least once a year, at the end of the fiscal year. (See Chapter 4 for more information on a fiscal year; and see File 11-12 at `www.dummies.com/go/nonprofitkitfd5e` for tips for reading financial statements.) Many organizations also prepare monthly or quarterly "in progress" versions of their annual financial statements.

Although good nonprofit accounting software can organize your financial information for you, preparing and interpreting financial statements is a special area of expertise that goes beyond the scope of this book. Many nonprofits seek outside professional help for this essential task. If hiring such assistance is beyond your organization's means, we recommend that you become acquainted with *Nonprofit Bookkeeping & Accounting For Dummies*, by Sharon Farris, and *Accounting For Dummies*, 6th Edition, by John A. Tracy, CPA (Wiley). If you do choose to prepare your own financial statements, you may want to hire an accountant at year's end to review them for accuracy.

The information in your annual financial statement resembles (but may not be identical to) the 990-EZ or 990 financial report that your organization is required to submit annually to the IRS. (You can find more information about the differences between these types of 990s and filling one out in Chapter 5.) You'll also include your financial statement in your board orientation packets and with requests for funding. Some organizations publish it in an annual report.

Seeing the value of an audit

At year's end, your organization may be required to or may choose to have its books audited by a certified public accountant (CPA) or firm. This service involves a formal study of the organization's policies and systems for managing its finances, a review of its financial statements, and commentary about the accuracy of those statements.

TIP

If you hire a CPA to conduct your audit or financial review, make sure the person has knowledge about or expertise in nonprofit organizations. Nonprofits use some accounting terms and bookkeeping methods that differ from for-profit businesses.

The accountant may issue an *unqualified opinion*, meaning that the statements appear to be accurate. In "auditor talk," the accountant didn't find a *material misstatement* — false or missing information. If the auditors have recommendations to make about how money is managed or how finances are recorded and reported, they may issue a *management letter* to the board about how the organization can improve its practices. These letters can contain a wide array of recommendations. For example, the auditors may tell the board that the organization didn't consistently collect original receipts before writing checks to reimburse staff; that staff travel costs increased dramatically without that increase being approved by the board (and recorded in board minutes); or that the organization's employees are owed so many vacation hours that the organization's work could be jeopardized if everyone were to take his or her deserved holiday.

Although some nonprofit organizations look upon a management letter as a scolding, its point is to help organizations safeguard their financial health by strengthening their policies and recordkeeping.

A representative of the auditing firm should present, formally and in person, any insights from the audit to the nonprofit board or audit committee. Often a draft version of the audit is first shared with the audit committee, providing it with an opportunity to respond before a final version is formally accepted by the board. Following this presentation and discussion — which may lead to a revision of the audit draft — the board should vote to formally accept the audit. If the audit or management letter contain recommendations for improving financial procedures or other matters, the nonprofit board should discuss them with the executive director and write a brief, formal letter acknowledging the points made and how the organization will respond.

TIP

The National Council of Nonprofits provides a helpful online Nonprofit Audit Guide and we've included a link in File 11-1 at `www.dummies.com/go/nonprofitkitfd5e`. You'll also find there the article "Get the Most Value from Your Audit" from the online journal *Blue Avocado*.

Knowing whether you need an audit

When do you need to conduct an audit? Use the following guidelines to help you determine this.

- **Many states require nonprofits that receive revenues over a specified amount to conduct audits.** This varies by state. Check with the office that regulates nonprofits in your state (usually the attorney general's office) to see whether it has specific audit guidelines.

» **Nonprofits that directly spend $750,000 or more in federal funds in a fiscal year are required to conduct an OMB A-133 audit — commonly called a Single Audit.** It has a fancy name because it's based on the Office of Management and Budget Circular A-133, and it takes a particularly close look at how government funds are tracked within the organization and whether the organization complies with federal laws and regulations.

» **Some other kinds of government programs also require audits of specific grants or contracts.** Other funders may also require audited statements from applicants.

Even if it isn't required, your board may decide that it's a good idea to have an outside CPA examine the books. Doing so provides reassurance that financial systems are healthy and the organization is working with accurate financial information.

If your organization is required to have an audit, some states require your board to form an audit committee and have rules about its composition. For example, in California, the board president, board treasurer, and paid staff members may *not* serve on this committee. The committee selects the audit firm, reviews a draft of the audit and any recommendations from the auditors, and, if necessary, investigates any practices that should be changed.

If you have an audit prepared for your organization, check out File 11-14 at `www.dummies.com/go/nonprofitkitfd5e`, which contains advice for reading an audit.

Although your organization can learn a great deal from an audit, the practice isn't appropriate or necessary for every nonprofit. The process is expensive and time consuming. A less-expensive option is having a CPA firm provide a formalized compilation or financial review of annual statements. Smaller nonprofits often go this route to cut down on costs but still provide some level of comfort and assurance to funders that the organization has the appropriate level of financial controls in place and that their financial statements are likely accurate.

Many nonprofits wonder whether they should seek pro bono audits from accounting firms or ask a board member's firm to audit their books. Although we're big fans of contributed services, a pro bono audit is a bad idea, and asking a board member for assistance is a very bad idea. An audit should be prepared independently of the organization's staff and board. It loses its value as the voice of unbiased, outside validation when provided as a gift.

Reading Your Financial Statements

Audited or not, prepared by you or by a bookkeeper or accountant, your organization needs to produce financial statements, and it's critically important for you and your board members to understand how to interpret them. So in this section, we show you everything you need to know, including how to identify the two parts of a statement and how to arm yourself with the right information.

Getting to know the parts of a standard financial statement

Nonprofit financial statements include two important substatements:

- **The *statement of financial position* (also called a balance sheet) provides an overview of what an organization is worth.** It outlines how much money is available in bank accounts and other investments; the value of property, furniture, and equipment; immediate bills; and other debts and liabilities. The basic formula at the heart of this statement is Assets – Liabilities = Equity.
- **The *statement of activities* outlines "Support and Revenue" and "Expenses and Other Losses," which tell you how much money the organization received in the past year, its sources for that income, and how it spent that income.** The basic formula at the heart of this statement is Revenues – Expenses = Net Income. Statements of activities divide an organization's income and expenses into three categories:
  - **Unrestricted funds** were available to spend in the fiscal year covered by the statement.
  - **Temporarily restricted funds** were promised or awarded in the year covered by the statement, but they were given to the organization for it to spend in the future or for a specific task that isn't yet complete.
  - **Permanently restricted funds** were given to the organization for permanent investments — such as endowments. Earnings on such money may be used toward the organization's costs (your board will set a policy about how much of these earnings may be used, subject to applicable state laws), but the amounts donated as permanently restricted gifts shouldn't be spent on current activities.

  The statement of activities also contains a statement of *functional expenses,* which details expenses spent on programs and those expenses that were exclusively for general and administrative costs and for fundraising costs.

WARNING

Many staff and board members are drawn to nonprofit organizations because they're knowledgeable about the services they provide, but their eyes glaze over when they're faced with financial statements. This lack of interest in finances is understandable but can be dangerous: The board and staff are stewards of the organization's resources, and the board accepts the responsibility of providing fiscal oversight. If your board lacks members with financial expertise, your board orientation could include training on reviewing and understanding nonprofit financial statements. BoardSource and other nonprofit service organizations also can help board members understand what they're reviewing and what questions they should ask to ensure financial integrity.

TIP

We include a sample of an audited financial statement at `www.dummies.com/go/nonprofitkitfd5e` in File 11-13 and some tips for reading an audit in File 11-14 that you can use to become familiar with which terms are used and how the information is organized.

Asking the right financial questions

At the year's end, when you've completed the financial statement, we suggest reviewing the following points, either as a board or within the board's finance committee. You can ask different questions about the statement of position (or balance sheet) and the statement of activities. The answers arm you with a clear understanding of your organization's financial health.

Looking at the statement of position, compare some of the numbers, asking:

- **Do the cash and cash equivalents exceed the accounts payable?** If so, the organization has enough cash in the bank to pay its immediate bills to vendors.
- **Are accounts payable increasing over time?** You'll see fluctuations in amounts owed from month to month and year to year, but if accounts payable are steadily rising, the nonprofit's bills are getting ahead of its capability to pay them.
- **Do the total current assets exceed the current liabilities?** If they do, the organization has enough money readily at hand to cover all its immediate obligations, such as bills and taxes, employee benefits, and loan payments.
- **How much of the assets are made up of property, equipment, or other durable goods the organization owns, and what is the nature of those assets?** Some organizations appear to be financially healthy because their total assets are high, but they may have trouble covering immediate expenses if all those assets are things that are hard to sell.

- **Does the current liabilities section show a high amount of payroll taxes payable?** If so, the organization may be delinquent in paying its taxes.
- **In the liabilities section, do you see an item for a refundable advance or deferred revenue?** If you see this item, it means money paying for a service was given to the organization before the organization provided that service. Make sure the organization has sufficient assets — preferably cash — to provide the services it has promised.
- **Has the revenue-to-date ratio changed?** To see whether income is coming into your organization in a steady and reliable way, divide the current year's revenue by the prior year's revenue at the same point in time. If your revenue is steady over time, the ratio figure you'll compute is close to 1. If the ratio is less than 1, money is coming into your nonprofit at a slower rate.
- **Are you using temporarily restricted funds for other purposes?** *Temporary restricted net assets* consist of money from foundations, donors, or customers for which the organization must still perform some kind of service. For example, if your nonprofit receives a foundation grant to produce a particular play and subscribers buy advanced tickets to that play, the money you receive from those sources is considered temporarily restricted until you produce the play. What if you need money immediately to pay the rent? Sometimes organizations borrow from their temporarily restricted funds for such a purpose. If you do so, the temporarily restricted funds should be restored as quickly as possible. Otherwise, you won't have the money you need to produce the play. To see if temporarily restricted net assets have been spent, divide cash plus accounts receivable by temporarily restricted net assets. The ratio should total 1 or higher.
- **Has the organization borrowed money?** If so, the amount appears in the liabilities section as a loan or line of credit. Compare its total unrestricted net assets to the total amount borrowed. If the numbers are similar, the organization may have cash at hand, but all of it must be repaid to a lender. This is not necessarily bad if the situation is temporary: Organizations may need to manage their cash flow with judicious borrowing.

The financial statement will be dated, and the answers to the questions you have just asked should give you a good picture of the health of your organization over its entire history leading up to that date. Now it's time to look at the statement of activities, which illustrates what happened in the past year. Two simple questions summarize that story. Look to the bottom of the page to see:

- **Is the change in net assets a negative or positive number?** If it's negative, the organization lost money in the previous year. If the loss is a high number, ask why it took place. If it's a modest amount, ask whether it's part of a downward trend or something that rarely occurs.

- **Is the number shown for total net assets a positive number?** If so, the organization is in a positive financial position (even if it lost money in the past year). That's good news!

In accrual accounting, if an organization receives a multi-year grant, the entire amount of that grant is shown as revenue in the first year. If that grant is a major part of the nonprofit's annual revenues, it may create a high positive net balance in the grant's first year that is followed by apparent losses in year two and year three. You may want to footnote this situation in presenting your finances to your board or funders. If your books are audited, you can ask the CPA firm to include a footnote about it in the audit.

Managing Financial Systems

Even in a small organization, you want to establish careful practices about how you handle money and financial documents. If you have a one-person office, creating all the following controls may be impossible, but you should try to implement as many of them as possible:

- Store checkbooks, savings passbooks, blank checks, financial records, and cash in a locked, secure place. If you're banking online, use secure passwords — multiword or nonconsecutive letters with numbers and symbols. Don't write your passwords down and don't allow your computer to remember them.
- Regularly back up financial records that you keep on computers, and store a copy off-site in a safe location.
- Assign to different people the functions of writing checks, signing checks, and reconciling bank statements. If you can arrange for three different people to perform these tasks, that's great. If not, maybe a board member can double-check the bank statements. Look for accuracy and for anything that looks fishy, such as payments to vendors you don't recognize . Embezzlement is rare but possible, and taking these steps is a way to detect it. Trust us, we've discovered such a case!
- Require two signatures on checks or bank transfers over a certain amount. We can't give you a standard amount for this practice: It depends on what qualifies as an unusually high transaction for your nonprofit. You can adopt this requirement as an internal policy and enforce it as a way of ensuring board oversight of large transactions. Although your bank probably monitors large or unusual transactions and may contact your organization if it notices unusual activity on your account, these days few banks accept responsibility for enforcing a two-signature policy on customers' checks.

» Retain in organized files, subject to a document retention/destruction policy, all paperwork that backs up your banking documents. These documents may include personnel time sheets, box office or other records for tickets sold, receipts, and invoices. You may scan and retain them as electronic documents, if you wish. If you do so, we highly recommend backing up the files and carefully protecting passwords.

» Keep an itemized list of any furniture or equipment you purchase or receive as donations (including computers), noting the date they were purchased or received and their value.

TIP

Blue Avocado (`www.blueavocado.org`) published a helpful article, "Treasurers of All-Volunteer Organizations: Eight Key Responsibilities," and we include a link to it at `www.dummies.com/go/nonprofitkitfd5e` in File 11-1.

» Looking at who you are and whom you serve

» Building your publicity through mass media

» Working the social media game

Chapter 12

Marketing: Spreading the Word about Your Good Work

You may think that marketing has no place in the nonprofit sector. After all, if an organization's work isn't driven by the dollar, and if its focus is on meeting a community need, why does it need to sell its services? If your motivation is to do good, isn't it crass to toot your own horn?

But wait. Marketing — the process of connecting consumers to services and products — is just as critical to the success of nonprofit organizations as it is to for-profit enterprises. Both nonprofit and commercial organizations depend on getting the word out, and the message from both is the same: "Here we are, and here's what we can do for you. Come and check us out. Use our services." For non-profit organizations, that message includes, "And please help us in our work by making a contribution."

You can go far if you have the "goods" — good planning, good stories, good services, good networking, and good luck. And, of course, persistence.

Check out File 12-1 at www.dummies.com/go/nonprofitkitfd5e for a list of web resources related to the topics we cover in this chapter.

Taking Care of the Basics

Before you begin tackling client surveys and media releases, you need to spend a little time making sure you have a few basic communication tools in place. These tools help you tell your clients, audiences, donors, and the general public who you are and what you do.

In this section, we provide some tips for the most basic communication tools that every nonprofit — large or small, new or old — should have.

Before we go much further, we want to introduce you to a few websites that can help you understand and use technology for nonprofit communications and management. Idealware (www.idealware.org) is the place to go for information and advice about which software will work best for your nonprofit. After you decide what you need, TechSoup (www.techsoup.org) provides free and discounted software packages to qualified nonprofit organizations. The Nonprofit Technology Network (www.nten.org) is an online community that helps nonprofits use technology tools to their best advantage.

Designing a logo and letterhead

Your letterhead should include your organization's name, address, phone number, fax number, and email and website addresses. Many organizations also list their board members on the letterhead. Doing so is a great way to highlight the board members' affiliations with your organization, but if your board is rapidly changing and growing, you may be reprinting the letterhead every few weeks. You may also want to include a tag line that briefly identifies what your organization does, like *Feeding the hungry in Tuborville.* Often the tag line is a phrase drawn from your mission statement and also appears on the home page of your website.

A logo is a graphic image that represents the mission of your organization. Many people remember pictures more vividly than words, so they may remember your logo more easily than they remember your organization's name. Try not to resort to standard computer font symbols or clip art when designing your logo. If you need to be frugal, ask around your community to see whether anyone knows a graphic artist or a printer with a good eye who can help you create something eye-catching and original. If you can't locate anyone, we recommend a simple solution: Choose an attractive typeface in which to present your name. You can get fancier later.

Whether it's complex or simple, as you're creating your logo, produce several drafts, spend time looking at those drafts, and show them to others, asking for feedback. You don't want to grow tired of your logo or have it confused with the logo of another group in your community.

Preparing an organization overview or brochure

An *organization overview* is a one-page description of your organization's mission and programs. You can enclose it with news releases, grant proposals, and fund-raising letters. Place copies in your lobby or reception area for visitors to read. Feature it prominently on your website.

As you grow, your organization may want to get a little bit fancier and produce a brochure with photographs or graphics. This may be a paper document or it may be a page on your website. An effective brochure clearly conveys the essence of your organization. You want it to be inviting and readable. Visual elements help to make a more attractive document and often enable you to tell your story in fewer words.

TIP

Keep the colors and design elements consistent among any printed materials and your website. You're creating a face and a voice for your organization, so try not to send out mixed messages.

Creating a website

Having a web presence is as important as being listed in the telephone book — maybe more important. (Who uses phone books anymore?) A basic website isn't difficult to create. You want to convey the key message of your agency on the home page and provide a few additional pages where people find out more about your organization's leaders and programs. Including strong photographs of your organization's work in action can make it more appealing to visitors. Most important, you want to let visitors know how they can learn more about, do more with, and become involved with you.

A few key features we highly recommend include:

» A "Get Involved" page or menu of pages that lets readers know what they can do to help your cause. Involvement invitations can range from removing blackberry bushes from hiking trails to taking and uploading pictures of the playground your nonprofit is building to sharing information with friends about the organization's work.

- A contact page with phone numbers and email addresses. Not only does a contact page suggest that you're available to provide more information, but it's also a means by which your organization can begin gathering email addresses for further communications about events, programs, and fundraising.
- If your organization provides services at a physical location, provide a map and transportation instructions (parking, bus routes, and so on) to help people find it.
- Clear information about how to make a contribution to your nonprofit — preferably a "donate now" feature for collecting donations. (See Chapter 14 for more information.)
- If you're collecting information from donors, volunteers, or others who come to your website, it's a good practice to develop and make available a privacy policy and an unsubscribe option on your website that connects them to a newsletter or other emailings.
- Icons for all your social media networks. It's an easy way for your website visitors to like or follow your organization. (We cover how to use social media to promote your organization at the end of this chapter.) You can also add the icons to the signature in your email and ask your staff to do the same.

In good time, you can add bells and whistles to your website: a gallery of photographs or short videos, articles about your field, a blog or discussion board, and much more. But keep in mind that as you add features to your website, you're also increasing the number and types of things that need to be maintained. If you need to learn about designing and building a website, have a look at *Web Design All-in-One For Dummies*, 2nd Edition, by Sue Jenkins, and *Building Websites All-in-One for Dummies*, 3rd Edition, by David Karlins and Doug Sahlin (Wiley).

TIP

In File 12-1 at `www.dummies.com/go/nonprofitkitfd5e`, we provide a link to the top 100 nonprofit websites of 2015. They're analyzed for their different qualities and strengths. Take a look for inspiration and also search for the top 100 nonprofit websites of previous or subsequent years.

TIP

If you're looking for assistance with creating a website, you may find free or low-cost help at a local college or university. Often students majoring in graphic arts, design, educational technology, and technical writing are required to complete internships or capstone projects as part of their degree programs. You may find a talented, eager designer by contacting his department or by reaching out to the campus's community-service learning program. Another option is to search online for a graphic artist through Volunteer Match (`www.volunteermatch.org`). If you do use such a volunteer, make sure that person leaves behind clear information about how you can continue to refresh the site in the future.

WARNING

An unattended website leaves a bad impression. Broken links and out-of-date information tell the reader that your organization isn't paying attention to communicating with its constituents. If you're going to have a website, assign someone in your organization the task of checking on and updating it regularly. If your budget allows, consider hiring a web design firm or web-savvy individual to create and maintain your site.

Producing annual reports and newsletters

Begin producing an annual report after your first year of operating. The report may be published as a letter, website feature, or even a booklet. Usually, the annual report includes a financial statement for the year, along with an overview of recent accomplishments. It may also include introductory letters from the board president and executive director of your organization. An annual report conveys the image of an accomplished organization that's transparent in its communications. The messages such reports convey build trust among people seeking organizations' services or considering making contributions.

A newsletter offers background stories and information about your organization. You can publish it as a feature page or slide show on your website, as a PDF document linked to your website, in email, or as a printed or photocopied document. Issuing a newsletter is a great way to keep in touch with your organization's constituents. It can take them behind the scenes of your organization. You may use a newsletter to profile members of your staff, board, and constituents; alert followers to upcoming events; summarize research; and announce news about your organization's work.

Taking care of customers — your most important marketing tool

If your constituents, audiences, patients, clients, volunteers, and board members aren't treated well, you can make major investments in advertising and public relations but still struggle to find new customers and keep your old ones. Small adjustments in customer service can make a big difference to your organization's image.

REMEMBER

If your agency is likely to receive calls from people requiring counseling or emergency assistance, make sure that everyone answering the phone is thoroughly trained in how to calm a caller and provide a referral.

MAKING CUSTOMER SERVICE A DAILY HABIT

The trick to providing exemplary customer service is to work at it in little ways all the time. Every staff member and volunteer needs to be aware of the importance of customer service. We offer five key areas to address if you wish to improve (or maintain) your service levels:

- **On the telephone:** Talk to all staff and volunteers about how to answer the phone politely and otherwise use it to maximum effect. Provide a cheat sheet with lists of extensions, instructions on how to forward calls, and any other useful information. Simple additions to a greeting, like "May I help you?" or telling the caller the receptionist's name can set a friendly, professional tone.

  Make a rule for yourself to return all calls within 24 hours or, if that's not possible, within one week. Let callers know, through your voicemail greeting, when they may expect to hear from you. If you're selecting a voicemail system, make sure it's user-friendly for your callers. We prefer a small number of transfers within the system and options for modifying standard recorded announcements with more personal and pertinent messages.

- **At the door:** Someone needs to (cheerfully) answer the door. If you're in a small office without a receptionist and the interruptions are frequent, you can rotate this task among staff members or volunteers. If visitors must use a buzzer or pass through a security system, try to balance the coldness of that experience with a friendly intercom greeting and pleasant foyer.

- **Before and after a sale or donation:** If you have tickets to sell, make it easy to buy. Accepting only cash is far too restrictive. The banker managing your business account can help prepare you to accept credit card orders. Also, consider selling tickets over the web; a number of web services assist small businesses and nonprofits with web sales and online donations. If a customer isn't satisfied with your service, invite the customer to give feedback and listen carefully. Offer a partial or complete refund. Doing so wins loyalty.

- **With a note:** Keep some nice stationery and a rough draft of a standard message handy so that writing thank-you notes is easy. Handwritten notes are more personal and often make a stronger impression than formal, typed letters or email messages. However, expressing your appreciation in a timely way is more important than the form of that appreciation. Use email if you can get to it more quickly. Write a thank-you note for a contribution within a few days of receiving it.

  If your organization receives *things* — like a theater group that receives play manuscripts or a natural-history museum that accepts scholarly papers for publication — have an acknowledgment email ready that confirms the item's receipt and states when the person submitting it can expect to hear from you.

- **In the details:** An old saying suggests that the devil is in the details, but that's also where you find the heart and soul of hospitality and service. Keep notes in a database about the interests, connections, and preferences of your board members, donors, volunteers, and constituents. Find out the names of your board and committee members' significant others and be prepared to greet them personally at events and on the phone.

Discovering Who You Are: First Steps to Marketing

You've put your basic tools in place and your programs are up and running, but you sense that something is wrong. You know that you're addressing a need and that your agency is highly regarded. Why aren't your programs full? How can you engage more people in the important work that you do?

If a marketing program is what you need (and we believe it is), the first step is understanding your current situation — what you do, whom you serve, who admires your work, and who is willing to support it. When you know the answers to these questions, you can begin to build marketing strategies that highlight your agency's strengths and are tailored to all your key audiences — the people who walk through your doors every day to receive your services, your board members and staff, your donors and volunteers, and the community at large. Blend in your budget constraints (we all have to do that!), and you're ready to roll.

YOUR A-TO-Z REFERENCES ON MARKETING AND PUBLIC RELATIONS

Whole books — hundreds and hundreds of books — have been written on the topic of marketing and the related field of public relations. Our job in this chapter is to take you on a quick flyover of marketing for nonprofit organizations so you get the big picture and know where to turn for details. For all the details on marketing plans, surveys, public relations, news releases, the use of electronic media, and related topics, we recommend four other books from Wiley: *Marketing Kit For Dummies,* by Alexander Hiam; *Small Business Marketing Kit For Dummies,* by Barbara Findlay Schenck; *Public Relations For Dummies,* by Eric Yaverbaum, Ilise Benun, Bob Bly, and Richard Kirshenbaum; and *E-Mail Marketing For Dummies,* by John Arnold.

Recognizing your current market

If you want to improve the way you reach the public, you first need to know how your current marketing works. Who are your constituents? How did they find out about your organization? Why do they make use of your programs? The following sections help you answer these kinds of questions.

Surveying your constituents to gain important info

You may never discover who reads about your organization in the newspaper or sees your sign every day on the bus, but some people — those with whom you directly communicate — can be identified. Start by defining your core group — your most important constituents — and work out from there.

Suppose that your organization is a small historical society that organizes exhibits and panel discussions at three libraries in your town, publishes a quarterly newsletter, and maintains a website featuring news and information about its collection. Your current constituents (or stakeholders, if you want to use a common nonprofit term), working from the core to the outer boundaries, include the following:

- Your board and staff (and their friends and relations)
- Your docents and volunteers
- Families and organizations that donate materials to your collection
- Local library staff and board members
- People attending your panel discussions
- Schools and other groups visiting your exhibits
- Scholars and other archivists writing to ask about your holdings
- Patrons of the three libraries
- Subscribers to your quarterly newsletter
- People visiting your organization's website

Drawing up this list of interested people is easy enough. But for marketing purposes, you need to know as much as possible about the characteristics, backgrounds, and interests of each group. Some things you can do to collect this sort of information include

- Creating a database of your supporters by gathering names and addresses from every possible source within your organization — items like checks from donors, subscription forms from online newsletter subscribers, sign-up sheets from volunteers, and email messages sent to the "contact us" address on your website. Enter these names and addresses in a database that can sort them by last name, type of contact, and date of entry; if you're planning to send traditional mail to them, sort them by zip code. Articles at Idealware (`www.idealware.org`) can help you choose a database software program.

TIP

Review the zip codes appearing most frequently on your list. If you're in the United States, you can visit the U.S. Census Bureau website (`www.census.gov`) and get demographic information about residents in those zip code areas.

- Asking the three libraries whether they collect demographic data when visitors apply for library cards and whether they can share that information with you. (Some public agencies offer information about their constituents on their websites.)
- Asking a few standard questions of schools or other groups when they call to sign up for a tour. Don't engage in a lengthy interview, but find out how they heard about your program, why they want to visit it, and whether they have other needs you may be able to address. You can gather similar information when collecting registrations online.
- Inserting brief, clearly worded and inviting surveys in the programs at your public events. At the beginning and end of an event, make a public pitch explaining why it's so important for people to respond to the surveys. Make pencils or pens available. Create incentives for completing the form, such as a free museum membership for a person whose survey is drawn at random. Make it easy for visitors to your website to subscribe to announcements or services, and send a brief survey to them by email. The higher your response rate, the more accurate and useful the survey information will be.

TIP

Your surveying may be *quantitative* (measuring the degree to which people do something or believe something) or *qualitative* (going deeper into understanding their beliefs and behaviors). Generally quantitative surveys are distributed and collected — either as paper documents or online — and qualitative surveys are presented by an interviewer in a guided conversation.

REMEMBER

Be aware that there's an art and a science to writing an effective survey: The way questions are worded can influence the answers you get, and you want to receive clear, candid responses. If you want help developing your survey, check with local colleges and universities for faculty members or graduate students who understand survey techniques and who may be willing to give you some guidance. You can also find survey subscription services and sample surveys on the web that give you ideas about wording questions. Three such services are SurveyMonkey (www.surveymonkey.com), Zoomerang (www.zoomerang.com), and SurveyGizmo (www.surveygizmo.com). Each offers somewhat different features and pricing models. All of them will distribute your surveys by email and tally the results for you.

TIP

In Files 12-2 and 12-3 at www.dummies.com/go/nonprofitkitfd5e, you can find two sample surveys that may suggest wording for your survey questions.

Organizing and interpreting the responses

After you have survey results, you can compile the responses by hand or use a spreadsheet program such as Microsoft Excel or Google Spreadsheet to tally responses. Of course, if you've used an online surveying service, it will compile the answers for you.

If your survey responders identify themselves and you want to keep track of information and opinions they share to inform your fundraising or volunteer-recruitment staff, you want to incorporate their information into the database you're developing of your supporters. Check Idealware (www.idealware.org) for recent reviews of Constituent Relationship Management software options. Such programs allow you to record every point of connection — from workshop attendance to dance contestant to donor — you may have with an individual. If you don't want to go that route, Microsoft Access and Filemaker Pro are commonly used tools that can be adapted to your needs.

You may discover that you serve several distinct groups of people. For example, low-income students may use the library after school and visit your exhibits while they're there. Middle-income mothers from the immediate neighborhood may bring their toddlers to the library for afternoon story time and take advantage of your programs. And wealthy older adults may volunteer as docents, serve on your board, and attend your organization's panel discussions.

With this valuable information, you may be able to recognize ways to reach more people who resemble the ones you're already serving. The more challenging task is to reach and entice new groups of people.

TIP

Do people you don't know gather at your organization's programs? Sponsor free drawings in which contestants compete for prizes by filling out forms with their names, email addresses, and phone numbers.

DESIGNING A USEFUL SURVEY

Developing a picture of your current constituents is critically important to creating a marketing plan, and a good survey can be your camera. If you're going to go to the effort to conduct a survey, it needs to have a purpose and must be clear and direct. Here are some general guidelines to keep in mind:

- If your survey is long, fewer people will complete it.
- You probably want to include some multiple-choice questions, but be sure to have one or two open-ended questions as well so people have a chance to speak their minds.
- Be careful not to ask *leading questions* — questions that suggest the answers you hope to hear.
- Explore different kinds of information by asking how convenient and appealing your events and services are and by asking your patrons what other kinds of programs they would enjoy.
- Ask your current constituents how they found out about you. The answers tell you which forms of your current marketing are effective and begin to suggest how you may best further spread the word about your programs.

TIP

An advantage of designing a survey online is that you can make it engaging by using the techniques of branching or piping. In such a survey, someone's answer to one question alters the next question she's asked. For example, if you were to say that you preferred ice cream to pie for dessert, you would next be asked if you preferred chocolate, vanilla, or spumoni.

If you discover that your target audience likes your program offerings but finds the times you offer them inconvenient, do you want to experiment with new times and formats? For instance, do people find Sunday afternoons (when the library is closed) to be more convenient? What other barriers inhibit their involvement? Maybe mothers with toddlers want to come to your lectures but need childcare. Perhaps you charge a modest admission fee for lectures but students find that charge to be too high.

One of your most difficult marketing tasks is analyzing the very basis of what you do and how you do it. You may feel that your historical society's close working relationship with libraries is its greatest asset, but the surveys may point out that those libraries are cold and musty during winter months. You may do better by taking over a neighborhood restaurant and creating a "warmer" atmosphere — even offering hot gingerbread and cider.

Using focus groups to gain detailed feedback

Surveys are useful for providing you with general information. You read them looking for an overview of what most people think and feel. Focus groups are another common tool for gathering information from constituents. Because they're based in conversations, they often give you more specific, nuanced responses. Perhaps you want to know what small business leaders think of your agency; a focus group of businesspeople can help you uncover their attitudes and needs.

When producing a focus group, remember to limit the number of questions asked to fewer than ten (five or six is ideal). Provide refreshments to create a convivial environment. The conversation will likely be more candid if it's led by a volunteer facilitator, not someone clearly identified with your organization. Seek permission from participants to record their discussions.

Defining whom you want to reach and how

Armed with information about the people who already know about your nonprofit, you're now ready to extend your reach by defining target groups you want to serve and discovering how best to reach them. In general, it's better to begin with your current constituents and work to expand within their demographic group or to others who are similar to them.

WARNING

As you shift your attention to reaching new groups, be cautious. You may alienate and lose current followers. The more different demographically your new target groups are from the people you now serve, the more difficult and expensive your marketing task may be.

If you run a local history archive and have audiences of low-income students, middle-income mothers of toddlers, and affluent docents and volunteers, logical new target audiences may include the following:

- Family members and classmates of the students using the libraries
- Mothers and toddlers from a wider geographic area surrounding the libraries
- Docents and volunteers who assist other local cultural institutions
- Friends and acquaintances of the docents

HOW MARKET RESEARCH CAN OPEN YOUR EYES

A small nonprofit alcoholism treatment center for women was started in San Francisco in the late 1970s in response to studies that found multiple programs to help alcoholic men but few focused specifically on women. No local service of this type focused on helping Spanish-speaking or immigrant women.

The program's mission was to provide comprehensive counseling and alcohol treatment services to low-income women, particularly those originally from Mexico or Central America. Four years into the nonprofit's history, its programs were full and effective, and it had won prestigious contracts from the city. The center had even launched a capital campaign to create a permanent home. It appeared to be very successful. With one exception. The women being served were mostly middle class and white. A few were African American. None came from the neighborhood where the program was based.

The board realized that the center needed to change its image and marketing strategy. Through interviews with women from the center's target population, the board and staff realized that the stigma associated with alcoholism was particularly strong among Mexican and Central American families, and women from these cultural groups were struggling with alcoholism in private.

The center began a multifaceted campaign to change this situation, beginning with cultural sensitivity training for counselors. When the center hired new staff, an aggressive effort was made to find Latinas to fill positions. The group published all brochures and other informational materials in Spanish and held news conferences for the Spanish-language press. The center hired a Latina outreach counselor to meet with community groups, churches, and schools and to develop connections and a system of referrals to the agency.

Within two years, more than one-third of the women served were low-income women from Mexico and Central America. The marketing aimed at this community was well worth the investment. Fulfilling the organization's mission depended on it.

Your marketing plan should then be tailored to reach these groups. To succeed, you may need to change both how you present your work and how you spread the message about that work. Consider some of the ideas in this list to reach the following groups:

» **The students' classmates and friends:**

- Contact local history teachers and work with them to link their lessons to archival materials in your collection. Invite them to bring their classes to see your exhibits and archival collections.

- Offer student internships and then use your interns as docents for the school tours.
- Invite the students who currently use the library space to help in planning, researching, and presenting an exhibit. Honor them for their involvement at the exhibit's opening and provide them with invitations for their friends and acquaintances.

» **More mothers and toddlers:**

- Advertise or place articles about your organization's work in local newsletters for parents of young children.
- Post fliers about your organization's work at parks, playgrounds, local stores that sell goods for small children, and other cultural institutions with children's events.
- Provide childcare at lectures and weekend programs.

» **Docents and volunteers who also help other organizations:**

- Advertise or place articles in your local volunteer center's newsletter.
- Contact local parent-teacher organizations and offer their members special tours or presentations.

MARKETING ON A SHOESTRING

Effective marketing doesn't have to be expensive. It can be based on multiple grassroots efforts that are small in scale and specifically targeted. *Guerrilla marketing* is the term for marketing that's cheap, creative, and effective. Jay Conrad Levinson's website (`www.gmarketing.com`) is a good place to find out more about low-cost marketing.

The guerrilla approach may use sandwich boards with provocative messages, cards passed out to people waiting in ticket lines, or enticing online offers. A famous guerrilla marketing example from the business world includes an Internet startup company that offered families $5,000 if they named a child after one of its products. The point of guerrilla marketing is to get customers talking about your organization and its products to develop powerful word-of-mouth advertising.

Another low-cost tip is to apply to Google AdWords for support from its nonprofit component. Qualified organizations can apply for up to $10,000 a month in keyword advertising. We haven't used this feature, but you may want to check it out at `www.google.com/grants`. The program isn't available for government entities, schools, or hospitals.

- Co-host an event with another local cultural organization, and combine your email lists to extend invitations. Invite guests to sign up for your newsletter or announcements to keep in touch with them in the future.

» **Friends and acquaintances of your current docents:**

- Hold a volunteer recognition party and provide each of your volunteers with ten or more invitations for friends and acquaintances.

REMEMBER

The cost of implementing new programs and spreading the word about them can add up quickly. Consider what resources you have available (including your time) before committing to big changes in your organization.

Using Mass Media to Reach Your Audience

What if you want to reach all the people, all the time? We're sorry, but you can't. You can, however, extend your reach beyond the contacts known by your friends and family by approaching mass media — newspapers, magazines, radio, television, and the web. Although some nonprofits pay for advertising to spread the word, most organizations rely on free publicity from the media.

Planning for effective publicity

Before contacting the media, first decide what story you want to tell and whom you want to reach with that story. What's distinctive and important about your organization's work? Why is it newsworthy? Be honest: Analyze your idea as if you're a news editor who has to choose among many different stories from many sources. How does yours stand up to the competition?

Ask yourself about your proposed story's relevance: Will it affect the lives of its readers? Is it timely? Does it contain conflict, intrigue, or prominent people? We uncovered some good advice about writing news releases from the American Land Rights Association, and you can find a link to the article in File 12-1 at `www.dummies.com/go/nonprofitkitfd5e`.

TIP

Reporters often have more story ideas than they have time to cover them. You can increase your chances of coverage if you write (or record or film) a basic document and vary it to address the different media interests that you appeal to, making use of your multiple story angles.

THE ART OF SHAPING A NEWS STORY

A few years ago we were involved as grant makers in supporting the creation of a mural on the exterior of a community center next to a small urban park. One of San Francisco's best-regarded mural artists led the project. She spent a great deal of time talking about images to use in her mural with the park's neighbors and people who used the building. People kept mentioning a young couple who had been killed in the park in a random, accidental shooting several years earlier. Although it wasn't her original subject idea, she included the couple's portraits in the painting surrounded by images representing peace and renewal.

The story of the mural's unveiling was presented to the media as a public memorial event for the neighbors and families of the young couple. Front-page, color images of the piece appeared in both the morning and evening papers the next day, and five television stations covered the event that evening. An acclaimed artist completing another mural wouldn't have been a story, but a neighborhood mourning two lost teenagers was hot news.

We hope your nonprofit never needs to tell a tragic story, but we give this example to show that not all good work is news. Tragedy is news. Drama is news. Breakthroughs are news. Surprises are news. Stories that connect to current trends are news.

Developing a media list

A *media list* is a compilation of names, addresses, email addresses, and phone numbers of contacts at local (and maybe national) newspapers, radio and television stations, magazines, and websites.

Your media list is a valuable tool that you refine and expand over time. Some metropolitan areas have press clubs and service organizations from which you can buy membership lists, providing a basis for your list. Cision (`us.cision.com`) is a comprehensive online media service. Its databases encompass contacts across traditional and social media platforms. It also offers educational programs and tools for communicating with government contacts. A full membership to Cision can be expensive, but one-time searches are reasonable.

WARNING

Journalism is a rapidly changing field with frequent layoffs and employee turnover, so it's important to update your media list often.

TIP

For all the nuances of media-list development and management, we recommend that you take a look at *Public Relations For Dummies*, by Eric Yaverbaum, Ilise Benun, Bob Bly, and Richard Kirshenbaum (Wiley). In general, your objective is to build two working lists, one that you use with practically every news release you send out and the other for specific opportunities for publicity. The latter group may include the following:

- Social and entertainment editors to whom you send news of your annual benefit gala
- Social and business editors to whom you announce new members and officers of your board
- Opinion page editors for letters to the editor
- Sunday magazine supplement editors for in-depth profiles of leaders in your field of work
- Features columnists for amusing anecdotes or unusual news
- Writers of blogs on topics related to your work

Understanding how the media works

Different media outlets require different amounts of lead time. In an ideal situation, you want to begin your efforts to reach the media four months — or even longer — in advance of that hoped-for coverage. The first two months are spent creating news releases and public service announcements and shooting photographs. Always check with the publication beforehand to find out how much lead time is necessary.

TIP

Many organizations create a media section on their website where they post news releases, background information, and high-resolution images with captions and photo-credit information. This tool enables your media contact to expand upon the information you send and download the photos he prefers for publication.

After you've done the initial legwork over the first couple of months, then you start distributing the materials:

- Most monthly magazines need to receive your news release and photographs at least two — preferably three — months in advance of publication (even earlier if they're published quarterly or bimonthly).
- At the same time that you're emailing your release to magazines, send advance notice to your most important daily and weekly outlets.
- We recommend sending public service announcements (see the section "Putting together public service announcements" later in this chapter) two to three months prior to the time when you hope they'll be used. Although most stations have set aside time for broadcasting nonprofits' announcements,

they need time to rotate through the many announcement materials they receive.

- News releases to daily and weekly papers should be sent four to six weeks in advance of the event you want covered. You may also send follow-up releases approximately ten days prior to the event. You certainly want to make follow-up phone calls.
- Releases inviting members of the media to attend a news conference or witness a special event or announcement may be sent three to ten days in advance of the event. If a news outlet is reluctant to cover your announcement, invite a reporter to visit your program and see your good works firsthand.

TIP

Take rejection graciously if your media contact decides not to use your story. You need to be able to go back to these people in the future. If you receive a rejection, ask the reporter for honest feedback about the kinds of stories that would better fit her needs.

Submitting materials to your media contacts

Each section of a newspaper and each part of a TV or radio program is made up of materials from multiple sources that are competing for time and space. You improve the odds of receiving attention in the media by providing clear, accurate, and provocative materials in time for consideration and possible use by reporters and broadcasters.

We recommend the following steps when you submit material to the media:

1. **Even if you've purchased a reputable media directory, call to confirm the most appropriate contact person for your story at the newspaper, radio station, TV station, or online outlet, and ask about the format that each wants you to use in your submission.**

   Most media outlets want your news release, photographs, and recorded materials to be sent via email. Paste the news release into the body of the email so the recipient doesn't have to mess with an attachment.

2. **Submit clear and accurate written materials and labeled, good-quality photographs.**

   The media generally want photographs to be in the form of high-resolution JPEGs, which work well in print.

3. **Call to see whether your materials have been received.**

   Take this opportunity to ask whether more information, further interview contacts, a different format, or additional photography are needed.

4. **If requested, submit additional information immediately and call to confirm its receipt and clarity.**

5. **If you don't receive a clear response (either "Yes, we'll cover it" or "Sorry, I don't see the story here") to your initial release, call again in a few days.**

6. **If a member of the media covers an event you've announced, provide a media packet (generally by providing a link to an electronic file).**

   A media packet commonly contains a news release, fact sheet, background information, and photographs. It briefs the reporter about your event and makes it easy for him to check facts when writing a story. If a reporter is covering your event, introduce yourself and be available to answer questions or introduce the reporter to key spokespeople, but don't be a pest. Let a reporter find his own story.

   You can help the reporter write a better story by recruiting ahead of time constituents who have benefitted from your program and are willing to be interviewed.

7. **If your situation changes and the news release is no longer accurate, immediately call in the change and, if necessary, revise and resubmit your original release.**

When making calls and sending emails to members of the media, recognize that they're busy and often working to meet deadlines. If you ask brief, clear questions that they can respond to quickly, you'll find it easy to engage them. For example, "How many photographs would you like?" is easier to answer than "What do you want a picture of?"

At `www.dummies.com/go/nonprofitkitfd5e`, you can find a sample news release (File 12-4), media alert (File 12-5), calendar release (File 12-6), photo caption (File 12-7), and photo permission form (File 12-8).

Getting your event into "What's Happening?" calendars

Getting your organization's events listed in newspaper calendars and broadcast on TV and radio can be critically important to attracting a crowd. Readers, viewers, and listeners use these calendars to help decide how they're going to spend their time on, say, a Thursday night. If your event is listed clearly and accurately, you're in the running.

TIP

Larger newspapers and many other media outlets assign the preparation of calendar sections to specific editors. When you put together your media contact list, make sure you identify the calendar editors. And to improve your access to this important source of publicity, contact the calendar specialists in advance and ask for instructions about how they prefer calendar listings to be formatted and how far in advance of the event they want to receive your information. Sometimes publications provide these explicit instructions — or the specific form you need to complete — on their websites.

Putting together public service announcements

Many radio and television stations allot a portion of their airtime to broadcasting public service announcements (PSAs) on behalf of nonprofit causes. Although these stations rarely give away their best viewing and listening hours to this free service, sometimes they do tack PSAs onto the end of a newscast or special program during prime time. But even an announcement played during the morning's wee hours can reach many people.

Public service announcements are brief. Most are 15, 30, or 60 seconds long. You may submit them as written text to be read by the stations' announcers, or you may submit them recorded or filmed on a CD or DVD or as an emailed podcast for direct broadcast. Many stations are more willing to use PSAs that are already recorded or filmed, but this isn't universally true. If you submit a prerecorded PSA, also include a print version of the text. Some stations string several announcements together in a general public announcement broadcast, and it's easier for them to work from text.

If you choose to submit a prerecorded PSA, make sure it's of broadcast quality with excellent sound and/or images. If you're the narrator, spit out your gum! If you mumble or if the images are blurry, stations can't use it. Using the voice of a celebrity — even a local luminary — for your prerecorded PSA can be particularly effective.

REMEMBER

The challenge to writing public service announcements is to clearly convey a lot of information in a few words. Write your message and test it against the clock. If you're rushing to finish it in time, it's probably too long. Ask someone else to read it back to you and time that person. You don't want to be the only one who can finish it in 15 seconds.

TIP

File 12-9 at `www.dummies.com/go/nonprofitkitfd5e` is a sample public service announcement addressed to a media outlet.

Using Social Media for Fun and Profit

How many people do you know without a Facebook account? Probably not many. Facebook, Instagram, LinkedIn, Twitter, and other platforms have penetrated our culture and enabled us to view photos of the children of our cousins and the pets of our friends and acquaintances. It doesn't matter if you think social media platforms are a positive step forward or the first step on a slippery slope toward the total destruction of personal privacy. Social media is here, and it's not going away.

It's not a surprise that nonprofit organizations are working to turn Facebook and other social media applications to their advantage. After all, what could be better than the opportunity to communicate easily with clients, volunteers, and donors — and to recruit more of the same? You can let people know what you're doing without the hit-and-miss efforts of trying to place articles in mass-media publications.

In this section, we give you an overview of the most popular social media tools and some tips for getting the most out of the time you spend on Facebook, Twitter, LinkedIn, and other sites that may become popular in the future.

Developing a social media policy

If you have staff and volunteers posting on your organization's social media platforms, you're wise to establish guidelines for what is and isn't appropriate to reveal to the public and to use a tone that builds a positive image for your nonprofit.

In your social media policy, include the following points:

REMEMBER

- **Be professional.** Blog posts, status updates, and tweets determine your organization's online face. Review your posts before posting. Spellchecker is your friend.

  Keep your personal social media play separate from your organization activities.

- **Post regularly.** Update regularly to keep your content fresh. We recommend creating a calendar and assigning responsibility for updates so your information is routinely refreshed.

- **Know what's appropriate to post.** Announcements, awards, milestones, and general news about the day-to-day activities of your organization are all worthy of posting. Rumors, gossip, bad news, and the like should never be circulated through social media. Here are some other tips about posting:

- Ask questions. Engage your audience in conversation.
- Respond to comments as soon as you can. Don't be defensive about negative comments.
- Photos and videos typically draw more comments and "likes" than text postings. You can add images to your website, or, if you have more images than your website can hold, open a Flickr account for still photos and upload videos to YouTube or Vimeo. Let people know when you've added new material by posting links on your social media pages.
- Get permission to post photos and names of people who work at your organization or clients who receive services. If you find good photographs on the Internet that you want to use, be conscientious about getting permission to use them. Providing a photo credit isn't enough — you need permission (from the photographer and the subject) to use the images.

» **Respect copyrighted material.** Ask permission to repost (if you can't just post a link) and give proper credit.

» **Prepare to respond to comments.** It's a good idea to develop policies for responding to social media comments. Engaging with the public is the essence of social media and many of the responses you receive will be positive ones, but you also need to be prepared for criticism or even offensive feedback.

» **Use good judgment.** Don't post anything you wouldn't want your mother to see or you wouldn't want to read on the front page of your local newspaper.

In addition to questions of taste and tone, you want to be sure that employees and volunteers understand that any 501(c)(3) organization must not take a stand on political campaigns for public office. And although you can't ban your staff and volunteers from using their own social media accounts for personal activities, remind them that their friends probably know where they work and that a tasteless or vulgar post can reflect badly on the organization.

TIP

We provide links in File 12-1 at www.dummies.com/go/nonprofitkitfd5e to information to help you develop your social media policy.

Planning your social media posts

Because social media can swallow up a lot of time, especially if you're distracted by cute cat videos, you should develop a plan before jumping into promoting your nonprofit through social media channels. Knowing what outcomes you want to achieve through social media helps you post or tweet information that is useful to both your organization and your followers.

Here is a list of questions you should ask yourself:

- What is your primary goal? Do you want to make more people aware of your services? Do you want to recruit volunteers? Do you want to raise money? Do you want to drive people to your website?
- How will you achieve that goal? After you determine your goal, set an objective. For example, you may decide to recruit ten new volunteers over the next three months.
- Whom will you target? If your nonprofit is a community theater, for instance, you want to look for people who enjoy the dramatic arts, either as audience members or performers.
- How much time can you spend implementing your plans? Can you spend an hour each day? Three hours a week? Do you have volunteers who can help?

We don't think it's necessary to create a formal plan document, but the more specific you can be about your goals and objectives, the more successful you're likely to be. We suggest *Social Media Marketing All-in-One For Dummies*, by Jan Zimmerman and Deborah Ng (Wiley), for good information about using social media marketing.

TIP

It's important to track visitors' activity on your website and social media networks so you can see how you're doing and make adjustments as necessary. Google Analytics (`www.google.com/analytics`) can help you do this. As you use these analytics and observe which features and posts attract the most attention on your social media pages, you can shape future content and campaigns to reach maximum numbers. You also want to keep an eye on qualitative information, such as whether influential people follow you on Twitter, whether your contacts share and re-tweet your material, and whether the comments you attract are thoughtful and well-informed.

Choosing your social media platforms

We doubt that anyone has been successful in developing a complete list of all social media applications because so many exist and because new ones are being launched almost on a daily basis. However, in the following sections, we list three that can't be ignored.

Facebook

For a company that was founded in 2004, it's amazing that Facebook can be called the granddaddy of social networking. If you do nothing else, you should create a Facebook page and begin collecting fans and likes. The company claims more than one billion users worldwide. And although it's unlikely you'll generate interest from more than a very tiny percentage of all users, it's the best place to find the niche for your nonprofit.

Facebook provides advice, including a resource guide and examples of how nonprofits are using the platform to advance their work. Visit `www.facebook.com/nonprofits` to find good ideas that you can adapt for your nonprofit's page.

Keep your organization's page separate from your personal account. Facebook provides a handbook to guide you through setting up a page for your nonprofit. *Facebook For Dummies*, by Carolyn Abram (Wiley), is a useful resource.

Facebook used to manage a "Facebook causes" feature that can now be found on a discrete platform, `www.causes.com`. Causes.com guides its users to organize boycotts, raise money, and build volunteer involvement.

Twitter

Twitter has fewer active users than Facebook, about 313 million according to estimates at the time of this writing, but it can be a powerful driver for traffic to your website and Facebook page. Your tweets are limited to 140 characters, so don't plan on publishing any essays on Twitter. On the other hand, it's perfect for writing a snappy headline and including a link to the essay or blog post on your website.

You build Twitter followers more quickly if your tweets have an authentic voice that sounds like a person who wants to have a conversation with followers — not someone who's reading a news release or shouting slogans.

Re-tweeting Twitter posts that you want your followers to see is an effective way to participate in the Twitter community. Don't forget that photos can be a part of your tweet and can be a fine way to promote a program or event. You can find a guide for nonprofit use of Twitter at `www.twitter4good.com`. Also, check out *Twitter For Dummies*, by Laura Fitton, Michael Gruen, and Leslie Poston (Wiley).

LinkedIn

LinkedIn is frequented by business folks and professionals, and who could be better to network with, especially when you're looking for potential board members or corporate connections? You want to create a company page for your organization. Remember, this is a professional network, so, you know, keep it professional.

LinkedIn For Dummies, by Joel Elad (Wiley), provides in-depth information about the ins and outs of LinkedIn.

LinkedIn groups can be a good way to exchange information with people who share your interests. If you're feeling adventurous, you can open your own group for your board members and volunteers. While you're at it, encourage those board members and volunteers to identify their participation in your nonprofit in their profiles. Declaring their affiliations with your organization can be a subtle but effective way to introduce their professional associates to your work.

Protecting your online reputation

Not all the information that people find about you online is generated by your organization. A number of online services rank and rate nonprofit organizations against such criteria as financial health, use of resources, transparency, and constituent feedback. Just as none of us can control gossip, you don't have the power to change the data these services use or their ways of interpreting that data. However, you can provide them with information, remain vigilant so you're aware of any critiques, and be proactive if your nonprofit is criticized.

TIP

One step in protecting your online reputation is regularly searching for the name of your nonprofit online. This practice can guide you to any weaknesses (or strengths) in your reputation and also alert you if someone is fraudulently using the name of your organization to raise money. Such scams are rare, but they do take place — particularly at times of disasters or other emergencies.

Here's a list of major sites that review nonprofits so you can provide them with up-to-date information:

- The Better Business Bureau's Wise Giving Alliance (`www.give.org`) rates more than 1,000 national nonprofits. Local BBBs also review nonprofits — approximately 10,000 of them. They use a range of standards to grade nonprofits. You can strengthen your rating by building your board of directors and making sure it meets more than three times a year, paying attention to the balance between your management costs and program expenses, being honest and transparent in fundraising materials, and providing options for donors who don't want their names to be disclosed. You can also gain a seal as a BBB Accredited Charity, for which you'll have to pay a fee on a sliding scale of $1,000 to $15,000 per year.
- Charity Navigator (`www.charitynavigator.com`) analyzes the finances, apparent effectiveness, and accountability and transparency of approximately 6,000 of the United States' largest charities with revenues of more than $1 million, and assigns 0- to 4-star ratings. If you are a new, emerging nonprofit,

you're unlikely to be included on this charity rating site, but it might be helpful to understand how nonprofits are assessed so you can plan for the future.

By studying a few sample profiles of nonprofits on Charity Navigator's website, you can discern how it analyzes nonprofits. Apart from the long-term hard work of growing your organization and containing overhead costs, your nonprofit can improve its rating by having conflict-of-interest and whistle-blower policies and publishing a donor privacy policy on your website.

- GuideStar (`www.guidestar.org`) offers an array of nonprofit information and is particularly valuable for posting three years of nonprofits' most recent 990 tax forms. You can strengthen your presence on GuideStar by writing the profile of your organization and checking to be certain that your most recent 990 forms are posted. The IRS provides the 990 forms to GuideStar directly, and occasionally there's a delay.

COLLECTING DONATIONS THROUGH CROWD FUNDING

Crowd funding (also called crowdsourcing) is a form of online fundraising that can benefit from the social networks you've developed. Kickstarter (`www.kickstarter.com`) and Indiegogo (`www.indiegogo.com`) are two of the best-known crowd-funding sites.

Kickstarter is all about projects and isn't specifically designed for nonprofits. You can't use Kickstarter to raise funds to support the ongoing costs of your nonprofit, for example. But if your work involves making a film or producing a play, Kickstarter may be the place to go. Generosity by Indiegogo (`www.generosity.com`) lets you raise funds for a cause, as long as your organization is recognized as a 501(c)(3) public charity. Fundly (`fundly.com`) is just for nonprofits and works well with Facebook. So if your organization has a lot of Facebook friends, and you're eager to try out crowd funding, Fundly may be a useful fundraising tool. For more details about these innovative funding initiatives, check out *Crowdsourcing For Dummies,* by David Alan Grier (Wiley).

Using crowd-funding platforms does involve some costs, so be sure to research carefully before you commit your organization to any one of them. Also, don't expect the contributions to flow into your organization's bank account without lots of work publicizing the campaign to your friends and followers. If you need a review of the principles of raising money from individuals, look at Chapter 14.

3 Fundraising Successfully

IN THIS PART . . .

Know where your money is coming from by writing a detailed fundraising plan and creating a fundraising budget.

Check out the many ways you can ask individuals to donate money to your organization. Many small donations can add up over time.

Go big and plan a special event that will draw in the dollars. The event may take a lot of work, but the payoff can make it worthwhile.

Nonprofits often receive grants as a source of income. Find out where to look for grants and how to apply for them.

Capital campaigns provide large amounts of funding for specific purposes — often property or buildings. Figure out how to establish a realistic plan and launch a campaign.

» Identifying the different types of contributions

» Asking the hard questions and preparing to take action

» Creating a great funding plan and putting it into action

» Fundraising as a startup project or organization

Chapter 13
Building a Fundraising Plan

If an organization is going to provide a public service, it needs money. Plain and simple. It may be run by volunteers and need just a little money, or it may need lots of money to pay for employees, office space, and formal research. If it's a brand-spanking-new organization, it needs startup money, or what's called *seed funding.*

Any nonprofit — whatever its purpose — grows out of someone's idea or passion (maybe yours!) to make the world better. Raising money is inviting others to share in that belief and passion.

Successful fundraising also is based on a plan, and a good plan recognizes an organization's likeliest sources of funding. After all, different kinds of causes appeal to different people or institutions. You'll find exceptions, but for most organizations, a good fundraising plan assumes the nonprofit will ask a variety of sources for support. No organization should put all its eggs in one basket and imagine its entire omelet will be coming from one generous chicken.

In this chapter, we show you how to create a plan for gathering your funding eggs from a number of nests. For those of you whose organizations are new, we offer some ideas for where to begin.

TIP

Check out File 13-1 at www.dummies.com/go/nonprofitkitfd5e for a list of web resources related to the topics we cover in this chapter.

Recognizing Who Can Raise Funds

Federal tax codes designate more than a dozen different kinds of nonprofit organizations (which we discuss in Chapter 2). This book focuses on those that are authorized under section 501(c)(3) of those tax codes and that qualify as public charities. Such nonprofits are exempt from paying some kinds of taxes, and their donors can take tax deductions for making contributions to them (unless they are foreign 501(c)(3) organizations or among those whose purpose is testing for public safety). We discuss grant-making foundations in Chapter 16. The IRS makes it easier for them to award their grants only to 501(c)(3) public charities. And if they want to award a grant to a for-profit entity or individual, they have to go through more steps to satisfy the IRS. Many individual donors, and foundations are motivated, in part, by that tax law. Therefore, this chapter and the fundraising chapters that follow are written about 501(c)(3) nonprofit organizations with public charity status.

TECHNICAL STUFF

Churches, if they operate solely for religious and educational purposes, and very small organizations (if their annual gross receipts are less than $5,000) that are performing a public service can receive contributions that are tax-deductible to their donors even though they may not have applied for and received 501(c)(3) status from the Internal Revenue Service (IRS).

WARNING

Forty-seven states and the District of Columbia require charities to file registration forms before engaging in fundraising solicitations. Each state defines its registration requirements differently. For example, some don't require universities, hospitals, or churches to register. You can see which states require such registrations at the website of the Council for Nonprofits (www.councilofnonprofits.org/tools-resources/charitable-solicitation-registration). If your state requires charities to register, make sure you do so before you start soliciting funds. After all, not filing the forms can cause your organization to lose permission to operate in your state or even, in rare cases, lead its officers and board members to be charged penalties. If you hire a fundraising consultant, both that consultant and your organization may need to file. If your organization raises money through games of chance (such as a raffle or casino night), you likely need to register for permission to do that as well.

Naming Possible Funding Sources

Before plunging ahead, we need to introduce some of the terms we use in this chapter to describe the different kinds of contributions an organization may seek. Here's what you need to know:

- **Grants:** *Grants* are formal contributions made to an organization by foundations, corporations, or government agencies, often to help the nonprofit address defined goals. Some grants, called *project* or *program grants,* are for trying out new ideas or enhancing existing programs. Others, called *general operating grants,* support the overall work of an organization.
- **Corporate contributions:** Some corporations create their own foundations (that award grants), and some award contributions directly out of their business coffers through corporate giving programs — often managed out of their public affairs, community relations, or marketing departments. Many corporations give *in-kind gifts* — contributions of goods and services — rather than or in addition to cash contributions.
- **Individual contributions:** Gifts to organizations from private individuals represent the largest portion of private money given to nonprofit organizations (72 percent in 2014). Individual donors may support specific activities or the nonprofit's general costs. You may seek these contributions through the mail, over the phone, through your website or social media, in face-to-face visits, or at special events. Common types of individual contributions include the following:
  - **Annual gifts:** A contribution written once a year to a charity is called, appropriately enough, an *annual gift.* The consistency of such gifts makes them of great value to the recipient.
  - **Major gifts:** As suggested in the name, a *major gift* is a large amount of money. For some organizations, a major gift may be $1,000, and for others it may be $500,000.
  - **Memberships:** Similar to an annual gift in some ways, a *membership* is a contribution made once a year. The difference is that you often make a membership gift in exchange for a benefit or service from the nonprofit, such as discounted admission to programs. Some nonprofits are structured so their members play a role in their governance, but that's a different, more formal relationship than calling a contributor a member because she paid to join. (See Chapter 4 for details on having members in your organization.)

REMEMBER

According to IRS regulations, if a donor receives a free gift in exchange for a contribution to your nonprofit organization, he cannot take a tax deduction for the value of that gift if its exceeds more than 2 percent of his contribution. He can only deduct the portion of the contribution that exceeds the fair market value of any items received. If that donor's contribution is more than $75, your nonprofit is required to provide a written disclosure of the value of the gift to the donor. In your thank-you note, you can write, "The estimated value of goods or services provided in return for your donation is $ (fill in the appropriate value)." Your nonprofit also should acknowledge in writing to a donor any gift of $250 or more. It is best to acknowledge contributions within seven to ten days.

- **Planned giving/bequests:** These gifts are contributions that donors make to nonprofit organizations through their wills or other legal documents specifying what happens to their money and property either during their lifetimes or after they die. Generally, the donor works with a trust officer at a bank or law firm to design her planned giving. Large nonprofits often employ staff members who specialize in providing donors with technical assistance in this area.
- **Crowdfunding:** Nonprofits may seek donations from individuals by setting up special online campaigns known as *crowdfunding.* The nonprofits contract with web services providing technology platforms that help them spread the word, collect the funds, and acknowledge the donors. Usually these campaigns focus on a service or product that the nonprofit wants to produce within a limited time frame for a specific amount of money. Crowdfunding also may be used to attract investors to for-profit businesses.
- **Special events:** From marathons to chicken dinners to online auctions, fundraising events generate income that supports organizations. Contributors can deduct from their taxes the portion of any event ticket that's above and beyond the cost of the meal or other goods received at the event. Individuals, corporations, and small businesses are the likeliest supporters of special events. We talk more about special events in Chapter 15.

Examining Your Potential

Different approaches to raising funds work best for different kinds of organizations. When you make a fundraising plan, you have to be both ambitious about your goals and realistic about what's likely to work for you. As you try to figure out how much money you can glean from each possible funding source, answer the following questions:

- **How far do your services reach?** Do a lot of people understand, care about, and benefit from your organization's cause? Do you focus on a small geographic area or work at a national or international scale? The answers to these questions tell you whether your nonprofit should be casting its net close to home or all over the country or world.
- **Are you one of a kind?** If you're unique, you may have a more difficult time explaining to potential donors who you are and what you do. But you may have an advantage when you appeal to foundations that like model programs and new approaches to solving community problems.
- **How urgent is your cause?** When a river floods a community or famine breaks out in a refugee camp, if your nonprofit is on the scene providing emergency shelter and food, you'll find that social media tools can be particularly effective in raising money from people who understand the importance of responding quickly.
- **Does your cause elicit strong feelings?** Even if it doesn't appeal to large numbers of people, a hot topic with a few passionate believers can still attract major gifts. However, organizations that focus on potentially controversial topics — such as family planning or eliminating the death penalty — may find that corporations and businesses are uncomfortable with having their names associated with the cause.
- **How well regarded are your leaders?** Most people feel better about supporting an organization when they believe in its leaders — whether they're famous research scientists or well-liked next-door neighbors. People give money to people they trust, so your organization's leaders (both staff and board) are critically important to its fundraising.
- **How well known is your organization?** Just as any commercial company with a well-known brand name has an easier time selling its products, a nonprofit with a widely recognized name often has an easier time attracting contributions from individuals. On the other hand, an agency with a low profile that's recognized by experts for doing good work may be more successful with foundations.
- **Whom do you know?** Your nonprofit's contacts are important to its ability to raise money, especially when seeking funds from individuals. Knowing somebody who may write a big check is great, but knowing a lot of people who may write small checks is just as beneficial. Crowdfunding is a good tool for securing multiple, smaller contributions.

- **What can you provide to a donor?** Many donors appreciate when they benefit from their gifts to you, so look at what your nonprofit can provide. Maybe your organization can offer contributors the best seats in a concert hall, print their names in the programs, and introduce them to the lion keeper. Organizations that are easily able to give tangible items and special recognition to donors may be particularly successful with membership drives or corporate campaigns.
- **Do you have fundraising expertise?** Do staff members, board members, or volunteers have experience with raising money? If not, can your organization afford to hire expert help?
- **Can you cover the needed fundraising costs?** Special events and *direct mail* (fundraising letters sent to large numbers of people) are expensive forms of fundraising. Grant writing takes time, but its cost is relatively low. We say more about fundraising costs in the later section "Estimate fundraising costs."
- **How does this year's fundraising climate compare to last year's?** Apart from the value and importance of your wonderful organization, donors give money according to their capabilities, and those capabilities change with the times. Corporate mergers, natural disasters, and economic downturns all affect how much money your organization can raise.

Drafting a Fundraising Plan

Setting out to raise funds is a lot easier when you have a guide to follow. After you analyze your nonprofit's fundraising potential, we recommend that you take the following steps to create a fundraising plan for your organization:

- **Set preliminary goals.** Ask yourself how much you need to raise to cover your organization's costs in the year ahead. Naming a clear and reasonable financial goal is the first step in your planning.
- **Ask whom you know.** Brainstorm lists of contacts — both the ones you and your board have now and the ones you want to have in the future. As you gather names, ask yourself the best way to approach those contacts — whether it's through formal proposals or a game of golf.
- **Research.** Your research phase asks both who might contribute and how much they might give. Using donor lists from other organizations, the Internet, and advice from others, find out as much as you can about your potential sources. Continually build your lists of prospects and then refine and edit it as you go.

» **Estimate costs.** List the tools you need for your fundraising — perhaps a simple fact sheet , a "Donate" button on your website, or pancake batter for a breakfast at the firehouse. Research the costs of those tools, the fees for consultants if you use them, and estimate the staff time devoted to fundraising. Also remember to invest in a good record-keeping system.

» **Get real.** Draft a schedule of what needs to be done when. Assign staff members, board members, or volunteers to different tasks and — this is important — agree to a system of checking in with one another to make sure you're all completing your assignments. (We address this step in the later section "Getting Down to Business: Moving to an Action Plan.")

TIP

In budgeting for fundraising, you will, of course, want to include the cost of registering to raise money in your state if required. If you're thinking of fundraising in a wider region or nationally, look into the costs of those registrations in other states. This can help you determine how broadly you want to focus your fundraising.

The following sections outline these steps in greater detail.

Setting a preliminary goal

A first step in your fundraising plan is to look at your estimated expenses for the year ahead. Spending this money should allow you to do the work your nonprofit has laid out in its organizational plan (see Chapter 7). For help connecting expenses to activities, see Chapter 11.

Have you ever daydreamed about winning a lottery and how you would spend your mega-millions? In that frame of mind, you might daydream a fundraising goal that depends on your nonprofit's receiving a seven-figure gift from a billionaire or the nation's largest foundation. That's great fun to imagine, but do you even know that billionaire or what that foundation supports? Your fundraising goal should be aspirational, but you also want it to be realistic.

Look at the competition and the amount of funds your organization has raised in the past. Set goals that allow you to cover your essential costs but that also seem reasonable based on the experience of your staff and board. This step grows easier over time as your organization develops a track record and solid donor relationships.

Asking whom you know

If your organization is brand-new, brainstorm with your founding board members the names of people and organizations they know who may support your cause. If your organization has been around for a while, begin by listing its previous contributors. Annotate your list by identifying which supporters are likely to contribute again. Then stretch your thinking to consider new prospects — people who know your work or who support other causes that are related to yours. You may even ask a few loyal donors to contribute to this brainstorming exercise.

After you begin this effort of generating names and ideas, you'll start to notice prospects all around you. A good fundraiser is ever vigilant, collecting names from news stories, athletic event programs, public television credits, donor display walls in buildings, and related organizations' websites and annual reports.

REMEMBER

This kind of brainstorming is best for identifying possible individual donors, but don't restrict your thinking at this stage. Go ahead and name all the prospects you've observed and thought of, including

- The foundations and government programs that fund other organizations like yours
- The corporations where your board members or volunteers work or those that make products that relate to your nonprofit's work
- The business associations or clubs that raise money for good causes
- The churches and places of worship in your neighborhood that offer community support and volunteers

Researching and refining your prospect list

Professional fundraisers will tell you that the three keys to raising money are research, research, and research. To see what they mean, consider this scenario: You know that your next-door neighbors' child plays music. After all, her band rehearses at all hours in the garage. That leads you to put the parents on your list as good prospects for your community music school. But you may also look harder at your neighbors' other interests or where they work. For example, their employer may sponsor local events (such as the music school's annual recital) or match its employees' charitable contributions. A little research can go a long way!

Professional list brokers and services can provide data about donors, but you don't need to follow that expensive, formal route. Local news media, the Internet, social media, and informal conversations can tell you a great deal about your neighbors' affiliations and interests. You just need to pay attention and keep notes about what you learn.

After you've brainstormed a list of individual donor prospects, go back through your list of names with board members, volunteers, and trusted associates. Mark which of your prospects seem to be highly likely, somewhat likely, and not very likely to donate to your efforts, based on what their interest appears to be in your organization's purpose, whether someone you know can contact them personally, and how much they've given to other nonprofits.

Researching institutional sources

If you want to raise money from foundations and government agencies, you'll likely begin by using the directories, databases, and websites specifically geared toward helping grant applicants. However, if your brainstorming session uncovers institutions as prospects, be sure to check out their websites or printed annual reports for information on their current guidelines and giving priorities. You can find details about institutions that don't have their own websites by subscribing to the Foundation Center's online directory at `www.foundationcenter.org`. Search for federal government agencies and programs at the Grants.gov website (`www.grants.gov`). See Chapter 16 for more information about conducting foundation research.

As you conduct your research, you're likely to uncover other prospects that you didn't think of when brainstorming, and you may also eliminate many that you discover are inappropriate.

Estimate how much your prospects will give

Identifying names of possible donors is just the beginning of the puzzle. You also need to estimate the amounts your prospective donors may give. For this estimate, you have to continue your research. For example, you may find clues by tracking down the approximate amounts of their contributions to other organizations that publish their donor lists.

For your foundation prospects, you can look up sample grants awarded to similar projects in the "Search Grants" section of *The Foundation Directory* online (find it at `fconline.foundationcenter.org`). (If you don't have access to a library that subscribes to this directory, you may want to purchase your own subscription.) You can also find grant lists for foundations by downloading their IRS Form 990 tax reports from the Foundation Center (`www.foundationcenter.org`) or GuideStar (`www.guidestar.org`) websites. We recommend using 990s as research tools for studying small and midsize foundations: PDFs of 990s for a large budget foundations can be enormous and unwieldy.

Estimating how much money you may receive requires research into each type of donor. You'll also want to set overall goals — perhaps attracting a certain number of new donors, sustaining many of the contributors your organization has had in the past, and convincing some previous donors to increase their gifts.

Estimating fundraising costs

As you probably already know, raising money costs money. No organization receives every grant or gift that it seeks. So before you put your fundraising plan into action, you need to make sure that your organization can afford its potential fundraising costs and that the possible returns outweigh those costs.

Fundraising costs should be a modest part of an organization's budget, but that said, they're real costs and organizations shouldn't feel that they have to hide them. The costs will be higher when the organization is starting up or launching a major new fundraising effort. These costs likely will decline after a few years. Although in 2016 Network for Good's research showed that the percentage of first-time donors who returned to give to a nonprofit in a second year had dropped to 27 percent for the average U.S. nonprofit, generally speaking, recruiting a new donor costs more than securing a second gift from a past supporter. Raising money through special events, mass mailings, and telemarketing are expensive fundraising methods. However, all of the above may be worthwhile if they bring visibility, new donors, or unrestricted support to your organization.

Consider the time and money you'll invest in fundraising

Here's a quick look at the costs — in terms of both time and money — of the fundraising activities your organization may consider using:

- **Grants and contracts:** Most of the costs of securing grants and contracts are labor costs for planning and writing the grant proposal. Sometimes you need to travel to meet in person with the agency awarding the money, so don't forget to budget that, too.

  Pay attention to each grant maker's requirements for reporting the results of a project and make sure you can afford to complete the report. Some grants require the collection and analysis of extensive data. Some require audited financial statements.

- **Corporate sponsors:** Businesses and corporations may require visible recognition for their contributions to nonprofit organizations. Such support is often called a *sponsorship,* and the nonprofit and corporation agree upfront to the type of acknowledgment that they expect — whether it's a corporate logo printed on volunteers' T-shirts or a 6-foot banner announcing a company's support of a children's playground. Costs of printing T-shirts, banners, and other promotions generally are incurred by the nonprofit organizations.

- **Individual contributions:** Costs related to securing individual contributions include salaried staff members' time spent compiling lists of possible donors and the tools used for making successful donation requests. Common tools include solicitation letters and emails, brochures, and printed reply envelopes. Adding capacity to your website so it can accept online contributions is another important investment. Don't forget the cost of registering to fund-raise in the states where you're targeting donors.

TIP

Often when you think of solicitation letters, you probably think of *direct mail* — hundreds of thousands of letters sent to purchased lists of potential donors. Although direct mail can be a very successful form of fundraising for some causes — medical emergencies, animal rescue, and civil rights, just to name a few — it's also very expensive. A new, emerging nonprofit can make good use of letter-writing by preparing a small mailing of appeal letters addressed people who already know its work well, asking its board members and volunteers to make personal appeals to people they know, or by using email rather than snail mail.

WARNING

Some organizations hire companies to handle telephone solicitation campaigns. Be cautious if you take this step: Some of these companies charge a high percentage of the money raised in exchange for providing this service. Others are reputable and valued by the nonprofits they serve.

- **Special events:** Special events can be a great way to introduce new people to your organization, but producing such events can be one of the more expensive ways to raise funds. Spending 50 percent or 60 percent of the income from an event to pay for costs is common. After all, printing, advertising, food, and entertainment all cost money. Also, special events are labor intensive, so make sure you have an experienced volunteer group or detail-oriented staff to work on special events if you go this route.
- **Planned giving:** If your nonprofit isn't familiar with tax laws regarding wills and estates, it may need to employ or hire on contract a planned giving expert (or attract one as a board member). Bringing this person on board can be expensive in the short term, but doing so can yield important long-term support for your agency.

WARNING

Planned giving works best for organizations that have been around for a long time and that show good prospects for continuing. Universities, museums, and churches come to mind. Small, new groups have a difficult time attracting bequests.

Keep good donor records

Before it starts to raise money, you need to set up a system for keeping timely records on all your nonprofit's donors. This system can be a Microsoft Excel spreadsheet or a specialized electronic database for donor development or customer relationship management. For foundation grants, be sure to note deadlines for applying for grants and submitting final reports. While you're at it, you'll want to develop your organization's policies for protecting the privacy of your donors' information. You can find guidance about this topic at the Association of Fundraising Professionals' website. We've included a link to its Code of Ethics in File 13-1 at `www.dummies.com/go/nonprofitkitfd5e`, and encourage you to search the association's site for additional information about donor privacy protection.

TIP

For individual donors, a good records system can help you keep information up-to-date about when and how you've contacted them, who in the organization knows them, how and why they gave their gifts, and whether you promised them any special recognition or invitations when they contributed. Because nonprofits hope to talk to or keep in touch with their donors over time, recording their spouses' and children's names, their interests, their business affiliations, and any other pertinent personal information also is worthwhile.

If you decide to use an electronic database system to manage your donor records, you can choose a program that focuses specifically on fundraising, or you choose one that keeps track of every connection you have with anyone and everyone, including clients, ticket buyers, professional peers, donors, and volunteers. The "tracking every kind of connection" software often is called a customer relationship management (CRM) system.

WARNING

Specialized CRM software can really increase your efficiency in record keeping, but when buying a system, consider the staff skills and the time you can devote to managing the program as well as your budget. Also check on setup charges and availability of technical backup support. For example, open source software — which is made available for free use by anyone — is readily available for you to download and use, but it may not come with consistent technical support if you run into problems.

You can find reviews and articles to help you choose among different kinds of donor, member, and CRM systems at `www.idealware.org`. To help you start your search, here's a quick look at three CRM system options:

» CivicCRM (`www.civiccrm.com`) is available license-free to nonprofits. Although it's an excellent tool, you'll likely need someone with technology skills to set it up for you. It integrates easily with Drupal and Joomla content management systems.

» Salesforce (www.salesforce.org/nonprofit) provides nonprofits with up to ten free licenses for its use. Many nonprofits take advantage of the free licenses but then purchase apps or modifications to shape it to their needs. The Salesforce Foundation website provides helpful short films and articles to guide users.

» SugarCRM (www.sugarcrm.com) offers a free, open-source community edition of its software along with editions that require reasonable monthly subscriptions.

Some nonprofits choose to prioritize a system that focuses exclusively on keeping track of donors and potential donors. A few options are:

» DonorPerfect (www.donorperfect.com) focuses on donors but has some support for members, volunteers, and other contacts. It offers an online version for a reasonable monthly fee.

» DonorSnap (www.donorsnap.com) is a good, reasonably priced option that offers technical and setup assistance. If you're using a QuickBooks system for bookkeeping, it integrates easily with it.

» eTapestry (www.etapestry.com) offers a 30-day free trial for trying out its system. You can subscribe at different levels and prices based on your estimated number of donor records. (The basic service accommodates up to 1,000 records.)

» GiftWorks (www.frontstream.com/nonprofit-education/constituent-management) is a reasonably priced software option with a reputation for being easy to use.

TIP

After you choose the system that's right for your organization, check out whether you want to join TechSoup (www.techsoup.org). TechSoup members can purchase donated or discounted software for their nonprofits. You may find what you want at a reasonable price.

Initially, when you urgently need grants and gifts to start up your organization, keeping good records may seem to be a time-wasting activity, but in the long run, thorough records allow you to raise more money. You can start with a simple approach and upgrade it over time, but every time you shift to a new system, be prepared to invest time in the change.

The point of investing in a donor management system is to save and develop the best information you can about the people who are connected to your organization. That information will do you no good if you aren't using it to actively build relationships with those entries in your database. Motivating the donors you already have is much less costly that recruiting new ones. You'll want to make sure that you're spreading the word about your organization and its fundraising messages across a range of platforms, such as Facebook, Twitter, YouTube, Instagram, blogs, radio, and print media. When thinking about calculating these costs, consider the time you'll invest. If you're more than a one-person band, build responsibility for building connections into everyone's job description.

Getting Down to Business: Moving to an Action Plan

A good fundraising plan includes the practical details that move it from being a list of goals, contacts, and costs to being a road map for your fundraising destination. The following steps help you incorporate these practical details:

1. **Assign tasks, gather tools, and make a calendar.**

   For each of the revenue areas in your plan (government, foundations, corporations, and/or individuals), indicate who's going to work on raising the money, how many prospective sources you need, and which tools you need to meet your goals (such as fundraising letters, photographs, web newsletters, social media campaigns, membership cards, or an online donation system). Then outline a general time frame for how long your fundraising efforts will take and research specific deadlines for government, foundation, and corporate grants.

   Include in your calendar a schedule for staff, board, and volunteers to check in with one another on progress made toward completing their assigned tasks.

2. **Link your cost estimates to each fundraising goal.**

   You've already estimated costs of the time and tools you'll need to raise funds (see the previous section "Estimating fundraising costs"). Now you need to create a fundraising cash flow outline that shows when you need upfront money and when you can expect to secure income from your efforts. Keep in mind that some foundations have a rather lengthy time from letter of intent to grant funding, sometimes up to 12 months. Individual donors may have one time of year (before the end of the year, for example) when they make charitable contributions.

3. **Put the fruits of your labor together in one document — this is your funding plan!**

   Your finished plan likely consists of the following elements:

   - Sources sought and fundraising goals (for example: foundation grants, $75,000, and special event revenues, $4,200)
   - Prospects identified (both current and prospective contributors), along with amounts you expect them to give

- Number of prospects you need to achieve your goals in each category
- List of who's responsible for making particular contacts or contributing other services to raising the funds
- Estimated costs of pursuing the contributions in each category
- Timeline and cash flow projection

TIP

Many agencies create an optimum fundraising plan as well as a bare-bones fundraising plan — one based on their hopes and one based on what they must secure to survive. During the course of the year, they rebalance and adjust their plans.

TIP

Check out the two sample fundraising plans (File 13-2 and File 13-3) and two sample fundraising budgets (File 13-4 and File 13-5) at www.dummies.com/go/nonprofitkitfd5e. One set is for a small school music organization, and the other is for a slightly larger neighborhood park improvement organization.

Planting the Seeds for a New Organization

One tried-and-true rule of fundraising is that people give money to people they know. They also give money to causes they care about. However, making a contribution is an expression of trust. That means the likeliest contributors to a new nonprofit organization are people and agencies that know and admire its founders and their work.

Hitting up people you know

We know of a few people who have launched their organizations with the help of major grants from a government agency or a large, national foundation, but many more of them start close to home with gifts from their founders, their founding board members, and the people who know them. They then build out from that inner circle of relationships, gradually creating networks of associations with the friends, family members, and business associates of their initial contributors.

We asked some people who founded organizations how and when they got their first contributions. One invited 20 people to her house and, after a convincing pitch and some good wine, got most of them to write checks. Another started a youth mentoring program out of his dorm room when he was a college student. He talked his college into paying him a work-study stipend to begin his project, and then he charged his volunteer mentors modest membership fees to cover basic costs. Another noticed a city department's neglect of small city parks after budget cutbacks and, by knocking on local politicians' doors, secured a city grant to involve volunteers in park cleanup days.

Many organizations start up with contributions from individuals, in part because individuals often make up their minds more quickly than businesses or foundations.

Branching out with special events

Although special events are one of the more expensive ways of raising money, you probably can see real advantages to including them in the fundraising plan for a new organization. Events create a way to inform dozens or hundreds of people about your organization at the same time. The best part is that those people then begin to spread the word about your good work. Events don't have to be elaborate. A gathering of a dozen people can be a good start.

Approaching foundations

When approaching foundations, you may think that you're at a disadvantage because your project doesn't have a track record. On the contrary, some foundations specifically like to support startup projects and organizations. If your agency is new, when you're using *The Foundation Directory* check the "types of support" listings to see whether the foundation you're looking at awards *seed funding* — that is, support for new activities. If the foundation makes grants in your field of interest and geographic area, it may be a good prospect for helping you launch your organization. Find out more information about conducting foundation research in Chapter 16.

Considering government grants

Government grants and contracts may provide significant underpinnings for a new effort, but they come with three distinct disadvantages for new nonprofit organizations:

- **In general, they take longer to secure.** The review and approval process may be slow, and many government agencies have only one annual deadline.
- **They often require grantees to comply with rules and regulations regarding permits, board policies, hiring practices, audits, and financial reporting.** Although their rules may be good rules, when you're just starting out, your organization may still be working out its systems and policies.
- **They sometimes require their funded organizations to spend money upfront and then submit invoices in order to be reimbursed for an agreed-upon amount of money.** If you have few sources of income, you may not have any funds available for this upfront spending.

» Writing a case statement

» Tracking down possible donors

» Asking for major gift contributions

» Soliciting donations via mail

» Securing funds online

» Raising money through telemarketing

Chapter 14
Raising Funds from Individual Donors

Everyone has had some experience with asking for money. Maybe you sold cookies for a scout troop when you were a child. Maybe you asked neighbors to sponsor you in a summer read-a-thon for your school library. Maybe you once had to call Mom and Dad when you were stranded at a bus station in Toledo and couldn't get home for Thanksgiving.

These moments can be awkward, but asking for money for yourself and asking for money for an organization you believe in are very different. Organizational fund-raising can feel good because it helps you do something important for a cause you care deeply about. The hard part is finding the right words to say, the means for conveying those words, and the audience who's receptive to those words.

Responding to the right cause, the right message, and the right person, individual donors give to many kinds of organizations. In fact, they represent the largest portion of private contributions in the United States. According to Giving USA, financial contributions from individual donors accounted for 72 percent of all charitable giving in 2014. This chapter shows you how to flex your own fundraising muscles and persuade individual donors to give to your organization.

REMEMBER

Often when people think of raising large amounts of money for organizations, they think of securing major grants from foundations or sponsorships from corporations. But we recommend that your nonprofit also include individual giving as a part of its fundraising plans. In total, individuals give more to nonprofits than do foundations or corporations, and most individuals allow you to use the funds contributed for whatever your greatest needs may be. That kind of flexibility adds considerable value to a contribution — even if it's just $20.

TIP

Check out File 14-1 at `www.dummies.com/go/nonprofitkitfd5e` for a list of web resources related to the topics we cover in this chapter.

Knowing Why People Give Helps in the Asking

The key rule of fundraising is "If you don't ask, you won't get." Taking the step of asking is critical, but so is knowing how to ask. According to marketing experts who study motivations for doing just about everything, people contribute to nonprofit organizations because they want to

- Feel generous
- Change the world
- Exercise compassion
- Have a sense of belonging to a group
- Feel a sense of well-being, of safety
- Be recognized

By understanding these underlying donor motivations, you can phrase your request in the most effective way. Appeals to new donors often ask them to "join" or "become part of" a movement or cause in order to touch upon the desire for belonging. Appeals also call upon donors' compassion and idealism, and they commonly link the needs of particular constituents served by a nonprofit to the well-being and security of an entire community. That's making the most of asking.

REMEMBER

When you find yourself hesitating to ask someone for a contribution, keep in mind that you're not begging, you're offering him an opportunity to be part of something worthwhile. Giving your organization a gift can make him feel good.

REGISTERING TO SOLICIT DONATIONS

Many states, the District of Columbia, and some local jurisdictions require organizations that solicit charitable contributions to file registration forms before soliciting. This registration is meant to protect the public from solicitors touting illegitimate causes. The regulations differ from state to state. Typically, the rules don't apply unless the nonprofit raises more than $250 a year (which we hope you will!). Check with your state's attorney general's office to be sure your nonprofit is in compliance with all laws. It's also important to check local laws when raising money through what is known as charitable gaming (bingo, raffles, pull tabs, and so on). Laws regulating these activities often differ from county to county and city to city.

What if you're raising money online? Guidelines adopted in 2001 by the National Association of State Charity Officials (NASCO) state that any nonprofit that uses fundraising tools to target donors in a specific state must register in that state. This registration requirement extends to any organization that receives contributions from a state through its website on a repeated and ongoing basis — whether or not it is specifically targeting donors in that state. Plus, the 990 tax form for nonprofits asks organizations where they are registered to fundraise. If you're seeking donations nationally, take all relevant states into account and remember to renew your registrations annually.

Stating Your Case

A *case statement* is a tool nonprofits often use when asking for a contribution. It's a short, compelling argument for supporting the nonprofit that can be presented as a one-page information sheet, a section of your website, or a glossy folder filled with factsheets, photographs, budgets, or charts. It should be professional, factual, and short enough that readers will read it all the way through.

REMEMBER

A good case statement can be used in many ways in fundraising:

- After talking about the agency with a potential donor, you leave behind a copy of your case statement for her to consider.
- If you need to phone a potential donor, you keep the case statement close at hand as a reminder of the key points to make about the agency.
- If you mail or email a fundraising letter, or write a blog about your nonprofit, you can borrow wording from the case statement to write that item or include a link to or copy of the case statement with the appeal.

Instructions for writing a case statement resemble a recipe for stew or soup. A few basic ingredients make any version of this dish delicious, and the cook can spice it up with whatever other quality ingredients she has at hand. Follow these steps:

1. **Make notes about the following subjects. Be selective; allot no more than 100 words to any of these items:**
   - The mission statement of the organization
   - The history of the organization
   - The services it offers
   - Data illustrating the organization's key accomplishments
   - Affidavits, reviews, or quotes from enthusiasts who have benefited from the organization's work
   - The organization's future plans
2. **Toss in no more than 50 words on each of these topics:**
   - The organization's philosophy or approach to providing service
   - The pressing needs the organization works to address
   - The ways the organization will measure or recognize its success
3. **Stir in whatever else you have on hand, including**
   - Compelling photographs of the organization's work
   - Quotes and testimonials from clients, partners, donors, and others
   - Tables or maps illustrating growth — for example, the increasing numbers of clients the organization serves or the expanding geographic area it serves
   - An overview of the organization's budget and finances, particularly if the budget is balanced and the finances are healthy
   - Anything else that shines a light on the agency, such as publications completed, awards received, and so on
4. **State the giving opportunities your nonprofit offers.**

   The giving options must be clear and not too complicated. Donors want to understand how their contributions can make a difference. Sometimes giving opportunities are linked to the cost of providing services, such as the following:
   - The cost of immunizing one child against a deadly disease
   - The cost of replacing one chair in a planetarium
   - The cost of rescuing and rehabilitating one stranded otter
   - The cost of planting one acre of trees in a reforestation area

Sometimes opportunities are linked to premiums or gifts the donor receives in return. For example:

- For a gift of up to $50, the donor receives a tee shirt as a token of thanks.
- For a gift of several hundred dollars, the donor's name appears on a brass plate on the chair the gift has paid to refurbish.
- For a gift of $1,000, the donor receives quarterly progress reports from the scientist whose research the gift is supporting.
- For a gift of several thousand dollars, the donor's name is etched on a tile in the lobby of the organization's new building.
- For a gift of several million dollars, the lobby is named after the donor.

REMEMBER

For a contribution of $250 or more, you should acknowledge the gift in writing and specify the date of the contribution and the amount. If you give that donor something of value in exchange for the contribution, you should mention the value of that premium in your acknowledgment letter (because the donor must deduct the fair market value of that gift if she itemizes her contribution to your organization in her tax filing). Check out File 14-2 and File 14-3 at `www.dummies.com/go/nonprofitkitfd5e` for examples of acknowledgment letters.

TIP

When writing your case statement, you don't need to follow the exact order in which we list the elements. Lead with the strongest points and leave out the less compelling items. Choose a strong writer to draft the piece and then test it on others — both within and outside your agency — to see whether it tells a clear, impressive story.

RECOGNIZING YOUR DONORS

Before acknowledging a donor in print, on a donor wall, or in another public manner, always ask him for permission and ask how he wants to be listed.

When choosing how to recognize your donors, you want to consider the cost of the recognition, the nature of the activity supported, and the way the donor recognition advances (or doesn't suit) your organization's purpose and mission. A simple, printed list of names is dignified and inexpensive. Although donor gifts cost money, having your organization's name printed on T-shirts or tote bags that your donors wear or carry throughout the community may be an effective way of promoting your work. However, if you spend too much on gifts, you can annoy donors or leave them wondering how much you really need their contribution.

When your case statement is written, produce it for distribution and make sure to match its look to your nonprofit and its intended audience. If your nonprofit is a modest, grassroots organization, your case statement should be simple and direct in its presentation. If it's an environmental organization, you may want to distribute the case statement online instead of using paper. If it's an internationally known opera company, your case statement should look dramatic and elegant.

Identifying Possible Donors

Your doorbell rings, and you open the door to greet an adorable child you've never met. The child is selling candy bars to raise funds for band instruments at school. You played in a band when you were young, so you buy a candy bar, wishing the little fundraiser success.

A few minutes later, the doorbell rings again, and you open the door to greet a second adorable child. You know this child because he lives next door. You remember when his mother brought him home as a newborn, and you watched him ride his first bicycle down the sidewalk. He's selling the same candy bars to benefit the soccer team at his school: They need equipment and uniforms. You buy *five* candy bars.

Rich or poor, most of us are influenced in our charitable giving by how well we know the people, either directly or indirectly, who are doing the asking. Mapping the personal connections of each member of your organization is a key first step in identifying possible donors.

Drawing circles of connections

Figure 14-1 represents a common brainstorming exercise that nonprofit organizations use to identify possible individual donors. This exercise is most effective if both staff and board members participate. Follow these steps:

1. **Identify the people who are closest to the organization — those within the "inner circle."**

   This inner circle includes staff and board members and active volunteers or clients who frequently use the organization's services.

2. **Identify those people who have the second-closest relationship to the organization.**

   This second circle may include family, close friends, and coworkers of those people in the inner circle, former staff and board members, neighbors of the organization and its inner circle, volunteers, and clients who sometimes use its services.

3. **Take one more step backward and identify the people who make up the third circle.**

   These folks may include grandparents or cousins of those people in the inner circle, old friends whom they haven't seen recently, friends of former staff and board members, and former or infrequent users of the organization's services.

4. **Identify the friends, relatives, and other associates of those who make up circles #2 and #3.**

5. **Search for your cause-related friends and associates.**

   Look for people who may not know anyone involved in your organization but who demonstrate an interest in the subject and the purpose it represents (as indicated by their memberships, their magazine subscriptions, or their contributions to similar organizations). Follow the local media and watch for people who have a connection to your cause because of their personal experiences. For example, maybe their child was born with the congenital disease your nonprofit is studying, or maybe they come from a forested part of the state and care about the preservation of old-growth trees.

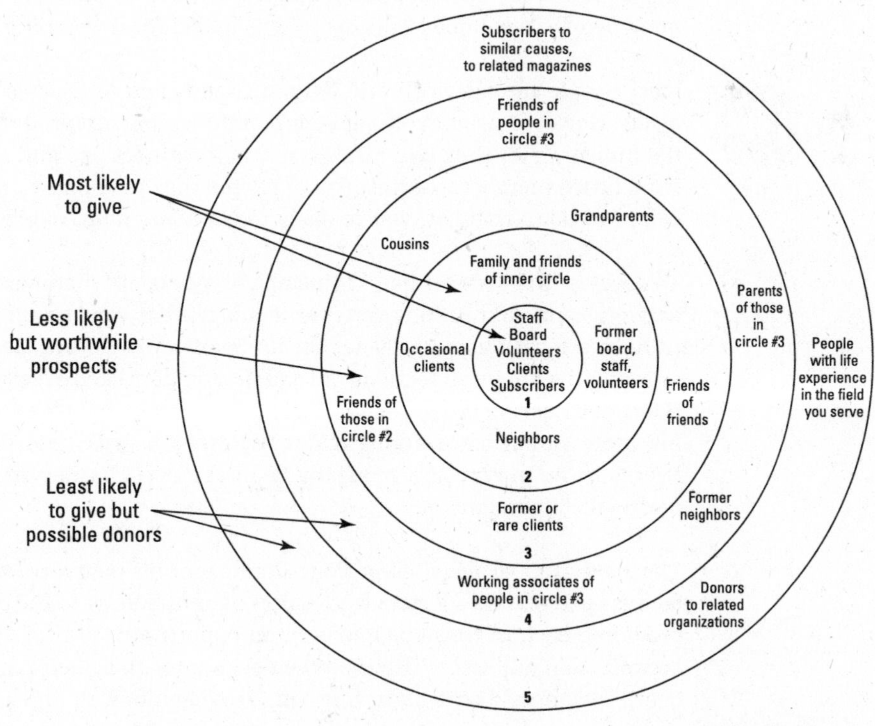

FIGURE 14-1: Drawing circles of connections helps you identify possible individual donors.

You can continue enlarging your circle of connections, but with every step you take from the center to the outside, the bond between the organization and the potential donors weakens. As that occurs, the cost of raising money from the people who inhabit those circles increases. Eventually, the cost of securing gifts from an outer circle becomes higher than the likely amount of income to be gained, and it's time to stop.

REMEMBER

How do you know when it's time to stop and back up? Most organizations try to keep the cost of their overall fundraising at or lower than 20 percent of their budgets. However, when you're starting out, particularly with some kinds of fundraising strategies, such as mailings or special events (see Chapter 15), your percentages may be higher in the first few years. These approaches generally begin to bring in significant contributions as you deepen your relationships with your donors and they renew their gifts. If you're still just "breaking even" on a fundraising activity in year three, it's time to back up and refocus your approach.

Getting a list of contacts from your board of directors

Every year, or before you start work on a special event or letter-writing campaign, ask each of your organization's board members to provide the names of ten or more people they know. This exercise is useful for developing a solicitation list.

TIP

Most people can sit down and list ten friends and associates off the top of their heads. However, when you ask your board members to produce a list in that way, it's unlikely that they will exhaust all their connections and relationships. If you hand board members starter lists of people they may know and then ask them to edit and add to that list, you're likely to get many more names.

To develop this "starter list," think about your board members and their probable networks and connections. Say, for instance, that your nonprofit is lucky enough to have a local business leader on its board of directors. Searching online and scouring the business sections of local newspapers can reveal a variety of connections, such as the board member's business partners and suppliers; members of his professional associations, clubs, or other boards; and his close neighbors. Everyone, no matter how modestly he or she lives, has a web of professional and personal connections.

The next step in developing your organization's network of connections is to address personalized letters from your board members to the people on their lists, providing each of those people with an opportunity to find out more about your organization and to contribute. When asking for their help, make it easy for your board members' contacts to respond: Provide a link to an easy-to-use donation page on your website or a simple form for them to fill out and a self-addressed, stamped envelope. If you send your query by email, make sure the subject line is compelling and the sender's name and email address are familiar to the recipient.

After you have developed your list of contacts, you can use it as the basis for a face-to-face individual donor campaign, as well as for mail and email campaigns and event invitations.

Growing a Major Gift

Face-to-face visits are the best way to secure larger contributions, what fundraising professionals call *major gifts* (see Chapter 13 for more information on major gifts). Rarely does one knock on a door, deliver a short speech, and depart with a check for a large amount. Generally, several contacts must take place before a major donor is ready to make a commitment. This process is often called "cultivation," and the comparison of planting the idea, tending the relationship, and harvesting the result when it's fully grown is an apt one. It takes time. In this section, we break down key steps and tips for growing a major gift.

TIP

Before you talk yourself into making a phone call or sending a letter rather than sitting down to talk to a donor face to face, remember that it's more difficult for donors to say no to someone sitting in front of them.

Deciding who should do the asking

If possible, the board member or staff person who knows the potential donor should make the visit. If that's not possible, the visiting team should be made up of two people who are peers of that person — perhaps a board member who's a local business leader and the executive director. Be careful not to overwhelm a potential donor with a large group. Two or three people are plenty.

Preparing to make your request

Gulp. It's time to ask someone for money. In preparing for this moment, you need to remember that you're not asking for something for yourself. You're inviting the potential donor to belong, to be part of something worthwhile. When planning how to describe the reason for the gift, step back for a moment and remember why volunteering for the organization is important to you. How did you become involved? What lives have you seen changed with the organization's work? Sally forth armed with your personal story and your case statement, which will remind you of the key points you want to mention.

Breaking the ice

Open the conversation with easy material. What did you find out about this person while conducting research, and how do you know her? Maybe your kids attend the same school or you're both baseball fans. Try to use something low key to open the conversation instead of forcing yourself on your "victim" with a heavy sales pitch. The key is creating rapport.

REMEMBER

Brief small talk can ease the conversation's start, but don't waste a potential donor's time. Let her know how you became involved in the organization, and then briefly give an overview of its attributes and current situation. Team members should take up different pieces of the conversation, remembering to let the potential donor talk, too, and paying close attention to the signals she sends. Keeping the meeting comfortable for the prospect is critical.

Adopting the right attitude

When soliciting gifts, be firm and positive, but not pushy. Pay attention to how the potential donor is acting or responding and step back if you discover he isn't feeling well or his business has taken a difficult turn. You'll also want to allow periods of silence: Don't rush to fill in every conversational pause; let your contact to have time to reflect and respond.

Just as you can see signs of when not to press the case, you also may notice signs of readiness when the discussion is going well. The potential donor may express interest in observing a program at the nonprofit. He may display pride in his familiarity with the field in which the nonprofit is working. Be a good listener as well as a good presenter.

TIP

Asking someone to contribute to an organization is easier when you've already made a gift yourself. Even if you can't make a large contribution, you'll feel more confident about asking others if you've already contributed — and the potential donor will find you more credible.

Timing the request: An inexact science

Many people don't believe in asking for a specific contribution at the first meeting. They believe in setting the stage — letting a prospective donor know that a campaign or special program is coming up and that the organization will be seeking her help in the future. At a second visit, they try to get this person to see a program in action or meet other board members at an informal gathering. The goal is building a relationship, inviting the potential donor to feel as if she wants to belong with the people leading your agency.

For some major gift campaigns, fundraisers use a feasibility study. In such a study, someone from outside the organization interviews board members and potential major donors, seeking their impressions of the organization and getting an idea of the size of the contribution they may make. This process both helps the organization to set realistic goals and warns the "research subjects" that the agency plans to come knocking at their doors.

REMEMBER

As with any cultivation, the timing of the harvest is critical. When the time is right, it's like recognizing a piece of fruit that's ready to be picked. Make a date for a follow-up visit. Plan a setting where the conversation can be congenial but focused. Don't rush (a common failing), but don't leave without asking. If the answer is disappointing, ask for feedback on your visit and try to figure out which aspects of your case the potential donor finds most (and least) compelling. Maybe the donor wants a naming opportunity that you didn't present, or maybe he feels loyalty to a similar organization. Or maybe the timing isn't right for the donor because he's just promised a major gift to another cause, but in a future year he would consider making a sizable gift. Sometimes prospective donors reconsider at a later time, so it's always good to seek permission to stay in touch.

Determining what to ask for

Before asking a potential donor for a contribution at one of your follow-up meetings, find out everything you can about the person's gifts to similar causes. (To do so, you can scour donor lists and ask board members who know this person.) This information helps you ask for an appropriate amount.

Many people believe that you should ask for an amount that's somewhat higher than what you expect to get, because making a request for a generous gift is flattering to the donor. However, you don't want to ask for so much money that the potential donor feels that whatever he gives will be a disappointment. Donors commonly say something like, "I can't do $50,000, but I'll consider $25,000 if that would be a help." (Add or subtract zeros as appropriate.)

REMEMBER

In most cases, when the donor suggests a level, you don't want to haggle; you may become annoying and cause him to back out of contributing. You *may* gently push for a higher amount by using the personal connections angle, however: "We were hoping you might become one of our $50,000 donors, along with Joe Schmoe and Sally Smiley, but whatever you can provide would be a big help."

Minding your manners

One of the most important parts of your major gift campaign is thanking your contributors. Do it very soon after any in-person visit is made, a pledge is made, or a gift is received. We recommend an immediate phone call or email and a handwritten note. If you acknowledge a donor thoughtfully and graciously, and follow up that thank-you note with newsletters, web posts, or other information about the organization's accomplishments, you're strengthening your relationship with that contributor and making her feel as if she's part of your organization. That's good manners and smart fundraising. Even if your in-person visit isn't successful, it's a good idea to send a short note to thank the prospect for her time.

Raising Money by Mail

Direct mail, the practice of soliciting donations through large-scale mailings, grows out of a sophisticated and fascinating area of fundraising. Your organization may be too small to make investing in a direct-mail campaign worthwhile, but you can use valuable tips from the "big guys" and develop small-scale letter-writing and emailing efforts that yield good results.

Taking the direct-mail route

If you want to try large-scale direct-mail fundraising, we strongly recommend that you hire a direct-mail consultant or firm to handle your campaign. Direct-mail fundraising can be an expensive investment, and you want the best, most up-to-date professional advice available.

Successful direct mail depends on the following:

- A cause that's meaningful to many thousands of people across your state or the country
- A compelling, well-presented letter that makes its reader believe that your nonprofit can make a difference
- A well-chosen mailing list, usually purchased from a list broker (opt-in email lists also may be purchased through these brokers)
- If you're using regular mail, nonprofit bulk-rate postage, which will save you a significant amount over commercial mailing rates

» Easy, clear ways your letter readers can respond to the request by using a return envelope, response card, or website donation button

» Testing your letter and list at a modest scale before sending the letter to hundreds of thousands of names

» A mailing schedule through which you solicit donors several times each year

TIP

When writing a fundraising letter, remember that many potential donors will read your letter's P.S. before they read the body of the letter, so include compelling information there, such as a testimonial from a client or information about a matching gift.

TIP

To secure a nonprofit bulk-mail permit, you must apply to the United States Postal Service, completing PS Form 3624. Check out File 14-1 at www.dummies.com/go/nonprofitkitfd5e for a link.

Successfully raising money through the mail depends on gradual development of a loyal cadre of donors who respond by mail. Fewer than 1 percent of the people you initially mail to may send contributions, but after they give, you add them to your *donor list* — and, according to research from Network for Good, an average of 27 percent of those donors will make a second contribution.

It has become more challenging to build loyalty among donors than in the past. It is just as important to apply the principles of cultivation and relationship-building with donors reached through mail or email as it is with those you meet face to face. This may mean sharing personalized news announcements from your after-school program director, sending invitations to parties or to online seminars — even links to webcam filming of newborn tiger cubs.

Donors who make repeat contributions are likely to stick with your organization for several years, making three or four (or more!) contributions in that time, and they may increase the size of their contributions. Adding to this value, you'll uncover prospects for major gifts and planned gifts (contributions from bequests) among these donors.

WARNING

Before your donor list develops into a significant and loyal resource, a direct-mail campaign on your organization's behalf may only break even on the cost of the initial letter-writing campaign. You may even lose money. That's why direct mail may not be appropriate for a small or startup organization. You need to be able to cover significant upfront costs and not expect a strong return on that investment for several years.

Trying your hand at a letter-writing campaign

Given the risks and costs of direct mail, we recommend that you borrow some of its techniques and work on a more affordable scale, using your organization's small but mighty contact list and your own writing skills. The key to success is targeting donors to whom the staff, board, and volunteers are already connected in some fashion and personalizing their letters.

REMEMBER

Your fundraising letter isn't just a letter; it also includes a mailing envelope, a reply envelope and card, and sometimes a brochure, a copy of a newspaper clipping, or a photograph. All these pieces should relate to one another and convey a clear, compelling message.

The outer envelope should be inviting to open. For instance, did you use a first-class stamp? Did you handwrite the address and add a special message? Is the envelope a different color and size than standard envelopes to draw attention to its importance? The letter itself should give readers enough information that they feel involved in the cause. Most of it should be dedicated to describing the problem that the organization is trying to solve. After that, it should discuss how things can be turned around for the better and the organization's specific method or program for doing so. Make the tone personal by using "I" and "you." The letter should close with a vision for how things will look if the plan succeeds. Keep paragraphs short and emphasize key points with underlined or bolded text.

Always make it easy for your donor to respond by offering return envelopes and clear information about how to give through your website. Always include your mailing address, website, and phone number on the letter in case a donor misplaces the envelope or other materials included in the mailing.

TIP

Your nonprofit can get a permit from the post office to offer postage-paid return envelopes. Although it's more expensive per delivery than first-class stamps, your nonprofit only ends up paying for the envelopes that are sent back (and we hope that those envelopes contain donations).

TIP

Check out File 14-4 at `www.dummies.com/go/nonprofitkitfd5e` for an example of a solicitation letter sent to individual donors.

Try out your letter first on your organization's internal lists of board contacts, clients, and donors. Then continually build your potential donor list by making it a habit to collect names, addresses, and emails at events you present and meetings and conferences you attend.

TREATING DONORS WELL FOR GREAT RESULTS

Good fundraisers recognize the importance of treating donors well. In fact, the Association of Fundraising Professionals has even developed a thoughtful Donor Bill of Rights that offers standards to follow (www.afpnet.org/Ethics/EnforcementDetail.cfm?ItemNumber=3359)), as well as an eDonor Bill of Rights (www.afpnet.org/Ethics/EnforcementDetail.cfm?ItemNumber=3285).

Good treatment of your donors includes allowing them to tell you if they don't want their addresses shared with other organizations. You can include a check box on your reply card where they can indicate their preference for keeping their information from being traded or sold. If you send email solicitations, be sure to include a safe unsubscribe option for donors who want to be removed from the mailing list.

Just as important as collecting names for your list is taking care of those names. You want to develop a good system for recording information about any donors in order to thank them, keep in touch with them, and ask for their support again in the future. You'll find that database programs to manage your donor information come in various degrees of complexity and prices (see Chapter 13). Choose the one that's best for your needs.

Raising Money the "E" Way (Easily and Electronically)

Just as your website and electronic-communications system are tools for serving your constituents and marketing your work, they are tools for raising money. Although giving in response to traditional mail has been declining, online giving is growing at a faster pace than overall giving. Blackbaud, which manages donor data for more than 5,000 varied U.S. charities, reports in "50 Fascinating Philanthropy Stats" that its clients saw overall giving increase 7.1 percent in 2015, but online giving increased at the higher rate of 9.2 percent. Although this growth in online giving suggests that it should be part of your fundraising plans, it still represents a modest portion (about 7 percent) of overall fundraising. We don't recommend that it be your only fundraising approach: It's a good tool when it's used in combination with other methods.

Given that more and more of us make online purchases and read electronic media, why is online giving so tepid? Part of the answer lies in who is giving online: They tend to be younger donors who make smaller gifts. (The same 2015 Blackbaud survey notes that the average online gift is $23.) Part of the answer is the importance of relationships: Many people find the Internet to be relatively impersonal and are unlikely to make a major contribution online. Part of the reason is that online donors tend to be younger and have less disposable income.

Online fundraising has worked very well in some contexts — especially disaster and emergency relief campaigns, for which donors want to respond quickly. People also respond well to online fundraising by organizations they know well, such as their college alumni associations.

Using email and related tools to build and maintain relationships

We cannot emphasize enough that fundraising is about building relationships with people. It's in a nonprofit's best interest to keep donors informed about what the organization is doing. It can't hurt to get to know the people who give money to your nonprofit, and the Internet can be of help with both of these tasks.

The most frequently used feature of the Internet is email. It's easy to use, and it's cheap and fast. Many nonprofit organizations produce monthly (or more frequent) online newsletters about their work. Distributing these bulletins to donors is a great way to keep them informed about how their contributions are making a difference.

TIP

If you're going to send bulk emails, you may want to use an email marketing service that can help you build your list, set up your message in an attractive format, and make it likelier that your message is received (and doesn't land in a spam folder). Some of these businesses allow nonprofits to send a certain number of messages without charge.

As a first step, obtain an organization email account and address that enables people to contact you (and enables you to begin building an online mailing list).

TIP

Your organizational emails, website, or other social media should connect and point to one another so that no matter how someone comes across your nonprofit online, he will find several ways to stay in touch.

When using email to raise funds, you'll want to compel readers to click through to your website, where they can read more about your organization and find out how to contribute. Remember to mention your online donation option in every piece of mail and email you send, and place your "Donate" button in a prominent place on your site.

Related online tools you want to have at hand include the following:

- **A website:** Your site should feature clear, timely information, calls to action that encourage readers to get involved, a "Contact Us" option linked to your email address, and an easy-to-use click-through "Donate" button that enables website visitors to make contributions online. If you have a wish list of equipment and supplies that people can donate, place a link to that list in a prominent location (perhaps accompanied by a link to your online wish registry at a website like Amazon.com [www.amazon.com], where donors can make a quick purchase that's delivered straight to your organization's door). You can read more about creating a website in Chapter 12.
- **A brief electronic newsletter:** This e-newsletter should feature an intriguing subject line and be highly readable. Imagine your typical reader: *If* she opens it, she'll scroll through it quickly. You want to write a compelling sentence or two about each topic, inviting the reader to click through to your organization's website and read more about it (noticing the "donate" button while she's there).
- **An organization blog:** Blogs are great tools for inviting comments and feedback. Tools for creating a blog are available for free at a number of websites.
- **Social-networking accounts:** Social marketing options include Facebook, LinkedIn, Google Groups, Twitter, Pinterest, Instagram, and many others. These social-networking tools remind people of your organization and urge them to take action on your behalf. Some of them provide features specifically for sharing information about organizations (such as Facebook Pages) or can be used as avenues for raising money. One example is to add a Facebook-compatible app — such as GoGetFunding (www.gogetfunding.com) or FundRazr (www.fundrazr.com) — to your nonprofit's Facebook profile.

TIP

At www.dummies.com/go/nonprofitkitfd5e, File 14-6 and File 14-7 provide sample e-solicitation letters, and File 14-8 shows how a nonprofit uses email correspondence to develop a relationship with a potential donor. File 14-9 offers tips on how to solicit donations through your website and social media outlets.

Building your email address lists

When corresponding with people who contact you through email or your website, consider them prospective donors. If you're emailing a new correspondent, ask whether he wants to be added to your mailing list. Make this invitation a habit and respect those who decline. Others will accept, and slowly and steadily you'll build your list.

Even though you can find robust email services that are free to use, many nonprofits choose to pay for commercial email services. Typically they provide templates that will make your email messages look better. You get less spam, less phishing, and more reliable delivery of your email. We always recommend checking product reviews and talking to others, but a few frequently used email services are Constant Contact (`www.constantcontact.com`), Google Apps for Nonprofits (`www.google.com/nonprofits/products`), MailChimp (`www.mailchimp.com`), and Vertical Response (`www.verticalresponse.com`).

REMEMBER

Email address lists, like mailing address lists, can be purchased, traded, or borrowed. A nonprofit may be willing to lend or trade its list with another nonprofit with a similar mission. When using a borrowed list, it's important to respect the privacy of the people listed. Don't pass it along to others without permission and always offer an "unsubscribe" option.

Using your website as a cultivation tool

A website can never replicate a face-to-face encounter with another person, but if you design your site to grab and hold a visitor's attention, you can succeed at attracting donors. Strategies may include:

- » Short profiles of clients, donors, volunteers, and staff members
- » Strong visual features — graphic-design elements, photographs, and short videos
- » Short films or slide shows of the organization's activities (One marine mammal rescue organization featured a story about stranded sea lion's rehabilitation.)
- » PDFs of studies your nonprofit has conducted or curricula it has developed and links to information and resources about the field in which it works
- » Specific fundraising campaigns such as "Be a hero," with a monthly gift to a search-and-rescue organization or "Adopt an acre," where you sponsor an acre of restored prairie grasslands
- » Opportunities to volunteer
- » Invitations to readers to provide contact information — and promises of specific, enticing rewards if they do so

» Blogs and chat rooms on subjects related to your work

» A clear and easy-to-use donation feature for making contributions and information about how donations will be used to make a difference

TIP

When incorporating photos and graphic material in your fundraising materials, be sure to secure permission and give appropriate credit to the photographer or artist.

The longer a website keeps visitors involved, the likelier it is to build their trust and interest.

Most websites include a "contact us" page. Make sure this option is user friendly and that someone on your staff is responsible for replying to queries quickly and courteously. Include opportunities for readers of the website to sign up for your electronic newsletter or other information and to leave an email address and/or telephone number.

TIP

You may build the most amazing website on the planet, but if readers aren't finding it, it's not fulfilling its role for your organization. One critical question is how to attract the notice of search engines. Some agencies sit back and wait for a web spider to find their page; some submit their page to search engines one by one, which can be time consuming; and some hire a service to submit their information to the leading search engines. See *SEO For Dummies*, 6th Edition, by Peter Kent, for information on getting your site noticed by search engines.

Gathering money online

If you have good technology support and a website that includes a system for securely collecting money, you can insert your own "donate now" button. However, many nonprofits find it easier to contract with a "donate now" service.

The number of options available can be daunting. When choosing one, think about the giving experience both from the point of view of your nonprofit and from the perspective of your donor.

From the perspective of your nonprofit, look carefully at how and for what an agency charges fees. Most online contribution services charge a setup fee, a monthly fee, and a per-contribution fee. Sometimes you can avoid the monthly fee, but when you do, you often pay a higher fee for each contribution you receive. Some less-expensive nonprofit technology services charge lower fees but send you your contributions in a lump sum, so you only receive the money, not the donor records. On the other hand, some services charge for keeping the donor records or for allowing you to upload them onto your computers. You also will want to check on ease of use, and number of types of credit cards accepted. Some services will allow you to add boxes on the donor form where your contributors can select premium gifts or choose to receive a newsletter.

You also want to consider the experience of making the donation from the point of view of your donors. With some services, such as PayPal and Network for Good's basic DonateNow, when a donor clicks on the Donate Now button on your website, she can immediately tell that she's making her gift through a different website. Check to see if the service you're using will allow you to modify the look of that donation form so that it matches the look and feel of your website and graphics. This can reassure nervous donors that they're in the right place.

We recommend checking Idealware's website (www.idealware.org) for helpful, updated articles about choosing online donation and credit-card processing tools.

REMEMBER

If you hire a marketing service to develop and maintain your organization's website, consider the breadth of the services you're purchasing. Will the agency design your website for you? Will it thank your donors? Will it integrate data about customers who pay to attend your workshops with information about those who contribute money? Will it manage a donor database for your online contributions? And is it a 501(c)(3) nonprofit organization or a for-profit entity? (If it is a nonprofit, it may be registered to raise funds in all states requiring such registrations and this could save your organization significant fees.)

Finding financial support from a crowd

Another method of raising money online, *crowdfunding* or *crowdsourcing*, involves seeking donations from people whom you reach through your email records or social media. Potential donors may learn only about your project or they may be directed to a page that features a dozen or more projects to choose among — one of which is yours.

Crowdfunding services provide individuals and enterprises with tools to conduct focused, time-sensitive campaigns to raise money. Someone seeking contributions must name a specific amount to be raised and a specific project to be launched or completed. Some crowdfunding services (such as Kickstarter) are "all or nothing," meaning fundraisers don't receive any of the pledged gifts until they meet their goals. Although this can seem to be a drawback, it also can motivate donors to give larger gifts or make additional contributions after their initial investments. Crowdfunding isn't exclusively for nonprofit organizations, but some platforms that include nonprofits among their clients are Fundly (www.fundly.com), Indiegogo (www.indiegogo.com), Kickstarter (for creative projects; www.kickstarter.com), and Kiva (for interest-free micro-loans; www.kiva.org).

Distributing your fundraising through volunteers

Technology provides novel ways for your organization's supporters to help raise money. In *distributed fundraising*, sometimes called "peer-to-peer fundraising," an individual supporter creates an online donation account for your organization, and through that account speaks directly to friends and family about your organization's work, urging them to contribute or volunteer. Often they do this in concert with a campaign you're promoting directly. This is a great thing to ask your board members or tech-savvy volunteers to do for you.

A peer-to-peer fundraiser can add a widget or badge that can be added to a website, blog, or social media page. They then distribute the page through their networks (for example, to subscribers to their blogs). Your volunteer fundraiser often uses an existing service to set up the page and collect contributions. Some of these services are Network for Good, Change.org, Changing the Present, FirstGiving, and Causes.com. Your volunteer will want to pay attention to the comparable setup fees and percentages of contributions collected by these services.

WARNING

While involving your known supporters and volunteers to help raise money for your nonprofit is a great idea, there is a risk in distributed fundraising if an individual who is not authorized or known by the nonprofit raises funds in the name of the organization. If this situation arises, you should contact the donor and the fundraising platform to stop the unauthorized solicitations.

Soliciting text-message donations

Many charities now accept donations made via mobile phone texts. Text message contributions have been particularly strong for natural disasters and other emergencies. In 2010, the American Red Cross raised some $12 million through its text-to-give campaign for Haitian earthquake relief. Mobile phone services waived texting fees for donors who gave. Charitable giving through text message platforms continues to grow steadily.

To set up this service, your nonprofit organization registers with a donations processor that, in turn, works with cellphone providers to add an agreed-upon contribution size to the user's cellphone bill.

A typical texted donation is modest in size, but a nonprofit may secure a new donor who is motivated by being able to take action immediately and by the relative ease of remembering a code word and phone number rather than a long website or street address. A text-to-give campaign works best when your nonprofit has an active social-media network or highly visible public event (such as a benefit concert) through which it can spread the word.

WARNING

Nonprofits should be aware of some disadvantages — the biggest one being that it's difficult to form a relationship with a texting donor. Your nonprofit and the donor will be working through a processing service such as mGive (`www.mgive.com`), and your organization won't gain access to the donor's information. Other disadvantages include:

- **Upfront costs:** Leasing a short code from the Common Short Codes Administration can cost $1,000 upfront for a "vanity" code word you choose or $500 for a randomly selected code. Furthermore, mobile vendors will charge a one-time set-up fee of $3,000 to $10,000 for a unique "vanity" short code and $1,500 for a shared short code.
- **Small contributions:** Nonprofits are legally permitted to ask only for modest gifts of $5 to $10, and cellphone carriers may limit the number of gifts donors can make monthly.
- **Fees charged:** The processing service will charge a modest fee for each donation.
- **Delays:** Contributions appear on the nonprofit's phone statements and are released after the phone bill is paid. It may take up to 90 days for the donor's gift to reach the nonprofit.

An alternative to texted donations is the use of QR codes (two-dimensional barcodes that appear on numerous products). Nonprofits can feature their QR codes on newsletters, letters, T-shirts, and invitations. Donors scan these codes with their mobile devices to connect with the nonprofits' websites and make contributions.

WARNING

With the growing use of mobile devices, QR codes would seem to be great means for engaging with their donors, but not all devices come equipped with QR code readers and potential users of the codes may have to download apps on their mobile devices first. They also may grow frustrated if the end-site downloads slowly or is hard to navigate.

Telemarketing: Dialing for Dollars

In the mid-1980s, direct mail was a wildly successful means of fundraising for many organizations. By the late 1980s, it had declined. Households were receiving too much direct mail, making recipients less inclined to read it, and the costs of printing and postage rose. That's when the dinner hour began to be interrupted by incessant telemarketing calls.

Now, with many people using Caller ID on their phones, and with cellphone numbers being difficult to collect for calling purposes, it's harder to succeed with telemarketing. Yet the phone is a powerful tool for connecting with others, and it's not yet time to abandon it in fundraising.

The advantages of telemarketing are:

- » It's hard to ignore. A human voice engages you in a conversation, and you must respond. It's not like a letter that you can throw into the recycling bin without the author noticing.
- » If you use trained volunteers to make the calls, you can manage a campaign at a very reasonable cost.
- » As with direct mail, you can use some of the elements that shape a large-scale, professional telemarketing project for a modest campaign.

The disadvantages? Donors grow annoyed and may stop contributing if called too frequently. Plus, if you don't have volunteers to make the calls, some commercial telemarketing firms take a very high percentage (as high as 75 percent!) of any funds raised, discouraging donors who want to see fundraising costs kept low.

WARNING

Citizen complaints about intrusive telemarketing have led to the creation of the National Do Not Call Registry, which makes it illegal for businesses to make telemarketing calls to people who register. In 2008, Congress passed legislation to make enrollment on the National Do Not Call Registry permanent. At present, nonprofit charitable organizations are exempt from these rules. However, if an individual asks to have her name removed from a nonprofit's call list, a third-party fundraiser must comply. Check the status of these regulations before embarking on a telemarketing campaign. Public policy articles on the Association of Fundraising Professionals website (`www.afpnet.org`) are a good place to check, as is the National Do Not Call Registry website (`www.donotcall.gov/faq/faqbusiness.aspx#ExemptOrg`).

The key steps to running a telemarketing campaign are writing a script, training volunteers, organizing follow-up calls, and — of course — thanking, cultivating, and upgrading donors over time. We explain everything you need to know in the following sections.

Writing a script

Every call should open with a clear, direct, personal greeting: "Hello, Mr. I'm-Getting-the-Person's-Name-Right, I'm Ms. Call-a-Lot, and I wanted to talk to you about the Scenic Overlook Preservation Fund Committee's work." Let your listener know right away that you're soliciting a contribution and calling on behalf of a particular nonprofit.

Having connected with the call recipient, the caller then tries to link that person's interests and behavior to the reason for the call:

- If she's a past donor, begin with a hearty thank you.
- If she's been involved in a related cause or effort, mention how important that work is.
- If she lives near the scenic overlook, mention how beautiful it is and the community's concern for the fragile surrounding environment.

You may notice a tricky moment in the call when you want the potential donor to relax and listen so you can tell your story and not be interrupted. To increase the caller's chances of keeping call recipients on the line, the script should be

- Engaging and information packed (to hold the listener's attention)
- Upbeat about the possibility of improvement or change
- Deeply concerned about the current situation
- Specific about the time frame in which things need to change
- Specific about the amount of money the caller hopes the listener can contribute

TIP

Check out File 14-10 at `www.dummies.com/go/nonprofitkitfd5e` for a set of telemarketing "do's" and "don'ts" and a sample telemarketing script.

Coaching your callers

Telemarketing can be an excellent board or volunteer group effort. If someone involved with your organization works in an office that has multiple telephone lines, see whether you can borrow the office for an evening. Early evening (6:00 to 8:30 p.m.) on weeknights is generally considered the best time to call, but — as you've probably noticed — telemarketers now frequently call during the day and on weekends as well.

Gather your volunteer team an hour before beginning. Feed them a good meal and give them a pep talk. Building camaraderie among the callers can relax anyone who's nervous. Setting a group goal for the evening and mapping it on a big chart can build morale.

Inform callers that they must deliver the message in a crisp, clear, and friendly voice. They shouldn't rush, but they also shouldn't leave holes in the conversation that the call recipient can close before the caller can ask for a contribution. They should ask for a specific contribution and confirm the amount. They also should tell potential donors that they're volunteers; this may make call recipients more likely to pay attention.

Provide your callers with information about each household they're calling, including a recommended gift request. You base your request on the potential donor's past contributions to your organization and others. Sometimes you're just guessing. That's okay so long as your callers are good listeners and deftly adjust the amount they're requesting in response to what they hear. If, after a gift is pledged, you feel that you asked for too little money, don't despair. You can ask the donor to upgrade his gift next time.

Collecting the pledges

If a pledge is made, the caller should thank the donor and try to get her to either provide credit-card information over the phone so the gift can be charged immediately or promise to return the contribution within a certain time. Ask the caller to confirm the spelling of the donor's name and address before saying goodbye.

REMEMBER

At the end of the evening, everyone present should write brief, personalized thank-you notes. For contacts who promised gifts but didn't charge them over the phone, send the thank-you notes with pledge forms (indicating the specific amount of the promised contribution) and return envelopes.

Every telemarketing campaign suffers from a percentage of unrealized promises, and callers want to keep that percentage as low as possible. You may ask a small number of your volunteers to reconvene for a short follow-up calling session a month after the initial campaign to jar loose any contributions that haven't yet been received.

» Recruiting influential people to support your event

» Creating a timeline for planning your event

» Generating media coverage and social-media buzz

Chapter 15
Making the Most of Special Events

From glamorous dances under the stars to pancake breakfasts at the local firehouse, social gatherings are part of the fundraising mix for most nonprofits. Special events don't just raise money. They're often credited with raising friends along with funds. New donors — who don't know the organization — may come forward because the event itself sounds fun and interesting, because of who invites them, or because someone they admire is being honored. Special events can be wonderful catalysts for attracting support from businesses that like to be recognized when they make contributions, or for attracting people who like to socialize and be seen.

At their best, special events raise substantial amounts of money, draw attention to an organization's good work, and attract new volunteers. If fundraising is about cultivation, a special event can be a greenhouse for nurturing growth. You hope to create a special event that people look forward to, an annual tradition. Still, a special event can be one of the most expensive ways to raise money. Expenses may eat up half or even more of the gross event income. Putting together a special event draws upon all your nonprofit management skills and can drain staff and board time and energy away from other important activities.

Bottom line? Special events need careful planning. In this chapter, we show you how to put together the leadership, timeline, and budget that can lead to a successful special event — and steer clear of obstacles.

TIP

Check out File 15-1 at `www.dummies.com/go/nonprofitkitfd5e` for a list of web resources related to the topics we cover in this chapter.

Thinking through the Whole Event

If you think a special event is in your future, we recommend sitting down with staff and board members and raising such questions as:

- **What would our organization's followers enjoy doing and how much would they pay to do it?**
- **Whom does our organization know who can provide event elements — donated goods, auction items, entertainment, or printing services?**
- **Whom do we know who can be honored at the event or serve on an event committee?** The event leadership and volunteer committee are critically important to attracting donors to the occasion.
- **When can we focus attention on a special event?** The last six weeks before an event are generally the most labor intensive. Look for a relatively clear six-week block of time in which you don't have grant deadlines or other commitments.
- **When can we hold an event without competing with our other fundraising drives?** If you hold annual fund drives in June and October, for example, why not plan your special event in the spring?

We offer some additional advice regarding these questions in the following sections.

Using your budget to guide decisions

Special events can be produced on bare-bones budgets, for princely prices, or for any amount in between. As with most kinds of investments, event planners expect a higher return in exchange for a higher investment. But don't exceed your means. In this section, we outline some ideas for tailoring an event to your organization's budget.

WARNING

When deciding on an event, we recommend avoiding anything that can make your guests uncomfortable or deter them from coming, such as:

- » Events that limit your guests' ability to come and go as they want — like a soiree on a boat in the middle of a lake.
- » Events at which your intended audience finds the attire, time, or place awkward.
- » Events that are designed to reach an audience that's completely unknown to your organization or its supporters. Do you like to go to a party where you don't know anybody?

TIP

After you've sketched out an idea for an event, go to your staff and board members and ask them directly: Would you attend this event? For this price? At this time of year? At this location? If your core followers and supporters aren't enthusiastic, the event isn't going to raise money. If the idea excites them and they're willing to play a role in the event planning, you're on your way to success.

Low-budget special events

Most nonprofit organizations have a wealth of talent simply waiting for a showcase. Some of the following suggestions won't work for your organization, but they can get you thinking about similar events that would be perfect for your nonprofit:

- » **Sign up neighborhood children for a summer read-a-thon that benefits the library or an after-school literacy program.** By sending forms home to parents in advance, you can secure their permission and help with collecting pledges.
- » **Offer a bake sale with a distinctive theme celebrating a cultural group or holiday.** Organize your volunteers around three primary activities: setting up and pricing, selling, and taking down the sale. Plan in advance what to do with items that don't sell.
- » **Hold a cocktail party in a board member's home, focused on a theme or honoring a special guest.** Although a cocktail party for 15 people may not raise much money, don't rule out this idea. If every one of your board members signs up to host such a party over the course of a year, the cumulative amount raised may pleasantly surprise you.

WARNING

Many nonprofit organizations produce raffles or prize drawings as low-cost activities to raise funds. If you plan to do so, proceed with caution. Such activities are defined as gambling or gaming in many states. Some states permit nonprofits to hold raffles, but in other states or jurisdictions, they may be illegal or may be presented only if you follow steps to secure a permit. You can find an overview of guidelines for hosting games of chance, raffles, and charitable auctions at the National Council of Nonprofits website (`www.councilofnonprofits.org`) and a quick guide to states' offices regulating nonprofits at the National Association of State Charity Officials website (`www.nasconet.org/documents/u-s-charity-offices`).

Mid-budget special events

If you can afford a few more features in your special event, you may want to do something like one of these ideas:

- **Identify an up-and-coming performer and ask for a donation of a performance in exchange for the promotion that your event will bring to him or her.** The club or theater rental is likely to be your highest cost. If you choose an unusual site that's donated to you, make sure it can accommodate your performer's needs — acoustics, sound system, lighting, electrical outlets, and the like.
- **Sponsor a daylong cleanup of a coastal area, park, or preserve.** Volunteers can be sponsored by having their friends sign up to pledge a certain gift amount to the organization for each pound or bag of trash that they remove. Offer prizes for the most unusual refuse items found and the most sponsorship sign-ups.
- **Produce an online auction.** Volunteers can help you reach out to friends to secure donations to be auctioned and draw attention to the event and auction items through social media. To build and hold interest in the auction, you'll want to post regular updates and news about the event (with pictures) on your website and social media pages. Don't forget that costs may include securing a seller's permit and remittance of sales taxes.

High-budget special events

If money is no object — at the front end, at least — these events may be of interest. All require more of an upfront investment:

- **Hire a major speaker or entertainer from a lecture bureau or theatrical agency and use that person as the focus of a dinner party or private concert.** Honor one or more business and community leaders at the event. Form an event committee of people who will invite their friends and people who may want to come for the sake of the honorees.

- **Present a dance featuring a live band.** Decorate festively. Make sure your crowd likes to dance and that your musicians' repertoire suits the moves that the crowd knows.
- **Organize a celebrity athletic event.** Celebrity golf, bowling, and ping pong — you name it — are all possible. Create teams pairing professional players with amateurs or local celebrities with donors. If you use professional athletes, liberally handicap the amateurs so everyone has a chance of winning. Provide trophies or certificates for many kinds of "achievements" — the longest drive, the bowling ball most often in the gutter, the quickest victory, and so on. Often these events end with celebratory dinners.

Sticking to your budget

The bottom line is very simple: Your total earnings from a special event must exceed your total cost — by a lot, you hope. But how do you get a handle on revenue and expenses? In this section, we offer some thoughts.

TIP

If you're staging an event for the first time, it's particularly important *early in the process* to ask your core supporters — board, volunteers, and event leadership — how much they intend to give. Because these people are the most likely to give generously, knowing their intentions helps you forecast the overall results.

Sometimes fundraising events take several years to garner high-level support, so budget exactly how much you intend to make in years one, two, and three. It's not uncommon to break even the first year of an event or to have small net proceeds from your first venture.

Another way to estimate the fundraising potential of an event is to check with organizations that produce similar events. If they've presented a program year after year and yours is a first-time outing, ask them where their income levels began. Try to objectively weigh your event's assets against theirs. Are your boards equally well connected? Is your special guest equally well known?

TIP

Check out `www.dummies.com/go/nonprofitkitfd5e` for three sample special events budgets — File 15-2 is a sample budget for a tribute dinner, File 15-3 is a sample budget for a concert or performance, and File 15-4 is a sample online auction budget.

Figuring the income side

Try to design your event so it generates income in more than one way. A rummage sale may also include the sale of baked goods. An auction may include advertising in a printed program along with tickets to the event and the income generated from the sale of the auction items. Standard event income categories include:

- Individual ticket sales
- Table or group sales (usually for parties of eight or ten)
- Benefactor, patron, and sponsor donations (for which donors receive special recognition in return for contributing higher amounts than a basic table or seat costs)
- Sponsorships of event participants (for instance, pledging to contribute a particular amount per mile run by a friend)
- Food and/or beverage sales
- Sales of goods and/or services
- Advertising sales (in printed programs, on banners, and so on)
- Purchasing a chance (raffle tickets, door prizes, and so on)

Capturing expenses — expected and unexpected

Unless a wonderful sponsor has offered to cover all your expenses, your event will cost money to produce. The general categories can include the following:

- Building/facility/location (space rental, site use permits, security guards, portable toilets, tents, cleanup costs)
- Advertising and promotion (save-the-date postcards, photography, posters, invitations, event programs, publicist costs, postage, event website with a ticket purchase feature)
- Production (lighting and sound equipment, technical labor, stage managers, auctioneers)
- Travel and per diem (for guest speakers, performers, or special guests)
- Insurance (for example, liability insurance in case someone gets hurt because of your organization's negligence, or shipping insurance to protect donated goods)
- Food and beverages (including permits for the sale or serving of alcohol, if necessary)
- Decor (flowers, rented tables and chairs, linens, fireworks, banners)
- Sales tax, which varies by state (Some states collect taxes on food and beverage sales, as well as auction items and other goods sold.)

» Miscellaneous (prizes, awards, talent treatment, name tags, signs, T-shirts)
» Office expenses (letter writing, mailing list and website management, detail coordination)
» All other staff expenses

TIP

Always inquire about nonprofit pricing, and let vendors know that your organization is tax-exempt.

WARNING

In spite of your careful planning, certain expenses can appear unexpectedly and cause you to exceed your budget. If you plan to serve food at your event, keep these tips in mind to avoid surprise charges:

» Confirm that all service and preparation charges are included in the catering budget.
» If you need to add additional meals at the last minute, find out whether your caterer charges extra. If meals that you ordered aren't eaten, you probably still need to pay for them. Check on your caterer's policy.
» If some of the wine that you've purchased isn't consumed, is the store willing to buy it back from you?
» If wine has been donated to your event, find out whether your caterer charges corkage fees for opening and serving it.

Soliciting in-kind gifts for your event

When you think about what your event will cost and how you can pay for it, think about the business contacts that your board, staff, and outside supporters have. Often, a business's contribution of *in-kind* (non-cash) materials is more generous than any cash contribution to your event. You may need to be flexible about timing and willing to drop things off or pick things up, but don't overlook in-kind gifts. Donated auction items are one example. Other examples include the following:

» Businesses in your community that have in-house printing equipment may be able to print your invitations and posters, saving you thousands of dollars.
» A florist may contribute a roomful of valuable centerpieces in exchange for special recognition in the dinner program.
» A donor may let you use her beautiful residence as your event site.

REMEMBER

You may not want or need all the in-kind gifts that are offered to you. In cases such as these, do look the gift horse in the mouth! Don't accept an in-kind contribution if it's not up to the standards that you need for your event. Also, consider the implications of accepting the gift — both your need to honor any restrictions or conditions the donor places on the in-kind gift and the appropriateness of associating with the gift and its donor. For instance, if your agency helps young people recover from drug or alcohol addiction, don't accept a sponsorship from an alcoholic beverage company. If you're afraid of offending the donor or hurting a relationship, you can ask to use the gift in another context and acknowledge the donor publicly for his generosity.

TECHNICAL STUFF

Your donors will want to claim tax deductions for their in-kind contributions. It's their responsibility to tell the Internal Revenue Service if they're seeking charitable contribution deductions for the gifts, and it's your responsibility to acknowledge the donation in a prompt thank-you letter. If they make a contribution of non-cash property worth more than $5,000, generally that item must be appraised before a value is assigned to it. For more information, see the IRS website at `www.irs.gov`.

Building your event committee

Strong volunteer leadership is the backbone of special-events fundraising. Some of the most important work you and your board will do as event organizers is to recruit a chair or co-chairs for an event committee. Choose people who are well connected and who bring different contacts to their committee work — perhaps one community volunteer, one business executive, and one local athletic star. Many organizations include a board member on the event committee, but they also use the event as an opportunity to recruit beyond their boards, bringing in new, short-term volunteers.

TIP

Go-getter volunteers can make your event a great success, but they don't have the same responsibilities as your board and staff to protect your organization's financial well-being and reputation. We recommend specifying in writing who is ultimately responsible for making decisions (likely your executive director, board, or board executive committee) and scheduling regular check-ins with volunteers to avoid poor decisions or misunderstandings.

We strongly recommend working closely with one or two volunteers who have planned fundraising events in the past and can help guide you and your committee through the process. Events are detail-heavy, and someone who has done this before can make sure that all details are addressed. Keep an event binder or shared online folder and document everything including meeting notes, invoices, lists of tasks, timelines, budgets, and attendees.

Your co-chairs' job is to build a network of support for the event. Usually they invite other well-connected people to join them as members of an event committee. Perhaps your board recruits three co-chairs, and those co-chairs recruit 25 event committee members. Your committee members and co-chairs then send personalized letters, emails, and invitations and make phone calls urging people to support and attend the event. If each of them brings 20 people, you'll have quite a crowd!

TIP

Check out www.dummies.com/go/nonprofitkitfd5e for two related sample letters. File 15-5 is written by an event co-chair, inviting someone to join an event committee. File 15-6 is written by a committee member to a potential donor.

Setting a date and location

You don't want to plan an event on Super Bowl Sunday or during your city's largest annual street fair. Apart from avoiding holidays and other obvious dates when you would be competing for your guests' attention, check around town to find out whether you're planning your event on the same date as one organized by another nonprofit agency. Be as thorough as possible in this date checking. Competing for the same audience on the same date — or even dates that are close to each other — hurts both organizations' results.

While the idea is not innovative, good hotels are generally excellent sites for special events. They often have several banquet rooms, and catering, podiums, sound systems, and parking valets are all available at the site. All these extras leave you with fewer details to manage.

Some of the best special events we've attended have been inside the nonprofit organizations that they supported or in sites thematically linked to the cause being supported — such as a historic building or maritime museum. Outdoor locations, such as vineyards, formal gardens, and mountaintops, can be beautiful but require backup plans for variable weather, and climbing rocky pathways or crossing uneven lawns may make your event inaccessible to some of your hoped-for supporters.

Another creative angle on choosing a location is to match it to an event theme. We've put on a great Halloween party in a former mortuary.

TIP

If you're running short on ideas, ask your friends and colleagues about interesting places they've attended events, or hold a brainstorming session with your board. Make sure your constituents will feel comfortable going to the location you're considering.

Setting Up Your Timeline

Although events vary greatly in size and complexity, and although we've pulled off adequately successful events in two or three weeks, we recommend working against a six-month schedule. This section outlines a scheduling checklist for a gala dinner. It can easily be modified to fit other types of events.

The first three months

The first three-month period is, not surprisingly, the slowest part of event planning. You may wait several weeks to hear back from an invited celebrity, and you may need several more weeks to find a replacement if you're turned down.

- Develop the plan and recruit the event's leadership — we recommend two or three co-chairs and an honoree.
- Secure entertainment and a location.
- Select a theme and a caterer.
- Send a save-the-date postcard and email announcement.
- Create an event web page and social-networking site.
- Apply for any necessary permits (for charitable gaming, selling auction items, serving alcoholic beverages, street closures, and so on).

Months four and five

As the event draws nearer, you let others know about the upcoming event and begin contacting volunteers.

- With the help of your event co-chairs, recruit an event committee and/or a core group of volunteers.
- Visit the site where the event will be held, checking out all the regulations and recommendations for its use.
- Solicit in-kind contributions of materials you need for the event.
- Send initial news releases announcing your speakers, celebrities, and/or event leadership. (See Chapter 12 for more details about a marketing and public relations program.)
- Call all potential committee members.

- Develop your invitation design. All the text on the invitation should be ready for final design, printing, and online posting by six weeks before the event.
- Select your menu and start working on decor ideas.
- Personalize invitations with the help of committee members.
- Mail and email the invitations.

Four weeks before the event

Spend the month before the event generating excitement and finalizing the details.

- Build interest with frequent website and social-media updates featuring photos and news bits.
- Email a second batch of news releases.
- Make phone calls to the media and to invitees to confirm coverage and travel plans.
- Design the printed program to be passed out at the event.
- Assist co-chairs and honorees with their speeches (if necessary).
- Gather the elements — baskets, banners, confetti, and so on — needed for decor.

TIP

In designing your event's promotional materials, you want the look of the invitation, program, event web page, signage, and audio-visual templates to be consistent. Schedule time to solicit logos from corporate sponsors and double-check spelling of donors' names and how they want to be recognized.

The week before the event

Don't get too stressed in the days before the event. You should have only a few last-minute details to attend to.

- Confirm the number of guests you expect. A few days before the party, call *everyone* who has made a reservation and confirm whether he or she is coming. A few people will have changed their plans, and you don't want to pay for uneaten meals or run out of food. When ordering food from your caterer, assume that under normal circumstances, 5 percent of your expected guests won't attend. If the event is free, 10 percent of them won't attend.

» Plan the table seating at the event (if necessary).
» Prepare place cards and table cards.
» Decorate the site.

ISSUING A MEMORABLE INVITATION

Make your invitation something your potential guests will open and remember. If it's a physical invitation, addressing the envelope by hand and using stamps rather than metered postage makes it look more personal, and an intriguing phrase or logo on the outside may lead to its being opened.

What do recipients see first when they open your invitation? Most invitations are made up of an outer folded piece, a reply card, and a reply envelope. Want some more attention? What if a compelling photograph slips out of the envelope? Or a small black cat, spider, or bat sticker falls out of your Halloween invitation?

In spite of our recommending these bells and whistles, we are firm believers in clarity. Make sure the reader of your invitation can easily see who's extending the invitation, what the event is, where and when it's being held, how much it costs, and how to respond to the invitation. If people have to search for these basics, the invitation will land in the recycling bin. And remember that people look forward to your event more if other people they know are going to be there. Make sure the names of the people on your event committee are clearly and prominently presented. Also list top sponsors (cash and in-kind) on the invitation so invitees can see what businesses are involved in the event. These tips are equally true for the electronic version of your invitation: Make it easy for your invitees to make reservations and send contributions.

Print the address, phone number, and/or website to which your guest should respond somewhere on the reply card, even though it's also printed on the reply envelope. Sometimes the pieces of an invitation become separated. You want your guests to readily know where to send their replies (and money!).

For informal and low-cost events, you may want to send only an e-invitation, using a service such as EventBrite or Paperless Post, which will distribute your invitation, collect reservations, send reminders to your guests, and urge them to invite others through social media. If you have good email lists of your donors and contacts, this approach can save you postage costs and reduce paper waste. A few disadvantages are that it can be challenging to include full credit to sponsors, and some people still find electronic invitations to be informal, making your event seem less important to recipients.

Check out www.dummies.com/go/nonprofitkitfd5e for additional special events timelines. File 15-7 is a sample timeline for a tribute dinner or luncheon and File 15-8 is a sample timeline for a concert or performance.

You may think that nobody can rain on your parade, but have emergency backup plans anyway. What if your performer is ill, a blizzard shuts down roadways, or your permits aren't approved on time? You need to quickly move, replace, reschedule, or cancel your program. The faster you can communicate any changes, the better your constituents feel about sticking with you and your cause.

Spreading the Word

If the mantra of fundraising is "If you don't ask, you won't get," the mantra for special-events fundraising is "People won't come if they don't know about it." Your invitation list is likely the most valuable tool you have at hand for reaching out to event guests, but don't overlook the role that social media and traditional media can play.

In recruiting your volunteer committee members, ask them which forms of social media they use and arm them with tools to spread the word to their networks. For example, every time you create a post on the event's web page, email them and ask them to share it. You also can arm them with good photographs and background information and urge them to use the materials to blog or post social-media updates.

Good photographs are essential to your announcements being noticed on social media. Often we think of hiring a photographer for the event itself, but photographic work done before the gala may be even more important.

If your event has a newsworthy angle, send news releases to the media in the hopes of their writing about it or broadcasting the news. (You can find information about writing news releases in Chapter 12.)

Finding a news angle

Think about all the angles you can exploit for publicity. Modify your basic news release to suit the content of appropriate newspaper and magazine section editors, producers of radio feature shows, and TV newsrooms. Follow up your emailed release with a phone call. A few possibilities include:

- **Entertainment section coverage of performers:** You may be able to arrange interviews between your special guests and local entertainment reporters.

- **Business section coverage of your event chair or honoree:** Many papers run a column highlighting the activities of local business and corporate leaders.
- **Feature stories about the community improvements that your agency has helped to bring about or human interest stories about individuals who have benefited from its services**
- **Food section coverage of your picnic lunch or the gourmet food trucks serving guests at your outdoor benefit concert**
- **Health advice connected to your event:** If your organization is sponsoring a 10K run, what's the current advice about dietary preparation for long-distance running? If local celebrities or unusual teams (for example, six cousins, or all the elementary school's teachers) have signed up, a local paper or news station might interview them in advance.
- **Society page coverage of your honorees, event committee members, or other guests at your event**
- **Radio broadcast of your honoree or guest speaker's speech**
- **Fashion page or television coverage of attire worn to your gala**

Getting a mention on radio or TV

The media need great gobs of news, announcements, and other content every day. You can get some media exposure if you, for example:

- Invite an anchor from the local news channel to serve as the emcee for the event and ask her to promote the event and/or provide an on-air mention about the event.
- Prepare a prerecorded public service announcement (PSA) for broadcast by radio stations. (See Chapter 12 for guidelines on PSAs.)
- Donate free tickets to your event to public television or radio stations to use as prizes of various kinds. If a station picks up on the idea, you get a free mention — maybe a number of free mentions.
- Invite live weather or traffic reporters to cover conditions from your location. It can be a novel way to draw attention to your event.

AFTER THE BALL IS OVER . . .

Write thank-you notes soon after the event to all your committee members and volunteers and make them as specific and personal as possible. They don't have to be long. Many people find handwritten notes of two or three lines to be much more sincere and memorable than boilerplate letters.

The thank-you letters you send to the attendees should clearly define how much money was donated and the cost of goods and services for the event. For example, if you host a dinner and the ticket cost is $50 per person and the food and beverage cost is $25 per person, the thank-you letter should thank the attendee for the $50 contribution and mention that the tax-deductible amount of the donation is $25. Thank-you letters for purchasing auction items should reference the amount paid for an item and the fair market value for the item as provided by the item's donor. It is a good idea to add language like, "Please consult your tax advisor for specific tax advice." Event donations can be tricky, and this can keep you out of the business of providing tax advice.

Thank-you letters also should be sent to any person or business that made an in-kind contribution to the event. If you held an auction at the event, you can note that the auction generated $10,000 that wouldn't have been possible without the generous donation from the XYZ Gift Emporium.

We also recommend holding a brief gathering for your key organizers and volunteers a week or two after the event. You can thank them personally again for their effort, but the real purpose is to get their ideas about what worked, what needs to be improved, and what should happen in the future. You may want to invite a few attendees to this meeting and solicit feedback from their perspective. Good records of this wrap-up meeting are a jumping-off point for planning the following year's program.

- » Tracking down private grant sources
- » Using directories and databases to research grant opportunities
- » Exploring options for public funding

Chapter 16
Finding the Grant Givers

Many of us are familiar with names of the nation's largest foundations. We can fantasize about receiving six- or even seven-figure grants from the likes of the Gates, Ford, or Robert Wood Johnson foundations. They have so much money — why wouldn't they support your nonprofit's work?

Foundations have missions just like other nonprofits, and the most likely foundations to give you a grant may not be the largest in the land. Your task is to become a sleuth and learn to identify the foundations whose priorities best match your nonprofit's goals. Spending time to write a good proposal matters — we discuss proposal writing in Chapter 17 — but finding a receptive audience for that proposal is just as important. It may even be more important.

This chapter shows you how to identify grant sources and figure out your approach to a grant giver.

TIP

Check out File 16-1 at www.dummies.com/go/nonprofitkitfd5e for a list of web resources related to the topics we cover in this chapter.

Planning a Foundation Grant Proposal

Grant sources can be grouped into two broad categories: private and public. Private sources are generally foundations and corporations, and public sources are based within some level of the government — city, county, state, or national. We write about public sources later in this chapter; here we provide the scoop on foundations and corporations.

When you look at the entire nonprofit sector, you'll notice that gifts from individuals provide the largest portion (more than 70 percent) of private contributions. Because they play a smaller role, many foundations see themselves as providing resources for innovation or program improvement. They're a good place to turn to when your nonprofit has a new idea, wants to expand to a new location, or wants to compare and evaluate different methods. Generalizing is wrong, however. Some foundations provide flexible dollars — general operating support — that your nonprofit can use for any of its expenses. In fact, we've seen a gradual trend toward more funding for "gen op."

Figuring out who's looking for whom

Sometimes foundations and government agencies actively seek grant proposals addressing particular regions or community needs. When doing so, they may issue a *request for proposals* (RFP) or a *program announcement.* Most foundations send their RFP announcements to *The Chronicle of Philanthropy* and The Foundation Center. At the *Chronicle of Philanthropy's* website (`www.philanthropy.com`), you can find these under "deadlines" in the "fundraising" menu. The Foundation Center announces RFPs in its *Philanthropy News Digest* (see "PND" at `www.foundationcenter.org`). You can sign up to receive a weekly newsletter from PND that will alert you to these announcements.

More often, you have to find foundations that have the broad missions and stated preferences that seem to match your organization's work and convince them — after studying their guidelines — that the issues your nonprofit addresses are compelling and that your organization has devised a good approach for tackling them. That process of identifying a funding source generally has two phases: developing a broad list of prospects, and then refining that list until you find the likeliest sources. We discuss both of these phases in this section.

TIP

Your time as a grant writer is best spent when you study the priorities and behaviors of the foundations you're approaching. *Shotgunning,* or sending the same proposal to a large number of foundations, is a waste of effort. *Targeting,* or focusing your attention on your most likely sources and writing to address their preferences, is much more effective.

Knowing whom you're dealing with: Different kinds of foundations

Foundations are formed and governed in different ways. These differences may affect the way you address them as a grant seeker. The four basic kinds of foundations include the following: independent foundations, corporate foundations, community foundations, and operating foundations. In this section, we describe how you can figure out which is which.

Independent foundations large and small

Most foundations in the United States are classified as *independent* — also called *private* — foundations. Some of the country's largest foundations, such as the Bill and Melinda Gates Foundation, the Ford Foundation, and the Lilly Endowment, Inc., are independent foundations, but many small- and medium-sized grant-making foundations also fall into this category.

Although they vary widely, the key characteristics that independent foundations share include the following:

- They're established with funds donated by individuals, families, or a small group of people.
- Their purposes and modes of grant making are set by their boards of directors, who may be family members (in which case, they're called *family foundations*) or their appointed representatives.
- Many (approximately 70 percent) of these foundations award grants primarily in the communities where they were formed, but others award grants in multiple locations.
- They must file a 990-PF tax form with the Internal Revenue Service (IRS) each year.

TIP

A 990 is the type of tax form filled out by a 501(c)(3) nonprofit or public charity (such as a community foundation), and a 990-PF is the type of tax form completed by a 501(c)(3) private or independent foundation.

Corporate foundations and corporate giving

Corporations and businesses may support nonprofits by creating corporate foundations, or they may award money directly from their company budgets. Some do both!

Whether the business manages its giving through a foundation or directly from company coffers, the staff member who coordinates grant making also may play another role within the corporation — such as public affairs director. You may notice that many corporate foundations and direct corporate-giving programs give away goods and services or provide technical assistance, as well as cash grants.

Common characteristics of corporate foundations include the following:

- They're established with funds that come from a business or corporation.
- Some of them maintain assets that they invest, and then they award grants from the money they earn on those investments. Many of them add to their grant-making budgets when their business is having a profitable year, but do not do so when business profits are weak. For this reason, their grant-making budgets can fluctuate dramatically.
- Often they're governed by a board made up of company executives.
- Most of them award grants in the communities where their employees live and work, and they see their grant giving as part of their role as good citizens.
- Sometimes corporate foundations award grants to organizations where their employees volunteer or contribute or that their employees recommend to them.
- As a type of private foundation, they must file a 990-PF with the IRS each year.

Companies' direct-giving programs share characteristics with corporate foundations, but they have a slightly different profile. Consider some of the following characteristics:

- All funds available for grants come directly from company budgets and depend on the business's available resources. In lean years, the business may not support charitable causes.
- Like corporate foundations, businesses with direct-giving programs tend to award grants in the communities where their employees live and work.
- Many corporate-giving programs want to receive public recognition for their gifts. They justify their charitable giving to their owners or stockholders as good public relations.
- Some companies are cautious about supporting controversial causes that could alienate their employees or stockholders.
- Gathering information about grants awarded through these programs may be difficult because the businesses don't need to specify their support of charities in their financial statements.

TIP

Companies often want their gifts to be visible. So, for example, you may not want to accept a sponsorship from a whiskey distillery if you're managing a youth program.

TIP

Many corporate-giving programs match contributions that their employees make to nonprofit organizations. When seeking information about your individual donors, ask whether they work for a company that will match their gifts.

In your neck of the woods: Community foundations

The first community foundation, the Cleveland Foundation, was created in 1914. Today, more than 780 such foundations exist nationwide. Community foundations have a different, advantageous tax status from other kinds of foundations. The IRS classifies them as *public charities.* Other characteristics that they share include:

- They're started with funds contributed by many sources, such as local trusts and individuals living in the communities they represent.
- Community foundations focus their grant making on the geographic areas in which they were established. Those areas may be as large as a state (Hawaii and Maine have community foundations), or they may be a county or city (Sonoma, California, or Chicago, Illinois).
- They're governed by boards of directors made up of people who represent the communities they serve.
- To keep their desirable public charity tax status, they must raise money as well as give it away. For that reason, even a very small community foundation has paid staff members.
- Donors to community foundations may contribute to general grant-making programs or may create donor-advised funds and play a role in recommending how to award grants derived from their contributions. The number of donor-advised funds in the United States is growing rapidly.

TIP

When reviewing a grants list for a community foundation, if you see a grant that doesn't seem to fit within the foundation's published guidelines or geographic focus area, realize that it may have been awarded by a donor from an advised fund.

Operating foundations: Locked doors for grant seekers

Operating foundations generally don't award grants. They're formed to provide an ongoing source of support for a specific nonprofit institution — often a hospital or museum. When you come across an operating foundation, skip right over it!

Using the Foundation Center to assemble a broad list of prospects

During the past 60 years, the public has gained increased access to information about funding sources. One milestone in this movement was the creation of the Foundation Center in 1956. Originally intended to collect information so foundations could learn about one another, the Foundation Center quickly became a leading source of information for grant seekers. We explain everything you need to know in the following sections.

Understanding the basics about the Foundation Center

TIP

The Foundation Center manages five libraries across the United States and a number of official cooperating collections within other libraries and nonprofits. (Check out File 16-1 at `www.dummies.com/go/nonprofitkitfd5e` for a list of these sites.) It also maintains databases of information on more than 108,000 foundations, corporate donors, and grant-making public charities in the United States and more than three million of their recent grants.

The Foundation Center maintains two websites — one that focuses on its data and news (`www.foundationcenter.org`) and another that provides access to classes, podcasts of public programs, webinars, and blogs (`www.grantspace.org`). The two sites can help you with your research and with honing your grant seeking skills. You can find links to them in File 16-1 at `www.dummies.com/go/nonprofitkitfd5e`.

The Foundation Center's most popular research tool, Foundation Directory Online, is available by subscription. The Foundation Center has begun building an international database so that foundations worldwide can learn about one another's work and collaborate more effectively. A new aspect of the Foundation Directory Online illustrates maps of global giving, which is also useful for U.S. organizations conducting international work.

If you don't have access to the online directory, you may work with one or all of the following books to aid your research:

- The *Foundation Directory* and the *Foundation Directory, Part 2,* are books featuring brief profiles of the 20,000 largest foundations in the United States. Most profiles include several sample grants awarded.
- The *National Directory of Corporate Giving* profiles more than 3,200 corporate foundations and 1,760 corporate-giving programs. In addition to describing companies' grant-making programs, it identifies their business products and their subsidiary companies.

Where does the Foundation Center find its information? It compiles data from foundations' annual IRS 990 or a 990-PF tax forms, which include summaries of the foundations' major activities and lists of all their grants for the year reported. The Foundation Center also surveys foundations annually, seeking more detailed information than sometimes appears on their tax forms. Increasingly, foundations are submitting their grants information electronically to the Foundation Center as soon as those grants are approved — likely several times each year.

TIP

If you can't find a nearby Foundation Center branch library or cooperating collection, check with reference librarians at your local public library or nearby college and university libraries. These places may carry the Foundation Center's publications or subscribe to the Foundation Directory Online.

TIP

If you find yourself working with the physical books published by the Foundation Center, don't be intimidated by their density. Each of them contains a helpful introduction and annotated sample listing to help you identify where you can find answers to your questions.

The number of foundations in the United States grows and shrinks; as we write this book, the estimated number is 108,000 (87,000 of which are active U.S. independent or community foundations). Reading profiles of all those funders is daunting (and a waste of your valuable time!). When using the Foundation Directory Online or the physical directories, to find the best options for your organization and project, you'll begin by asking whether a foundation satisfies the following four basic criteria:

- **Geography:** Does this foundation award grants in the area where your nonprofit is based?
- **Fields of interest:** Does this foundation award grants in the subject area of your organization's work?
- **Type of support:** Does this foundation award the kind of grant you want?
- **Application process:** Is this foundation open to reviewing proposals from grant seekers, or does it only give to preselected organizations? Some 72 percent of all U.S. foundations don't accept applications.

If you answered yes to all four, good work! You've found an entry for your broad list of prospects. Strive to come up with a list of about 15 or 20 foundations.

TIP

If you have subscribed to the Foundation Directory Online, you can note information about the foundations that interest you in Workspace, and return to that dashboard to track deadlines and your progress toward applying.

If you find too few choices, broaden your search (for example, looking for funders in your state — not just in your city). If you find too many choices, narrow your options, perhaps by adding a keyword search. But, wait, we're jumping ahead in the process. First you need to be more familiar with these databases and search fields.

Discovering the center's directories and databases

The Foundation Center organizes its information in several databases. As you're creating your broad list of foundation prospects, you'll be most interested in four of them — all of which are available through the Foundation Directory Online:

- **Search Grantmakers:** Brief profiles of all U.S. foundations, corporate giving programs, and public charities, along with a growing number of non-U.S. grant makers
- **Search Companies:** Brief profiles of companies that sponsor giving programs — either through corporate foundations or directly from their company coffers
- **Search Grants:** Indexes and brief descriptions of recently awarded grants
- **Search 990s:** Foundations' 990 and 990-PF tax forms that you may search using keywords

Another tool, Power Search, allows grant seekers to conduct keyword searches across all four databases listed here.

REMEMBER

The information contained in the Foundation Directory Online is updated constantly, and the *Foundation Directory* (the book) is published once a year with a midyear supplement.

TIP

The Foundation Center's resources aren't your only options. FoundationSearch (`www.foundationsearch.com`) is a commercial product that features 100,000 foundation profiles for study, and NOZAsearch (`www.grantsearch.com`) offers an extensive database of charitable donations. You also can conduct an Internet search. Only about 10 percent of U.S.-based foundations have websites, but many of the larger ones do. Some states and regions publish their own guides. Another research approach is to investigate who is funding other nonprofits in your area. Compile a list of organizations with a similar focus to yours and investigate their donor lists in their annual reports, newsletters, and 990s.

Effectively searching the center's information

Here are a few pointers to help you make the best use of the Foundation Center tools:

- **Practice searching** using the free tutorials on the Foundation Center's Grant Space website (`http://grantspace.org/skills/find-foundation-corporate-donors`) or included in the Help pages of the Foundation Directory Online.
- **Take advantage of search options.** The Foundation Directory Online offers options to search by grant-maker location; fields of interest; transaction type (for example scholarships, grants, or loans); geographic focus; and trustees, officers, and donors; as well as keywords. If you're using the printed directories, you'll find indexes that allow you to search by similar fields.
- **Use the indices.** The Foundation Directory Online uses a consistent vocabulary in its databases. When filling in the search form, click the small word *index* so that you, too, are using this vocabulary. It will greatly improve the effectiveness of your search.
- **Eliminate the ones that do not consider unsolicited proposals.** More than 70 percent of U.S. foundations don't open their doors and mailboxes to any applicant that chooses to apply. Instead, they study their fields of interest and go out and find grantees. One way to avoid wasting your time on these foundations is by selecting the box that reads, "Exclude grantmakers not accepting applications" when entering your search terms in the Foundation Directory Online. If you're working with the *Foundation Directory* books for your research, check the "limitations" section of foundations' profiles to see if they say "applications not accepted" or "gives only to preselected organizations." You also can find this information on 990s or 990-PFs by checking Part 15, Question 2.

Say, for example, that you're based in Alaska and you want a grant for "capacity building" to strengthen your organization's internal operations. You can choose Search Grantmakers and in the search form, in the Support Strategy field, select Capacity Building and Technical Assistance. You then will want to choose the state of Alaska as the recipient location.

When we chose these parameters, five foundations appeared as prospects. We then chose the same search terms in the Search Grants portion of the site and were able to scan more than 300 examples of capacity building grants that had been awarded to nonprofits in Alaska. We then chose the Power Search option and entered the keywords "Capacity Building AND Alaska," which uncovered many more items for exploration.

TIP

Earlier, we mention that you'll want to use the Foundation Center's controlled vocabulary for conducting your search. For many years, the Center used the Internal Revenue Service's system for classifying nonprofits — something called the National Taxonomy of Exempt Entities. Not all grant seekers found those terms to be easy to understand, so the Foundation Center invited its community of users to suggest more relevant terms for a new Philanthropy Classification

System. You can find a link to the list of terms on GrantSpace: We've included the link in File 16-1 at `www.dummies.com/go/nonprofitkitfd5e`.

- **Determine whether you or your board members have any contacts who serve on foundation boards.** You can gather this information using the Trustees, Officers, and Donors search box in the "Search Grantmakers" section of the Foundation Directory Online.
- **Don't forget that corporations also award grants.** *The National Directory of Corporate Giving* and the "Search Companies" section of the Foundation Directory Online present corporate foundations and giving programs that are managed directly by corporations. Corporations frequently give in the communities where their employees live and work and their profiles include information about where subsidiaries are located, as well as their corporate headquarters. Many companies award materials or services — often products they produce. You may find that these in-kind gifts are more valuable to your organization than the cash grants they also award.

TIP

You may want to consider how the company's policies and practices regarding such issues as domestic-partner benefits, staff diversity, energy consumption, or greenhouse gas emissions align with your nonprofit's values. The Search Companies profiles in the Foundation Directory Online provides links to an array of these Corporate Social Responsibility measures.

TIP

For a step-by-step guide to the foundation research process, check out File 16-2 at `www.dummies.com/go/nonprofitkitfd5e`. For each entry on your broad list, jot down notes and questions on a *foundation prospect evaluation sheet.* You can find a sample evaluation sheet in File 16-3.

If you now have your broad list of prospects in hand (or on your computer), it's time for cross-checking and deeper study that will help you to narrow your broad list down to the five or six best potential sources.

Digging deeper to narrow your prospects

At this phase in your research, you truly become a sleuth, taking your broad list of prospects in hand and finding out more about each of them. See what you can discover from the following sources:

- **Using Workspace:** If you're using the Foundation Directory Online, when you find a promising prospect, click Workspace in the upper-right corner of the funder's profile. Workspace will save the foundation's information in a dashboard for you. It includes an assessment tool that invites you to choose from a list of criteria to determine the strength of the match between your project and the foundation.

- **Websites and annual reports:** Some foundations (approximately 10 percent of them) produce websites, and some others publish printed annual reports with detailed information about their grant-making guidelines and grantees. These publications can be particularly helpful for grant seekers. They often include introductory letters from the foundation's leaders that provide insight into a foundation's philosophy or current direction. Foundation websites are easy to find with a simple Internet search. You can find announcements of new annual reports in the *Chronicle of Philanthropy* (`www.philanthropy.com`), explore printed annual reports in some libraries, or call the foundations and ask to have the reports sent to you.
- **Blogs and tweets:** Some foundation leaders post blogs or tweets about their work and grantees. Some have Facebook pages. You can find their social-media identities through their websites and "follow" or "like" them to receive ongoing updates.
- **Form 990s:** Foundation 990 forms may be the only place to find a list of every single grant a particular foundation has awarded. Foundations are required to make their three most recent 990 or 990-PF reports available to the public. If they don't post them on their websites, you can find them through the Foundation Center (`www.foundationcenter.org`), NOZA 990-PF Database (`www.grantsmart.com`), and GuideStar (`www.guidestar.org`) websites. We've provided links in File 16-1 at `www.dummies.com/go/nonprofitkitfd5e`. Some state agencies also provide online access to 990s.

Studying a foundation's 990 may intrigue you, but PDF files for the nation's largest foundations are enormous. We recommend against downloading a 990 from a large foundation that produces a website or annual report containing grants lists. Reviewing 990s or 990-PFs is more useful when you're studying small foundations whose grants lists may not be readily available.

- **The foundation's contact person:** If a foundation listing includes a contact person and phone number or an email address — and if it doesn't say you shouldn't call or write — feel free to contact that person to confirm information and ask any questions you may have. We recommend that you do so only after you've read as much as you can find about that foundation. This way you can be focused, clear, and knowledgeable as you seek additional information.

Going for a Government Grant

Before you start daydreaming about massive state or federal funding, compare the pros and cons of private funding and public funding. Table 16-1 sorts things out for you.

If, after reviewing the info in Table 16-1, you decide you're up to the task of seeking government grants, read on.

TABLE 16-1 Comparing Private and Public Grant Sources

Private Sources	Public Sources
The purpose of the grant-giving entity is set by donors and trustees.	The purpose of the grant-giving agency is set by legislation.
Most funders award grants to 501(c)(3) nonprofit organizations, some give to public agencies, and others give fellowships and awards to individuals.	Most agencies award grants, contracts, and loans to nonprofit and for-profit entities, to individuals, and to other government entities.
Most grants are awarded for one year, and, in general, grants are smaller than those awarded by public sources.	Many multiyear grants are awarded for large amounts of money.
The application process requires a limited number of contacts with individuals within the funding agency.	The application process is bureaucratic and may require applicants to work with staff at multiple levels of government.
Grants are sometimes awarded in response to simple two-page letters with backup materials.	Many proposals are lengthy (as long as 100 pages).
Agency files are private; applicants may not have access to reviewer comments or successful grant proposals submitted by other organizations. Review panels and board meetings may be closed to the public.	Agency records must be public. Applicants may visit many review panels; reviewer comments and successful proposals submitted by other nonprofits may be available.
Foundation priorities may change quickly according to trustees' interests.	Agency priorities may change abruptly when new legislation is passed or budgets are revised.
Developing personal contacts with trustees or with foundation staff may enhance your chances of securing a grant.	Developing personal contacts with congressional aides for state or national representatives may enhance your application's chances.
After accepting a grant, the recipient is usually required to file a brief final narrative and financial report.	After accepting a grant, the recipient may be required to follow specified contracting and hiring practices and bookkeeping and auditing procedures. It may have to prove other compliance with federal regulations. Grants can be relatively costly to manage.

Federal grants

To pursue government funds, you need to study a different set of research guides from those that help you with foundations and corporations. The resource that's most likely to help you is online at www.grants.gov. This tool is more than a directory of information: At Grants.gov, you can both find and apply to federal grant programs.

Determining the pros and cons of Grants.gov

Grants.gov is a perfect model of bureaucratic writing, and the site has so much information that it can be overwhelming. But the burden of this complexity is far outweighed by the tool's usefulness: Grants.gov is the only source you'll need for information about all federal grants. It incorporates more than 1,000 programs representing $500 billion in annual grant awards. If you take the time to read the introductory information in the Learn Grants menu found on the home page, you'll find that the site is relatively easy to search.

Supplementing your search with the CFDA

Federal agencies provide 15 different types of funding, including program grants, insurance, loans, and "direct payments for specified use." For information about all 15 types of federal assistance, turn to another federal publication, the *Catalog of Federal Domestic Assistance* (CFDA), which can be found online at `www.cfda.gov`.

We steer you to Grants.gov because, if grant support is what you're seeking, it's more tailored to your needs and, perhaps, less cumbersome. However, the CFDA website is a powerful tool that you also may want to explore. The two websites use the same catalog numbering system to identify funding opportunities.

Beginning your search with Grants.gov

REMEMBER

As you begin your search, pay attention to whether your organization is eligible to apply to a particular program. Some kinds of federal grants are available only to state and local governments, federally recognized Native American tribal governments, and other specified groups — not necessarily nonprofit organizations.

When opening the home page of Grants.gov, you're invited to search by

- **Opportunity Status:** You'll want to choose Posted, which means the funding program has been announced and is open for applications.
- **Funding Instrument Type:** You're probably interested in a grant, but a few other forms of financial support are also listed.
- **Eligibility:** This means the type of entity you represent — perhaps a nonprofit with 501(c)(3) status, a for-profit business, or a city or township.
- **Category:** You'll choose among broad subject areas such as Agriculture or Health. This is useful if you know you're interested in reviewing the programs

offered by the Institute of Museum and Library Services, the Department of Homeland Security, or another specific agency.

- **Keyword(s)**
- **Funding Opportunity Number and/or CFDA:** If you happen to know the number assigned to a program that interests you or the Catalog of Federal Domestic Assistance number assigned to it, you can enter that number and — *voilà!* — the program description will appear.

Grants.gov can tell you which agencies and offices manage the funds for different grant opportunities. After identifying federal programs of interest, experienced grant seekers usually call, write, or email those offices to confirm the information. That's because even though Grants.gov information is updated frequently, it also goes out of date quickly.

We outline more information about searching for federal grants and contracts and suggest further steps to take in your research in File 16-4 at www.dummies.com/go/nonprofitkitfd5e.

Before you apply for a federal grant, you need to register with Grants.gov. The process, which involves several steps outlined in File 16-4, can take up to two weeks. Begin this process as soon as you decide to apply for federal funding so delays don't cause you to miss a critical deadline.

Nonfederal government grants

Information about state, county, and municipal grants (which state or local tax dollars generate money for) can be trickier to find than those awarded through federal agencies. That's because the information isn't always compiled in resource guides. Some, but not all, state and city government offices do a good job of informing their constituents about funding opportunities.

During your research process, check the websites of state and local agencies related to your project (such as agencies overseeing education or child services) and visit with local government officials and congressional office staff members. Also, check to see whether your governor's website includes information about available grant programs.

» Writing the different sections of the proposal

» Varying your proposal for specific situations

» Saying thank you and reporting after receiving a grant

Chapter 17

Writing a Grant Proposal

Even though grant seeking is highly competitive, writing a good grant proposal isn't particularly mysterious or difficult. First you identify a problem and make it compelling, then you set goals for solving that problem, and last you propose a possible solution and a way to test the results. In the course of laying out the plan, you also want to assert the special strengths of your organization and a sensible financial plan for the project, both in the near term and later. You present your proposal in a readable form that tells a story and conveys your vision in terms both inspiring and practical.

Some proposals are quite long and elaborate. Some are just a few pages. Some are presented on paper, but many are submitted online. Your research into funding sources (which we describe in Chapter 16) will uncover the preferences of different foundations and government programs, so you can figure out the best mode of presentation, level of detail, and method for submitting your proposal. You'll encounter many variations, but this chapter shows you the basics of how to write the grant proposal.

TIP

Check out File 17-1 at www.dummies.com/go/nonprofitkitfd5e for a list of web resources related to the topics we cover in this chapter.

The Windup: Completing Pre-Proposal Tasks

Generally, a grant writer develops a proposal by talking with staff members, volunteers, or the board about a project idea. Before setting fingers to keyboard, the writer should investigate the following:

- What is the demonstrated need in the community for the work you intend to do?
- Who are the constituents who will benefit from your efforts? (Find out everything you can about them.)
- What are others doing in this field?
- What particular strengths does your nonprofit bring to the project?
- Specifically, what do you want to accomplish?
- What will it cost to do it?

TIP

Sometimes a funding source announces a specific initiative for which it's inviting proposals — called a request for proposals (an "RFP") or a program announcement. When you respond to an RFP, you're trying to convince the foundation or government agency that your nonprofit is the best one for the job.

Asking for permission to ask

Many funding sources screen proposal ideas before they invite extensive, detailed documents. This enables them to encourage only truly promising requests, saving both themselves and grant seekers the time and effort that goes into reviewing and writing longer proposals.

When you encounter a request for a *letter of inquiry* (LOI), boil down the essence of the proposal — all eight areas covered in the later section "The Pitch: Writing Your Proposal" — into a readable, compelling letter. The letter doesn't ask for a grant directly, but it asks for permission to submit more detailed information. Most letters of inquiry are two or three pages long. They may be made up of short answers to a foundation's online forms. Follow the foundation's stated preferences.

TIP

At `www.dummies.com/go/nonprofitkitfd5e`, File 17-2 is a sample letter of inquiry, and Files 17-3, 17-4, and 17-5 show samples of different types of full proposals. File 17-8 illustrates a foundation's budget form.

Passing the screening questionnaire

Some funding agencies screen potential applicants by asking them to respond to short questionnaires that are found on the funders' websites. Questions often determine specific eligibility — such as whether the nonprofit organization serves a particular geographic area or the size of its budget. You may have to provide your nonprofit's federal EIN so that the foundation can confirm with the Internal Revenue Service that you are a qualified 501(c)(3) nonprofit organization. In many cases, if you "pass" one of these surveys, you'll receive login information that will permit you to submit more information — either a letter of inquiry or a full proposal. Usually, you've learned that you're eligible to apply but you still don't know if your application will be a competitive one.

WARNING

Don't pass off the letter of inquiry or questionnaire as an inconsequential hurdle: First impressions can be lasting impressions.

The Pitch: Writing Your Proposal

Deep down, every good proposal is based on a well-considered plan, and designing that plan is a creative, even fun, part of the task. The differences from funder to funder are in how to pitch that plan and the order in which its parts are assembled. But the parts themselves are fairly standard:

- **Cover letter and summary:** Provide a brief overview of the entire project and its costs, linking the project to the interests of the funding source. Most online applications don't invite cover letters, but they do ask for project summaries.
- **Introduction or background:** Present the strengths and qualifications of the nonprofit organization.
- **Statement of need:** Describe the situation that the proposed project will try to improve or eradicate.
- **Goals, objectives, and outcomes:** Outline a vision for success in both the near future and the long term.
- **Methods:** Describe the project activities, who will manage them, the time frame for the project, and why the proposed approach is the best.
- **Evaluation:** Explain how the organization will measure whether it met its objectives, and outcomes.
- **Budget:** Present the project's costs and sources of income.

» **Sustainability or other funding:** Explain how additional funds will be raised if a grant is awarded but doesn't cover all costs. If the program is expected to continue beyond the grant period, how will future costs be covered?

TIP

Your proposal should be compelling to the foundation you're addressing, but be careful not to overstate what your nonprofit can and will do. If a grant is awarded, those promises could become binding in the foundation's grant agreement.

In this section, we cover each of these parts in detail.

REMEMBER

Just like other nonprofit organizations, foundations have mission statements and goals. As a grant writer, you're working to convince your reader of a clear connection between the foundation's mission and your organization's work.

Starting out with the cover letter and executive summary

Technically, the *cover letter* isn't part of the proposal narrative. It's attached to the top of the proposal, where it serves as an introduction to the document's primary points. One of its key roles is to convey how the proposal addresses the foundation's stated priorities. Cover letters often require a signature from the executive director and/or board president.

The letter mentions any contact the organization has had with the funding source. For example, you may say, "When we were introduced at the new teen center's ribbon-cutting ceremony last week, I was pleased to hear you talk about how important playing basketball was for you when you were growing up. Here at Backboard Nation, we're working to refurbish basketball courts in parks and playgrounds across the city."

Although you always want to lay out the basics of the request — how much money is needed and for what — you also can use the letter to say something about your personal connection to the cause. The letter should close with clear, specific information about the contact person to whom the funding source should direct any questions.

The body of the actual grant proposal begins with an executive summary (similar to an abstract for a report) containing an overview of its key ideas. You'll usually begin the summary section with a one-sentence overview of the project and how much money is being requested in the proposal. Next you include the primary ideas from every section of the proposal (boiled down to one or two sentences per section). Finally, close the executive summary with the prognosis for future programming and funding for the program.

TIP

Don't use the exact same wording to describe your project idea in the cover letter and in the project summary. You don't want to bore your reader, and you don't want to miss the opportunity to write in a more personal manner in the cover letter.

Introducing your agency

Some proposals contain *introductions.* Others call this section *background information.* However you label it, this part of the proposal describes the nonprofit organization that's seeking money.

Usually this section begins with a brief history of the nonprofit, its mission, and its major accomplishments. Then you describe the current programs as well as the constituents. After including these standard ingredients, you can draw upon whatever other credentials recommend the organization for the work you're about to propose. Here are some things you can mention:

- Media coverage
- Citations and awards
- The credentials and/or experience of the nonprofit's leadership
- Other agencies that refer clients to the nonprofit
- Major grants received from other sources
- Results of recent evaluations of your nonprofit's work

Although you're not yet describing the project idea, the items you introduce here should back up your nonprofit's qualifications to do the proposed work. For example, suppose that an after-school program for teenagers offers multiple programs — athletics, arts, and youth-led volunteer work. If the grant proposal seeks money to expand the volunteer program, the writer can tell the story of how that volunteer work began and evolved, where the teens have provided services, and who has praised their contributions.

TIP

If your organization is brand new and you don't have accomplishments or accolades to describe in this section, tell the story of how and why the organization was created, its mission and purpose, and the qualifications of its founders and board members.

If yours is the only organization that offers this program, mention it here. Funders like to know who else is addressing the same issues and how you differentiate yourself as an attractive source to receive financial support.

You may be tempted to write on and on (and on and on and on) about an organization's history, philosophy, or mission. Don't drag this section down by including too many details, using too much eloquent verbiage, or overstating the truth. One or two clear, factual paragraphs on these subjects are plenty.

Shaping the problem

A grant writer begins to shape the argument behind the proposal plan in a *statement of need* — sometimes called a *problem statement.*

This section brings forward the reasons behind the program for which your organization is seeking money (or behind all the organization's work if you're seeking general operating support). It does so by describing the needs of the constituents you will serve. It should incorporate some data or statements from experts to back up the needs or problems it describes. And it should capture its reader's attention — making him want to read on.

You shouldn't describe your "need" as a lack of money but rather as a situation in the lives of your constituents, whether they're tree frogs, retired adults, or former race horses. Don't start to solve the problem in your statement of need.

Setting goals, objectives, and outcomes

After you've described a need, the next step is to introduce what the nonprofit intends to achieve if it takes on the proposed project. You haven't yet described the project activities, but this section jumps ahead to show what can be accomplished if the project is a success. Why? First you pique the reader's interest and concern (in the statement of need), next you show him the possibility of a better future (in goals, objectives, and outcomes), and then you explain how to achieve that vision (in the methods section).

Goals, objectives, and outcomes are related but different terms. You can find out more about them in Chapter 7, which discusses planning. The following points give a brief overview of these terms:

- *Goals* are broad, general results. They may be somewhat lofty.
- *Objectives* should be specific, measurable results. What degree of change do you want to achieve in what time frame, involving how many people (or tree frogs or race horses)?

» *Outcomes* relate to your objectives, answering the question "So what?" So what if you provide anti-smoking classes to students entering high-school, reaching every first-year teen in four school districts? Your objective may be the number of those teens who pass an exit test, demonstrating that they learned from the classes. The outcome may be that, three years later, a lower-than-average percentage of the students who completed the classes begins smoking in high school.

REMEMBER

You want the outcomes to be significant, but you also have to be careful not to overstate how much the project can claim or measure. If a grant proposal says a nonprofit intends to achieve a long-term outcome, the organization should be prepared to follow up with participants in the future.

Presenting (ta-da!) your project idea

At last it's time to explain the project idea. Who will do what to whom over what period of time? If you're writing your proposal for a research project, generally this section is called *procedures.* For most other types of projects, it's called *methods* or *methodology.*

Although this section contains the idea that inspired the organization to seek funding, sometimes the writing is dull. The content is similar to writing a list of instructions, and writers often get wrapped up in describing each step in minute detail.

TIP

To avoid the dullness trap, think about constructing an argument from beginning to end. A good methods section opens with an overview of the approach and then leads the reader through key phases in the project's development. It includes enough detail so that the reader can imagine the project clearly but not so much that she sinks into the daily grind. Other techniques to preserve vitality include the use of charts or graphs to break up and complement the descriptive text.

Many projects require a few months of preparation before they can be launched. Factor in time for hiring and training staff, purchasing and installing equipment, identifying research subjects, or performing other necessary preliminary steps.

REMEMBER

Because this section contains all the project details, you may accidentally leave out or forget important information. These two topics are often overlooked:

» **Hiring:** If you need to hire new staff to do the project work, be sure to discuss the hiring process, job descriptions, and qualifications. See Chapter 10 for ideas on hiring paid staff.

» **Marketing:** Just because you create something great doesn't mean that anyone will show up to take advantage of it. Your proposal must explain how your organization plans to spread the word about this new project and entice the target population into becoming involved. Chapter 12 provides tips on marketing your program to the public.

The proposal must explain what the organization will do and why the organization is taking that approach — the rationale. The reason may be that nobody has ever done it this way before. The reason may be that the organization has tested the approach through a pilot program and knows that it works. The reason may be that another organization in another part of the country has tested this method.

Some funding sources ask you to fill out a table with columns in which you describe your goals, the *inputs* (resources you'll put into the project), short-term objectives, long-term outcomes, activities, and intended results. These tables go by several names. Often they're called *logic models.*

TIP

We introduced logic models in Chapter 8. Take a look at File 8-3 at `www.dummies.com/go/nonprofitkitfd5e` to see an example of a logic model.

WARNING

In an effort to raise as much money as possible, you may be tempted to change your programs to match a particular foundation's interests. Your board and management should carefully assess new project ideas to make sure they fit the organization's mission. Otherwise, the nonprofit may drift away from addressing its purpose.

Explaining how results will be measured

If you've carefully shaped the goals, objectives, and outcomes section of the proposal, the desired project results are clear. The *evaluation* section, then, explains how the organization will measure whether it met those goals, objectives, and outcomes.

This section says who will conduct the evaluation and why that person or consulting group is right for the job; what information is already known about the situation or population served; what instrument(s) will be used to measure the project's results; and how the finished report will be used.

Different kinds of evaluation are appropriate at different stages of a project. A nonprofit with a brand-new effort may want to spend the first year analyzing its internal efficiency, balance of responsibilities, and general productivity before beginning an in-depth study of whether outcomes are being achieved. In that first year, the most important question may be, "How can we make it run better?"

Some organizations evaluate all their projects with project staff. Who's better suited to understand the details and nuances of the work? Who else can grasp the goals and objectives of the project so quickly? Of course, the opposing view is that you get biased results by asking staff members whose ideas may have shaped the programs and whose livelihoods may depend on continuation of programs. Yet, outside consultants are more expensive, and they, too, are employed by the nonprofit organization and may bring some biases to the evaluation.

Here are some common questions addressed in evaluations (progressing from basic to complex):

- What activities did you complete? Were project funds spent according to the proposal plan?
- How many people are being served and by how many units of service (hours of counseling, copies of publications, and so on)?
- How do consumers of the organization's services rate the quality of those services?
- How does this organization's work compare to industry standards for effective work of this type?
- What changes has the program made in the lives of the constituents served?

TIP

Often a nonprofit can discover a lot about the effectiveness of its work by analyzing information that it already gathers. For example, an organization may have on file intake and exit interviews with staff and clients, videotapes of activities, or comments submitted to its website.

If your organization plans to gather data from new sources (such as surveys, interviews, and focus groups), the proposal should explain what those sources are, who will design the instruments to be used, who will gather the information, and who will analyze the results.

Talking about the budget

It's time for your proposal to present information about project costs and the other funding that's available. If your organization offers only one service, the organization's budget is the same as the *program budget* or *project budget.* If the proposal seeks funds for an activity that is one of many things the nonprofit does, it's presented as a smaller piece of the overall budget.

TIP

Chapter 11 discusses how to compute the indirect costs that all programs within an organization share. Different foundations and government agencies have different attitudes about covering the indirect costs associated with projects. In late 2013, the federal Office of Management and Budget (OMB) issued an overhaul of federal grant policies that made it clear that nonprofits' indirect costs were legitimate and should be reimbursed. Not every state or local government agency has adopted a similar stance, but this OMB decision has been influential and appreciated by nonprofits. Check with the organization to which you're applying to see whether it limits how much can be charged to indirect costs or to management and fundraising expenses.

REMEMBER

If you have a federal grant, your organization will need to negotiate an appropriate indirect cost rate based on your nonprofit's size and structure. After you've negotiated a rate for your organization, it will apply to any government grant from any federal agency.

The project budget should be clear, reasonable, and well considered. Keep a worksheet of budget assumptions with the grant file in case anyone asks months later how you computed the numbers it contains. Foundations don't want to believe that an organization is inflating the budget, nor do they want to think that the applicant is trying to make his proposal more competitive by requesting less than he really needs. Either approach — budgeting too high or too low in relationship to the cost of a project — can raise a red flag for funders.

Unless the foundation is willing to consider a proposal for 100 percent of the project costs, the proposal budget should include both income and expenses. Some foundations ask for the information in a particular format. For those foundations that don't require a special budget format, here are some general standards that apply to the presentation:

- Income should head the budget, followed by expenses.
- If the organization has multiple income sources, contributed sources should be listed apart from earned sources. Within contributed sources, a writer usually lists grants first, followed by corporate gifts and contributions from individuals.
- All expenses related to personnel costs — salaries, benefits, and consulting fees — come first in the expense half of the budget. Usually personnel costs are subtotaled.
- Non-personnel costs — rent, materials and supplies, printing, and so on — follow the personnel costs and are also subtotaled.

» Some kinds of project budgets allow for contingency funds of a certain percentage of the project's projected costs. These funds are usually listed last among expenses.

» Budget footnotes can explain anything that may be difficult for a reader to understand.

REMEMBER

The budget is like a spine for the rest of the proposal and supports every aspect of the plan. The costs of every activity in the methods and evaluation sections should be included.

Keep in mind that some project budgets seem simple and straightforward when they really are more complex. Say, for example, an agency wants to purchase a piece of equipment. It researches the cost and writes it down as the one and only budget item. This approach may be foolhardy: Other costs may include shipping and installation, insurance, training of staff, maintenance, and supplies.

TIP

Check out File 17-8 at www.dummies.com/go/nonprofitkitfd5e for a sample foundation budget form. You also can find a sample proposals and their accompanying budgets at the same site.

Showing where the rest of the money comes from: The sustainability section

Willingness to cover the full cost of a project varies from foundation to foundation and agency to agency. Some like to "own" a project and have their name strongly associated with it; therefore, they may be willing to cover full costs. Many like to be one of several supporters so the nonprofit isn't entirely dependent on them. That way, the project can go on even if the funder can't continue to support it in a future year.

In Chapter 13, we introduce the age-old concept of not putting all your eggs in one basket. Between many foundations' reluctance to pay for 100 percent of a project's costs and their preference for offering short-term support, if you seek a grant for a new project, you want to know how you can cover the balance of the costs in the present and all the costs when the initial project grants have been spent. Foundations reviewing your grant proposals want to know the same information.

Your *sustainability plan* — addressing these needs for additional and future funding — should be clear and reasonable. Because a nonprofit can't assume it will receive every grant for which it applies, a foundation understands if a nonprofit has applied for more grant funding than the program costs. If a nonprofit lists another foundation as a possible source of income for a project, it doesn't necessarily have to have received that grant, but the prospect should be plausible. The size of the grant should be appropriate to others awarded by the agency, and the focus of the foundation should be aligned with the project request.

When listing sources of potential funding, note whether the funding is *proposed* (application has been submitted), *committed* (grant has been awarded or promised, but funds have not been received), or *received* (money is in hand). Funders like to have a general idea of where you are in the fundraising process.

WARNING

Foundation staff members talk with colleagues at other foundations. Never lie about having submitted a proposal or having received a grant from another foundation. That lie can undermine your request (and future requests, too)!

Generally, foundations hope to see nonprofit organizations growing less dependent on grants over time and are happy to see proposal budgets projecting future increases in earned revenue or individual contributions. Here are some possible sustaining sources you may describe in your proposal:

- A government contract to continue a valuable service after it has been developed and tested
- The sale of publications, recordings, or services based on the project
- A membership drive
- A major donor campaign

Some proposals don't need to address these concerns. They seek support for projects that begin and end in short periods of time. For example, a grant proposal may support the publication of a study of an agency's work. After the agency edits, designs, and posts the report on its website, future funding isn't needed. For some other kinds of projects, future funding is very important. For example, you wouldn't want to start a recreation center for low-income youths and have to close it after a few years. The lack of program continuity can have a negative effect on clients.

Writing the P.S.: The appendix

A proposal usually needs an appendix or attachments. Four key items that are routinely included in the appendix (and often identified by foundations as required enclosures) are

- Proof of 501(c)(3) status from the IRS (in the form of a determination letter). (This may not be requested if you've provided an EIN in response to an online application questionnaire.)
- List of the board of directors (and of any advisory boards)
- Current year's organization budget
- Prior year's financial statement

Other common appendix items include the following:

- A list of major grants received in recent years
- An organizational chart outlining staff and board roles
- Copies of media clippings about the agency
- Job descriptions and/or résumés of key staff
- Samples of evaluations or reports
- A copy of a long-range plan
- Agency brochures and program announcements
- Letters of support

WARNING

Don't include attachments unless the grant instructions clearly ask for them. Funders receive many applications at a time and don't want to sort through information that isn't relevant to the proposal.

Throwing Special Pitches for Special Situations

If you prepare a grant proposal following the format we describe in the previous sections, you'll have in hand a very useful document, suitable for submission to many foundations. However, grant writing isn't a field in which one size fits all or even one approach fits all. So in the next sections, we describe some common situations that call for variations on your basic proposal.

TIPS FOR ONLINE APPLICATIONS

More and more foundations are inviting or requiring applicants to apply online. Online application systems vary widely, but a few things to keep in mind are:

- Well before any deadlines, check the funder's website to see if you need to create an account prior to submitting a proposal. This procedure can take a few days.
- Online applications can be tricky. Some don't work on certain web browsers. Others get overloaded and time out your submission.
- Plan ahead by gathering documents you want to include with your proposal and by setting up your electronic signature if one is requested on the application.
- In the foundation's instructions for applicants, pay attention to whether you can save your work in the online system and come back to finish it later. If you can't, draft, edit, and save your proposal before you start working on the foundation's online form.
- Check to see whether the funder specifies a word count or character count. Track the length of any drafts you write so you don't have difficulty fitting your sections into the online form. And don't exceed the length! It may not be obvious to you as the applicant, but often the funder can't see any words or sentences that run longer than the prescribed length.
- Some online applications consist of a series of questions that aren't revealed until you've logged into the system and begun filling out the form. When writing such a proposal, set aside a generous amount of time to concentrate on each answer. Don't skip a question just because you've have answered it. Also, even though your prose is broken up by the system's form, read through your answers as if they were a continuous document. Paying attention to the flow of your prose will make your proposal more interesting and clearer.
- If the funder's online system allows you to upload a PDF of your proposal, we recommend that you do so. We've had problems with using the Track Changes function in Microsoft Word to mark edits, and — even if we've saved a final, clean version of the proposal — having those edits reappear when the document is uploaded.
- Always look back over your drafted proposal before pushing the "upload" or "send" button. If possible, save it for a day to read with fresh eyes. Some application systems allow you to save drafts online before submitting them.
- Submit the proposal at least 24 hours before it's due to allow time to address any technical difficulties. It's best to prepare ahead so if you encounter difficulties, you still have time to try other approaches or contact the funder for more information.
- Don't forget to press Submit! Trust us, this happens. In most systems when it arises, the foundations are aware that you've prepared something, but they can't read it.

Trolling for corporate grants or sponsors

Proposals addressed to corporations and company-sponsored foundations are generally brief. A two-page letter or email is an excellent approach.

Here's what to do in your proposal:

- Ask for a specific contribution early in the letter. If you've had prior contact with the funder, mention it.
- If applicable, mention the involvement of any company employees on your board or volunteer committees.
- Describe the need or problem to be addressed.
- Explain what your organization plans to do if the grant is awarded.
- Provide information about your nonprofit organization and its strengths and accomplishments.
- Include appropriate budget data. If the budget is more than half a page long, include it as an attachment.
- Discuss how the project will be sustained in the future.
- Describe how your organization can acknowledge the gift publicly and provide visibility to the corporation.
- Make a strong, compelling closing statement.

Some nonprofits seek sponsorships from corporations to help underwrite special programs, such as marathons, museum exhibitions, or beach clean-ups. Approaching a corporation about a sponsorship resembles proposing a business agreement.

Your organization may, for example, suggest to a local business that if it sponsors your charity's annual walk-a-thon, its name will be included on banners and T-shirts that will be seen by 50,000 people.

REMEMBER

In seeking sponsorships from corporations, keep in mind the difference between a qualified sponsorship and advertising. In a qualified sponsorship, an individual or company makes a sponsorship payment to a 501(c)(3) nonprofit without an arrangement or expectation that the payer will receive a benefit in return. The nonprofit might acknowledge that sponsorship through use of the donor's name or logo or slogans, but may not specifically endorse that donor's products. The non-profit also might provide goods or services of insubstantial value to the company in recognition of the gift. The sponsorship relationship is considered advertising when the nonprofit specifically endorses purchase of the company's products or services. Revenue to the nonprofit from advertising sales likely is unrelated business income and potentially subject to taxes on that income. The IRS website (www.irs.gov) provides additional details about such unrelated business income taxes (UBIT).

A sponsorship pitch usually starts with a phone call that is followed by a brief, confirming email. You can attach to that email a list of possible sponsorship benefits and the level of contribution expected for each.

At www.dummies.com/go/nonprofitkitfd5e, you'll find sample company foundation guidelines and a sample letter proposal to a corporation.

Seeking general operating support

If you're seeking funds for general operating support, your proposal needs to make an argument for the work of the entire nonprofit rather than for a specific project. In this type of request, some of the information about current activities, which is often included in the introduction or background information sections, should be moved to the methods section. The grant-making organization judges the application based on overall organizational strength and the nonprofit's role in its field.

If you're tailoring a proposal for general operating support, we recommend the following:

- Prepare an introduction that covers the agency's purpose, goals, and current programs. Describe its leadership (board and staff) and its history.
- In the statement of need, describe current challenges the organization faces as it works to fulfill its mission.
- When preparing the section on goals, objectives, and outcomes, address the external goals (how the nonprofit plans to serve its constituents) and internal goals (such as training the board of directors to fundraise).
- Use the methods section to describe the agency's planned activities for the year ahead.
- In the evaluation section, describe various means the agency uses to assess and improve its programs.

- Include the entire annual budget for the organization in lieu of a project budget.
- In the sustainability section, briefly describe fundraising or earned income areas the organization is working to increase.

Seed money: Proposing to form a new nonprofit

Some foundations specialize in "seed funding" for new projects and new organizations. A seed proposal has two key ingredients: careful assessment of the problem to be addressed and special qualifications its founders bring to its creation. Here's a quick outline of a proposal for a new endeavor:

- Background information introduces the people who are creating the organization, their vision, how they identified the idea, and steps they've taken to realize their vision.
- The problem statement thoughtfully presents what the founders have observed and learned about the needs to be addressed.
- Although the goals may be lofty, stated objectives and outcomes should be reasonable considering the developing state of the organization.
- Methods present plans for the first year or two of activities. They include discussion of how the organization will be structured and how services will be offered.
- Evaluation plans may "go easy" during the organization's initial phases. Founders may be testing basic ideas for their feasibility and efficiency for a year or two before studying a program in depth.
- The budget is likely to be the entire organizational budget. Some seed grant funders are willing to cover such startup costs as equipment purchases or deposits for renting an office.
- The sustainability section should outline basic plans for supporting the nonprofit in the future (unless it addresses a discrete problem that may be solved within a few years).

TIP

The foundation may be willing to support a feasibility study that tests the viability of the seed project — how distinctive it is and who its likeliest sources of support might be.

The Homerun: Following Through after You Receive Funding

As is true with any good fundraising, your grant proposal is part of a relationship that you want to develop and maintain over time. If a funder doesn't want to be bothered, he or she will generally tell you. In most cases, we recommend:

- **Thank them.** Thank the foundation at the time the grant is awarded and acknowledge arrival of the check or wire transfer when the money arrives.
- **Recognize the gift.** For most kinds of grant awards, list the funder in your printed materials. It's always a good practice to check with funders about whether and how they want to be listed by sending them a brief letter or email with the text you plan to use. Some foundations tell you in their award letters how to acknowledge them.
- **Keep them informed.** Your funder should hear from you at times when you're not requesting money. Don't overwhelm your funder, but sending her a copy of good media coverage or inviting her to see your program in action are good ways to build a relationship.
- **Let them hear bad news from you first.** If your organization is forced to move, a key staff member decides to leave, or you lose a contract you need to do the project you proposed, make certain that your funder hears the bad news from you, not as a rumor. Let the funder know the steps you're taking to resolve the situation. Your organization's integrity is one of its most important assets.
- **Prepare to submit reports on time.** Most grant awards will tell you when and how to submit interim and final reports. Schedule time in your calendar to gather the information you need and write that report as soon as you know the due date. The foundation will likely send a reminder as the due date approaches, but being prepared and submitting reports on time makes a good impression and sets the stage for a possible future grant request.
- **Share what you've learned.** If your funder provides reporting guidelines, follow them. If not, look back at the goals, objectives, and outcomes that you identified in your grant proposal and write about what steps you took to accomplish them and whether you were successful. Of course, you and the funder hope that the project has met its objectives and initial outcomes are promising, but if it hasn't been successful, sharing what you learned — your thoughtful analysis of what went wrong and how you might adjust the program in the future — will be deeply appreciated.

  Remember the beginning of this chapter where we mention that foundations have missions just as other kinds of nonprofits do? They can't fulfill their missions if they aren't learning alongside their grantees.

» Figuring out where gifts will come from

» Understanding the pros and cons of capital campaigns

Chapter 18
Capital Campaigns: Finding Lasting Resources

As a nonprofit organization matures, its staff and board may become ambitious to obtain resources that can stabilize the organization. One option is to raise *capital* — money to make specific, usually long-term, financial investments.

Some organizations raise capital funds for buying property or building facilities. Other raise such funds for *ventures* — money they put aside for major program innovations. Still others raise capital for cash reserves (funds available to grow and adapt in the future) or *endowments* — money they permanently invest, with the investment returns providing a source of annual income. If you're the director of a nonprofit taking on a capital campaign, imagine yourself as suddenly having two jobs: leading your organization and its programs and leading its capital project. Can you picture yourself wearing these two hats? Exhausting thought, isn't it? However, if the campaign project goes well, it can attract new donors and attention to your nonprofit. That benefit, combined with new resources to invest in your work, can make those two hats worth wearing.

REMEMBER

As varied as their purposes are, most capital campaigns share a set of fundraising activities. Not every organization is positioned to make those strategies work. Read ahead and proceed cautiously before stepping onto the capital campaign trail.

TIP

Check out File 18-1 at www.dummies.com/go/nonprofitkitfd5e for a list of web resources related to the topics we cover in this chapter.

Beginning the Funding Plan

Suppose your adult literacy program wants to share its innovative teaching materials broadly through a series of books, but setting up an online publication and distribution program requires a big initial investment. Maybe your nonprofit school for children with hearing loss balances its budget each year, but it depends heavily on a small number of income sources. Having annual investment returns from an endowment may allow it to provide scholarships for children who can't pay full tuition rates. Or say you've found the building of your organization's dreams. It's just around the corner from where you currently serve clients, and it's bigger and better than your current building. Better yet, it's for sale at a reasonable price. You need money to buy it and money to make repairs.

REMEMBER

If funds are contributed to your organization for an endowment, that money must be set aside and invested in accordance with state laws regarding endowments. It is not "liquid" — if your nonprofit encounters cash flow problems, it should not borrow from the endowment to pay its bills.

In each of these cases, you need to raise a significant amount of money that's above and beyond your normal annual fundraising. Where do you begin? Capital fundraising, of course! However, keep in mind that it requires you to analyze your organization's situation and plan carefully and thoroughly. Not every organization is positioned to succeed at capital fundraising. You need to do the following:

- **Preplan** by gathering information about technical assistance available to you, reviewing your donor records, and creating a rough campaign budget. If you're raising funds for an endowment, do you have access to good asset management advice that can ensure it is invested wisely?
- **Develop** your case by brainstorming with staff and board members.
- **Test** the project's feasibility by meeting with a select group of potential contributors and gauging their interest. Consider the timing of your campaign: Will you be competing with other campaigns?
- **Analyze** what you find out from the interviews against your initial goal and then refine your goal and plan.

WARNING

An unsuccessful capital campaign can cost your organization money, hurt its reputation, and demoralize your staff and board. If your research suggests that you won't succeed, do *not* be foolhardy and forge ahead anyway.

Preplanning your campaign

After your board decides it may want to raise funds for a capital project, seek outside assistance in the form of workshops, classes, or consulting. Convincing yourself that your capital campaign can work is easy, so test your vision on others. Possible resources include the following:

- A program officer at a local foundation or staff in your city's economic-development office or regional community loan fund may be able to suggest nearby resources.
- The Nonprofit Finance Fund (`www.nonprofitfinancefund.org`) may offer programs in your area, including low-cost consultations, tools to analyze your financial position, workshops about raising capital, and loans for capital projects.
- Your board may contain members with professional experience in financial lending and investing, architecture, construction, real estate, city government, and law; all can provide valuable assistance. If your board doesn't include such professional skills, its members may know professionals who do.

You need to begin raising funds from individuals and institutions that know your organization and care deeply about its work. Assess your organization's primary sources of contributed income. How detailed and accurate are your donor records? Who's been donating money to your cause and at what level? Critically important in this scan of supporters is identifying a person or a small team of people with resources, dedication, and good connections who would be willing to lead your campaign.

REMEMBER

When identifying possible resources for your capital campaign, don't overlook loans, which often play an important role in facility projects, both as standard mortgages and to help organizations complete projects on time. Often donors of large gifts don't write checks for the full amount of their pledges immediately, so organizations borrow some of the money they need to finish construction while waiting to receive those gifts.

If you're interested in loans, find the name of the federal community development investment fund in your region. (Visit the Coalition of Community Development Financial Institutions at `www.cdfi.org` to view its list of members in all 50 states and the District of Columbia.) Also investigate whether any local foundations make program-related investments (low-interest loans) to nonprofits or guarantee bank loans for nonprofits.

Developing a rough budget

Before you test your project to find out whether it's feasible, you need assistance figuring out what it will cost. A common mistake organizations make is forgetting that raising money costs money. They need to dedicate staff time and invest in marketing materials and events. Often they hire outside consultants or new fundraising staff to assist them.

For example, if you want revenues from an endowment to cover five $10,000 scholarships, you need to raise at least $1 million. Your board may set a different policy, but most nonprofit organizations figure they can expect to earn between 3 percent and 5 percent on their endowments and spend at that rate each year without depleting the money they've put aside. When interest rates are very low, the board may decide to spend a smaller percentage earned on the endowment, which requires you to increase your fundraising goal, but we're going to stick with the $1 million goal for this example. If you estimate that meeting that goal will take you a year and a half, you may want to budget for

» A skilled fundraising consultant who will spend an average of ten hours per week on your project

» Printed materials and a project web page describing your organization and the fundraising campaign

» Funds to cover the cost of meeting with prospective donors — travel and meals

» Money to invest in recognizing and thanking donors, including a celebratory event at the culmination of the campaign

Budgeting for building projects is complex. If you identify a plot of land you want to build on or an existing building you want to renovate, you need to begin with rough estimates and develop a specific budget later — after you secure architectural drawings, construction permits from your city or county, and bids from contractors.

As you begin developing a budget, find out about the construction rules you need to follow. A local architect or city appraiser/inspector should be able to tell you what enhancements you're required to include in your building. For example, the Americans with Disabilities Act (ADA) establishes an important set of requirements. Buildings for public use must offer ramps and elevators for wheelchair users; appropriately placed plumbing, railings, and equipment; and large-print signs or audible signals. You may think to budget for the obvious, more-visible modifications that are needed, but an experienced professional can point out more subtle requirements that you may overlook.

REMEMBER

Just because a building is currently in use doesn't mean you can move right in and begin operating your programs there. Building codes are closely tied to the building's function. Rules for classrooms, for example, sometimes change according to the age of the children. And rooms in which large numbers of people gather are likely to have stricter safety requirements than those rooms used by individuals as offices. Building codes often change over time, and when a building changes hands or its use changes, it must then comply with the most recent requirements.

TIP

At www.dummies.com/go/nonprofitkitfd5e, you can find a checklist from the Nonprofit Finance Fund on hard and soft costs (File 18-2), and three sample capital project budgets (Files 18-3, 18-4, and 18-5).

Testing feasibility

Remember that tried-and-true method for buying a used car? The one where you circle the vehicle, kicking all the tires? An organization's capital campaign starts with that kind of tire kicking, except that it's called a *feasibility study,* which is research that tests the hypothesis that you can raise the amount of money you need. Four feasibility study test points are

- **Tire One:** How much money does this organization have the capacity to raise?
- **Tire Two:** For this particular need?
- **Tire Three:** At this location and at this time?
- **Tire Four:** Is that enough money to meet the campaign's goals?

The accuracy of the feasibility study is important to the capital campaign. Whoever conducts it has to have enthusiasm for the project, yet be able to listen carefully to the direct and indirect messages conveyed in the interviews. The person conducting the study must interview key leaders in the organization along with its current supporters and others whose support of the project would be critical to its success.

REMEMBER

Nonprofits often use a consultant for feasibility studies for this reason: She's a step removed from the organization, and therefore interview subjects are more likely to be frank with her.

Leading the way with the board

Whether or not an organization's board includes wealthy individuals, all its members should contribute to the campaign. In fact, the full board's willingness to support the campaign is essential to its success. Potential donors who aren't board members are likely to ask whether you have 100 percent board participation.

REMEMBER

Board members are expected to lead the way. Their contributions may not be the largest, but they should be made in the campaign's earliest phases.

In addition, the size of a donor's gift (in relation to what the donor can afford) is some indication of enthusiasm. If a penniless playwright who serves on the board gives $1,000 toward a capital campaign for a theater, he's making as clear a sign of enthusiasm as a wealthy banker on the board who makes a gift of $1 million. If the wealthy banker makes the $1,000 gift, it suggests either that the board has one malcontent or that the project is unsupportable.

Conducting feasibility study interviews

A typical feasibility study interview opens with an overview of the proposed project, emphasizing why the nonprofit selected the project and how it can enhance the organization's programs and better serve its constituents. Next, the interviewer describes how much money the project needs and how, in general, the organization plans to raise it. If any major donors are already involved, the interviewer mentions their levels of support. Finally, the interviewer suggests a possible donation to the potential contributor and makes note of whether that person is likely to give to the campaign and how much the contribution may be.

The interview should be a conversation rather than a one-sided presentation. The more the interviewer can engage the interviewee in discussion, the more the interviewer finds out about how outsiders view the organization and about the fundraising potential.

In these conversations, most contributors aren't making promises, but they're suggesting their probable levels of support. The study, therefore, is important for two reasons:

- It gathers information that helps the organization estimate whether its capital campaign goal is feasible.
- It begins the process of "cultivating" contributions to and leadership for the campaign.

TIP

At `www.dummies.com/go/nonprofitkitfd5e`, you can find an example of feasibility study interview questions (File 18-6) and an overview of the parts of a finished study (File 18-7).

Analyzing the results of your study

To continue with the used-car metaphor, a feasibility study is like finding out — after talking to your parents, spouse, or the bank — how much you can afford to pay for a car. Your parents may say that they'll give you money for a small sedan but not for a red sports car, and the same is true for a feasibility study. Board members, community leaders, and potential donors may tell you that they would support the venture or endowment if it were smaller, or the building if it didn't need a new foundation.

If the feasibility study suggests that you can't afford a "Champagne" version of your plans, you may have to settle for a serviceable "house wine" edition. If your plans were already of the house-wine variety, you may want to break the campaign into phases. If yours is a building renovation project, over the first two years you can make major safety and structural changes or renovate one of two floors. Then, maybe three or four years later you can launch the second phase of a capital campaign to cover the upper-floor renovation costs or pay off the mortgage.

You may also find out that you need to wait. Perhaps your potential major donors recently gave to another campaign, and although they like your project, they aren't currently in a position to make a major gift to your organization. You may be able to come back to them in two years and receive major contributions. Or maybe potential donors voiced concerns about your organization. Maybe your board has dwindled in size or your executive director just accepted another job. When you organize, recruit, and hire anew, your campaign can proceed.

And you may find out that undertaking the project isn't a good idea at all. Period. At that point, it's time to start over and plan to address your goals in a different way.

Developing a Case Statement

In Chapter 14 on raising money from individuals, we outline steps for writing a case statement. A capital campaign cries out for a *case statement* — a brief, eloquently stated argument on behalf of the capital project. Sometimes these case statements are fancy brochures with profiles of scholarship recipients, drawings of a planned building, or stories about clients of the future. Sometimes they're just simple PDF statements.

A capital campaign case statement should incorporate the following elements:

- A mission statement and brief history of the organization
- A list of major accomplishments
- Compelling information about the constituents served
- A vision of how the mission can be served better as a result of the capital project
- A vision for the results of the capital investment
- The campaign's leadership and goals
- Naming and giving opportunities

After you've put together your case statement, you can show it to prospective donors and use it to build your pyramid of gifts (see the next section).

Building the Pyramid of Gifts

You've sketched out your budget. You've set your sights on a reasonable goal. Your next most important need is to identify *campaign leadership* — a team or committee of volunteers who are willing to cheerlead on behalf of the campaign, make personal visits and calls, and sign letters asking for contributions. These leaders may be board members, but we recommend including some people from outside the board if appropriate enthusiasts can be identified. Not only do their efforts ease your board's workload, but their contacts also expand the pool of possible donors, and their involvement illustrates broader community support for the campaign.

Each campaign develops its own strategies, and each organization has distinctive strengths to call upon, but most of them use the same diagram, which is based on conventional wisdom. It's called a gift table, and it looks like a pyramid (see Figure 18-1).

Remember the old saying, "A handful of people do most of the work"? A capital campaign works that way. Only a few donors are able to contribute large gifts. They go at the top of the pyramid. Smaller gifts from many other donors fill in the lower levels as the campaign moves forward and the pyramid takes shape.

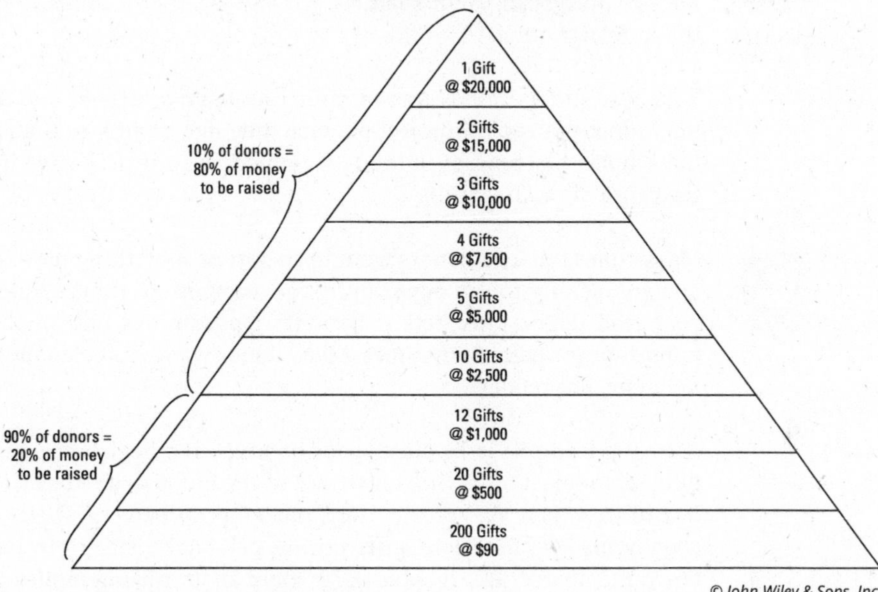

FIGURE 18-1: A sample gift table showing 80 percent of donations coming from 10 percent of donors.

Starting at the top

We know you don't start at the top when building a pyramid, but that's where you begin with a capital campaign. Similarly, we structured the sample gift table in Figure 18-1 starting at the top, using the following scheme:

- Ten percent or more of the money comes from a single gift, known as the *lead gift.* This large donation crowns the tip of the pyramid.
- Fifteen percent of the money comes from the next two largest gifts.
- Fifteen percent of the money comes from the next three largest gifts.
- Overall, 80 percent of the money comes from 10 percent of the donors. In other words, if you receive 100 contributions, ten of those people will provide 80 percent of the money that you must raise.

Say you're driving across town and pass the Janice Knickerbocker Symphony Hall, the Engin Uralman Medical Center, and the Jerome Kestenberg Museum of Natural History. What do Janice, Engin, and Jerome have in common? Generally, those people whose names adorn buildings made the contribution of 10 percent or more of the campaign total. They're called the *lead donors.*

REMEMBER

When you conduct your feasibility study, one thing you're trying to determine is the size of the campaign's largest gift and whether it totals 10 percent or more of the campaign total.

Your lead gift becomes one of your campaign's attractions. Some contributors may support a capital project because they like or admire the person after whom the scholarship fund or building is named, even if they have little connection to the cause or organization.

TIP

Many times the lead donors want to be part of something new and transformative and are looking for an opportunity for recognition. However, always clarify with your lead donor prospects if naming opportunities and public recognition are important to them. Sometimes a lead donor wants to be anonymous; you'll want to honor that request.

After the lead gift is in place, your organization begins seeking the other major gifts that make up the first one-third of the funds to be raised. (In our example in Figure 18-1, just six contributions make up 40 percent of the goal.) Individuals usually pledge these lead gifts during personal visits from nonprofit staff and board members. (See Chapter 14 for more about raising money from individuals.)

Continuing down the pyramid's structure to the widening middle, an organization usually moves beyond board members and major donors to seek grants from foundations and corporations.

Applying for grants as part of your capital campaign

Grant proposals for capital projects follow the general outline of a standard grant proposal (see Chapter 17) with some variations and additions. The grant writer needs to describe the organization's goals and activities and its constituents' needs, and convincingly describe how a stronger organization with stronger capital assets will better serve the organization's goals and clients. Although your proposal may focus on any kind of capital investment project, we discuss a standard building project in the following outline of a campaign grant proposal:

- The introduction includes general information about current facilities and leads up to a discussion of the need for a new or renovated building.
- The problem statement describes the needs of clients or potential clients and how the organization's current building (or lack of a building) hinders their satisfaction of those needs.

- The goals and outcomes section discusses aspirations that the organization holds for serving its clients and its goals for the capital project (the building's dimensions and amenities).
- The methods section briefly touches on how you're delivering services but primarily focuses on how you're conducting the capital campaign, how construction will proceed, and the activities you're undertaking to connect current and future clients to the building. Organizations often include a discussion of the results of the feasibility study in this section as a rationale for how they set goals and how they planned the campaign.
- The evaluation section focuses both on whether the building project will meet its goals (such as achieving all city fire and safety code standards) and on how the improved building serves clients.
- The budget usually includes two major sections: hard costs (for land or building purchase or construction) and soft costs (for financing, fundraising, promotion, and so on).
- The sustainability section discusses how the capital campaign is progressing and where the organization anticipates raising the necessary funds to complete it. This section also provides information about how the finished project will affect the organization's operating costs and how any additional expenses will be covered in the future. It may include a statement that completing the building project will help ensure the organization's future stability.

Ending the quiet phase and moving into the public phase

Early on, when you're conducting the capital campaign among trustees, close friends of the nonprofit, past donors, and foundations, the campaign is in what's called the "quiet phase." When the organization has raised 75 percent to 80 percent of the money it needs, the fundraising style changes. The managers are growing confident that the campaign can succeed, and they announce it to the general public through a press conference, tour, gala party, or cornerstone-setting event.

This is the time to seek smaller donations from lots of people — neighbors, friends of friends, and grandparents. You can often raise these contributions through special events, mailings to individuals, and proposals addressed to smaller foundations and businesses.

Smaller gifts from many people construct the base of the pyramid. Don't discount these gifts. They're important financially to close out the campaign and also to build a feeling of participation among all your donors.

Other focused fundraising drives are structured along the gift-table pattern. Whether you're raising money for your child's school, an election, or a community fair, a gift table can help shape your plans.

TIP

Check out File 18-8 at www.dummies.com/go/nonprofitkitfd5e for a sample capital campaign gift table.

REMEMBER

Funds raised for a capital campaign must be used for the purposes described in the solicitation — whether it is a case statement or grant proposal. Many people don't recognize that the capital contributions can't just be treated as unrestricted funds if not enough money is raised to engage in the capital project.

Realizing the Benefits and Risks of Capital Campaigns

Although you may describe your request for capital support as a one-time need to potential supporters, many campaign donors continue to give after you finish the campaign project. They've been introduced to the agency, they've left their names in its lobby or attached to a scholarship fund, and they want to be sure that it succeeds over time. In the best situations, capital campaigns strengthen the nonprofit organization's programs both by enabling it to improve services and by broadening its donor base. A capital project also can benefit staff morale because it improves working conditions.

In recent years, some thoughtful foundations, service organizations, and consulting groups have advocated for better "capitalization" of nonprofit organizations. Their point is that donors — particularly foundations — have encouraged nonprofits to come up with break-even financial results year after year. Although breaking even is much better than going into debt, emphasizing it as a virtue means that nonprofit organizations rarely put money aside for an unexpected crisis or infrastructure investment.

No for-profit business would thrive under these circumstances. And nonprofit organizations, many of which are formed to tackle important social and educational needs, should be just as innovative as businesses — maybe even more so. A campaign to raise "working capital" for your nonprofit that you can use to innovate or to weather a financial shortfall may be harder to explain to donors than a campaign to build a building, but it may be just as important to your organization's vitality.

TIP

At www.dummies.com/go/nonprofitkitfd5e, we include links in File 18-1 to some key papers about capitalization that may help your nonprofit make the case for a campaign to raise working capital.

Although capital projects are meant to enhance your organization's programs and vitality, capital campaigns also have their drawbacks:

- Capital campaigns may detract from organization's fundraising for operations. If you ask a donor to contribute to a building project, he may not contribute to the organization's ongoing programs in the same year.
- Capital campaigns may double, triple, or quadruple an organization's fundraising expenses while they're being conducted.
- Campaigns that don't succeed or that drag on for a long time can damage an organization's reputation. Because buildings tend to be visible entities, the public may be more aware of an organization's slow-moving construction project than of a problem with its programs or services. If the campaign doesn't succeed, it is important to discuss the situation with its donors, and, of course, honor the terms under which the gifts were made. Donors may want their contributions to be returned, or — if taking the tax deduction is important to them — they may choose to alter the terms and purposes of their gifts.
- Organizations often have turnover in their fundraising staffs after a capital campaign. Employees may stick around to achieve the campaign goal, but a heavy workload may cause burnout.

In short, capital projects offer opportunities and pitfalls, buy-in and burnout, and new donor development and loss of current annual fund donors. But when completed, they often pay for concrete, lasting benefits and are worthy of celebration.

4 The Part of Tens

IN THIS PART . . .

Find out steps you can take if your organization runs into problems.

Check out some proven ideas for increasing your fundraising dollars.

Discover what risks your organization faces and get advice for protecting it.

IN THIS CHAPTER

» Making choices to keep your nonprofit afloat

» Knowing when to close your doors

Chapter 19

Ten Tips for Adapting in Hard Times

It's likely you'll come across problems in the course of managing your nonprofit organization. Everyone does. But sometimes those problems are so severe that you must change the way you do your work to address them.

You may face a catastrophe — a natural disaster, the destruction of a venue, or the serious illness or death of a staff or board member. Or shifts in legislation or foundation policies may cause you to lose important grants or contracts. Although some crises are sudden and fierce, others may build gradually. Several modest deficits may add up to significant debt. The needs of your constituents may change. A competitor may win away your clients or staff members.

The choices you make in challenging circumstances depend, in part, on how much time you have to plan. Is the change you're facing sudden, or has it been gradual? Is it long-term or temporary? Thinking about a crisis or dry spell when you're currently successful isn't fun, but it's necessary in order to survive. So, in this chapter, we provide ten tips for making choices in tough times.

REMEMBER

Transforming or even closing a nonprofit in hard times draws on the same courage, leadership, and resourcefulness you used to create it. You need to be clear and decisive; include others in making and executing your plans; ask for help when needed; and base your decisions on accurate information.

Recognize the Need for Change Before It's Too Late

Recognizing a sudden crisis is easy, but often the most difficult aspect of managing a nonprofit is seeing a gradual decline and the need to change. Likely, you work so hard and feel such dedication that you can't believe your organization is floundering.

REMEMBER

Financial problems have clear consequences (so be on the lookout for them): When deficits exceed 10 percent of an operating budget, the organization will have trouble paying bills within 30 or 60 days. When the deficit exceeds 20 percent of the budget, the pressure intensifies. The inability to make payroll or timely tax payments signals severe problems.

A loan or emergency fundraising drive may solve a short-term problem, but an organization must consider whether its business model is working. Do programs consistently cost more money than the organization's earned income and fund-raising capacity? It can't keep borrowing funds or crying out for help without alienating its friends and supporters.

Communicate When Making Hard Decisions

A colleague of ours once asked, "What do you mean by 'hard decisions'? Are they hard because they're unpopular? Or are they hard because we must choose among options and the best choice is unclear?" The answer is both.

If you're facing an unpopular decision — laying off staff, for example — it's better to communicate clearly and move forward decisively. The longer you delay, the greater harm you may do to your organization and to those employees. In most cases, that communication should be both internal (to board and staff) and external (to donors and constituents). If the change is going to affect your scale of operations, you may even want to announce it to the press.

Step Back and Regroup

The appropriate first action when you recognize that your organization faces a crisis may be to step back and take a breath. You need to ask which programs best address your mission and are of highest quality, which are most expensive, and which are best at earning revenue or attracting contributions. That analysis requires time and effort on the part of the staff and board. You can make small adjustments, such as reducing the hours that a program operates or cutting back on costlier activities. You may even consider suspending all programs temporarily to give yourself time to plan. Not every nonprofit can do so without harming constituents, but many can. A hiatus can allow you to focus full attention on your options while reducing overhead costs.

REMEMBER

If you do take a break, communicate your reasons to your members, clients, and funders and keep everyone informed of your progress. It's critical to your long-term health as a nonprofit that your supporters know you care about them. Also, even if your income and expenses are negligible, be sure to file your 990 tax form with the IRS.

Set a Manageable Fundraising Goal

When raising money, you want to present an upbeat story and can-do attitude, right? And when challenges arise, you may feel the need to hide from donors. When a financial shortfall is creating the organization's crisis, sometimes people stop fundraising. But if donors care about your nonprofit (and why would they have given you money in the first place if they didn't care?), many will still care when times are tough.

We recommend setting a fundraising goal for something concrete — perhaps for two months of office rent or finishing a youth basketball tournament. Make that goal realistic yet meaningful, ask for help with meeting it, and report back on your progress. When you meet that goal, it gives you a positive story to share that helps rebuild your confidence and enthusiasm among your supporters.

Collaborate with Others in Your Field

If the field in which you work has seen a decline in funding or an increase in competition for staff or clients, it may be time to identify others who provide similar programs and begin to work together. Who shares your organization's values and approach? Can you serve clients better by combining services?

Collaboration enables you to bring different strengths, resources, and knowledge to a task. It assumes that you can work with greater depth and understanding through a partnership. You also may save some money.

We recommend that you try a temporary, modest project with your partner before launching a large-scale combination of services. Shared goals and values and shared fields of interest matter, but personal compatibility is important, as well. Also, as you step into a partnership arrangement, make sure that more than one leader in each organization is participating in the conversations and negotiations. Collaborations often falter if the key contact person at one of the two organizations leaves. A team can ensure continuity.

Share a Back Office

Maybe program collaboration won't work for your organization, but other types of collaboration may be successful. For example, why not approach another nonprofit about sharing a staff member? By hiring together, you may be able to offer a stable, full-time job with a competitive salary rather than a part-time gig. In turn, your organizations can engage more experienced and qualified employees.

Of course, you'll want to be careful in setting up this arrangement. Two organizations hiring the same person half-time may diminish that person's benefits. One way around this problem is for one of the organizations to serve as the paymaster, with the other one contributing to the cost of the employee, but this works only if clear lines are drawn that specify when the person is working for which organization. It's a good idea to talk to an employment attorney to devise a clear shared employee arrangement that benefits all parties.

When sharing an employee, both organizations need to give that person clear direction about setting priorities and balancing attention between them.

Sharing things instead of people also can reduce your operating costs. Consider whether you can save money by purchasing equipment, securing office space, or combining supply orders with others. And don't assume that you can share these things only with organizations that resemble yours. In a study of nonprofits in the Puget Sound region, researchers found that some of the best shared-space experiences were among unlike nonprofit organizations, such as a theater school that operated in the late afternoons and evenings and a day-care center that operated early in the day.

Place a Program within Another Agency

One question you should continually ask is how to best meet the needs of the people you serve. Your clients are your reason for existence, after all. If your nonprofit is stretched too thin to achieve everything it's trying to do, why not spin off a program, placing it within another reputable organization that shares your values? You may find that another organization, operating at a different scale or involving people with different skills and knowledge, is positioned to do a better job of managing that program.

We understand that it's difficult to give up a piece of your agency's work, but making this choice is the responsible thing to do for your clients. Plus, it may enable you to focus on the services that best address your nonprofit's mission. Of course, placing a program with another agency also will likely reduce your costs.

Merge with Another Nonprofit

One challenging but responsible choice for your nonprofit may be to merge with another organization. Mergers in nonprofits are similar to those in the business world. Generally, in a merger, one organization dominates the other.

Mergers require thoughtful planning that involves teams of people from both nonprofits. Plan in two phases: First, thoroughly discuss your missions, goals, and program philosophies. If those aren't aligned, the merger won't work. Keep those goals foremost in mind as you move into the second phase of planning, which involves choosing a name, selecting staff, consolidating boards, and arranging a new structure. If problems arise in the second phase of planning, return to the primary conversation and ask whether your missions are aligned and your constituents will be well served by the merger.

TIP

We strongly recommend that, when executing a merger, you use a facilitator to lead you through the plans and decisions, and work with an attorney to determine questions to be raised and risks assessed prior to agreeing to a merger. That same attorney can draw up your final agreements if the merger is prudent. Also be open to the idea of bringing in new staff and board members who will be committed to the merged organization without loyalties to either of the original entities.

Close with Dignity If Necessary

Nonprofit organizations generally resist closure, maybe because they're usually founded by people with a vision and a desire to serve the public. Yet, sometimes closing is the right choice.

If your nonprofit's financial health fails and its program quality declines, it's best if you accept the idea of closure before it's a last-gasp necessity. Shutting down a nonprofit involves several stages — some of them formally defined and some of them merely good practices. We suggest you do the following:

- **Take care of your employees.** Warn them about the impending closure and, if possible, provide them with job counseling and severance pay.
- **Take care of your clients.** Work with other agencies to make sure that client needs will be met in the future. If you have subscribers or members, offer refunds or work with other nonprofits to honor their benefits.
- **Tell your donors and professional partners.** They should hear about your planned closure from you rather than through rumor or through the media. Clear communication of your story must be presented before rumors begin.
- **Pay your debts or negotiate settlements of your obligations before closing.** Doing so protects your reputation and the reputation of your board members and also engenders trust in nonprofits as good business partners.
- **Document your work.** What's your nonprofit's legacy of services to the field? What has it learned? How can your knowledge benefit others? We recommend gathering your information and placing it in a library, school, historical society, or online archive.
- **Celebrate and recognize your staff, board, and volunteers.** The people who have contributed their time, labor, and expertise to your nonprofit deserve applause and thanks.

Complete the Closing Paperwork

If you decide to close your nonprofit, it isn't enough to lock your doors and put up a Gone Fishin' sign. You also need to take a few formal steps. One of the first is to have a board meeting at which the board votes to dissolve the organization and records that decision in the board minutes. Another is to take an inventory of all your assets — money, furniture, client lists, web domains, costumes, fish tanks, and so on — and pass them on, sell them, or return them appropriately, honoring your donors' intentions if some things were contributed for specific purposes. If you have assets to distribute, you must pass them on to a nonprofit with a similar mission.

You'll likely want some technical assistance from a consultant or attorney to complete the paperwork, because you have certain state and federal requirements to meet:

- **State requirements:** Procedures and documents vary by state. For example, in California you need to make an initial filing related to the planned dissolution with the attorney general. After the attorney general issues a waiver of objections to the dissolution plan, you file a certificate of dissolution with the secretary of state. Check with the state agency through which you incorporated on the appropriate steps to take.
- **Federal requirements:** Just as you turned to the Internal Revenue Service (IRS) to create your nonprofit, you return to the IRS to shut it down. You must file a final Form 990 tax return (see Chapter 5 for details about the 990) within four months and 15 days of your organization's formal dissolution date. As you do so, check the Terminated box in Part B on the first page of the return. You'll also need to answer yes in Part IV about whether the organization is terminated, liquidated, or dissolved, and file Schedule N, which asks for a list of assets, their value, and where they were distributed.

» Getting prepared to raise funds

Chapter 20
Ten Tips for Raising Money

Raising money is essential to managing a successful nonprofit organization. In fact, nothing is easier than knowing you should be raising money — and nothing is more difficult than asking for it. This chapter contains our top ten tips for that all-important task of raising funds.

Ask

One of fundraising's oldest adages is, "If you don't ask, you won't get."

Developing fundraising plans, compiling lists of names and email addresses for potential donors, and designing invitations to fundraising events are labor-intensive tasks. But those tasks aren't the things that slow you down in fundraising. Instead, many people pause when picking up the telephone or ringing the doorbell — in other words, when it comes to asking for money. Then, when it's a little too late for the prospective donor to make a decision, write a check, or forward a proposal to a board meeting, these fundraisers make their move and stumble over their own procrastination.

We repeat, "If you don't ask (and ask at the right time), you won't get."

Hit Up People You Know

Some fundraisers believe that the entire money-raising game is in knowing people with money and power and working those contacts — charming them to bend their wills and write those checks. To be honest, if yours is a good cause, that approach isn't bad.

But what if you don't know wealthy people? Does that mean you can't raise money? No, it doesn't. Begin with people you know. Don't be afraid to ask your friends and associates. From a donor's point of view, saying no to someone you know is more difficult than saying no to a stranger.

Tell Your Story

The best way to write an effective fundraising letter or make a successful presentation to potential donors is to tell a story. You don't have to explain how your organization was founded and everything it has done since then (although that history may be worth a brief mention). Instead, the best stories and presentations focus on the constituents you serve and how they benefit from your efforts. Such stories are hopeful. They paint a picture of a better future and describe what "better" looks like in clear, specific terms.

REMEMBER

Pace the story so it has a bit of drama (but don't stoop to melodrama or hyperbole). Include facts — clear evidence of your success. Recognize and discuss the complexity of the field in which you work, but don't drone on and on about technical matters that will cause your audience to zone out.

Show How You're Improving Lives

In grant-writing terms, this piece of advice would be worded as "clearly describe your outcomes." Keep in mind that outcomes are different from outputs. *Outputs* is the word used for the quantity of work a nonprofit organization produces — the number of meals served, shelter beds offered, workshops led, miles covered, or acres planted. *Outcomes* is the word used for the changes that occur as a result of those outputs. Well-defined outcomes are the hallmark of a good grant proposal, fundraising letter, web page, or pitch.

In other words, providing training about nutrition to a group of 50 seniors isn't enough if those seniors don't change their eating patterns and live longer, healthier lives. Removing toxins from a lake isn't enough if its fish population and ecosystem aren't revived. Exposing 500 children to formal music lessons isn't enough if none of them can read a simple score later. Don't leave potential donors thinking, "Sounds nice, but so what?"

Make the Numbers Sparklingly Clear

Effective requests for money include information about how much is needed to achieve change or test an idea. Make sure to present any data you cite in clear terms. In most cases, telling your reader or listener how much a needed change costs — the cost per child to participate in a special classroom for a year, the cost per injured sea mammal rescued, the cost per well in a remote village — helps to make your point.

REMEMBER

Nothing undermines a well-written proposal or case statement faster or more thoroughly than a confusing budget or a muddled financial statement. Double-check your presentation to make sure every activity in your proposal is represented in the budget and that every item in your budget can be traced easily to the work outlined in your proposal. If some items may be confusing to your reader, include budget notes. In all cases, check your math.

Research, Research, Research

Earlier in this chapter, we tell you to ask people you know for money. We're not taking back that advice, but at some point you need to move beyond your immediate circle of acquaintances. That's where doing your homework pays off.

Before you send a fundraising letter, submit a proposal, or visit with a corporate giving director, find out as much as you can about the prospective contributor. Do you have anything in common on a personal level? Maybe the foundation director you're meeting recently published an article. If you read it, you have a conversation topic to break the ice. This advice holds true even when your approach is a membership mailing or email appeal: You want to know as much as you can about the people whose names are on the lists you develop, borrow, or purchase.

More important, you want to find out as much as you can about your potential donor's giving behavior. Does this person give small amounts of money to a wide array of organizations or generous gifts to a few selected agencies? Does the foundation like to be the only contributor to a given project, or does it prefer to support an activity along with others? Does the corporate giving program prefer a low-key style, or does it like to have the company's involvement highlighted?

TIP

You can turn to many sources for this information; just be sure to make research a habit. For instance, you can do any of the following:

» Refer to the Foundation Center library and its published and online resources (www.foundationcenter.org). Don't overlook its Philanthropy News Digest ("PND" on its website), which announces research published and grants awarded by foundations and their grantees.

» Look up the foundation at the Council on Foundations' website (www.cof.org) to see whether it has recently issued a report or been featured in an article.

» For personal information about individuals, conduct an Internet search; look at LinkedIn bios (www.linkedin.com); peruse local newspapers; and contact college alumni associations.

» Follow business and social news along with obituaries to keep track of people's families, professional developments, and affiliations.

» Pay close attention to contributor lists when you attend events at other nonprofit organizations.

Know Your Donors' Point of View

Have you ever heard the old saying, "To catch a trout, think like a trout"? Well, this saying also applies to raising money for your nonprofit. The point of conducting research is to be able to talk or write about your organization in ways that are compelling to your listener or reader. You don't want to warp your message or change your mission, but you do want to think about it (and talk or write about it) in ways that respect your audience's point of view. To do so, you need to think about your organization as if you were a prospective donor yourself.

Donors have different personal giving styles, so you need to put yourself in donors' shoes when deciding how to raise money from particular individuals or companies. You don't want to offer to hold a tribute dinner in honor of someone who prefers to contribute anonymously, for instance. And you don't want to downplay a gift from a contributor who relishes public acknowledgment. Some donors prefer the sociability of supporting a cause through a special event, others respond to email appeals, and still others like to see as much of their money as possible going directly to the service being provided. (If possible, let them see that service with their own eyes!)

TIP

Whenever possible, ask donors how they prefer to be recognized and also ask them why they donate to your organization. Their answers may surprise you and will help you craft your future message and approach.

Many people forget that foundations are nonprofit organizations with mission statements and that their job is to support proposals that further their missions. Your job as a grant seeker is to measure how your goals align with their purposes. For example, many corporate giving programs and company-sponsored foundations work to improve the communities in which their employees live and work. When these organizations discover that their employees are involved in or contribute to a given cause, they may be more inclined to support that cause.

Government grant-making programs are created through legislation. If you find a program that seems well suited to your organization's work, reviewing the legislation or, when possible, attending public hearings about the program is worth your time. Doing so helps you fully understand the program's context and intentions.

Build a Donor Pyramid

Earlier in this chapter, we suggest starting by asking people you know. Imagine your fundraising approach as starting with those people at the peak of a pyramid and working your way down to the broader base. After asking people you know, try to enlist those donors in asking their contacts to support your organization as well. Some are likely to respond (in part because people they know also are contributors). Then ask those new donors to ask their friends. If each contributor leads you to two additional contributors, you steadily build your pyramid.

Distributed fundraising platforms such as `www.causes.com` are one easy way for your donors to raise money for you. (See Chapter 14 for more about distributed fundraising.)

Make It Easy to Respond

You've written a brilliant appeal letter. You've created a compelling website. You've delivered a stirring speech to a roomful of prospective donors. You've got them hooked. They want to contribute. But they're glancing around the room with confused looks on their faces. You blew it. You didn't give them an easy way to respond.

Always suggest a specific amount for donors to consider contributing. Connect that amount to what you need and to their potential giving levels (which you can estimate from your research). And always make immediate giving easy for them: Distribute an addressed envelope and reply card with each mailing, enable a reader to click through your email to the Donate Now button on your website, or set up a labeled box by the exit where donors can leave contributions. Give potential donors pens, stamps, email addresses, pledge cards, and any other tools to help them respond when you have their attention. Also, if you're targeting donors in multiple states, remember to register to fundraise in each state that requires it.

Keep Good Records

After you begin attracting contributors, your donor records become your most valuable fundraising tools. Individuals who give to your organization once are likely to continue giving for three or more years. If you thank them, address them as if they're part of your organization, and generally treat them well, the size of their gifts is likely to increase over time.

Working with foundations, corporations, and government sources is a different story. In their case, you want to keep clear records of your original project goals and outcomes, project budget, and due dates for any required reports. Although these sources may not be willing or able to support your organization year after year, their future support is more likely if you're a conscientious grantee who submits reports on time and keeps clear records.

You can create a simple database by recording names, addresses, phone numbers, email addresses, patterns of giving, and personal information (such as whether a donor knows one of your board members or whether he or she is married and has children). Consider investing in *donor management software*. Several good ones are available. Some are pricey, others are free. Find out more about donor software that's right for your organization by searching articles on Idealware's website (`www.idealware.org`).

However you keep it, guard your donor database carefully and invest the time necessary to keep it up to date. It's one of your organization's most valuable resources.

» Filing official paperwork and taxes

» Protecting your reputation

» Insuring your future

Chapter 21

Ten Tips for Protecting Your Nonprofit

Protecting your nonprofit organization encompasses tasks that can range from purchasing sandbags to filing government forms. These activities should be driven by an ongoing, thoughtful assessment of your organization's risks.

Some risks that nonprofits face vary according to where they're based and what they do. Is your office on the banks of a river or in tornado country? Do you work with preschool-age children or with injured raptors? Other items are universal: protecting the safety of staff and clients, protecting the organization's assets, protecting its tax-exempt status, and protecting its reputation. Although no nonprofit manager can anticipate every problem that arises, it's wise to put systems and precautions in place.

TIP

Check out File 21-1 at www.dummies.com/go/nonprofitkitfd5e for a list of web resources related to the topics we cover in this chapter.

Assessing Your Risks

Evaluating risks should be a routine part of your planning and operations. We recommend that you begin with a big-picture analysis of your nonprofit and its potential vulnerabilities. Your list should include the following:

- Location and facility
- Activities and programs
- Beneficiaries of your programs
- Management and financial systems
- Contracts and collaborations
- Employees
- Board members and volunteers
- Communications and reputation

You may note that your organization works with young children and uses organization-owned vans to transport them; its board has limited financial management knowledge; its employees and programs must be licensed; and its computer systems are old and prone to crashing. Each of these observations makes the organization vulnerable.

Planning for Emergencies

Blizzard belt, floodplain, earthquake fault line, hurricane path, gas leak: Likely all the nation's nonprofits are vulnerable to some kind of natural or man-made disaster. Your emergency plan should have three elements:

- Preventing problems by eliminating as many potential risks as possible.
- Preparing for emergencies by devising procedures and assigning roles for enacting them.
- Training everyone on your staff to respond to several scenarios. Drill and practice your responses.

We recommend that your nonprofit put together an emergency planning team of three or four volunteers, staff, and board members. It's wise to include people on the committee who represent different kinds of expertise and who would be level-headed in an emergency. Sketch out different response scenarios. Three common options are evacuation, lockdown, and shelter-in-place.

Your emergency team should

- **Prepare to ensure the whereabouts and safety of staff members and their families.** Phone lists that include information for staff members, their local family members, and an out-of-state contact should be kept up to date and in a safe place. The same kind of information should be kept current for clients.
- **Plan evacuation routes, identifying nearby shelters, fire stations, or hazards.**
- **Collect and store emergency supplies — sufficient food and water for 72 hours along with flashlights, batteries, a radio, and a first-aid kit.** Remember that cellphones may not work in an emergency. Encourage staff members to keep a few days' worth of prescription drugs (if you must stay in place) and a pair of comfortable shoes at work (if you must evacuate).
- **Safeguard your records.** Are your computers backed up regularly? If you depend on data collection, do you send it away regularly to a safe location? Also, if your accounting system is online, you may want to lock away some checks that can be handwritten in an emergency.
- **Plan to take action.** How will your board take action to respond to an emergency if calling a meeting and securing a quorum is difficult? Some states permit nonprofits to add allowances to their bylaws that outline how decisions can be made under emergency circumstances.

TIP

You can find useful tools for emergency planning at `www.ready.gov` (the federal government's all-hazards approach to planning). San Francisco CARD (`www.sfcard.org`) offers a template for an emergency plan that focuses on the San Francisco Bay Area but is adaptable to other settings.

Filing Annual Federal Forms

Each year, the federal government requires 501(c)(3) nonprofit organizations to submit a *Return of Organization Exempt from Income Tax*, commonly known as Form 990. The complexity of that document varies according to the size of your organization:

- The 990-N or e-postcard is for organizations with gross receipts that normally are $50,000 or less.

» The 990 Short Form or 990-EZ is for organizations that normally have annual gross receipts above $50,000 but less than $200,000 and total assets of less than $500,000.

» The 990 is for organizations that are larger than those permitted to file the 990-N or the 990-EZ. Smaller organizations that qualify to file the 990-N or 990-EZ may choose to make the greater disclosures required and file the 990.

REMEMBER

Your 990 tax form needs to be postmarked or filed electronically on the 15th day of the fifth month after the close of your financial year unless it falls on a weekend or holiday. You may request an extension if needed. Failing to file can jeopardize your nonprofit status.

Filing Annual State Forms

Each state has its own laws and regulations for how and when a nonprofit must report its activities. In some states, your nonprofit will receive a sales-tax exemption. To find your state laws, visit the directory of state offices on the IRS website at www.irs.gov/charities-non-profits/state-links.

If you change your articles of incorporation or bylaws during the year, you'll need to follow the laws of your state (in the case of articles of incorporation) or the rules described in your bylaws to register the change.

Furthermore, as of this writing, 36 states, the District of Columbia, and some local governments require nonprofits to register and pay an annual fee if they are fundraising in the state. The number of states requiring this fundraising registration changes from time to time. See www.multistatefiling.org for updated information.

Paying Employment Taxes

Although nonprofit organizations don't need to pay federal taxes on revenue related to their exempt purpose, and — in some states and jurisdictions — don't need to pay local property taxes or sales taxes, they must pay employment taxes, including the employer's portion of employment and Social Security taxes and unemployment insurance. In most cases, these taxes are paid quarterly to state tax boards and the Internal Revenue Service. Not every state requires nonprofits to pay unemployment taxes but every nonprofit must pay federal unemployment taxes. Failing to pay these taxes leads to stiff penalties and interest charges. Unpaid employment taxes are one of the few nonprofit liabilities for which the organization's board members may be personally liable. Check with your local state unemployment office or department of revenue for filing and payment requirements for your state.

Reporting Payments to Consultants

If your organization hires a consultant to design its website, conduct an evaluation, or manage other specialized tasks, and if that consultant is paid $600 or more during one year, your nonprofit must report these payments by filing the IRS Form 1099 shortly after the end of each calendar year. Some states also require that this information be reported. You can find more information at `www.irs.gov/charities-non-profits/state-links`.

Maintaining Transparency

Transparency is a buzzword in the nonprofit arena these days, and although it may seem like jargon, it represents an important set of standards, namely: Nonprofits should be aboveboard and honest in their delivery of services, personnel matters, fundraising, and finances. One smart thing to do is to review IRS rules for public disclosure by visiting `www.irs.gov`. Also, it's useful to review the recommended principles and practices for nonprofits in 23 different states, compiled by the National Council of Nonprofits at `www.councilofnonprofits.org`. The National Council of Nonprofits also has summarized a set of Standards for Excellence. Recommended transparency practices include the following:

- Be honest with donors about how their contributions are used and offer them the opportunity to opt out of being identified as contributors apart from being listed on Schedule B of the 501(c)(3) nonprofit's 990 tax form.
- Have a conflict-of-interest policy for staff and board members.
- Have your IRS Form 1023 exemption application available for disclosure upon request.
- Post your most recent 990 or financial statement on your website.
- Adopt an internal complaint and whistle-blower policy in accordance with laws in your state.

Responding to Negative Press

Even good, conscientious nonprofit organizations can draw negative media attention. The negative coverage may be based on accurate or inaccurate information, but even untrue accusations can be damaging. When bad press arises:

- Don't speak to the media or members of the public until you've fully investigated the complaint or critique and are ready to give a clear statement. You don't want to delay a response for a long time because your silence can appear to be an admission of guilt, but you also don't want to retract an initial statement. It's fine to say, "We are issuing a statement on Friday morning" when asked to comment.
- Decide on a small number of people (perhaps your executive director and board chair) who are authorized to speak to the media. Don't permit others to comment. It's important to present a clear, consistent message.
- Answer the media's questions directly and truthfully, but also be sure to insert a positive message about how you're rectifying the situation, if the complaint is valid, or the value of your organization's work.
- Acknowledge and address the controversy on your website, in a press release, and — if it's a major concern — in a letter to your funders and donors. In all communications, indicate that you're taking the criticism seriously.

Protecting Your Online Reputation

Information reported online is important to your organization's reputation. Negative comments collected on your website or social media pages can be damaging because of how quickly and easily they can be transmitted to many people.

If someone is disappointed with your services, apologize (if appropriate) and invite them to try again, either for no charge or at a discounted fee. In doing so, be careful of an admission of responsibility and liability, especially if the organization wasn't responsible and should not be liable. Sometimes people may offer what they consider to be constructive criticism. Treat these comments with respect and thank them for their advice. Generally, it's best to ignore angry, vengeful comments. Responding often makes it worse.

Determining Insurance Needs

A nonprofit's insurance needs vary according to, among other things, the following:

- Whether it offers services within its own (or a rented) piece of property
- Whether it has employees (we discuss employee benefits in Chapter 10)
- Whether it makes use of volunteers (see Chapter 9)
- Whether it has direct contact with clients, students, patients, or audiences and the nature of that direct contact
- Whether it involves professional services
- Whether it creates and publishes materials

Most of the claims reported by nonprofit organizations are accidents and injuries related to falls at nonprofit locations or special events or are related to the use of automobiles. All nonprofit organizations should purchase general liability (sometimes called "slip and fall") insurance. You need to purchase specialized kinds of liability insurance if your organization has significant direct client contact that involves a potential risk. Health clinics and therapy programs need such insurance to protect them if a client is hurt or handled inappropriately. So do programs offering activities with possible physical dangers, such as rock climbing, sailing, or horseback riding.

If your organization occasionally produces events or conducts work in locations other than its central office or building, double-check to make sure you're covered for these off-site events. You may need to purchase a rider to your regular liability insurance policy to cover such situations.

You should also purchase "non-owned/hired" auto insurance in case an employee or volunteer is involved in an automobile accident.

In most states, workers' compensation insurance is required by law. If an employee is injured in the course of performing her job, this insurance helps protect both the employee and the employer. The cost of your workers' compensation insurance is calculated according to the number of employees, the number of hours they spend on the job, and the nature of the work they do.

Other kinds of insurance to consider are:

- Property insurance for damage to property and equipment owned or leased by the nonprofit.

- Employee dishonesty insurance or fidelity bonds to protect against embezzlement.
- Directors and officers (D&O) insurance to protect directors', officers', and employees' personal assets from lawsuits against the nonprofit organization. Corporate laws vary by state, but in most cases directors and officers are protected (even without insurance) if they're providing reasonable oversight of the nonprofit. That "reasonable" oversight involves attending meetings, asking questions, keeping well informed about the organization, and acting in good faith in the best interests of the organization. However, D&O insurance generally covers the cost of a legal defense against lawsuits. We encourage nonprofits to shop around among different insurance companies because D&O insurance products vary widely.

5 Appendixes

IN THIS PART . . .

The nonprofit world has its own lingo. Flip to the glossary when you encounter an unfamiliar word.

The website that supplements this book includes many helpful documents and links to useful websites. Appendix B lists what's available.

Appendix A
Glossary

annual report: A report published by a nonprofit organization, foundation, or corporation describing its activities and providing an overview of its finances. A foundation's annual report usually lists the year's grants. Many states require a 501(c)(3) nonprofit to provide an annual financial report to members or board members.

articles of incorporation: A document filed with an appropriate state office by persons establishing a corporation. Generally, this filing is the first legal step in forming a nonprofit corporation.

bylaws: Rules governing a nonprofit organization's operation. Bylaws often outline the methods for selecting directors, forming committees, and conducting meetings.

capacity building: A general term used to describe activities that help a nonprofit organization strengthen its internal operations so that it can do its job better. Examples include staff and board training, computer and financial-management systems upgrades, and consultant assistance for planning.

capital campaign: An organized drive to raise funds to finance an organization's capital needs — buildings, equipment, renovation projects, land acquisitions, or endowments.

capital support: Funds provided to a capital project.

case statement: A brief, compelling document about an organization's plans, accomplishments, and vision.

challenge grant: A grant made on the condition that other funds must be secured before it will be paid — usually on a matching basis and within a defined time period.

charitable contribution: A gift of goods, money, or property to a nonprofit organization.

charity: The word encompasses religion, education, assistance to the government, promotion of health and the arts, relief from poverty, and other purposes benefiting the community. Nonprofit organizations formed to further one of these purposes generally are recognized as exempt from federal income taxes under Section 501(c)(3) of the Internal Revenue Code.

community foundation: A grant-making organization receiving its funds from multiple public sources, focusing its giving on a defined geographic area, and being managed by an appointed, representative board of directors. Technically, it's classified by the Internal Revenue Service as a public charity, not a private foundation.

community fund: An organized community program making annual appeals to the general public for funds that are usually disbursed to charitable organizations rather than retained in an endowment. Sometimes called a federated giving program.

corporate foundation: A private foundation that derives its grant-making funds primarily from the contributions of a profit-making business.

corporate giving program: A grant-making program established and managed by a profit-making company. Unlike a corporate foundation's grant making, a corporate giving program's gifts of money, goods, and services go directly from the company to grantees.

demonstration grant: A grant made to experiment with an innovative project or program that may serve as a model for others.

donor-advised fund: A fund held by a community foundation or other charitable organization for which the donor, or a committee appointed by the donor reasonably expects to have advisory privileges over the distribution or investments of the assets — including through awarding grants and contributions to 501(c)(3) nonprofit organizations.

donor-designated fund: A restricted fund — often held at a community foundation — for which the donor has specified how the proceeds from the fund will be used.

endowment fund: Endowment funds generally are invested to provide income for long-term operation of an organization. Some are meant to be kept permanently.

excise tax: An annual tax of net investment income that private foundations must pay to the Internal Revenue Service.

family foundation: Not a legal term, but commonly used to describe foundations that are managed by family members related to the person or persons from whom the foundation's funds were derived.

federated campaign or federated giving program: Raising funds for an organization that will redistribute them as grants to nonprofit organizations. This fundraising often is led by volunteer groups within clubs and workplaces. United Way and the Combined Federal Campaign are two examples.

fiscal sponsor: A nonprofit 501(c)(3) organization that formally agrees to sponsor a project led by an individual or group from outside the organization that does not have nonprofit status. If the outside individual or group receives grants or contributions to conduct an activity, those funds are accepted, on behalf of the project, by the fiscal sponsor.

501(c)(3) nonprofit: A term describing the Internal Revenue Service's designation for organizations whose income isn't used for the benefit or private gain of stockholders, directors, or other owners. A nonprofit organization's income is used to support its operations and further its stated mission. It may sometimes be referred to as a nonprofit organization (NPO).

foundation: A nongovernmental, nonprofit organization with funds and a program managed by its own trustees and directors, established to further social, educational, religious, or charitable activities by making grants. A private foundation may receive its funds from an individual, family, corporation, or other group consisting of a limited number of members.

gift table: A structured plan for the number and size of contributions needed to meet a fundraising campaign's goals.

grantee: An individual or organization receiving a grant.

grantor: An individual or organization awarding a grant.

independent sector: The portion of the economy that includes all 501(c)(3) and 501(c)(4) tax-exempt organizations as defined by the Internal Revenue Service, all religious institutions, all social-responsibility programs of corporations, and all people who give time and money to serve charitable purposes. It's also called the voluntary sector, the charitable sector, the third sector, or the nonprofit sector.

in-kind contribution: A donation of goods or services (not of money).

lobbying: Efforts to influence legislation by shaping the opinions of legislators, legislative staff, and government administrators directly involved in drafting legislative policies. The Internal Revenue Code sets limits on lobbying by organizations that are exempt from tax under Section 501(c)(3).

logic model: A graphic illustration that outlines an organization's or program's resources, activities, outputs, and outcomes, and shows, thereby, how it intends to achieve change. Sometimes called a theory of change.

matching gifts program: A grant or contributions program that matches employees' or directors' gifts made to qualifying nonprofit organizations. Each employer or foundation sets specific guidelines.

matching grant: A grant or gift made with the understanding that the amount contributed will be matched with revenues from another source on a one-for-one basis or according to another defined formula.

mission statement: A succinct statement of the purpose and key activities of a nonprofit organization.

outcome evaluation: An assessment of whether a project achieved the desired long-term results.

program officer: A staff member of a foundation or corporate giving program who may review grant requests, recommend policy, manage a budget, or process applications for review by a board or committee. Other titles, such as program director or program consultant, also are used.

program-related investment: A low-interest loan or other investment made by a foundation or corporate giving program to another organization for a project that would accomplish one of the foundation's exempt purposes.

proposal: A written application, often with supporting documents, submitted to a foundation or corporate giving program when requesting a grant.

public charity: A type of organization classified under Section 501(c)(3) of the Internal Revenue Code. A public charity normally receives a substantial part of its income from the general public or government. The public support of a public charity must be fairly broad, not limited to a few families or individuals.

restricted funds: Assets or income whose use is restricted by a donor for a specific purpose or activity.

tax exempt: A classification granted by the Internal Revenue Service to qualified non-profit organizations that frees them from the requirement of paying taxes on their income. Private foundations, including endowed company foundations, are tax-exempt; however, they must pay a modest excise tax on net investment income. All 501(c)(3) and 501(c)(4) organizations are tax-exempt, but only contributions to a 501(c)(3) are tax-deductible to the donor.

unrelated business income: Income from a trade or business, regularly carried on, and not substantially related to furthering the exempt purpose of the organization. Such income may, therefore, be taxable.

unsolicited proposal: A proposal sent to a foundation without the foundation's invitation or prior knowledge. Some foundations don't accept unsolicited proposals.

» Perusing a list of the digital files

Appendix B
About the Online Content

Throughout this book, we refer to specific files that help you set up and run your nonprofit organization. We've placed those files at www.dummies.com/go/nonprofitkitfd5e. The following list summarizes the files that you'll find on the website:

- **File 1-1:** Chapter 1 Web Resources
- **File 1-2:** Nonprofit Activity Types
- **File 2-1:** Chapter 2 Web Resources
- **File 2-2:** Sample Fiscal Sponsorship Contract
- **File 3-1:** Chapter 3 Web Resources
- **File 4-1:** Chapter 4 Web Resources
- **File 4-2:** Checklist for Forming a Nonprofit Organization
- **File 4-3:** Writing Organizational Bylaws
- **File 5-1:** Chapter 5 Web Resources
- **File 6-1:** Chapter 6 Web Resources
- **File 6-2:** Sample Grid for Planning Board Recruitment

- **File 6-3:** Board Officer Position Descriptions
- **File 6-4:** Sample Board Contract
- **File 6-5:** Sample Outline of Board Meeting Minutes
- **File 6-6:** Sample Simple Board Meeting Agenda
- **File 6-7:** Sample Formal Board Meeting Agenda
- **File 7-1:** Chapter 7 Web Resources
- **File 7-2:** Sample Planning Retreat Agenda
- **File 7-3:** Sample SWOT Analysis
- **File 7-4:** Matrix Map: A Tool for Bottom-Line Decision Making (included with permission from Jeanne Bell, CEO, CompassPoint Nonprofit Services, www.compasspoint.org)
- **File 7-5:** Sample Outline of a Strategic Plan
- **File 7-6:** Sample Needs Assessment Questionnaire
- **File 7-7:** List of Required Spaces (included with permission from the Nonprofit Finance Fund, © Nonprofit Finance Fund, www.nonprofitfinancefund.org. All rights reserved.)
- **File 7-8:** Planned Change and Facilities (included with permission from the Nonprofit Finance Fund, © Nonprofit Finance Fund, www.nonprofitfinancefund.org. All rights reserved.)
- **File 7-9:** Sample Planning Interview Protocol
- **File 8-1:** Chapter 8 Web Resources
- **File 8-2:** Evaluation: Some Definitions of Terms
- **File 8-3:** Sample Logic Model (included with permission from Harvard Family Research Center, © 1999 President and Fellows of Harvard College. Published by Harvard Family Research Project, Harvard Graduate School of Education. All rights reserved. No part of this publication may be reproduced without permission of the publisher. Since 1983, HFRP has helped stakeholders develop and evaluate strategies to promote the well-being of children, youth, families, and communities. HFRP's work focuses on early childhood education, out-of-school time programming, family and community support in education, complementary learning, and evaluation. Visit www.hfrp.org to access hundreds of resources with practical information for evaluators, practitioners, researchers, and policymakers.)
- **File 9-1:** Chapter 9 Web Resources

- **File 9-2:** Sample Volunteer Job Description, Food to the Table Delivery Volunteer
- **File 9-3:** Sample Volunteer Job Description, Take My Hand Adult Mentor
- **File 9-4:** Sample Volunteer Application Form
- **File 9-5:** Sample Volunteer Agreement Form
- **File 10-1:** Chapter 10 Web Resources
- **File 10-2:** Sample Job Description: Executive Director
- **File 10-3:** Sample Job Description: Development Director
- **File 10-4:** Sample Job Description: Office Administrator
- **File 10-5:** Sample Reference Checking Form
- **File 10-6:** Sample Hire Letter
- **File 11-1:** Chapter 11 Web Resources
- **File 11-2:** Sample Organization Budget: Photography Workshops
- **File 11-3:** Sample Organization Budget: Midsize Sheltered Workshop for Persons with Disabilities
- **File 11-4:** Sample Organization Budget: Education/Recreation Program for Homeless Children
- **File 11-5:** Total Project Costs: Tutoring Program
- **File 11-6:** Tracking Actual vs. Budgeted Revenue and Expenses
- **File 11-7:** Tracking Actual Income and Expenses
- **File 11-8:** Five-Year Financial Trend Line (included with permission from LarsonAllen Nonprofit and Government Group)
- **File 11-9:** Monthly Information Every Nonprofit Needs to Know (included with permission from LarsonAllen Nonprofit and Government Group)
- **File 11-10:** Sample Cash-Flow Projection
- **File 11-11:** Cash-Flow Projection Form from the Emergency Loan Fund (modified for general use; included with permission from Colin Lacon, president and CEO, Northern California Grantmakers, `www.ncg.org`)
- **File 11-12:** Using your Financial Statement to Answer Key Questions
- **File 11-13:** Sample Audited Financial Statement
- **File 11-14:** If You're Reading an Audit
- **File 12-1:** Chapter 12 Web Resources

- **File 12-2:** Sample Audience Survey Form: Neighborhood Historical Society
- **File 12-3:** Sample Participant/Volunteer Survey
- **File 12-4:** Sample News Release (modified version, included with permission from Carla Befera)
- **File 12-5:** Sample Media Alert
- **File 12-6:** Sample Calendar Release
- **File 12-7:** Sample Photo Caption
- **File 12-8:** Sample Photography/Video Permission Form
- **File 12-9:** Sample Public Service Announcement
- **File 13-1:** Chapter 13 Web Resources
- **File 13-2:** Fundraising Plan: High School Music Awards Program
- **File 13-3:** Sample Fundraising Plan: Neighborhood Park Coalition
- **File 13-4:** Sample Fundraising Budget: High School Music Awards Program
- **File 13-5:** Sample Fundraising Budget: Neighborhood Park Coalition
- **File 14-1:** Chapter 14 Web Resources
- **File 14-2:** Sample Donor Acknowledgement Letter: Contribution of $250 or More, No Good or Service Provided
- **File 14-3:** Sample Donor Acknowledgement Letter: Contribution of $250 or More, Good or Service Provided
- **File 14-4:** Sample Solicitation Letter
- **File 14-5:** Tips for Writing an Email Fundraising Letter
- **File 14-6:** Sample E-Solicitation Letter (reprinted with permission from Graywolf Press, Minneapolis, MN, `www.graywolfpress.org`)
- **File 14-7:** Sample E-Solicitation Letter (reprinted with permission from Graywolf Press, Minneapolis, MN, `www.graywolfpress.org`)
- **File 14-8:** Sample Email for Building a Potential Donor Relationship
- **File 14-9:** Using Social Media and Your Website to Enrich Your Fundraising Appeal
- **File 14-10:** Telemarketing Do's and Don'ts and Sample Telemarketing Scripts
- **File 14-11:** Fundraising Ethics: Respecting and Protecting Your Donors
- **File 15-1:** Chapter 15 Web Resources
- **File 15-2:** Sample Budget for a Tribute Dinner

- **File 15-3:** Sample Special Events Budget for a Concert or Performance
- **File 15-4:** Sample Online Auction Budget
- **File 15-5:** Sample Special Event Committee Invitation Letter
- **File 15-6:** Sample Special Events Solicitation Letter
- **File 15-7:** Sample Special Events Timeline: Tribute Dinner or Luncheon
- **File 15-8:** Sample Special Events Timeline: Concert or Performance
- **File 15-9:** Sample Special Events Invitation
- **File 15-10:** Tips for Organizing an Online Benefit Auction
- **File 16-1:** Chapter 16 Web Resources
- **File 16-2:** Foundation Research Overview
- **File 16-3:** Foundation Prospect Evaluation and Tracking Sheet
- **File 16-4:** Federal Grants Research
- **File 17-1:** Chapter 17 Web Resources
- **File 17-2:** Sample Letter of Inquiry (included with permission from the San Francisco Mime Troupe)
- **File 17-3:** Sample Letter Proposal
- **File 17-4:** Sample Fictional Grant Proposal Addressed to the Randall A. Wolf Family Foundation Requesting Support for Tri-City Hot Meals' new program (included with permission from Kristie Kwong)
- **File 17-5:** Sample Full Proposal Submitted to the Walter & Elise Haas Fund for Dancers' Group Presents and Bay Area Dance Week (included with permission from Dancers' Group and grant writer Julie Kanter)
- **File 17-6:** Sample Project Budget Submitted to the Walter & Elise Haas Fund for Dancers' Group Presents and Bay Area Dance Week (included with permission from Dancers' Group and grant writer Julie Kanter)
- **File 17-7:** Sample Federal Grant Request for Operating Support (included with permission from UCSF AIDS Health Project)
- **File 17-8:** Sample Foundation Budget Form
- **File 17-9:** Steps to Take for Finishing and Submitting Your Grant Request
- **File 17-10:** Sample Company-Sponsored Foundation Guidelines and Application Information
- **File 17-11:** Sample Letter Proposal to a Company-Sponsored Foundation
- **File 18-1:** Chapter 18 Web Resources

» **File 18-2:** Hard and Soft Costs for Capital Improvement Project Checklist (included with permission from the Nonprofit Finance Fund, © Nonprofit Finance Fund, www.nonprofitfinancefund.org. All rights reserved.)

» **File 18-3:** Sample Capital Campaign Budget, Building a Youth Center within an Existing Building

» **File 18-4:** Sample Capital Project Budget, Building a Small Theater in an Existing Building

» **File 18-5:** Sample Capital Campaign Budget, Constructing a Childcare Center in a Former School Building

» **File 18-6:** Sample Feasibility Study Questionnaire

» **File 18-7:** Sample Capital Campaign Feasibility Study

» **File 18-8:** Capital Campaign Gift Table

» **File 18-9:** Case for Support: Sample Women's Audio Mission Case Statement for Capital Project (included with permission from the Women's Audio Mission)

» **File 18-10:** Case Statement Background Information: Women's Audio Mission Background Information (included with permission from the Women's Audio Mission)

» **File 18-11:** Why WAM? Excerpt from Capital Campaign Case Statement (included with permission from the Women's Audio Mission)

» **File 21-1:** Chapter 21 Web Resources

Index

Numerics

A

B

C

D

E

F

G

H

I

J

K

L

M

N

O

P

Q

R

S

T

U

V

W

Y

Z

About the Authors

Stan Hutton: Stan became involved in the nonprofit world after co-founding a nonprofit organization in San Francisco. Since that time, he has worked as a nonprofit manager, fundraiser, consultant, and writer. He served as executive director of the Easter Seals Society of San Francisco and a fundraiser for Cogswell College. For five years, he wrote for and managed a website about nonprofits for About.com. He has worked at the San Francisco Study Center and the Executive Service Corps of San Francisco. He currently is a senior program officer at the Clarence E. Heller Charitable Foundation.

Frances N. Phillips: Frances is program director for the Arts and the Creative Work Fund at the Walter and Elise Haas Fund in San Francisco. She also teaches creative writing at San Francisco State University, and taught grant writing at SFSU for more than 25 years. Previously, Frances worked as executive director of Intersection for the Arts and of the Poetry Center and American Poetry Archives at San Francisco State University, and as a partner in the public relations and fundraising firm Horne, McClatchy & Associates. Frances is an advisory board member for Kelsey Street Press, and a policy council member of the California Alliance for Arts Education. She also is co-editor of the *Grantmakers in the Arts Reader.*

Dedication

To Alice, Tristan, and Isaac.

Authors' Acknowledgments

This book started with our agent, Marina Watts, who had the idea and approached us to write it. Many thanks to her, to project editor Elizabeth Kuball, to acquisitions editor Tracy Boggier, and to technical editor Gene Takagi, who guided us through shaping and revising our manuscript for this edition. We also thank our colleagues who gave us permission to include examples of their work on the website that accompanies this book. Finally, we are grateful to the staff at Foundation Center West, particularly Michele Ragland Dilworth, Natasha Isajlovic, and Sarah Jo Neubauer; Marcelle Hinand, President of M. Hinand Consulting; Julie Taylor formerly of Chitresh Das Dance Company; and Jennifer Wong of the California Alliance for Arts Education for their advice.

We owe much of our knowledge of the nonprofit sector to a remarkable array of professionals with whom we have worked. It is impossible to list everyone who has helped and inspired us, but we would be remiss if we didn't name a few. Lori Horne, Jean McClatchy, and Ginny Rubin showed Frances the satisfaction of raising money for good causes. She is grateful to the boards of the Poetry Center at San Francisco State and Intersection for the Arts for giving her a chance to be a nonprofit executive director, and to her current and past colleagues at the Walter and Elise Haas Fund — particularly Bruce Sievers and Pamela David. Stan wants to add his thanks to Arthur Compton, Louise Brown, Neil Housewright, John Darby, Jan Masaoka, and Bruce Hirsch, all of whom have given him opportunities to work in the nonprofit sector and provided guidance along the way.

Publisher's Acknowledgments

Senior Acquisitions Editor: Tracy Boggier

Project Editor: Elizabeth Kuball

Copy Editor: Elizabeth Kuball

Technical Editor: Gene Takagi

Production Editor: Vasanth Koilraj

Cover Photos: © Antonov Roman / Shutterstock

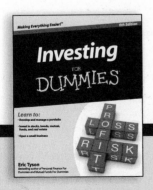

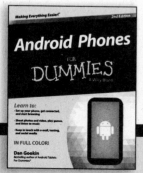